AF695219

Water Dynamics in Plant Production

2nd Edition

FSC
www.fsc.org
MIX
Paper from responsible sources
FSC® C013604

Psalm 65, 9–10

You care for the land and water it;
you greatly enrich it.
The streams of God are filled with water
to provide the people with corn,
for that is how you prepare the land.
You drench its furrows
and level its ridges;
You soften it with showers
and bless its crops.

Water Dynamics in Plant Production

2nd Edition

Wilfried Ehlers

University of Göttingen, Germany

and

Michael Goss

University of Guelph, Canada

www.cabi.org

CABI is a trading name of CAB International

CABI
Nosworthy Way
Wallingford
Oxfordshire OX10 8DE
UK

Tel: +44 (0)1491 832111
Fax: +44 (0)1491 833508
E-mail: info@cabi.org
Website: www.cabi.org

CABI
745 Atlantic Avenue
8th Floor
Boston, MA 02111
USA

Tel: +1 (617) 682 9015
E-mail: cabi-nao@cabi.org

A catalogue record for this book is available from the British Library, London, UK.

Library of Congress Cataloging-in-Publication Data

Ehlers, Wilfried, author.
Water dynamics in plant production / Wilfried Ehlers and Michael Goss. -- 2nd edition.
pages cm
Includes bibliographical references and index.
ISBN 978-1-78064-381-6 (hbk : alk. paper) -- ISBN 978-1-78064-382-3 (pbk : alk. paper) 1. Crops and water. 2. Plant-water relationships. I. Goss, M. J., author. II. Title.
S494.5.W3E37 2015
633--dc23

2015023023

ISBN-13: 978 1 78064 381 6 (hardback)
978 1 78064 382 3 (paperback)

Commissioning editors: Vicki Bonham/Joris Roulleau
Editorial assistant: Emma McCann
Production editor: Tracy Head

Typeset by SPi, Pondicherry, India.
Printed and bound in the UK by CPI Group (UK) Ltd, Croydon, CR0 4YY.

Contents

This book is enhanced with supplementary resources. To access the discussion and multiple choice questions, please visit: http://www.cabi.org/openresources/43823

Preface to 2nd Edition

Since the publication of the first edition some 12 years ago, *water* has become an even more topical subject. Flooding has disrupted life in the UK, Europe, North America, the Indian subcontinent, West Africa and Asia, whereas drought has dominated in Australia, East Africa, South and Central America, and the south-western states of the USA. Conflicts between the use of water for irrigated agriculture and for maintaining natural ecosystems, particularly wetlands, are more widespread and the decline in water reserves of major groundwater aquifers continues.

Climate change has become more evident, with greater volatility in the climate, as evidenced by the number of extreme weather events. In the northern hemisphere the last three decades are considered to have been the warmest for 1400 years – an interval of time that *includes* the Medieval Warm Period – despite the current slowdown in global temperature rise. But the present hiatus is predicted to end and the rise is expected to continue uninterrupted for the rest of the century.

These changes are taking place as increases in world population means that more people than ever need to be fed. So understanding the relationships between water use and crop production has relevance to a wide range of key subjects important to food security, infrastructure development, environmental sustainability and preventing conflict between countries sharing international rivers. Water use by crops depends on the instantaneous volume of water in the rooting zone, the rate at which it can move to the root surfaces and into the plant as well as the rate of evaporation from the substomatal cavities in the leaves and stems. It is the dynamics of water through the hydrologic cycle and the associated mechanisms that the plant employs to optimize its growth potential that forms the focus of the book. In this respect it offers a multidisciplinary introduction to the fundamentals and applications of water dynamics in managed and natural ecosystems. Such holistic approaches are needed but the pressures on researchers are such that the vast majority of new papers are narrowly focused and contribute small steps to the development of topics. By taking a somewhat historical approach, we have attempted to show the key steps in the development of the various areas covered and how new information is enhancing knowledge and its application.

We are particularly pleased that the new edition meets our original intention of a text in a soft-cover format aimed at advanced undergraduate students, with a hard-cover version for those using it for instruction and as a reference source. Support material is also available online (http://www.cabi.org/openresources/43823).

A number of people have contributed greatly to the appearance of this second edition. Commissioning editors Victoria Bonham and Joris Roulleau have been hugely supportive throughout the project. Both editorial assistant Emma McCann and production editor Tracy Head have provided vital help at critical times. We thank Dr Mary Beth Kirkham and Dr Christine Watson for their reviews of the text. We are very grateful to Dr Terry Gillespie, who for a second time provided essential feedback for the final version. Our special thanks go to Dr Ole Wendroth, who not only reviewed every chapter but also provided food for thought on each one. Marie-Christine Ordnung gave her husband, WE, tremendous support and was patience and encouragement personified as we disrupted schedules and mealtimes in our attempts to meet deadlines. She alone created the impossibly jolly termite!

Wilfried Ehlers and **Michael Goss**
Göttingen and Guelph, October 2015

Preface to 1st Edition

The source of life is water. Life began in the oceans, which represent the largest stock of water on Earth. Much less water is stored below the land surface in the form of fresh groundwater, amounting to not quite 0.8% of the earth's total water reserves, while lakes and rivers combined only contribute a further 0.007%. Therefore terrestrial life depends primarily on the global water cycle. This cycle makes the land productive by the infusion of fresh water precipitation, originating from the salt water of the oceans. However, in many regions the precipitation does not provide a sufficient or reliable source for the sustained presence of plants and animals. In fact, in all regions, precipitation proves to be highly variable in time and space, and human activities that have led to global warming have also increased the variability and intensity of rainfall.

Assimilation and biomass production in natural plant communities are intimately linked to water use through transpiration. The same is true for agricultural crops. However, extreme weather events like drought and torrential storms threaten agricultural enterprises and the well-being of an ever increasing world population. Three-quarters of the renewable fresh water resources used by mankind are consumed in irrigated agriculture, but such practices are at risk in several regions for varying reasons. These include climate change, weather variability, decline in groundwater reserves owing to over-utilization, and the degradation of soil and water quality. The 21st century has been referred to as the 'century of water'. At the world food summit held in Rome in 1996, water was identified as the major threat to food security. The global water crisis is predicted to intensify within the coming decades. Areas of acute water shortage are expected to spread, particularly to large regions of Africa and the Middle East.

In an age characterized by an increasing demand for fresh water and at the same time by an actual decline in reliable water resources for both rainfed and irrigated agriculture, conservation of water is essential. Agricultural water management must aim to eliminate unproductive water losses and optimize transpirational water use. The goal is to achieve optimum economical yields per unit of water used without compromising the environment. The development of an understanding how to approach such a goal is a central theme of this book.

Terrestrial plants obtain their water supply from the soil. They use their root system to access the water held within the soil profile, and transpire the water into the atmosphere. Plants provide, therefore, the most important link between the liquid phase of water in the soil and the gaseous form – water vapour – in the atmosphere. We focus on the main causes and processes that govern water movement through this continuum between soil, plant and the atmosphere. Flows and exchange processes that take place when water enters and leaves cropped land are explained. Moreover, the responses of plants to a decline in the water supply are highlighted. Both topics, water movement and responses to water stress, are essential for exploring practices in soil and crop management that enhance the efficiency of water use in plant production.

Crop plants are grown under a wide range of climates. For that reason, there is a need for a range of management strategies. These include both matching the water supply with the given soil and climate as well as the provision of additional resources through irrigation. We have used information from case studies, dealing with management systems in various parts of the world. We illustrate the importance of the underpinning processes and show how knowledge of the processes can guide development of better practices.

There are a number of textbooks on plant physiology, ecology, plant nutrition, soil physics and irrigation science that deal with individual topics of soil–plant–water relations, but none has attempted to integrate current knowledge across the continuum for agricultural crops. There are some books available at an advanced level, but most of them consist of a compilation of individual contributions. They were prepared largely in support of advancing the science, and as such are not very convenient as teaching aids. Our book is intended for university and college students and those starting postgraduate studies. Our treatment of the subject matter is certainly not exhaustive, and in part it aims to raise questions in the mind of the reader that will encourage a more detailed inquiry into a fascinating area of study.

The preparation of the book was the inspiration of Wilfried Ehlers, and we have drawn heavily on material of his earlier book *Wasser in Boden und Pflanze* (*Water in Soil and Plant*), published in 1996 by Eugen Ulmer GmbH & Co. in Stuttgart. Mr Roland Ulmer generously waived his copyright to encourage our project for an English-speaking audience. We are most grateful to Dr Murray Brown and Dr Terrie Gillespie of the Department of Land Resource Science at the University of Guelph, who read the complete text and provided a detailed critique. Their comments were very valuable in preparing the final script. We are indebted to our publisher in England, Mr Tim Hardwick, who supported our work and showed tolerance when we were somewhat elastic with our deadlines. Our special thanks go to Mrs Anita Bartlitz, who skilfully converted our sketches into accurate and precise diagrams, and created the annotations in a far less familiar language than her native German. Wilfried thanks his wife Marie-Christine Ordnung for her continual support and patience over the years. We also wish to acknowledge the contribution of Marie-Christine Ordnung and Amarilis de Varennes, who provided ideal venues to work, food to eat and the encouragement to complete our endeavour.

Wilfried Ehlers and **Michael Goss**
Göttingen and Guelph, December 2002

Abbreviations

Abbreviation	Name	Units used
A	energy flux fixed by assimilation	$J\ cm^{-2}\ day^{-1}$
A	cross sectional area	cm^2
ABA	abscisic acid	
ADP	adenosine diphosphate	
ATP	adenosine triphosphate	
a	year (annus)	
asl	above (mean) sea level	
BD	bulk density	$g\ cm^{-3}$
b; b*	regression coefficient	
CAM	crassulacean acid metabolism	
CER	CO_2 exchange rate	$\mu mol\ cm^{-2}\ s^{-1}$
CFC	chlorofluorocarbons	
CGR	crop growth rate	g DM m^{-2} soil surface day^{-1}
CWB	climatic water balance	mm
CWD	climatic water deficit	mm
CWSI	crop water stress index	
c	concentration	$g\ cm^{-3}$; $cm^3\ cm^{-3}$; $mol\ mol^{-1}$
c_i	water vapour concentration inside leaf	$g\ cm^{-3}$; $cm^3\ cm^{-3}$; $mol\ mol^{-1}$
c'_i	CO_2 concentration inside leaf	$g\ cm^{-3}$; $cm^3\ cm^{-3}$; $mol\ mol^{-1}$
c_o	water vapour concentration outside leaf	$g\ cm^{-3}$; $cm^3\ cm^{-3}$; $mol\ mol^{-1}$
c'_o	CO_2 concentration outside leaf	$g\ cm^{-3}$; $cm^3\ cm^{-3}$; $mol\ mol^{-1}$
c_{xs}	specific heat capacity of the xylem sap	$cal\ g^{-1}\ {}^{\circ}C^{-1}$
D	(subsoil) drainage	mm
D	diffusion coefficient	$cm^2\ s^{-1}$
DAS	days after sowing	
DM	dry matter	
DW	dry weight	g
d	regression coefficient	
d	density of water	$g\ cm^{-3}$

Continued

Continued.

Abbreviation	Name	Units used
d_w	density of sapwood tissue	g cm^{-3}
dz	thickness of soil layer	cm
E	evaporation	mm
E	evaporation rate	mm day^{-1}; g cm^{-2} day^{-1}
E_a	ventilation–humidity term	g cm^{-2} day^{-1}
E_a	actual soil evaporation	mm
E_f	actual evaporation rate of fallow soil	mm day^{-1}
E_p	potential evaporation rate	mm day^{-1}; g cm^{-2} day^{-1}
E_p^*	potential soil evaporation rate under crops	mm day^{-1}
E_{pf}	potential evaporation rate of fallow soil	mm day^{-1}
ET	evapotranspiration	mm
ET_c	crop evapotranspiration rate	mm day^{-1}
ET_o	reference crop evapotranspiration rate	mm day^{-1}
ET_p	potential evapotranspiration rate	mm day^{-1}; g cm^{-2} day^{-1}
ETE	evapotranspiration efficiency	kg DM m^{-3} water
ETR	evapotranspiration ratio	l water kg^{-1} DM
e	actual vapour pressure	mbar; Pa
e_s	saturated vapour pressure of the air	mbar; Pa
FC	field capacity	cm^3 H_2O cm^{-3} soil; vol.%
FC_{av}	available field capacity	cm^3 H_2O cm^{-3} soil; vol.%
FW	fresh weight	g
f_v	volume flow of sap	cm^3 day^{-1}
G	energy flux to heat the soil	J cm^{-2} day^{-1}
GHG	greenhouse gas	
G_r	relative rate of cell enlargement	day^{-1}
g	acceleration of gravity	cm sec^{-2}
g_{min}	minimum value of stomatal conductance	g H_2O m^{-2} s^{-1}
g_s	stomatal conductance	g H_2O m^{-2} s^{-1}
H	energy flux to heat the air	J cm^{-2} day^{-1}
HI	harvest index	
h	height of capillary rise	cm
h	plant height	cm
h	hour	
I	interception	mm
IAA	indole-3-acetic acid	
IR	irrigation water	mm
IR_{WUE}	water use efficiency of irrigation water	kg DM m^{-3} water
i	infiltration rate	cm day^{-1}
J	diffusion flux or rate	cm^3 cm^{-2} day^{-1}
j	plant-specific ratio from equation 15.2	mm H_2O per MJ m^{-2}
K	hydraulic conductivity	cm day^{-1}
K'	constant	
K_c	crop-specific coefficient	
K_s	saturated hydraulic conductivity	cm day^{-1}
K_u	unsaturated hydraulic conductivity	cm day^{-1}
k	von-Karman constant	≅ 0.41

Continued

Continued.

Abbreviation	Name	Units used
k	factor of Bierhuizen-Slatyer equation	Pa
k*	constant	
k_s	thermal conductivity of stem tissue	cal °C^{-1} cm^{-1} day^{-1}
L	depth of soil profile	cm
L	latent heat of vaporization	cal g^{-1}; J g^{-1}
L	length of root	cm
L_v	root length density	cm cm^{-3}
LAD	leaf area duration	
LAI	leaf area index	m^2 m^{-2}
LAR	leaf area ratio	m^2 g^{-1}
LE	latent heat flux (the product of L × E)	J cm^{-2} day^{-1}
LWU	water uptake rate within soil layer	mm day^{-1}
LWU_{max}	maximum LWU	mm day^{-1}
ly	langley; radiation density	1 cal cm^{-2}
M	soil resistance to deformation	MPa
M	moisture content of sapwood tissue	cm^3 g^{-1}
m	crop specific factor	
NADP	nicotinamide adenine dinucleotide phosphate	
NAR	net assimilation rate	g DM m^{-2} leaf surface day^{-1}
n	site-specific factor	
n_s	number of moles of solutes	
O	osmotic potential	cm; bar; MPa
P	biomass dry matter	g; t ha^{-1}
P	precipitation	mm
P	pressure potential	cm; bar; MPa
P	pressure; turgor pressure	MPa
P_a	ambient air pressure	mbar
P_c	pressure applied to chamber	MPa
P_L	net photosynthesis rate of the leaf	g cm^{-2} s^{-1}
P_n	normalized precipitation	mm
PAR	photosynthetically active radiation	mol of photons m^{-2} s^{-1}; J m^2 day^{-1}
PEP	phosphoenolpyruvate	
Pi	inorganic phosphate	
PSE	precipitation storage efficiency	% of precipitation
PWP	permanent wilting point	cm^3 H_2O cm^{-3} soil; vol.%
p	partial pressure	bar; MPa
pF	*pondus/potentia* free energy	
p_i	CO_2 partial pressure inside leaf	bar; MPa
p_o	CO_2 partial pressure outside leaf	bar; MPa
Q	volume flow rate	cm^3 day^{-1}
q	water flux or water flux density	cm^3 cm^{-2} day^{-1}; cm day^{-1}
q_c	capillary water flux	cm day^{-1}
q_c	heat dissipated by conduction	cal day^{-1}
q_f	mass flow of heat in xylem sap	cal day^{-1}
q_h	heat generated by a heater	cal day^{-1}
q_r	water flux through roots	cm day^{-1}

Continued

Continued.

Abbreviation	Name	Units used
q_r	rainfall rate	cm day^{-1}
q_t	total water flux	cm day^{-1}
R	hydraulic resistance	day; bar day cm^{-1}
R	runoff	mm
R	universal gas constant	8.314 J K^{-1} mol^{-1}
R*	diffusion resistance	bar day cm^{-1}
R_L	long-wave back radiation	J cm^{-2} day^{-1}
R_N	net radiation	J cm^{-2} day^{-1}
$R_N{}^*$	net radiation below canopy	J cm^{-2} day^{-1}
R_T	total or global radiation	J cm^{-2} day^{-1}
RH	relative humidity of the air $((e/e_s) \times 100)$	%
RL	root length in a soil layer	cm cm^{-2} soil surface area
RuBisCO	ribulose-1,5-bisphosphate carboxylase/ oxygenase	
RuBP	ribulose-1,5-bisphosphate	
RUE	radiation use efficiency	g DM MJ^{-1} PAR
RWC	relative water content	
r	radius	cm
r	reflectivity coefficient or albedo	
r_l	diffusion resistance for H_2O vapour in intercellular spaces and stomates	s cm^{-1}
r'_l	diffusion resistance for CO_2 in intercellular spaces and stomates	s cm^{-1}
r_o	diffusion resistance for H_2O vapour at boundary layer	s cm^{-1}
r'_o	diffusion resistance for CO_2 at boundary layer	s cm^{-1}
r'_m	diffusion resistance for CO_2 in mesophyll	s cm^{-1}
r_1	radius of the root	cm
r_2	radius of soil cylinder around root	cm
S	sorptivity	cm $day^{-1/2}$
S	suction force of the cell	MPa
SD	saturation deficit of the air e_s - e, Δe	mbar; kPa
SLA	specific leaf area	m^2 g^{-1}
SRL	sum of or total root length in soil	cm cm^{-2} soil surface
T	temperature	°C
T	transpiration of the crop	mm
T	water tension	cm; bar; MPa
T*	absolute temperature	K
T_A	air temperature	°C
T_L	leaf temperature	°C
T_L	transpiration rate at the leaf surface	g cm^{-2} s^{-1}
T_W	transpiration of weeds	mm
TDR	time-domain reflectometry	
TE	transpiration efficiency	kg DM m^{-3} water
TE_G	transpiration efficiency related to grain	kg DM m^{-3} water

Continued

Continued.

Abbreviation	Name	Units used
$(TE)_L$	transpiration efficiency of the leaf	mg CO_2 g^{-1} H_2O
TR	transpiration ratio	l water kg^{-1} DM
$(TR)_L$	transpiration ratio of the leaf	g H_2O mg^{-1} CO_2
TW	weight at turgidity	g
t	time	s; day
U	wind speed, wind run	km day^{-1}
UR	specific root water uptake rate	cm^3 water cm^{-1} root day^{-1}
UV	ultraviolet	
V	cell volume	cm^3
$\bar{V}$	partial molal volume of water	cm^3 mol^{-1}
V_a	volume of apoplastic water	cm^3
V_t	total volume of water in leaf	cm^3
VOC	volatile organic compounds	
W	cell wall pressure, turgor pressure	MPa
W_{extr}	extractable soil water	mm
W_{rc}	crop water requirement	mm
WU	water uptake rate	cm^3 water cm^{-3} soil day^{-1}
WU_{max}	maximum water uptake rate	cm^3 water cm^{-3} soil day^{-1}
WUE	water use efficiency	kg DM m^{-3} water
WUE_G	water use efficiency related to grain	kg DM m^{-3} water
X_{ABA}	concentration of abscisic acid	mol kg^{-1}; mol l^{-1}
x	space coordinate	cm
$\bar{x}$	mean of x	
Y	minimum turgor pressure	MPa
Y_d	yield under dryland conditions	t ha^{-1}
Y_G	economic or grain yield	t ha^{-1}
Y_i	yield attained by irrigation	t ha^{-1}
Z	gravitational potential	cm; bar; MPa
z	space coordinate in vertical direction	cm
z_r	rooting depth	cm
$z_{r\,eff}$	effective rooting depth	cm
z_0	roughness length	cm
α	contact angle	degree
β	ratio partial pressure to concentration	bar $(cm^3\ cm^{-3})^{-1}$
γ	surface tension of water	N cm^{-1}
γ	psychrometric constant	mbar °C^{-1}
	(at 1.013 bar air pressure and 20°C)	0.667 mbar °C^{-1}
Δ	difference	
Δ	slope of the function e_s versus T	mbar °C^{-1}
$\Delta^{13}C$	discrimination of carbon isotope ^{13}C	
Δe	saturation deficit of the air	mbar; kPa
$(\Delta e)_o$	standard saturation deficit	mbar; kPa
Δ MWD	change in mean weight diameter	mm
ΔS	change in soil water storage	mm
Δx	distance along the x coordinate	cm
Δz	distance along the z coordinate	cm
ε	cell wall extensibility	day^{-1} MPa^{-1}

Continued

Continued.

Abbreviation	Name	Units used
ε	ratio molecular weights of water to air	0.622
η	viscosity of water	poise; g cm^{-1} s^{-1}
θ	volumetric soil water content	cm^3 H_2O cm^{-3} soil; vol.%
π	potential osmotic pressure	MPa
ρ	density of air (at 1.013 bar air pressure and 20°C)	g cm^{-3} (1.205×10^{-3} g cm^{-3})
ρ	absolute humidity of the air	g water m^{-3} air
ρ_{XS}	density of xylem sap	g cm^{-3}
τ	dew point	°C
ϕ	total water potential	cm; dyn cm^{-2}; erg g^{-1}; bar; MPa
ϕ_L	leaf water potential	MPa
Ψ	matric potential	cm; bar; MPa
Ψ_r	matric potential at root surface	cm
Ψ_s	soil matric potential	cm

1 The Role of Water in Plant Life

1.1 Functions of Water in the Plant

In some ways the life of plants is much more directly dependent on *water* than is life in the animal kingdom. One reason for this is that higher plants differ from animals because nearly all are nutritionally self-sufficient, or *autotrophic*. Water serves as a hydrogen donor and thereby as a building block for *carbohydrates*, which are synthesized by plants making use of sunlight (Box 1.1). Another inorganic building block used by plants in the synthesis of organic primary products is *carbon dioxide* (CO_2), which plants can only take up from the atmosphere at the same time that they return water vapour to the atmosphere. This exchange of gases is necessary because, during their evolution, plants never developed a membrane that was permeable to CO_2 but impervious to water vapour. For the exchange of these two gases there are special openings in the leaf epidermis called *stomates*. The contra-flow *gas exchange* of water vapour and CO_2 that takes place between the inside of a leaf and the atmosphere through the stomates is therefore unavoidable. If the exchange of gases is to be maintained for the production of dry matter, growth and development, plants require a continual supply of water in liquid form. This is particularly true for plants other than succulents and halophytes, since the internal store of water is normally very small relative to the daily loss. The steady use of water demands its continual uptake.

There is another reason why plant life is immediately dependent on water. In contrast with animals, land plants live permanently in one place, so they have to remove water from the *soil water reservoir* in their immediate vicinity. Plant life depends essentially on water that is stored within the soil and is available for extraction. For extraction of water, plants rely on their root systems, which continue to grow through most of their life. The quality of the soil as a store of water accessible to roots depends on texture and structure.

Box 1.1. Light and water – prerequisites of photosynthesis

In the so-called light reaction of photosynthesis water is split into oxygen, protons and electrons:

$$2H_2O \rightarrow O_2 + 4H^+ + 4e^-$$

The oxygen produced moves out of the plant by diffusion preferentially through the stomates. At the same time during this light reaction nicotinamide adenine dinucleotide phosphate (NADP) is reduced to NADPH and in a coupled process adenosine diphosphate (ADP) is phosphorylated by use of inorganic phosphate (Pi), forming the 'energy-rich' adenosine triphosphate (ATP). NADPH and ATP as well as enzymes bring about the fixation of CO_2 in the so-called dark reaction. In this reaction CO_2 is reduced, ATP is dephosphorylated and NADPH is oxidized. The CO_2 gets assimilated, and organic compounds can then be built (Fig. B1.1).

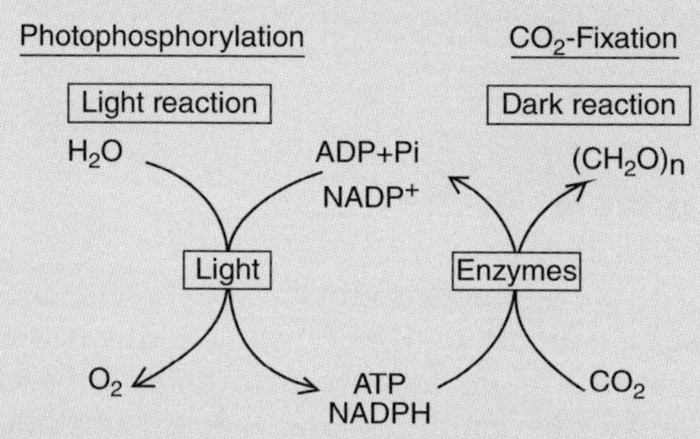

Fig. B1.1. Light and dark reactions during photosynthesis (after Gardner *et al.*, 1985).

The enzyme involved in the primary process of CO_2 assimilation is known as ribulose-1,5-bisphosphate (RuBP) carboxylase. That is true for the C_3 plants (see Section 1.2), but for the C_4 plants it is another enzyme, named phosphoenolpyruvate (PEP) carboxylase. The latter enzyme is also involved in CO_2 assimilation by certain succulent plants. These plants have the capability of crassulacean acid metabolism (CAM).

The daily *throughput of water*, that is the removal of water from the soil by roots, its movement through the plant in liquid phase, and its final transfer to the atmosphere in the vapour phase, can amount to a considerable quantity in comparison with the mass of the plants involved. In the middle of June, 1 week before heading, the dry weight of an oat crop amounted to 400 g m^{-2} (4 t ha^{-1}). The daily water use at this stage of development was equivalent to 6 mm of precipitation, which in this case came from the soil storage (Ehlers *et al.*, 1980a). These 6 mm of water throughput convert into 6 l or about 6 kg of water m^{-2} or 60,000 l (approximately 60 t) ha^{-1}. Hence, relative to the dry mass of the standing crop, 15 times more water was returned to the atmosphere on a daily basis. Assuming that 85% of the shoot mass was water, the oat crop contained 2270 g water m^{-2}. Compared with this store of water in the shoot, 2.6 times more water was extracted from the soil and passed on to the atmosphere. This transfer of water by plants to the atmosphere in the form of vapour is called *transpiration*.

Clearly we can say that the demands of plants cannot be satisfied with a small amount of water. Certainly it can be said that nature allows plants to be prodigal with this resource. The amount of water that is transpired daily by plants is generally 1–10 times more than the water stored in them. Compared with the amount needed for cell division and cell enlargement, the amount is 10–100 times more, and finally compared to the needs for photosynthesis it is 100–1000 times greater. So that prompts the question: 'Why is all the water taken up?'

Water is an *important constituent* of all plants. Roots, stems and leaves of herbaceous plants are 70–95% water. In contrast, water comprises only 50% of ligneous tissues and 5–15% of dormant seeds.

Water is the *basis of life* for a single cell and for the aggregate of cells that combine to form the structure of higher plants. It not only influences the processes and activities of cell organelles, but can also determine the final appearance of a plant. As a *chemical agent* it takes part in many biochemical reactions, for instance in assimilation (Box 1.1) and respiration. It is a *solvent* for salts and complex molecules, and mediates chemical reactions. Water is the medium of transport for nutrient elements and organic molecules from the soil to the root and the *means of transport* of salts and assimilates within the plant. Stimulation and motion of organelles and cell structures, cell division and elongation are examples of processes controlled by hormones and growth substances, with water being the carrier of these messengers, enabling the *regulatory system* of the plant.

Other functions of water are much more apparent. Water confers shape and solidity to plant tissues. If a previously sufficient supply of water is disrupted, herbaceous plants and plant organs that lack supporting sclerenchyma will lose their *strength* and wilt. The hydrostatic pressure in cells is dependent on their water content, and permits *cell enlargement* against pressure from outside, which originates either from the tension of the surrounding tissue or from the surrounding soil. Root tips experience a confining pressure when penetrating soil because the particles, held together by cohesive and adhesive forces, have to be pushed apart to allow the root to penetrate. The large *heat capacity* of water greatly dampens the daily fluctuations in temperature that a plant leaf might otherwise undergo, due to the considerable amount of energy required to raise the temperature of water. Energy is also required to convert liquid water to a vapour that transpires from leaves causing *cooling due to evaporation*. Without these temperature compensating effects, plants would warm up much more and eventually die from overheating. Interestingly, because of these effects, transpiration rates can be estimated from surface temperatures, obtained by infrared thermography, either immediately within the crop stand (Section 14.1) or using remote sensing from aeroplanes or satellites.

1.2 Adaptation Strategies of Plants to Overcome Water Shortage

Depending on the amount and distribution of rainfall and the probability of occurrence, the regions of the world vary greatly in the supply of water. The support to plant life ranges from great abundance to extreme poverty. Plants have developed various strategies to counter the problems of temporal or spatial water shortage.

According to the presence and supply of water, ecologists divide terrestrial plants into hygrophytes, mesophytes and xerophytes. *Hygrophytes* are plants that thrive in generally humid habitats, where there is no shortage of water throughout the growing season. In temperate zones, in addition to these plants with a humid biotype, there are many shade-loving herbaceous forest species that also belong in this category.

At the opposite end of the spectrum are the *xerophytes*. These plants are adapted to water shortage, which may occur regularly and may persist over long periods of time. Anatomical and physiological specialization has taken place to meet the requirements of these plants so that they can survive extended periods of water scarcity. To this group belong succulent plants that establish an internal water reservoir for use during dry spells, thereby postponing desiccation. Another group of xerophytic plants are able to endure considerable water loss from their tissues without losing their ability to survive.

Mesophytes fit in between these two extremes. Many plants from temperate climates belong to this group, but the cultivated plants from those regions are also included. The latter cannot endure an extreme form of arid climate without being irrigated. However, they are well prepared for short periods of water shortage. When water supply falls short, they can reduce their transpiration rate dramatically and modify other processes.

Weather patterns may result in temporal and spatial shortages in water supply with varying intensity. As mentioned earlier, rainfall is a principal factor responsible for limitations in the plant water supply. But water shortages depend also on a second climatic variable that sets the rate of transpiration and thus how rapidly the water stored in the soil becomes limited by root extraction if it is not replenished by rainfall. This variable is the evaporative demand of the atmosphere. It is principally dependent on the humidity and temperature of the air in contact and interchange with the transpiring plant stand. Dry and warm air induces a large demand and humid and cool air a small demand. Rainfall and demand determine the actual soil water storage and are prominent causes for drought. 'Drought occurs when the amount of plant-available water in the soil due to precipitation is not enough to meet the atmospheric demand for evapotranspiration' (Woli *et al.*, 2012). The term evapotranspiration will be differentiated from transpiration in later chapters.

How do plants react to drought, and what kind of strategies have they developed to achieve *drought resistance*? The principal stress that all plants undergo as a result of a severe water deficiency associated with drought is a deficit of water within their tissues. Strategies to evade deadly water deficits are quite varied. A definition, such as that given in Fig. 1.1, may appear to be arbitrary but it serves the purpose of clarifying the facts in a particular case. Plants also combine several of the possible strategies.

Those plants that are adapted to *drought escape* will germinate from dormant seeds only when there is abundant rainfall. Afterwards they can manage with a limited supply of water because they can terminate vegetative growth and become reproductive after a very short life cycle of just a few weeks, even ending with mature seed. Subsequent dry periods are escaped through seed dormancy. Another strategy is *drought avoidance*. Here plants may avoid or at least retard desiccation

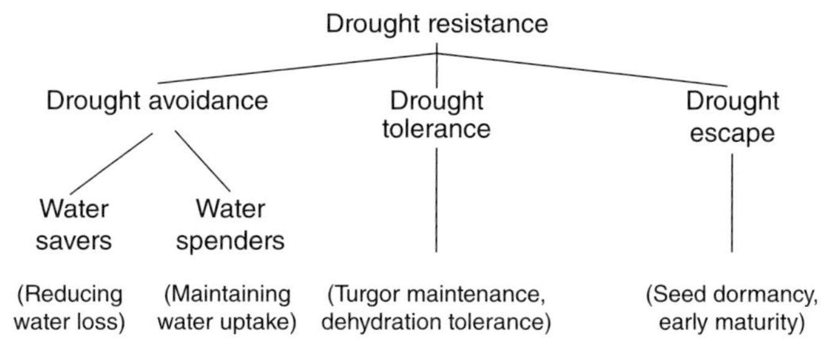

Fig. 1.1. The different forms of drought resistance (after Levitt, 1980).

of their tissues by increasing water uptake, reducing water loss, or by enhancing the internal storage of water. Like the first group these plants maintain a water balance that is largely in equilibrium. They belong to the hydrostable or homoiohydric species. A third strategy, *drought tolerance*, has also to be mentioned (Fig. 1.1). Plants relying on this strategy are able to tolerate a certain level of tissue desiccation. During phases of desiccation they limit their vital functions quite considerably. The plants are said to be hydrolabile or poikilohydric (Larcher, 1994).

The various strategies of adaptation can be observed most strikingly in arid deserts. Within the group of plants that avoid drought are those that have adopted the strategy of *water savers*. Many of these plants are succulents and can save a large volume of water within parenchymatous tissue when the very short periods of rainfall occur. This stored water can be used during longer-lasting periods of drought by exercising very thrifty water exchange. Quite a number of species in the family *Cactaceae* belong to this group. Cacti, as well as plants of the families *Crassulaceae*, *Agavaceae*, *Asclepidiaceae* and others are representatives of a group that demonstrate *crassulacean acid metabolism* (CAM). These *CAM plants* effect a unique physiological adaptation to water shortage. During periods of high radiation and air temperature (i.e. during the day) stomates of plants with succulent leaves or stems will remain closed. During the night, however, they will be opened for CO_2 assimilation and accumulation in the form of organic acids, which during the daytime supply CO_2 again for producing carbohydrates by photosynthesis (see Box 1.1).

There are also water savers among *C_3* and *C_4 plants* (see below). In many cases the plants possess distinct anatomical features such as: (i) stomates that are deeply sunk into the epidermis; (ii) thick and leathery or fleshy leaves; (iii) small leaves; (iv) leaves with waxy coatings over the cuticle; and (v) leaves with a felt-like cover of fine hairs. Some of the water savers restrict water loss during dry periods by rolling or folding their leaves, thereby reducing both the area of the leaf that intercepts radiation and the area through which transpiration occurs. Finally, extreme desiccation can be avoided or at least postponed by premature leaf drop. The rapid regrowth of leaves after rainfall allows a considerable adaptability to variations in the state of their water supply. Yet other plants respond to drought by having a leaf area that is relatively small, or having lateral branches that instead of bearing leaves are transformed into spine-like spurs.

C_3 plants are so-called because the first identifiable metabolic product of CO_2 fixation is a molecule with a chain of three carbon atoms. This compound is 3-phosphoglyceric acid. It is formed by an instantaneous disintegration of an unstable molecule with six carbon atoms, which is generated by catalysis of the enzyme RuBP carboxylase (Box 1.1). C_4 plants, on the other hand, form a molecule with a four-carbon chain, oxaloacetic acid, which results from the carboxylation of PEP – a reaction supported by the corresponding enzyme.

Deep rooting plants like the North American mesquite (*Prosopis juliflora*) belong to the group of drought avoiding plants that follow the strategy of *water spenders*. Mesquite is a leguminous tree from the Mojave Desert, with roots extending to 20–30 m depth, thereby giving the plant access to a comparatively large water reservoir. Some deep rooting plants from the steppe raise water during the night from deep layers to more shallow ones, where the water is released from the roots into the surrounding soil (Caldwell and Richards, 1989). This 'hydraulic lift' enables plants to make use of a larger water supply during the day for transpiration and for CO_2 assimilation. Neighbouring plants with shallow roots can also make use of the water brought up from depth (Caldwell *et al.*, 1998). A much less cooperative plant is the creosote bush (*Larrea divaricata*) of the Californian deserts. It counters competition for water from neighbouring plants by secretion of toxins from its roots.

Among those plants that have a strongly developed *drought tolerance*, sometimes called the 'genuine xerophytes', are numerous algae, lichens, mosses and ferns. However, the same tolerance may be found to a certain degree in some angiosperms. These plants are commonly referred to as 'resurrection plants'. They are able to withstand periods of desiccation of their protoplasm without too much injury, even though vital processes are slowed down. After rewetting, metabolic processes can resume promptly. Species classed as drought tolerating plants are to be found in the families of the *Myrothamnaceae*, *Lamiaceae* (*Labiatae*), *Scrophulariaceae* and *Poaceae* (*Gramineae*). Within this group of xerophytes is *Borya sphaerocephala*, a perennial species of the lily family (*Liliaceae*) from Western Australia. Its roots are restricted to shallow soil troughs on top of granite, and the plant becomes greatly dehydrated during the 6 months of the dry season. At that time the plant enters into a state of dormancy, and the whorled lineate leaves change hue from green to orange-red, becoming prickly and brittle. With the onset of autumn rain, the leaves revert to green and become smooth and pliable.

When desiccation develops slowly over time, many plants are able to accumulate inorganic ions or organic compounds, such as sugars, alcohols and amino acids, in their tissues. These materials are osmotically active and draw water into the cells. This capability of solute accumulation is termed *osmotic adjustment* (Moinuddin *et al.*, 2005). The solutes are concentrated in the cytoplasm and vacuoles, but the water content of the cells is maintained at a more or less stable level. By osmotic adjustment plants guard against a loss of turgidity. This adjustment will allow the plant to survive periods of drought more vigorously and for longer periods of time, and can allow the extraction of an additional amount of water from the soil.

Finally, plants adapted to *drought escape* will avoid long-lasting periods of desiccation by terminating their short life cycle before the onset of drought. These plants will germinate only after sufficient rainfall, but then will reach the flowering stage after just a short period of development, which takes place at a fast rate. When in bloom, the desert is truly alive, garnished with a dense and colourful plant cover.

Among *cultivated plants*, the short-lived two-rowed barley (*Hordeum vulgare*) is a drought escaper. Peanut (*Arachis hypogaea*) and cowpea (*Vigna unguiculata*) are classed in this group along with C_4 plants from the various species of millet. All of these crops reach maturity, although annual precipitation may not exceed 250–300 mm (Rehm and Espig, 1976, 1991; Andreae, 1977; Eastin *et al.*, 1983). Sorghum (*Sorghum bicolor*) is considered as a crop species characterized by a strongly developed drought tolerance compared with other crops. Some cultivars of soybean (*Glycine max*) are capable of osmotic adjustment, and the same is true of other grain and forage legumes, of sugarbeet (*Beta vulgaris*) and various other crops (Ludlow and Muchow, 1990). The succulent sisal (*Agave sisalana*) is a water saver, and members of the water spenders include sainfoin or esparcet (*Onobrychis viciaefolia*). This is a perennial deep-rooted forage legume, adapted to calcareous soils and native to Mediterranean regions, but now cultivated to some extent in more temperate zones.

1.3 Water and Net Primary Production

Thus far we have stressed that water has a unique physiological importance in the life of plants – for CO_2 assimilation, for biochemical transformations and for the transmission of impulses and signals. Furthermore, it was made clear that during the course of evolution plants have developed many strategies to adapt to situations of water shortage. From all this it may seem reasonable to conclude that there ought to be a more or less well defined relationship between water use and the amount of dry matter produced.

To explore the possible relationship, the *net primary production* (i.e. gross primary production minus respiration) together with the total biomass are compiled in Table 1.1 for different types of ecosystem. Production is expressed in terms of unit area and time.

These data indicate that water supply not only plays a major part in determining net primary production and *biomass* (rainforest – savannah – desert), but it also accounts for the impact of other environmental factors such as *temperature*

Table 1.1. Mean annual net primary production and biomass for various types of terrestrial ecosystems (after Whittaker, 1975).

Type of ecosystem	Net primary production ($g\ m^{-2}\ year^{-1}$)	Biomass ($kg\ m^{-2}$)
Tropical rainforest	2200	45
Tropical seasonal forest	1600	35
Temperate deciduous forest	1200	30
Boreal forest (taiga)	800	20
Savannah	900	4
Temperate grassland	600	1.6
Tundra	140	0.6
Desert	90	0.7
Extreme desert	3	0.02

(monsoon forest – deciduous forest – boreal forest). An upper limit of productivity is determined by the *radiation*, which is not used very efficiently by any of the plant communities. Utilization of radiation and hence the level of net primary production may be reduced by factors that are variable in time and space, such as water and temperature that were just mentioned, but also by the supply and availability of *mineral nutrients*. All these factors influence plant growth and can regulate net primary production either through the *net assimilation rate* (NAR, rate of growth per unit of leaf area) or by constraining growth. The constraint to growth may be such that only a relatively small biomass is formed, and yet this represents the maximum possible, taking account of the full set of prevailing conditions (Table 1.1). A small biomass will result in a small *leaf area index* (LAI, total green area of one side of a leaf as a ratio of one unit of soil surface area). Therefore, the leaf canopy will not intercept all of the incoming radiation. Rather, some part of the radiation will reach the soil surface and not be used for photosynthesis. A small LAI is the second cause of reduced productivity. The actual net primary production, based not on 1 year's growth but on shorter time intervals of days or weeks, represents the growth rate of the plant stand, the *crop growth rate* (CGR) in arable farming. The CGR is the rate of growth per unit of soil surface area.

$$\mathrm{CGR} = \mathrm{NAR} \times \mathrm{LAI} \tag{1.1}$$

Equation 1.1 establishes that the productivity of a crop stand is dependent on the photosynthetic net productivity of the single leaf and of the size of the total leaf canopy.

Figure 1.2A represents a relationship between net primary production of terrestrial forests and annual precipitation as a rough index of the level of available water. The dry matter produced includes the above-ground material but not the root system. The relationship is not very exact, suggesting that there is likely to be dependency on some other environmental factors such as temperature (Fig. 1.2B).

One of the difficulties in making a quantitative demonstration of the significance of water for net primary production is how to measure *the quantity of water used* by a plant stand. Figure 1.3 gives some results of *lysimeter* studies on peanut. In this example the net primary production is shown in terms of the marketable product, the seed, rather than the total dry matter less that of the roots.

In this example with peanut, the rest of the environmental factors influencing plant weight, such as radiation, temperature and nutrients were kept at a constant level, which represented the optimum conditions for the two locations. Just the one limiting factor, water supply, was varied systematically using supplemental irrigation. In this experiment, pests and diseases were controlled so that they did not act as yield-reducing factors. Under these conditions, the relationship between water use and yield, the 'production function', became very evident. None the less, the values of the function differed between the sites. In Georgia a larger yield was obtained per unit of water used than was found in Florida. It is possible that the explanation of the difference lies in the evaporative demand of the atmosphere at the two sites (see Section 1.2). That idea will be explored further in following chapters.

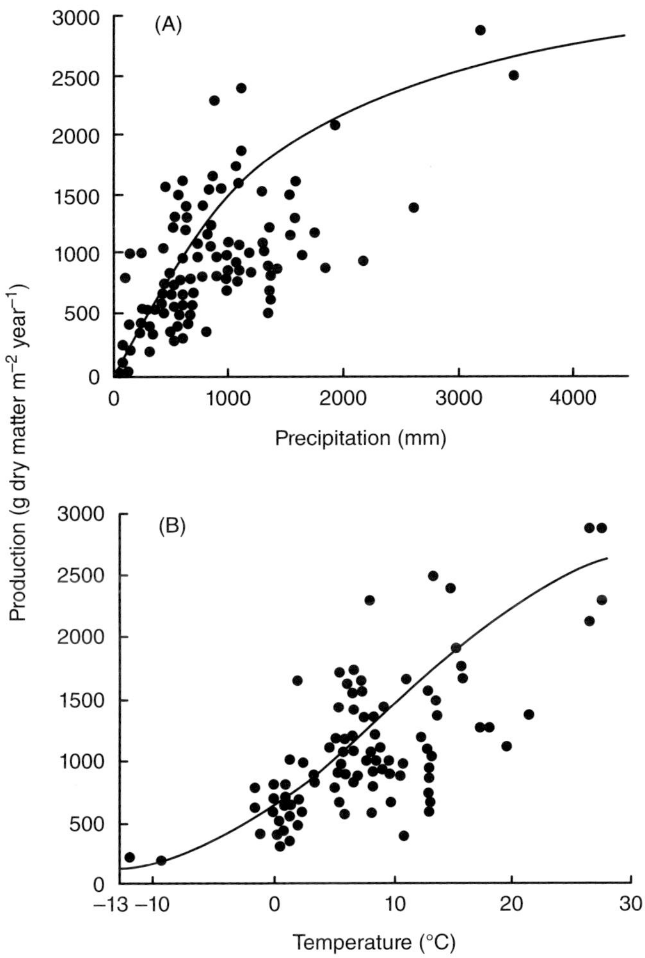

Fig. 1.2. Relationship between net primary production of terrestrial forests and annual precipitation (A) and mean annual temperature (B) (based on Begon *et al.*, 1991).

1.4 Water and Type of Vegetation

As already shown, water is a determining factor in the productivity of various types of ecosystems and for plants that are cultivated for their usefulness as food, fuel or fibre. However, the effect of water on productivity is also most likely to be governed by the condition of another climate variable: temperature. Temperature is important in terms of its magnitude and duration, which can be explained in terms of the direct effects of temperature on the rate of *gross photosynthesis, respiration* and *net photosynthesis* (Fig. 1.4A); but there can be effects on *growth rate* (Fig. 1.4B) and on phenological development.

The effect of *water supply* on dry matter production may be optimized at an appropriate *temperature*, but may decrease under warmer or cooler conditions.

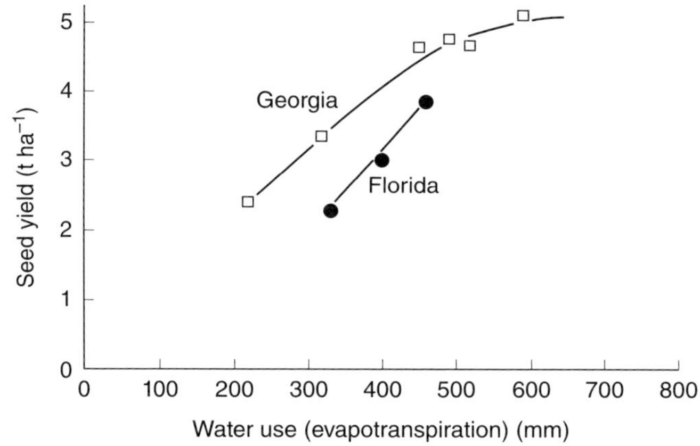

Fig. 1.3. Seed yield of peanut as related to water use. The water use includes the transpiration of the crop and the evaporation from the soil. Lysimeter studies in Georgia and Florida, cultivar is 'Florunner' (after Boote and Ketring, 1990).

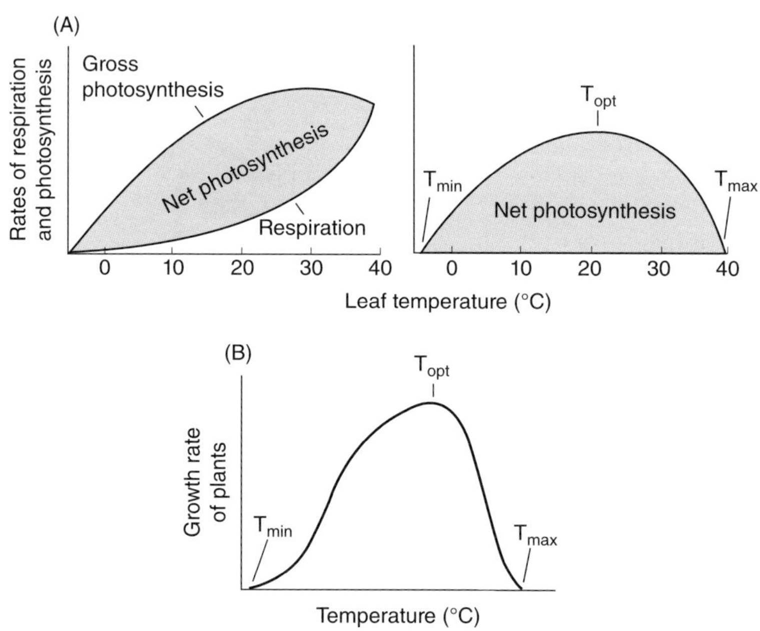

Fig. 1.4. Influence of temperature on the rates of gross photosynthesis, respiration and net photosynthesis (A) as well as on growth rate (B). The three cardinal points for temperature are the minimum, optimum and maximum values (T_{min}, T_{opt} and T_{max}, respectively). (Schematic after Pisek *et al*., 1973 for (A) and Fitter and Hay, 1981 for (B).)

However, the role of water in production can also be affected indirectly by temperature. The higher the temperature experienced by a community, the more water has to be available to the plants (see Section 1.2) if they are to realize the potential for net primary production because more water will be *transpired*. The

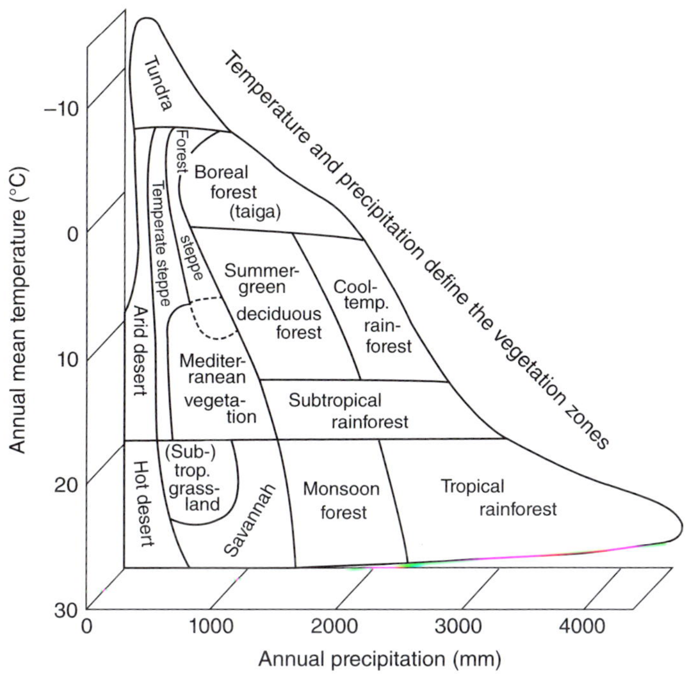

Fig. 1.5. Mean annual temperature and annual precipitation govern the types of terrestrial vegetation (based on Strasburger *et al.*, 1991).

significance of the interaction between temperature and the amount of precipitation for the delineation and classification of *climate types* was already recognized by the middle of the 19th century. Linsser introduced the concept of the 'humidity factor' in 1869, while living in St Petersburg, but it was not until 1915 that it gained significance. When it did, it was mainly in the field of pedology, where it became known as 'Lang's rain factor'. This factor is a function of the annual precipitation and the mean annual temperature (Blüthgen, 1966). As a first approximation, this factor characterizes the proportion of precipitation and evaporation, and thereby indicates the *aridity* or *humidity* at a location. For describing climates the linking of precipitation and the mean annual temperature has been further modified and adapted. The famous classification system for climate types proposed by Köppen, and the one proposed by Thornthwaite, are both based on rain–temperature relationships (Blüthgen, 1966).

Figure 1.5 depicts, in a simplified way, how temperature and water supply act together to influence the form of various *terrestrial vegetation types*. In tundra and desert, dryness will restrict the growth of forests. With increasing precipitation, the steppe and savannah will change to forest in temperate and tropical latitudes. In the tropics, forests develop where the annual rainfall exceeds about 1400 mm, but in temperate zones they may occur when rainfall is only about 600 mm.

2 The Role of Water in Soil

2.1 Soil Genesis and Soil Functions

Water is of primary importance for *soil genesis*. Without the action of water, soils would not develop. Soils originate from parent rock. The first step towards soil formation is the *weathering* of these rocks. Water contributes to the processes of weathering through physical and chemical actions.

Physical weathering splits rocks and minerals, but their chemical composition is not basically changed. Water is the main agent and works through frost action, but in arid regions differential thermal expansion of minerals can also split rocks. The fragments formed are transported by surface water from higher elevations downhill into valleys. Here the fine debris is deposited in alluvial fans, burying the former valley bottoms and levelling the topographical features.

Chemical weathering breaks down minerals by hydration, hydrolysis and dissolution. The disruptive force of water is greatly augmented by protons or hydronium ions (H_3O^+) that are derived from organic and inorganic acids. In this way even the sparingly soluble silicates are finally broken down. Increasing the temperature accelerates the kinetics of destruction. For example, in the humid tropics the decomposition of minerals is usually well advanced at much greater depths below the soil surface than occurs in temperate regions. As a result, many tropical soils are strongly developed, but at the same time they may only provide an inadequate supply of essential plant nutrients. In contrast, weathering and soil genesis in Arctic regions have penetrated into the mineral stratum to only a shallow depth.

In the course of *soil genesis*, soil-specific minerals are formed, but at later stages of development these may themselves undergo decay. Disintegration of solid particles liberates solutes, such as various cations, silicic acid, and iron and aluminium compounds. Water transports these solutes together with colloidal solids deeper into the soil body or even beyond the soil into deeper strata. Disintegration, displacement, precipitation and leaching are essential parts of the *pedogenic processes* supported by water. In temperate-humid regions, with loess deposits for instance, the first step towards soil formation is the decalcification of the parent material. Calcium ions will be leached and the pH will decrease. As a result, 'primary' minerals decay and 'secondary' silicate minerals are formed, which belong to the 'clay' fraction. These particles are small, having an average diameter of less than 2 μm. During clay formation the soil colour turns brown. Thereafter the clay starts to migrate within the soil from an upper layer to a lower layer, a soil forming process called 'lessivage' (Fig. 2.1). Another process is 'podzolization'. Here clay disintegrates by chemical weathering, and the sesquioxides of iron and aluminium released are leached along with soluble humic acids to deeper layers (Fig. 2.1). Characteristic *soil horizons* are formed that are diagnostic for distinct *soil types*, soil units (Fig. 2.1) or soil orders, the name being dependent on the classification system in use.

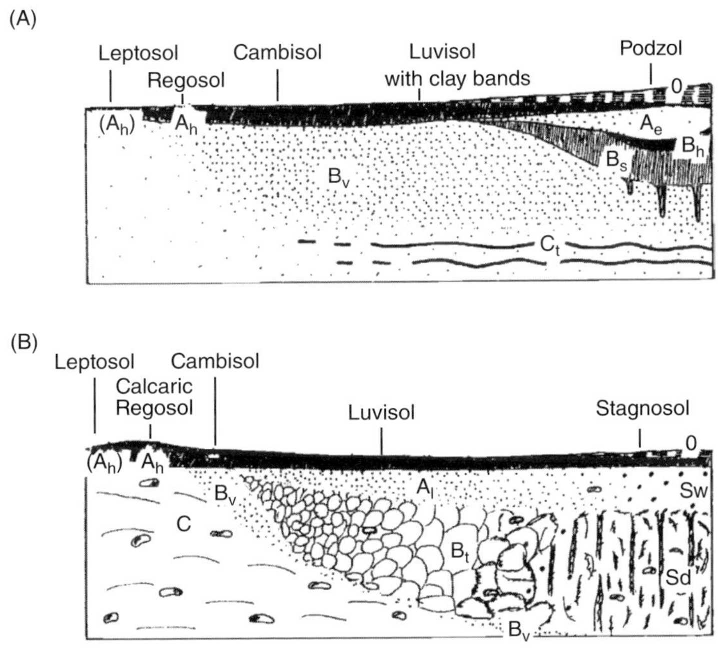

Fig. 2.1. Water is the cause and driving force of pedogenic processes. Examples are given for north-west Germany. (A) Braunification, clay migration or lessivage with generation of clay bands. Thereafter podzolization occurs with formation of an iron hardpan or ortstein. Parent material: dune sand. (B) Decalcification, braunification and clay formation, lessivage and gleyification with consolidation. Parent material: glacial till from the late Pleistocene (after Scheffer and Schachtschabel, 1989).

Designation of horizons: O, organic horizon above mineral soil; A_h, mineral horizon adjacent to soil surface with accumulated organic matter ('h' from humus); A_e, horizon bleached by acids ('e' from eluvial); A_l, horizon with loss of clay ('l' from lessivage); B_v, horizon of the subsoil with weathering *in situ* ('v' from *verwittert* = weathered); B_s, illuvial horizon enriched with sesquioxides; B_h, horizon with illuvial concentration of organic matter; B_t, illuvial horizon with clay enrichment ('t' derived from *Ton* = clay); C_t, horizon of parent loose rock with separated clay bands; Sw, bleached horizon with perched water table, water conducting ('S' derived from *Stauwasser* = stagnant water; 'w' from *wasserleitend* = water conducting); Sd, strongly mottled horizon with perched water, causing waterlogging by impeded drainage ('d' from *dicht* = dense, impermeable).

In arid regions, however, water flows in an upward direction to the soil surface during prevailing rain-free periods. Here soil water is lost by evaporation, and solutes from the subsoil are deposited near the surface. As a consequence, soils become saline, barren and desolate. In some regions of Western Australia soils turned saline after trees were cleared for cultivation. Removal of the transpiring trees caused a rise in groundwater and movement of the associated salts into the surface soil.

Soil genesis is accompanied by the *formation of soil structure*, which is essentially dependent on soil water. Water causes the clay minerals to swell and shrink, and the soil matrix becomes subdivided by planes of weakness or by visible fissures. Also, ice formed by frost can separate the soil matrix into aggregates of characteristic size and form. On the other hand, the gleyic horizons of Stagnosols (Fig. 2.1) become cohesive and compact due to waterlogging.

Without water there would be no transport processes involving the mobilizing of materials within the soil. Water in the soil is seldom in a state of equilibrium. Usually it is in constant motion within the soil profile. The reason is that the energy state of water, its *potential*, is generally not the same but varies between different locations within the profile. Differences in the potential from one place to another cause water to flow, and the water movement tends to cause a change in water potential. Moreover, water flow may alter the proportion of *water-filled pore space*. As a result, water indirectly affects some other transport processes in the soil such as *gas diffusion* and *heat conduction*. When water flows it may transport solutes and suspended fine material, sometimes over such long distances that even groundwater may become contaminated by ions, bacteria or pesticides carried with the soil water. When solutes become the focus of attention, the term 'soil water' is replaced by the term *soil solution*.

It is difficult to measure the water use of a plant community and hence quantify exactly how much water remains in the soil profile and is available to contribute to drainage and the recharge of groundwater. This same uncertainty limits our ability to determine how much material is transported from the soil profile by leaching. The determination, for instance, of the *concentration* of nitrate or environmental toxins in the solution extracted from the soil is relatively easy. Much less certain is the methodology required to quantify the *rates of seepage*. Up to now there is no generally applicable and easy method available for the direct measurement of the flow rate – the 'flux' of water in '*unsaturated*' *soil*. Soils for agricultural production and rooted by plants are normally unsaturated. That means the pore space is only partially filled with water. The rest is filled by air. Figure 2.2 shows the variation in the concentration of nitrate nitrogen as a function of depth and time under various crops. The seepage rate was evaluated by laborious measurements in the field, supplemented by mathematical modelling. Based on nitrate concentration and seepage rate the nitrate leaching was determined.

Water is one of the agents contributing to soil genesis. It is the agent for the transport of material in the soil profile. And water is an agent that can destroy the soil. The kinetic energy of rain can break down aggregates at the soil surface. A water front can penetrate into surface-exposed aggregates from all sides, putting entrapped air under such pressure that the aggregates are blown apart. Suspended clay and silt particles are then deposited at the soil surface, forming a skin-like cover. Large pores become clogged, reducing the rate at which rainwater enters the soil pore space below. The soil's *capability to accept rainfall* is reduced. In the case of sloping land, surface water displaces soil particles from the upper to the lower areas. The transposition process may occur in a more imperceptible form over a large area, called *sheet erosion*, or more conspicuously in a localized area as *gully erosion*.

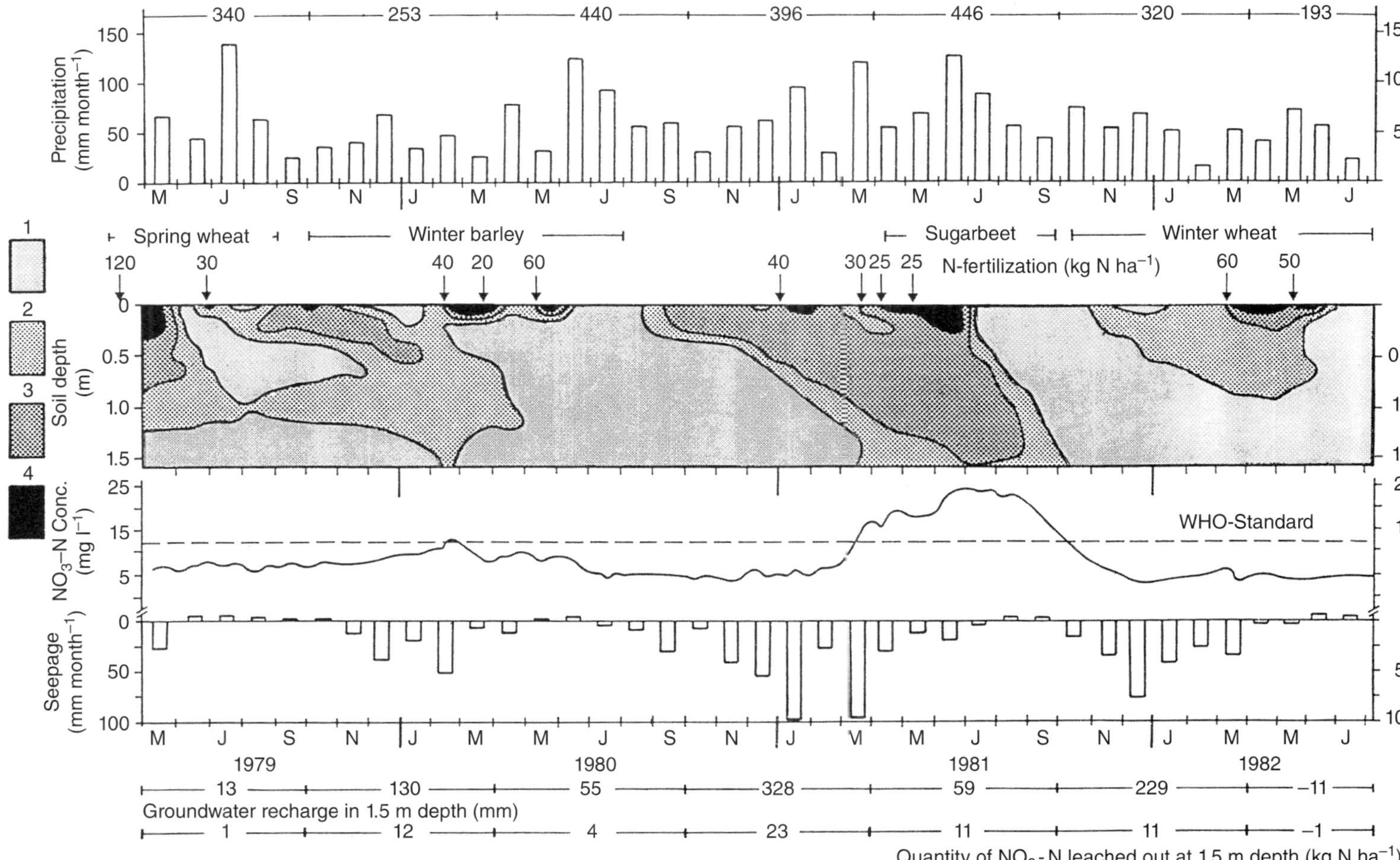

Fig. 2.2. Concentration of nitrate nitrogen in the soil solution (1: 0 to 11; 2: 12 to 22; 3: 23 to 45; 4: >46 mg NO_3–N l^{-1}), seepage water loss and leaching of nitrate nitrogen from 1.5 m soil depth under various crops. The top line shows totals of precipitation (mm) for the periods indicated. WHO-Standard, World Health Organization maximum acceptable concentration in drinking water. The soil is a Luvisol derived from loess (after Strebel *et al.*, 1989).

2.2 Soil Fauna and Vegetation Cover

The soil is a *porous body* built up from inorganic and organic solid particles with pore spaces in between. Many of the soil's physical properties and the transport phenomena within the soil may be described by application of physical laws, a description that may be imprecise and approximate. However, the soil represents much more than a medium that obeys physical laws. The soil is the *habitat* of plants and animals, which are organized to a greater or lesser extent. Soil water supports soil organisms, for instance mediating the transport of nutrients to flora and fauna while protecting the organisms from desiccation.

Plants and animals colonize the soil. Therefore they are subjected to its properties and have to conform to its natural laws. But at the same time they react to the soil and modify their habitat. The development of immature soils is accompanied by: (i) humification; (ii) weathering of minerals; (iii) release of nutrients; (iv) a vigorous development of plant cover; (v) stronger rooting and soil colonization; and (vi) an intensified nutrient cycling. Such soils become more porous. Soil animals of different sizes work through the ground, mixing inorganic and organic particles, creating connecting pores, and stabilizing aggregates within soil horizons and near the land surface. In the course of this development the soil's quality as a *habitat for plants* improves. The development of the soil and that of the plant association go along with and depend on each other. The soil is effectively protected against erosion by a dense vegetative cover, by roots and by a litter layer. Soil aggregates exposed at the soil surface are much better protected from slaking, being stabilized by humus due to physico-chemical linkages as well as by microbial activity. Stabilized soil surfaces allow rainwater to enter the soil at high rates, a very important characteristic during intense rain storms. Rainwater accumulation in the soil will in turn stimulate plant growth.

The soil (Fig. 2.1) and the plant cover will mutually develop as a function of time. It may take hundreds of years for both to reach a mature state. The intimate dependence is generally not apparent at first sight from local soil associations. But it is obvious by close inspection of 'banded' landscapes in arid and semi-arid regions of Africa, Australia and America (Valentin *et al.*, 1999). Banded vegetation ('tiger bush') is comprised of alternating bands of vegetation and bare soil, aligned along the contours of more gentle slopes. The bare interbands, 10–185 m wide, have a smooth, crusted surface, generating runoff of rainwater, which is collected by the vegetated stripes, which are 5 m to nearly 50 m in width. Here the soil surface is rougher and covered by litter, thus favouring termite activity. It has been concluded from a straight comparison that the differences between soils of the vegetated bands and the bare interbands are 'induced by positive feedbacks from vegetation, faunal and hydrological differences' (Valentin *et al.*, 1999).

The favourable state of soil conditions under vegetation cover will not last forever. Climatic change, leaching of calcium carbonate, progress in weathering, acidification, or human impacts can contribute to soil *degradation* and loss of its properties as a habitat (Fig. 2.1).

In summary, this chapter leads to the conclusion that water is crucial for soil. Water affects its development and properties, both directly, as a physical and a chemical agent, and indirectly, by enabling its colonization by plants and animals.

3 The Interdependence of Soil Water and Vegetation

3.1 The Significance of the Soil for Water Storage

The space in the soil between the solid particles is made up of pores of different effective sizes. The pore space is the medium that stores and conducts water. A sample of dry soil will soak up water like a sponge. At first that might be a comparatively large quantity of water per unit of time. But as the process goes on, the rate of water uptake will decrease and finally it will drop to zero. At that stage the forces in the soil sample inducing water attraction are neutralized.

These forces originate in the surfaces of fine soil particles belonging to the clay fraction. In many soil layers alumino-silicates predominate in this fraction. They adsorb water by 'short range' forces, like the London–van der Waals forces. Also active over a wider range are electrostatic forces, which derive from the negatively charged surfaces of clay minerals and from the adsorbed positively charged counter ions. Beside these *adsorptive forces* water is bound by *capillary forces*. These forces result, on the one hand, from the adhesion of water molecules at the surfaces of solid particles and on the other hand from the cohesion between the water molecules themselves. Capillary forces bind water at the spots where granular particles touch each other to form a 'pore neck'. These forces also bind water in any of the soil's capillary voids and, commonly, form a concave curvature at the boundary between the water and air, called the meniscus (Fig. 3.1).

At its maximum the sponge or the soil sample can soak up so much water that the total pore space is filled up. But in reality, during filling up, some air will normally become enclosed within some pore wedges and voids (Fig. 3.1). Prolonged rain will cause water saturation of the soil. But the water, having filled the coarse pores, will drain from the soil profile for some time after the rain has ceased. The volume of water that is retained 'against gravity' by the soil after 2 days of drainage without rain is frequently referred to as '*field capacity*'. It will be shown later that the concept of field capacity has to be taken as a subsidiary one, convenient for

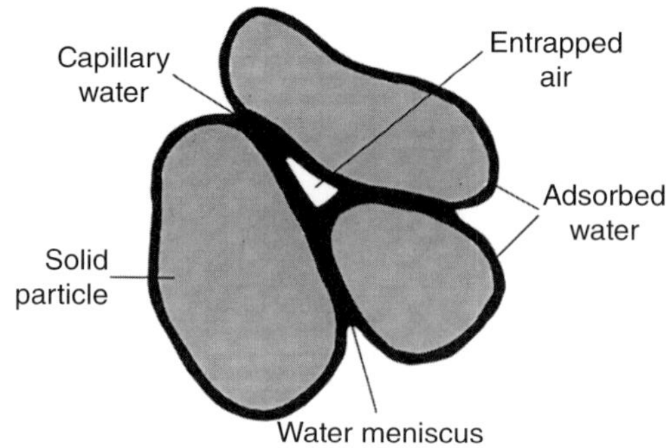

Fig. 3.1. Part of the soil water is bound by adsorptive forces, part by capillary forces (based on Brady, 1990).

some storage calculations. The reason is that even after 2 days the drainage of soil water will not have ceased but will go on for some time. There is another characteristic value of soil water content on the 'dry' side of the scale. This value is called the '*permanent wilting point*'. It is the water content that is envisaged as being the limit of extraction by plants as the remaining water is held by strong adsorptive forces (Savage *et al.*, 1996). This strongly adsorbed water is known as 'residual' water because it is not available to plants.

The difference between field capacity as a water content on the wet side and the permanent wilting point on the dry side is the water content potentially 'available' to plants. The available water content changes with the soil textural class, a soil grouping based on the composition of sand, silt and clay (Fig. 3.2). The dynamic aspects of soil water availability are discussed in later chapters.

3.2 Transpiration and Seepage of Water with Different Types of Vegetation

All the water that is transpired by natural communities comes from precipitation, if the locations are such that the groundwater is well below the rooting zone. Before plants can use precipitation, it will be stored temporarily in the pore space of the soil in a plant-available form. By considering the precipitation as intake into the soil and the water transpired as an output from soil, the concept of a *water balance* has been developed. The surplus rainwater that does not add to the plant-available water in the soil reservoir, and therefore does not contribute to meeting the actual transpiration, will drain through the subsoil towards the groundwater. But depending on local conditions part of the precipitation may not infiltrate the soil profile but flows away as surface runoff. It therefore follows that the soil water balance comprises the input quantity of precipitation and various output quantities. The balance acts as a feedback system with plants, which have a great need for water, taking priority.

The water balance equation reads:

$$P = E + I + T + D + R + \Delta S \text{ (mm)} \quad (3.1)$$

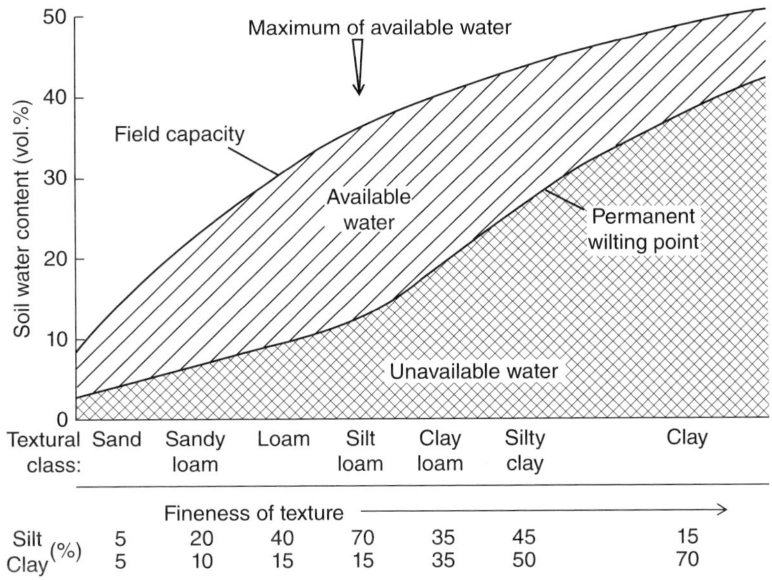

Fig. 3.2. Depending on soil texture, which is determined by the particle-size distribution, soils will vary in water content at field capacity and at the permanent wilting point. Both characteristic values enclose the plant-available water content. Silt loam soil contains the maximum of available water. The water at the permanent wilting point is not available to plants. The fineness of texture increases with the silt and clay content, presented as approximate percentages.

The P stands for precipitation, E for evaporation (the ‘unproductive’ loss of water from the soil surface to the atmosphere), I for interception (the evaporative loss to the atmosphere of precipitation that was retained on leaf surfaces), T for transpiration (the ‘productive’ loss of water by the plants to the atmosphere via stomates), D for subsoil drainage, R for runoff and ΔS for change in soil water storage. This change may be positive (gain of soil water) or may be negative (loss of moisture from the soil profile). All the terms listed are considered for specified time intervals (i.e. days, months or years).

Quite frequently the terms E and T are not distinguished. ET, the *evapotranspiration*, is often used, as it is quite troublesome to separate E and T when measuring the balance in the field. Also the *interception*, I, is usually not recorded as a separate term, and therefore is often included in ET. In some cases, water may rise from deeper soil layers or from groundwater into the rooting zone by capillary action. In such a case the output quantity loss of water by deep drainage or seepage has to be replaced as an input quantity. This is the capillary rise, which is an upward transport of water and a source of input.

As an example of a complete water balance, the record from a German mountainous region is shown below (Benecke and van der Ploeg, 1976). In the Solling mountain region, 500 m in altitude, two neighbouring stands of mature trees were investigated. One stand represented a 125-year-old beech (*Fagus sylvatica*) population, the other one a 100-year-old plantation of spruce (*Picea abies*). To determine the water balance the authors used meteorological and soil physical techniques to measure the components as well as computer simulation (Fig. 3.3). The interception of the trees was determined by comparing precipitation in the open with that within the stand. Precipitation was partitioned within the stand into stemflow, canopy drip and through-fall, the latter being the part that fell through the gaps in the canopy.

Compared with the lowland of north-west Germany (600–800 mm) the Solling is a region with high rainfall (1064 mm of annual precipitation in 1969). During the winter months from January to March, the ET of both tree species was small (Fig. 3.3). Probably it was only a matter of soil evaporation, as the beech trees had shed their leaves. The rates of seepage were quite considerable and attained a maximum in April. In May beech usually comes into leaf, and temperatures increase. Therefore during spring the ET of both tree species increased. In the summer months of July and August the ET of the beech stand surpassed that of spruce. These ET relations changed again in autumn during the leaf fall of the beech, allowing the spruce stand to maintain a higher ET. One can easily recognize the control of the water balance in favour of the trees as mentioned above. In summer time, when plants transpire at high rates and soil water gets depleted, the seepage is greatly reduced.

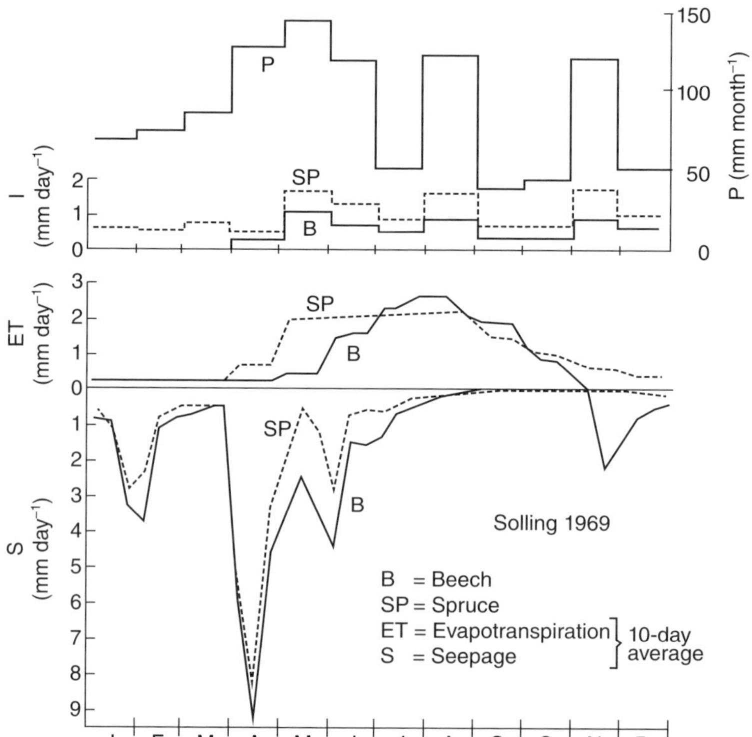

Fig. 3.3. Temporal course of evapotranspiration (ET), seepage (S), interception (I) and precipitation (P) in mature stands of beech (B) and spruce (SP) in the Solling mountains, north-west Germany (after Benecke and van der Ploeg, 1976).

On a yearly basis the ET of spruce was much larger than of beech, because of the evergreen needles. The same was true with the interception I (Fig. 3.3). Because of higher evapotranspiration and interception, the water losses from seepage were smaller in the spruce stand than in the beech stand (Table 3.1).

At a location with identical climatic and soil characteristics, the seepage will be controlled by the water use of the existing plant community. On the other hand, there will be differences between types of soil, when precipitation and vegetation are the same. Deep drainage or seepage and therefore *groundwater recharge* will be greater on sandy soils than on soils with larger silt and clay content like silt loams derived from loess (Table 3.2). Because of their small field capacity, the water storage capacity of sandy soils will be replenished much more quickly than that of soils with a loamy or clayey texture with a greater field capacity (Fig. 3.2). Soils that can store more water offer the possibility of greater water extraction by the roots, thus reducing percolation losses. Due to the freer drainage through coarser textured soils, water shortage may affect plant growth much more. Groundwater recharge will also depend on the form of land use. Greater seepage rates were found in arable fields compared with pasture land and coniferous forests (Table 3.2). Therefore the yearly ET follows the trend of: coniferous forest > pasture > arable land.

The results in Table 3.2 indicate that arable cropland contributes more to groundwater recharge than does a forest stand. This appears to contradict a widely held opinion that forests are good for groundwater recharge. But our conclusion about recharge from cropland is true only if the comparison is based on the same amount of precipitation and a similar soil. We have to consider that forests usually are associated with more mountainous regions where rainfall is greater. Because of terrain steepness and shallow depth to bedrock, large amounts of precipitation

Table 3.1. Components of the annual water balance (measured in mm) within a beech (*Fagus sylvatica*) and a spruce (*Picea abies*) stand in the Solling mountains, north-west Germany, 1969. There was no surface runoff (after Benecke and van der Ploeg, 1976).

Tree species	Precipitation	Evapotranspiration	Interception	Drainage	ΔS[a]
Beech	1064	307	152	583	+22
Spruce		384	320	384	–24

[a]Change in soil water storage.

Table 3.2. Annual groundwater recharge (in mm) as influenced by the form of land use and soil textural class for two levels of annual precipitation (after Strebel *et al.*, 1984).

		Land use		
Textural class	Precipitation	Arable	Grassland	Coniferous forest
Sand	600	200	190	130
	1000	500	450	330
Silt loam	600	100	90	50
(from loess)	1000	390	350	210

and a short period for vegetative growth, crop husbandry may be possible here only with restrictions, perhaps in the form of ley farming. This is a system whereby annual crops are cultivated only after fodder crops have been grown for a period of at least 2 years. But arable farming may also be completely out of the question. Taking these circumstances into account, the larger amounts of seepage water actually measured under forests are much more understandable. Soils of forests and pasture land are characterized by permanent rooting and surface protection by litter and leaves. These features and the humus accumulation in the surface layer mean that these soils are very well protected against erosion. Consequently, large rates of soil water infiltration strongly reduce runoff, and the waves of water discharge from mountainous areas become damped and the danger of flooding is averted. In contrast, when forests are cleared erosion increases. Then the soils no longer act like a sponge to soak up the water, and surface runoff rapidly takes place after the onset of rain, leading to flash floods, erosion and even landslides.

4 Properties and Energy State of Water

4.1 Physical–Chemical Properties

The chemical composition of the water molecule is quite simple. Nevertheless, water has very special and notable features because of its stereochemical configuration. The H–O–H atoms are not arranged in a straight line, but combine to form an angle of about 105°. Each hydrogen atom shares its single electron with the oxygen atom, which for its part contributes one electron to form a *shared pair of electrons* (Fig. 4.1). In this way the O atom attains eight electrons in the outer shell through the combination with its own *free pairs of electrons*. This arrangement represents a stable configuration. At the same time the first shell of electrons of each H atom is stabilized because the space available for two electrons is filled by the shared electrons (Fig. 4.1).

The shared pair of electrons is the basis of a covalent bond, also called an *atomic bond*. This results in the three atoms being held tightly together. Admittedly the water molecule is electrically neutral as a whole, but the two shared pairs of electrons build up a surplus of negative charge (δ–) at the oxygen atom. The reason is that the pairs of electrons revolve around the oxygen atom at a shorter distance than around the hydrogen atom. This shift of electrons towards the O atom explains also the positive surplus of charge (δ+) at the two H atoms (Fig. 4.1).

The *polarity* (strictly speaking the bipolarity) of the molecule induces a bond with neighbouring water molecules by hydrogen bridges. The molecules are coupled together by a link-up between the δ– oxygen side of one molecule and one of the two δ+ hydrogen sides of another molecule. Because of molecular motion the hydrogen bond will last only for a moment. Then the bond will break, but it will be re-established instantly with other molecules. As a result water molecules are interlinked, more or less as if in a crystal, but still constituting a liquid.

The polarity causes some unusual properties of water. The liquid is quite a good *solvent* for ionic substances like salts. But because of its own small molecular

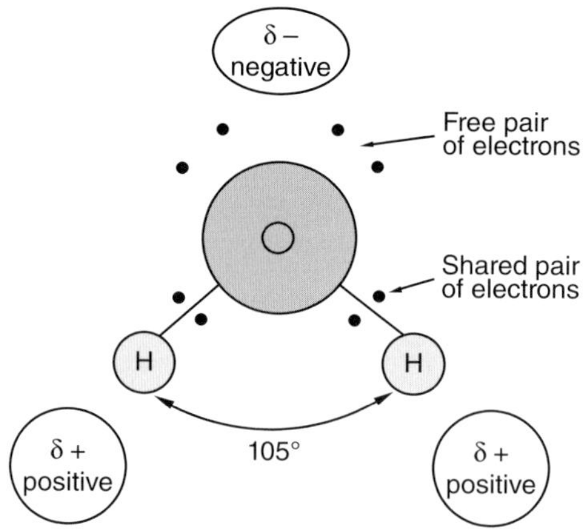

Fig. 4.1. The bipolar structure of the water molecule. The molecule is a dipole that is electrically neutral as a whole, but is characterized by opposite charges at the periphery. There are two centres of positive (δ+) and one centre of negative (δ–) charge.

size water will also dissolve quite a large number of other substances that occur in nature and which may be composed of larger molecules of various sizes. The polarity causes water molecules to be attracted by charged, ionic molecules and by non-electrolytes with a surplus of either positive or negative charges. Hydration shells are formed, loosening the bonds between the molecules of such substances, thus causing them to separate and dissolve.

Because of the bonding between water molecules the *specific heat* of water is large compared with other liquids. The calorie, a unit of heat once in common usage (1 cal = 4.19 joule, J), was defined as the amount needed to warm up 1 g of water from 14.5°C to 15.5°C. This relatively large energy input is required to increase the degree of freedom between water molecules that are interlinked by hydrogen bonds. The propensity for hydrogen bonding is an important property causing the large specific heat of water and is one of the main reasons why wet soils warm slowly under the radiant energy of the sun in spring. It also helps to explain how plants are protected from overheating. The *heat of vaporization* for water is also large because of the requirement to overcome the strength of the hydrogen bonding between water molecules. A comparatively big input of energy is necessary to convert the interlinked molecules of liquid water to an independent form: that of the vapour phase. When water evaporates from the leaf, the required energy is removed from the surroundings. This energy removal results in the cooling effect that is so important to the plant.

There are some other features of water that need to be mentioned, which also result from polarity and hydrogen bonding. They are the high boiling point and the viscosity of water.

Due to polarity, water molecules hold strongly together. The binding forces between the water molecules, called *cohesion*, are strong. But water molecules will not just link each other together. They will also be taken up by solid surfaces, for instance by cell walls or by soil particles. Between water molecules and solids the forces of *adhesion* are active. Cohesion makes it possible for more distant water molecules to adhere to soil solids. Cohesion and adhesion are responsible for the water retention capacity of soils. And in combination with the viscosity of water, these forces influence the water movement through the pores of soils and plants.

Another property of water, its *surface tension*, is also related to the cohesion force. At the water surface a molecule will be attracted by neighbouring molecules from all sides, but not from the direction of the vapour phase. That is why the surface molecule will be pulled into the liquid by molecules below the surface. A net force is acting on the water molecules at the boundary surface, which is directed inwards. The effect is as if an invisible, thin and elastic membrane covers the water surface. This phenomenon is called surface tension. The convex curvature of a water drop shows that the 'skin' pulls the drop together, thus creating a small surface area.

Adhesion and cohesion forces cause water to rise in glass tubes (Fig. 4.2) and in soil pores (Fig. 4.3). Water is pulled upwards by *capillary attraction*. Adhesion, attaching water molecules to the solid surface, and cohesion, which joins the water molecules together, combine to allow the water to rise. The adhesion causes the rise and the cohesion makes all the water molecules follow the upward pull. The water surface is curved but, in contrast to the water drop, the shape of the curvature is concave. The concave curvature indicates the presence of a pressure below the water surface being smaller than the surrounding pressure. The normal atmospheric pressure is around 1 bar (= 0.1 MPa). Therefore the concave shape signifies a subatmospheric or 'negative' pressure, which is caused by the weight of the hanging water column in the tube. The water will rise to the height where the adhesion and cohesion forces are counterbalanced by the weight of the water column.

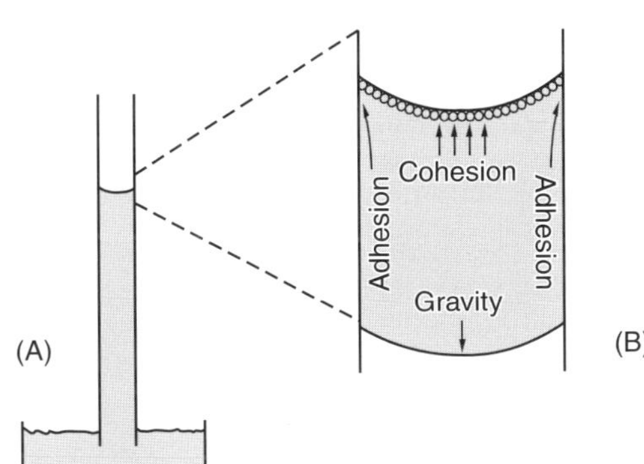

Fig. 4.2. The phenomenon of capillarity. (A) A glass tube is inserted into a water-filled bowl, and the water ascends in the tube. (B) Forces of adhesion are active at the circle of contact between liquid and solid phase, pulling the water column upwards. Forces of cohesion tie the water molecules together. Due to the force of gravity there is a subatmospheric pressure in the water column, signified by the concave curvature of the water surface (after Brady, 1990).

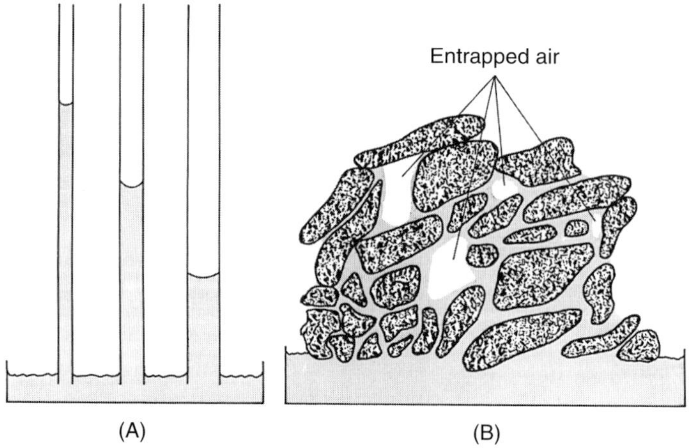

Fig. 4.3. In glass tubes (A) the height of capillary rise is largely determined by the diameter of the tube. In soils (B) the height of capillary rise is quite irregular, as size and continuity of pores can change. In addition, hollow spaces can occur between aggregates (after Brady, 1990).

In narrow tubes water will rise higher than in wider tubes (Fig. 4.3). The height h (cm) of capillary rise is given by:

$$h = 2\gamma(rdg)^{-1}\cos\alpha \qquad (4.1)$$

In this equation γ is the surface tension of water (N cm^{-1}), r is the radius of the tube (cm), d is the density of water (g cm^{-3}), g is the acceleration of gravity (cm s^{-2}) and α is the contact angle formed by the curved water surface and the solid surface.

By inserting the appropriate values and by assuming a contact angle of 0° as between water and quartz, one obtains a simple equation, the *capillary rise equation*:

$$h = 0.15r^{-1} \quad \text{(h and r in cm)} \qquad (4.2)$$

For the purpose of demonstration we will assume that in the xylem of a plant a tracheid acts as the water conducting tube. This is a wood vessel of perhaps 100 μm in diameter. According to Eqn 4.2 the capillary rise will be no more than 30 cm. Therefore the phenomenon of capillary rise can by no means explain the transport of water from the roots to the canopy of a large tree! Later on we will return to this point.

The soil does not contain perfectly cylindrical capillaries. Yet the water will enter and be retained in the very fine tortuous and winding spaces between the solid particles. In plant cells the walls are made up of very fine fibrils, which are soaked in water, just like a wick. This protects the walls from drying out even at those spots where water transfers from the liquid to the vapour phase. This is the case with the cell walls of the spongy mesophyll in leaves, which form large intercellular spaces. One of the most striking of these spaces is called the substomatal cavity or the air chamber, located just below a leaf stomate.

4.2 The Concept of Water Potential and the Darcy Equation

When the concept of field capacity was discussed in Section 3.1, we mentioned that the soil's pore space may be filled by rainwater during a heavy shower but that much of the surplus of stored water will drain out in a short period of time (2 days) after the rain has stopped. Therefore a change in the water content of soil is a regular occurrence. It is caused by the flow of water through the pores of the soil. Without this flow there could be no seepage of water or evapotranspiration (Fig. 3.3). The flow of water is a basic transport phenomenon in soils and plants. Before plants can deliver and return an appreciable amount of water from parenchymatous leaf cells through the stomates to the atmosphere in the form of vapour, the liquid water has to flow from the soil to the roots. From there it has to pass across the root cortex, arriving at the stele, from where it will flow through the vascular tissue of the stem to the leaves.

Why does the flow occur? What is the *driving force* for the water movement? It is certainly not a difference in water content, as can be seen from the following experiment. Two containers, one filled with a sandy soil and the other one filled with a clayey soil, are brought into contact. Each of the soils contains 20% by volume (20 vol.%) of water (Fig. 4.4). The question is: will water flow or not? The answer is: water will move from the sandy to the clayey soil. After some time, when the water redistribution has ended, the clayey soil will have the larger water content. Therefore we can conclude that differences in water content have not been the cause of water flow. Another quantity must have been operating to make the water move. This quantity will be described in some more detail.

Physicists distinguish between properties of *capacity* and *intensity*. We are familiar with one example from the science of electricity. The quantity of charge stored in a battery is a matter of capacity, but the rate with which the charge moves (the current) is dependent on the intensity, the potential difference (voltage) between the battery terminals. A quantity of electrical charge (coulomb) will be conducted as a difference in potential becomes effective. The electric current, I (ampere = coulomb s^{-1}), is proportional to the potential difference, V (volt). The factor of proportionality is the electrical conductivity (R^{-1}). The reciprocal (R) is the electrical resistance (ohm = volt $ampere^{-1}$).

It is *Ohm's law* that combines the three quantities in the transport equation for electric current:

$$I = 1/R \cdot V \tag{4.3}$$

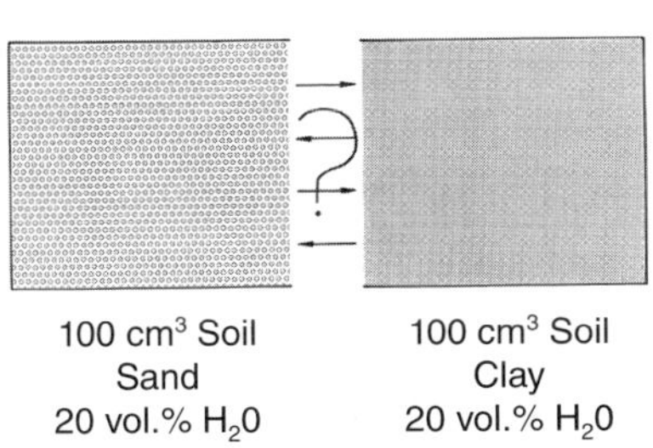

Fig. 4.4. Water flow between a sandy and a clayey soil, containing identical volumetric water contents. The question is: will water flow between the two soils of different textural class when the containers are brought into contact?

The potential difference (V) is the driving force for the electric current (I) that is conducted. The resistance (R) is a property of the conducting material (i.e. a metallic wire). R will increase with the length and will decrease with the cross-sectional area of the wire. It will also depend on the specific resistance, which is a constant for the metal concerned.

In the earlier example with soil water, it is the difference in potential of the water (a property of intensity) that causes an amount of water (a property of capacity) to flow from the sandy to the clayey soil. In the same way a difference in potential causes river water to flow downhill from the mountainside to the water meadow in the valley floor below.

The potential expresses thermodynamically the free energy that is associated with the water. The potential describes therefore the ability of water to do work. The free energy of the electrically neutral water depends on its concentration, the ambient pressure, the temperature and gravity. The potential is never set in absolute terms, but it is always seen as a relative quantity and is therefore compared to some normal or reference state or condition. The unit of the potential described above can be expressed as energy per mole of a substance, the substance here being water. The potential defined in this way is the *chemical potential* of water.

Such a definition of the potential is not very convenient for solving practical problems. In this case we need to consider the energy not per unit mole, but per unit of mass, volume or weight. Doing so changes the chemical potential of water to the *soil water potential* with more convenient units. The soil water potential is the quantity of intensity that makes the quantity of capacity, some part of the stored water in a soil section A, to flow in the direction of section B (Fig. 4.4). The relationship between the quantity of capacity and the quantity of intensity with respect to water flow was first described by Darcy in the 19th century. Henry Darcy (1803–1858) was a French hydraulic engineer, constructing wells and pipes. In his hometown of Dijon he built a central pure water supply system for the public in just 18 months. It was the first one in a European city (Philip, 1995). Similar to Eqn 4.3, Darcy's empirical equation reads:

$$q = K(\Delta\phi/\Delta z) \tag{4.4}$$

The term q is the *flux density* (cm^3 H_2O per cm^2 cross-sectional area per unit of time) or just the *flux* of water, K is reciprocal to a hydraulic resistance, named the *hydraulic conductivity* (K). The water potential is ϕ and z is a coordinate in the three-dimensional space, usually the one in the upward direction.

The *Darcy equation* states that a flux will originate only when there is a difference in water potential, namely $\Delta\phi$. The potential drops over the distance of Δz, which is, for instance, the distance between two soil layers or soil horizons. The quotient $\Delta\phi/\Delta z$ is called the *hydraulic gradient*. In contrast to diffusion processes water flux is a *mass flow* process, also termed convection.

The unit of z is centimetres (cm). What then is the unit of ϕ? Relating the free energy of water to unit mass, one obtains the unit J g^{-1} or erg g^{-1}. When taking unit volume, the unit of ϕ is J cm^{-3}. Because 1 J is 1 N m, one obtains the unit N cm^{-2} or dyn cm^{-2}, which are units of a pressure. When relating the energy to unit weight one obtains, for example, erg (g cm $s^{-2})^{-1}$ or erg dyn^{-1}. As an erg is a dyn cm, we obtain a unit for ϕ that is a length in cm – the head of water in a column, which makes calculations easy.

When choosing cm as the unit for ϕ, the resulting unit of K will be cm per unit of time. When choosing the unit of pressure for ϕ (i.e. N cm^{-2}), K takes on the unit of cm^4 N^{-1} per unit time. When using the unit Pa for the pressure, K will be in cm^2 Pa^{-1} per unit time. These examples help us realize that the unit of K depends on the chosen unit of ϕ and that it is straightforward to relate the water potential to unit weight. Doing so is simple as we can calculate with K as a length per unit time.

The water potential ϕ is the *total potential* of water. It is the sum of various component potentials. These component potentials are the *gravitational potential* Z, the *matric potential* Ψ, the *osmotic potential* O and the *pressure potential* P. O is caused by the presence of dissolved substances in the water. P is due to compressive forces acting on the water, for example the pressure of walls on living plant cells.

Water movement will always be driven by the hydraulic gradient $\Delta\phi/\Delta z$, and in the soil ϕ is composed of the gravitational potential Z and the matric potential Ψ. The physical cause of the matric potential Ψ is the adsorption of water molecules at the surfaces of inorganic and organic soil particles that constitute the soil matrix. The adsorptive forces represented by Ψ cause the ascent of water in dry soil.

Let us imagine water from a supply reservoir is rising through dry soil in a glass cylinder. The rise is caused by capillary action in the upward direction, the direction of the z coordinate (Fig. 4.5). After some time the ascent of water ends. Water is now in a state of equilibrium.

If we take the lower edge of the soil column as a reference level for the gravitational potential Z, it is easy to accept that a unit of water lifted above the reference level has a higher gravitational potential than the water in the reservoir. When relating the potential to the unit weight of water, the height of capillary rise will directly indicate the quantity of the potential in the unit of a length (i.e. in cm). The higher the capillary rise of water in the direction of the *z*-axis, the larger is the value of Z.

But now the transport rule applies: water will always flow from a site of higher potential to one of lower potential. Nevertheless there will not be any additional water movement due to capillary flow after attainment of equilibrium state, although

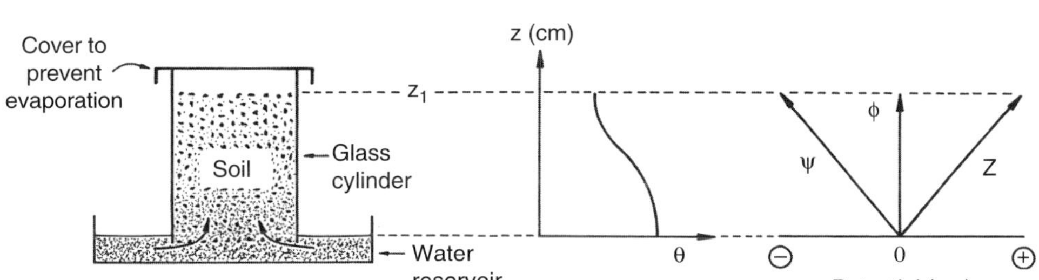

Fig. 4.5. Potential distribution within a soil column, where water has attained a state of equilibrium. The system is open to atmospheric pressure so that air replaced by water can escape from the pores without an air pressure increase. The term ϕ stands for the total potential, Z for the gravitational potential and ψ for the matric potential. The distribution of the volumetric water content θ in the soil column is also shown.

there are obvious differences in Z. The explanation is that it is not gravity, but the total potential that is decisive for water flow, and that is the same everywhere within the soil column. From this realization, it follows that the gravitational potential, (Z), at each height, is counterbalanced by another potential of equal size but opposite in sign (Fig. 4.5). This potential is always negative. It is the matric potential Ψ. At the level where the surface of free water or the water table is located, Ψ is zero. Below the water table, Ψ is replaced by the hydrostatic pressure potential, which is positive and increases with increasing depth below the water table. But with increasing height above the water table, Ψ will linearly decrease to the same extent as Z increases.

The osmotic potential (O), which is also always negative is only expressed when two water bodies are separated by a semipermeable membrane. As there is not a semipermeable membrane in the body of the soil water column, O is not important for water movement in the soil. Moreover, in soils of humid regions O is usually quite small numerically (only slightly negative). But in saline soils for instance, O may be much more negative and important due to the elevated concentration of solutes. In that case O may decide the rate of water uptake by the roots, usually reducing the ability of roots to extract water. In any case the osmotic potential is an important component of the total water potential in plants. Here also the pressure potential P is important.

Now we can express the composition of the total water potential by the following equation:

$$\phi = \Psi + Z + O + P \text{ (soil–plant system)} \tag{4.5}$$

In the soils of regions with humid climate the equation reduces to:

$$\phi = \Psi + Z \text{ (soil)} \tag{4.6}$$

In the plant these two component potentials are quite often not taken into consideration, as the two remaining component potentials O and P are of much higher order of magnitude. In such a case the equation reads:

$$\phi = O + P \text{ (plant cell)} \tag{4.7}$$

When considering the distribution of potentials within a tall tree, we will refer back to Eqn 4.5 with all the component potentials (see below).

According to the reference level of Z (for instance, the surface of free water in Fig. 4.5), the value of Z at a certain position in the soil or in the soil–plant system will change, and can attain positive or negative values. By definition, the matric potential Ψ is zero at the height of a 'free' water surface. Above the surface of free water Ψ is always negative. Also the osmotic potential O is always negative, whereas the pressure potential P is mainly positive as it is in turgid cells. But in drying cells it may possibly attain negative values. In the vascular tissue (the xylem of the plant) P is usually negative. But an exception is also recognized here, when plants develop a positive root pressure. We will develop these ideas in more detail over the next four chapters.

5 Water Storage and Movement in Soil

5.1 Fundamentals and Principles

Capillary attraction and the adsorption of water by surface forces of soil particles are expressed by the *matric potential*. The less water present in the pore space of the soil, the more negative is the matric potential. Alternatively, one can say that the smaller the water content, the larger is the positive value of the soil water *tension* or suction – expressions for a pressure less than that of the atmosphere. The relation between (volumetric) water content and water tension is described by the *soil moisture characteristic* curve, also known as the pF curve or the water retention function.

Schofield (1935) suggested the term pF by analogy with Soerensen's logarithmic scale of hydrogen ion concentration, the pH. The p derives from the Latin words *pondus* or *potentia*, the F from thermodynamic free energy. The pF expresses the tension of water, retained in the soil. It is the base 10 logarithm of the tension, which is measured as a head of water in centimetres (Fig. 5.1).

From Fig. 5.1 one can estimate the quantity of plant-available water as a fixed quantity with the unit of vol.%. This percentage is related to the *total* volume of the soil. The unit *vol.%* corresponds numerically to the notation '*millimetre of water* stored *per decimetre depth* of the soil profile'. When reporting the quantity of plant-available water this way, it is envisaged that the availability is identified from two 'fixed points'. At one end there is the *permanent wilting point* (PWP). At this point the residual water is held so firmly ($\Psi = -15{,}000$ cm or pF 4.2) at the surfaces of clay minerals that it cannot be extracted by plant roots. This residual water is therefore unavailable to plants. Of all the various textural classes, clay soils retain the largest percentage of *unavailable water* (Figs 3.2 and 5.1).

The other fixed point is the *field capacity* (FC). FC (Fig. 5.1) refers to the water content that is retained within the soil profile by adsorptive and capillary forces, after surplus rain or irrigation water has drained away. The water is retained

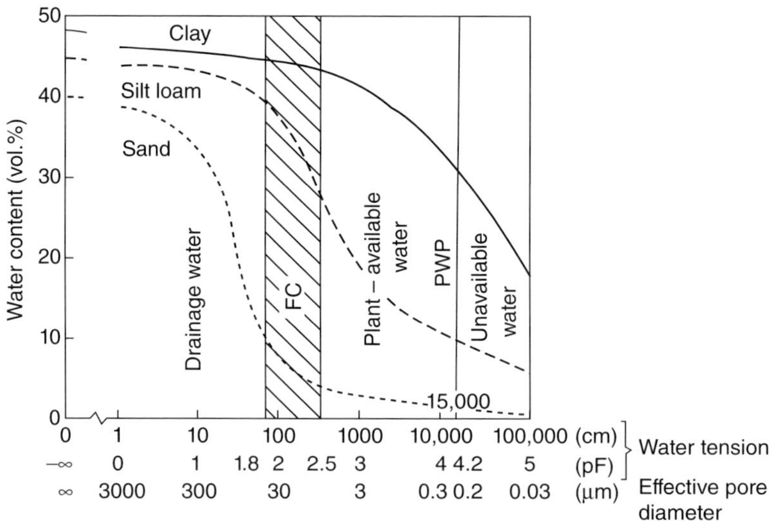

Fig. 5.1. The relationship between volumetric water content and water tension in a sand, a silt loam and a clay soil. The shape of the soil moisture characteristic curve is modified by the soil textural class. The quantity of plant-available water is fixed by the field capacity (FC) and the permanent wilting point (PWP). The supply of plant-available water is least in the sandy soil and largest in the silt loam soil. The water tension corresponds to an 'effective pore diameter' (see Eqn 4.2, the capillary rise equation, containing radius r). Therefore the shape of the curve is indicative of the pore size distribution. For instance, the sandy soil drains quite effectively between pF 1 (10 cm tension) and pF 1.8 (about 60 cm tension). It follows that the large pores with an effective diameter between 300 μm and 50 μm constitute a significant proportion of the total pore space, equivalent to 22 vol.%. The silt loam soil, on the other hand, contains a large percentage of medium-sized pores, which drain between pF 2 (30 μm) and pF 3 (3 μm). Finally, the clay soil holds about 12 vol.% of fine pores within the range of pF 3 (3 μm) and pF 4.2 (0.2 μm) (based on Scheffer and Schachtschabel, 1989).

'against the gravity force'. Usually the water content at FC is measured 2 days after the rain has ceased, assuming that by then the gravitational surplus water will have drained away. But according to Fig. 5.1 FC is not a fixed limit. Rather it may change between the tensions of 60 and 300 cm, corresponding to pF 1.8–2.5. In sandy soils the tension taken for FC is usually lower than in loamy or clayey soils. Despite the comparatively low water tension (pF 1.8), the water content at FC in sandy soils is clearly much less than in loam and clay soils, where the assumed tensions (up to pF 2.5) are greater (Fig. 5.1). The largest water content at FC is to be found in clay soils (Fig. 3.2).

With increasing clay content, FC and PWP will not change to the same extent (Fig. 3.2). FC tends to be close to the maximum in silt loam soils. But the PWP steadily increases with clay content. The difference between FC and the PWP represents the quantity of plant-available water, which tends to a maximum in silty and clayey loams. This maximum value is in the range of 22–25 vol.% of water (Figs 3.2 and 5.1).

The division of soil water into gravitational water, plant-available water and unavailable water is based on a *static* way of looking at things. In reality there is no reason to distinguish between these forms of soil water. Nevertheless, the concept is helpful in the development of ideas for estimating the quantity of plant-available water in spring or after an application of water by irrigation. It is also helpful for calculating the timely onset of irrigation during a period of drought, an approach widely used in more humid areas.

For the latter purpose let us assume a uniform, 'effective' rooting depth of 60 cm for all the soils shown in Fig. 5.1. In the 60-cm soil profile the sandy soil will store 54 mm of plant-available water (a volumetric water content of 10% at FC and 1% at PWP – see Fig. 5.1 – so 9 mm per 10 cm soil depth; therefore, over the 60-cm profile, 9 mm dm^{-1} × 6 dm = 54 mm). For the silt loam we get 150 mm, and for the clay soil there are about 84 mm. Let us further assume weather conditions with relatively high evaporative demand, approaching 6 mm day^{-1}. We suppose that under these conditions irrigation needs to start whenever 60% of the available water supply has been used and only 40% of stored water remains. The reason to start irrigation at this level of water storage is the finding, based on experience, that otherwise the dry matter production of crop plants will be impaired. Based on these considerations and information, we can calculate the period after which irrigation ought to restart. For the sandy soil the *time interval* is just 5 days (54 mm × 0.6 = 32.4 mm; 32.4 mm/6 mm day^{-1} = 5.4 days). For the silt loam we obtain a value of 15 days. The period for the clay soil is intermediate at 8 days.

When there is sufficient water in the soil, the actual rate of evapotranspiration (i.e. the combination of evaporation from the soil and transpiration by the crop) will more or less meet the evaporative demand. Under these circumstances the decrease in soil water storage can be calculated from the daily record of a climatic variable, called potential evapotranspiration, which is the basis of the *climatic water balance record*. In the case of rainfall, the amount that the soil moisture deficit is replenished can be estimated by calculation, using the equivalent height of precipitation. The store of available soil water is gradually filled up until FC has been attained. Any rainfall beyond FC is not taken into consideration. It is assumed that this surplus rain seeps away.

The climatic water balance, which is a static approach, provides a very simple technique of accounting. This technique allows us to control timing of irrigation and determine the amount of irrigation water required, just by recording the input and output of water. Measurement of soil water is not necessary.

Although the static approach is simple, it does not lead to an understanding of the physical foundations of the soil's hydrologic balance. There are no physical reasons for separating qualitatively different forms of soil water. For example, when a soil has become water saturated during a rainfall event and then for some time after rainfall (e.g. 2 days), large pores containing water will empty, the drainage will not stop even though FC has been reached. That is because not all of the remaining soil water will be retained 'against gravity'. Instead, the water will keep on flowing but the flow rate will decrease as the soil slowly dries and the tension increases. This implies that increasingly smaller pores are water-filled and will maintain the flow process. According to Eqn 4.4 the water flux gets smaller, as first the hydraulic conductivity decreases with the reduction in the soil water content, and secondly the hydraulic gradient also decreases. The gradient declines due to redistribution and drainage of water within the soil profile. When the gradient reaches zero, drainage finally stops.

About 40 years ago Hillel and van Bavel (1976) used a computer-assisted simulation model to calculate the water content and water tension distribution within the soil profile as a function of time, after the soil had been saturated with water to depth. The change in water tension and content resulted only from drainage. Any water loss by evaporation was deliberately excluded. The authors applied their simulation model to three soils of different textural classes, a sand, a loam and a clay. As expected, the water content decreased rapidly during the first 2 days of drainage, as shown in Fig. 5.2 for the 40-cm soil depth. The change was most obvious in the sandy soil. After 2 days of drainage the remaining soil water content was 45 vol.%, 29 vol.% and 16 vol.% for the clay, loam and sand soil, respectively.

These values of water content are similar to the data for FC that can be derived from Fig. 5.1. The values of water tension calculated for those particular water contents and for the three soil types were 130 cm, 110 cm and 80 cm or pF 2.11, 2.04 and 1.90. These tensions also fall well into the range presented for FC in Fig. 5.1. But, contrary to the static view, even after 2 days, seepage and water loss continued (Fig. 5.2). Between the second and tenth day, water losses were greatest from the loam soil (8 vol.%), whereas losses from the clay and sand were 3.5% and 4%, respectively. During that time interval the tensions continued to increase, reaching 180 cm in loam and clay and 130 cm in sand.

In contrast to the static concept, this *dynamic approach* leads to the conclusion that water movement in the soil is not an exceptional phenomenon, but the normal condition. Usually in the soil profile the total water potential is not the same at all locations but will vary between places. This inequality will cause water movement, a water flux. And the transport rule applies: *water always flows from*

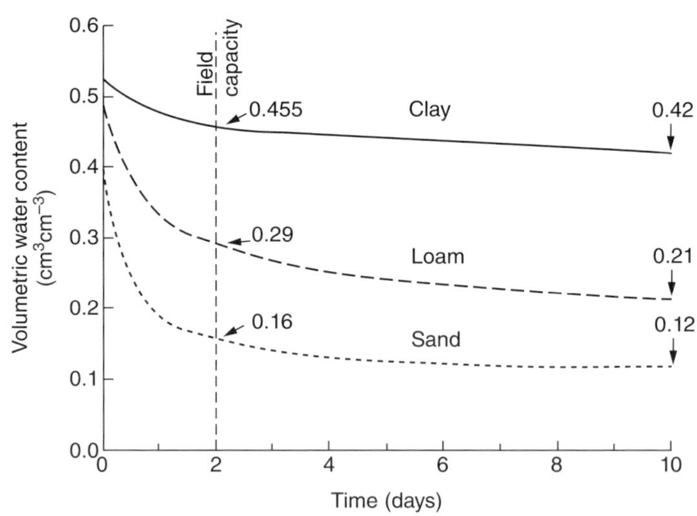

Fig. 5.2. Change of volumetric water content at the 40-cm soil depth of a clay, a loam and a sand soil as a function of time. Before drainage started, the soils had been saturated by water. Evaporation was prevented. Results of a simulation study (Hillel and van Bavel, 1976).

places of higher potential to places of lower potential. Differences in the total potential between two places cause the water to flow and the water flux at a given matric potential is proportional to the level of the hydraulic gradient (Eqn 4.4). For the water flow between two places it is not simply the particular matric potential that is crucial, but rather the magnitude of the total potential.

These more theoretical, abstract principles of water flow will now be illustrated by presenting some practical but simple examples. The illustration starts with Example 1 (pp. 35–36) and Example 2 (pp. 36–38).

Having become familiar with the arithmetic handling of matric and gravitational potential for identifying the direction of water flow, we now have to come back to Eqn 4.4. This equation tells us that the hydraulic gradient determines not only how much water flows but in which direction it moves. In addition to the gradient, the hydraulic conductivity (K) also has a determining influence on the

Example 1. In a salt-free soil the level of total potential is determined by two component potentials, the matric potential and the gravitational potential (Eqn 4.6). How can the gravitational potential be measured? What is its level at a certain spot in the soil profile?

Before answering these questions two things have to be decided. We have to fix a reference level for the gravitational potential, Z. We also have to decide on the direction of increasing values of Z. We will show later that this is equivalent to defining the positive direction of the *z*-axis. Please note that *z* pertains to a space coordinate, while Z is the gravitational potential.

Case A. Reference level for Z is the soil surface, the *z*-axis is directed upwards. What is the level of the gravitational potential Z at points P_1 and P_2? How large is the difference in gravitational potential between P_1 and P_2?

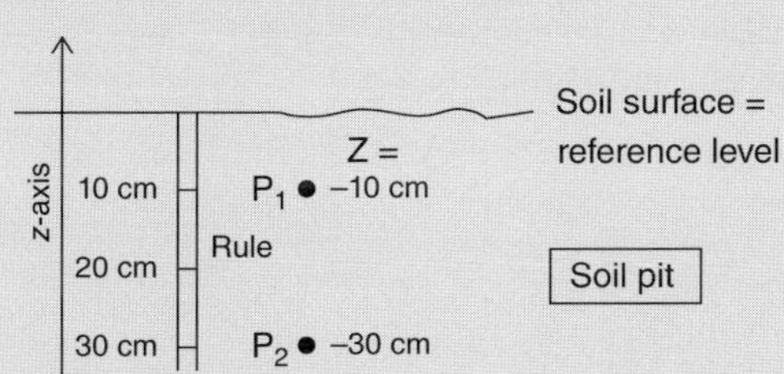

We will always define the positive *z*-axis in an upward direction. (One can also define the axis positively in a downward direction. By doing so, Eqn 4.6 will change to: $\phi = \Psi - Z$.) The reference level for the gravitational potential Z is the soil surface (case A). We use the *folding rule* to measure Z as the vertical distance to the reference level. At P_1 the distance is 10 cm, and Z_1, which is the gravitational potential at P_1, is by inference −10 cm. At P_2 the gravitational potential Z_2 is −30 cm. The difference in potential between P_1 and P_2 is:

$$\Delta Z = Z_1 - Z_2 = -10\,\text{cm} - (-30\,\text{cm}) = +20\,\text{cm}$$

Case B. The reference level for Z has now been shifted from the soil surface to the depth of P_2. The questions formulated in case A will be posed once more.

Continued

Example 1. Continued.

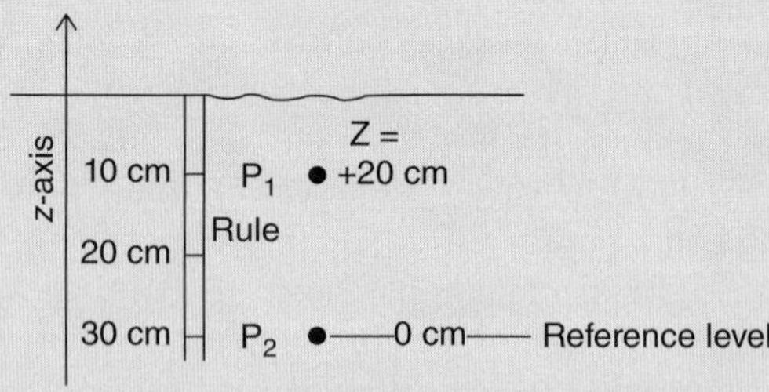

By shifting the reference level for the gravitational potential, the absolute value of the potential will be changed. Now Z_1 is +20 cm and Z_2 is 0 cm.

$$\Delta Z = Z_1 - Z_2 = +20\,\text{cm} - 0\,\text{cm} = +20\,\text{cm}$$

We notice that shifting the reference level changes the absolute value of the gravitational potential. However, the difference in the gravitational potential remains the same. For the movement of water in the soil only the difference is important.

Example 2. For the determination of the total potential (ϕ), in addition to the gravitational potential (Z), the matric potential (Ψ) also has to be known. The level of the matric potential (negative sign!) is determined using *tensiometers* (Box 5.1, pp. 41–45). The reference level for the matric potential is the free water surface (Fig. 4.5). The negative matric potential is numerically identical to the water tension. Tension is always designated by a positive sign.

Case A. Reference level for Z: soil surface, the positive *z*-axis directed upwards. At point P_1 in the soil profile the water tension (T) is 70 cm, and at P_2 it is 50 cm. The questions to be answered are: (i) Will water flow between P_1 and P_2, and if so, in what direction?; and (ii) How big is the hydraulic gradient (Eqn 4.4)?

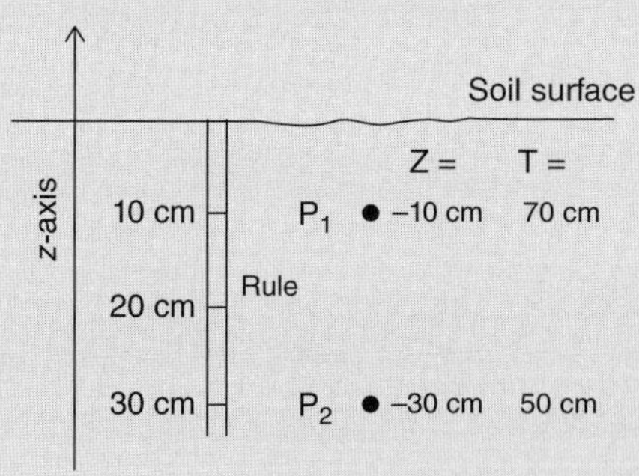

To answer the first question, the total potential ϕ must be calculated at P_1 and P_2 (Eqn 4.6):

$$\phi_1 = \Psi_1 + Z_1 = -70\,\text{cm} + (-10\,\text{cm}) = -80\,\text{cm}$$

$$\phi_2 = \Psi_2 + Z_2 = -50\,\text{cm} + (-30\,\text{cm}) = -80\,\text{cm}$$

The answer, therefore, is that in spite of unequal matric potentials there is no water flow between P_1 and P_2, as the total potentials at the two points are identical.

Continued

Example 2. Continued.

The answer to the second question, on the size of the hydraulic gradient, is as follows. The driving force of water flow (Eqn 4.4) is zero, as shown by the following calculation (all data in the unit cm):

$$\frac{\Delta\phi}{\Delta z} = \frac{(\Psi_1 + Z_1) - (\Psi_2 + Z_2)}{z_1 - z_2} = \frac{(-70 + (-10)) - (-50 + (-30))}{-10 - (-30)}$$

$$= \frac{-80 - (-80)}{20} = 0\,(\text{cm/cm})$$

Case B. Like case A, but at P_2 the tension measured is 70 cm as at P_1.

$$\phi_1 = \Psi_1 + Z_1 = -70\,\text{cm} + (-10\,\text{cm}) = -80\,\text{cm}$$

$$\phi_2 = \Psi_2 + Z_2 = -70\,\text{cm} + (-30\,\text{cm}) = -100\,\text{cm}$$

Although the water tension is identical at P_1 and P_2, the water will drain from P_1 to P_2. Now the hydraulic gradient is calculated:

$$\frac{\Delta\phi}{\Delta z} = \frac{-80 - (-100)}{-10 - (-30)} = \frac{20}{20} = 1\,(\text{cm cm}^{-1})$$

The hydraulic gradient is 1. As there is no gradient in the matric potential, the hydraulic gradient results just from differences in the gravitational potential at P_1 and P_2. The water seeps due to gravity. Such a case is found quite frequently in deeper layers of a homogeneous soil profile after a fairly long period of rain and subsequent drainage.

Case C. A period of rain is followed by a longer spell without precipitation. In addition to a loss of water by seepage from the lower part of the profile the upper part loses water by evaporation. We shall have a look at the upper part of the soil profile.

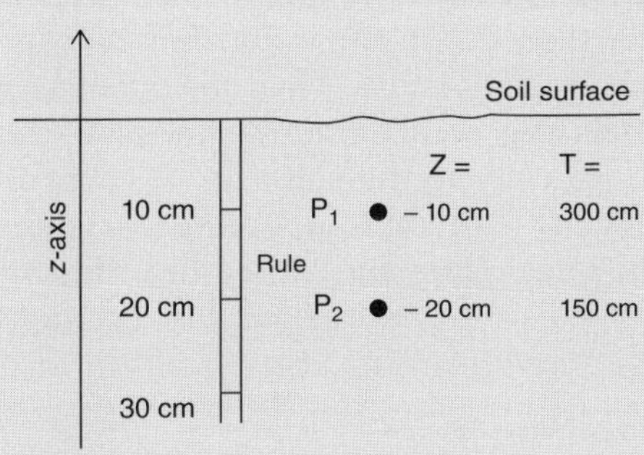

$$\frac{\Delta\phi}{\Delta z} = \frac{\phi_1 - \phi_2}{z_1 - z_2} = \frac{(-300 + (-10)) - (-150 + (-20))}{-10 - (-20)}$$

$$= \frac{-310 - (-170)}{10} = \frac{-140}{10} = -14\,(\text{cm cm}^{-1})$$

Continued

Example 2. Continued.

We notice that by fixing the positive *z*-axis in an upward direction, the calculated gradient is *negative* for the upward water movement caused by evaporation. The gradient will be *positive* on the other hand, when the water flow is directed downward such as during drainage. When taking the *z*-axis positively downwards and now taking account of the fact that $\phi = \Psi - Z$, then one should obtain numerically the same gradient, but with an opposite sign.

After all it does not matter whether the hydraulic gradient is calculated according to the formulas given or whether the sequence of the terms is changed as below:

$$\frac{\Delta\phi}{\Delta z} = \frac{\phi_2 - \phi_1}{z_2 - z_1}$$

The result will not be modified by the rearrangement (in the numerator as well as in the denominator!) of the numerals. *The best the reader can do to become clear with this issue is to calculate some examples with real figures according to the rules given. In doing so the reader will also learn the skills of handling algebraic signs.*

soil water 'flux'. Soil physics distinguishes between '*saturated*' and '*unsaturated*' conductivity, named K_s and K_u. These shorthand symbols signify that the soil is able to conduct water through its pore space when it is saturated as well as when it is only partly saturated. First of all we will start with the idea that water can flow only through pores that are filled with water. With increasing dryness due to water loss, the cross-sectional area of the soil capable of conducting water declines. (Conversely, the cross-sectional area of pores that contain and conduct air, increases.) Hence the soil's ability to conduct water will fall; that ability falls sharply, much more than just corresponding to the decrease of the wetted cross-sectional area. This is because the effective diameter of the remaining water-filled pores declines too (Fig. 5.1). The diameter predominantly governs the water flow rate through the pores. According to *Hagen–Poiseuille's law*, which is defined in a later chapter (Eqn 7.12), the water flow rate through a pore will decline to one-16th of its original value, when the diameter of the conducting pore is reduced just by half. In general terms, the law states that the pore water flow rate varies with pore diameter to the power of 4. Therefore soil dryness causes the hydraulic conductivity to decrease according to a power law (Fig. 5.3). Rewetting the soil helps refill larger pores and causes the conductivity to increase again.

We can rewrite Eqn 4.4, introducing two changes that now can be easily understood:

$$q = -K(\Psi)\cdot(\Delta\phi/\Delta z) \tag{5.1}$$

The term $K(\Psi)$ (read as: K as a function of Ψ) means the hydraulic conductivity K is a function of the matric potential Ψ. As Ψ changes with the volumetric water content θ of the soil (Fig. 5.1), K may of course also be regarded as a function of θ (therefore $K(\theta)$. The minus sign has been introduced for obtaining the result: a positively marked water flux is flowing in the direction of the positive *z*-axis, no matter if the axis is set positively upwards or downwards. The use of Eqn 5.1 is illustrated in Example 3.

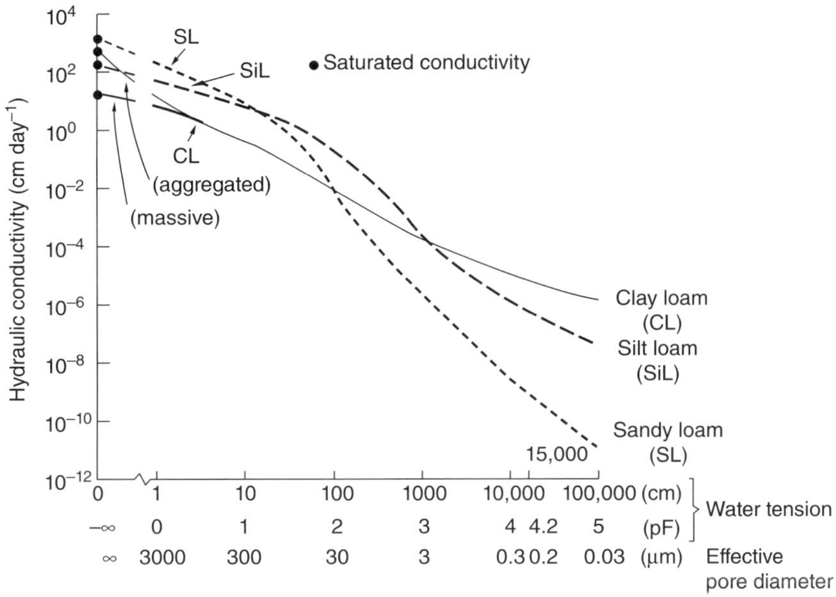

Fig. 5.3. Hydraulic conductivity functions for a sandy loam, a silt loam and a clay loam soil. The respective values for saturated hydraulic conductivity are presented by a dot. In an aggregated clay soil the saturated conductivity is greater than in a massive clay soil.

Example 3. Using Example 2, case C, we would like to know how much water will flow from P_2 to P_1 during 1 day. We shall suppose that water content and water tension will not be changed 'significantly' by the water flow within the time period of 1 day. First, a mean water tension is calculated for the flow distance between P_2 and P_1. It amounts to 225 cm (mean of 150 cm and 300 cm). From a 'hydraulic conductivity function' of the soil in question – as depicted for instance in Fig. 5.3 – we can read a conductivity at the mean tension of – let's say – 10^{-2} cm day^{-1}. The water flux q is calculated by:

$$q = -K(\Delta\phi / \Delta z) = -0.01\left(\text{cm day}^{-1}\right) \times -14\left(\text{cm cm}^{-1}\right) = 0.14\left(\text{cm day}^{-1}\right)$$

Therefore during 1 day 0.14 cm of water is flowing from point P_2 to P_1, or 1.4 mm. When the potentials involved have the unit of cm (see above), the hydraulic gradient will be dimensionless (cm cm^{-1}). At the same time K will obtain the dimension of a velocity with the units of cm day^{-1}. Seemingly, q also has the dimension of a velocity, but that is not the case. Rather it is that the unit cm day^{-1} is the short form of the unit cm^3 water per cm^2 cross-sectional area of the soil per day, the correct unit of the water flux. The actual *flow velocity* of the water through the soil matrix, the pore water velocity, is much greater than we calculate for q. That depends primarily on the fact that only part of the cross-sectional area of the soil is porous and not solid. Only the porous part is able to conduct water, provided the pores are wetted, whereas at the same time non-wetted pores support aeration. A second reason for the larger flow velocity is the fact that the water molecules will not move forward in the soil from point P_1 to point P_2 in the form of a straight line. Instead they have to flow around the mineral and organic solid particles. That causes the actual distance of water movement to be tortuous, and hence longer than the linear distance measured between P_1 and P_2.

Figure 5.3 shows some typical *hydraulic conductivity functions* for a sandy loam, a silt loam and a clay loam soil. It is quite obvious that the conductivity in the wet range is greatest for the sandy loam soil. The reason is the high proportion of large pores with an effective diameter of >50 μm, which can be derived from the shape of the corresponding soil moisture characteristic curves (Fig. 5.1). Compared with the silt and the clay loam soil, the unsaturated conductivity of the sandy loam soil declines sharply with increasing water tension and the diminishing effective diameter of the water conducting pores. The silty loam soil has a comparatively large conductivity over a broad range of water tension (see Fig. 5.1), because of the sizeable percentage of medium-sized pores with an effective diameter between 3 μm and 30 μm. At water saturation the conductivity of the clay loam soil (and clay soils) may be quite large, due to the presence of planar voids and cracks between aggregates that have been formed earlier by shrinkage during periods of drying. They will drain at very small tensions, if they do not close by swelling. In clayey soils that do not develop structural units and have not formed aggregates, the saturated conductivity will be relatively small (see data for clay loam soil in Fig. 5.3). The large percentage of fine pores (<3 μm in diameter) explains the relatively high level of the unsaturated hydraulic conductivity of clay soils in the 'dry' range at pF values >3 (Fig. 5.1).

Example 4 introduces water flow causing a change in soil water content. From that example we can formulate the *principle of mass conservation*: the divergence of the fluxes of a material causes a change in the content of that material. Related to Example 4 we can indicate:

A divergence of water fluxes causes a change in water content:

$$dq/dz = d\theta/dt \tag{5.2}$$

If the difference in water fluxes is not defined by $q_2 - q_1$ but by $q_1 - q_2$, the algebraic sign will change:

$$dq/dz = -d\theta/dt \tag{5.3}$$

Equation 5.2 is also known as the *equation of continuity*. The equation can be combined with the Darcy equation (Eqn 5.1), thus obtaining the so-called *Richards equation*:

$$d\theta/dt = d(K(\Psi) \cdot d\phi/dz)/dz \tag{5.4}$$

With the aid of modern, high-speed computers, Eqn 5.4 can be handled even by non-mathematicians. The flow processes, varying in space and time, are split up into an enormous number of small portions. This is achieved by breaking down the two variables – space and time – into discrete steps. Over a short distance (e.g. 1 cm) soil water fluxes are brought about by gradients in water potential. These fluxes can cause, within a short time period (e.g. 1 min), a change in water content within an elemental soil volume. The computer program will calculate the fluxes and the accompanying temporal and spatial changes in water tension, in water content and in the conductivity, brought about by these

Example 4. *Stationary flow conditions*, as indicated in Example 3, are seldom to be found in soils. Under such *steady state* conditions water will flow through a soil volume without changing the soil's water content θ (Fig. 5.4A). *Transient* or *non-stationary flow conditions* will prevail whenever water flow causes change in the soil water content. When, for instance, more water is flowing into the soil layer from 10 cm to 20 cm depth than flowing out (the level of the two fluxes is symbolized by the length of the arrows), the water content in the layer will increase (Fig. 5.4B).

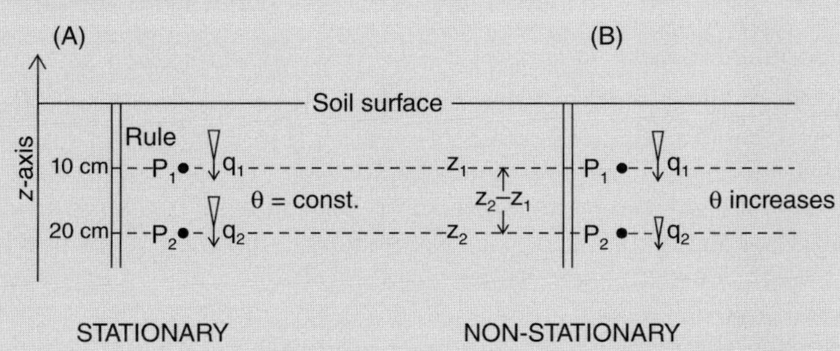

$$q_2 - q_1 = 0$$

$$q_2 - q_1 = \int_{z_2}^{z_1} \frac{d\theta}{dt} dz$$

Fig. 5.4. Stationary (A) and non-stationary or transient (B) flow conditions in soil.

As an example of non-stationary flow conditions the following figures may serve: q_2 = –1 cm day^{-1}; q_1 = –2 cm day^{-1}. The unit cm day^{-1} is shorthand for cm^3 cm^{-2} day^{-1}. It follows that:

$$-1-(-2)\ \text{cm}^3\ \text{cm}^{-2}\ \text{day}^{-1} = \int_{-20\,\text{cm}}^{-10\,\text{cm}} \left(0.1\ \text{cm}^3\ \text{cm}^{-3}\ \text{day}^{-1}\right) \cdot dz$$

$$1\text{cm}^3\ \text{cm}^{-2}\ \text{day}^{-1} = 10 \cdot \left(0.1\ \text{cm}^3\ \text{cm}^{-3}\text{day}^{-1}\right) \cdot 1\text{cm}$$

$$1\text{cm day}^{-1} = 1\text{cm day}^{-1}$$

The difference in water fluxes at the two boundaries of 20 cm and 10 cm depth is 1 cm^3 cm^{-2} day^{-1} = 1 cm day^{-1}. As a result, the water content as an average of the soil layer from 10 cm to 20 cm depth increases daily by 0.1 cm^3 cm^{-3}. That means that on average each of the 1-cm-deep layers will undergo a change in water content of 0.1 cm day^{-1} = 1 mm day^{-1}. For the 10-cm layer it adds up to 1 cm day^{-1}. In other words, the difference in fluxes at 10 cm and 20 cm (1 cm day^{-1}) is numerically the same as the change in water content in the 10–20-cm layer (1 cm day^{-1}).

Box 5.1. Measuring soil water

Water potential

The soil water potential, strictly speaking the *matric potential* of soil water, is measured by *tensiometers* (Richards, 1928, 1942). The matric potential corresponds to a subatmospheric pressure (value has a negative sign) that will be recorded as a water suction or *water tension* (value with positive sign). An indispensable part of the tensiometer is a porous cup (Fig. B5.1) made from ceramic with very fine pores. The cup retains water within these fine

Continued

Box 5.1. Continued

pores against air pressure acting upon the cup. Under conditions of increasing air pressure the cup continues to hold the water until a critical pressure is reached, the so-called air entry value, which depends on the diameter of the largest pore within the material (Eqn 4.2). Therefore below this value the cup will stop any air exchange between tensiometer and surroundings, even when there is a difference in air pressure. The cup is permanently linked to a manometer via a water-filled tube through an airtight connection. Such a manometer is either a commercial vacuum gauge (Fig. B5.1 middle) or a mercury manometer (Fig. B5.1 right).

For an understanding of the principle of measurement, we can imagine the tensiometer being inserted into a soil at a certain degree of dryness. Some water will pass from the tensiometer into the surrounding soil by flowing through the pores of the cup. The loss of water causes a small negative pressure within the tensiometer. The water flow through the cup into the soil will come to an end when the matric potential in the soil and the negative pressure in the tensiometer reach the same value. At this stage of equilibrium the sub-atmospheric pressure in the device, indicated by the manometer, corresponds to the matric potential in the soil. During rainfall, when water is infiltrating and wetting the soil down to the installation depth of the cup (Fig. B5.1), the tensiometer will absorb water from the soil. The matric potential will increase, which will be shown by the development of a less negative pressure within the tensiometer.

Depending on the length of the connecting tube of the tensiometer, porous cups can be positioned at various soil depths. Before insertion, it is necessary to auger a tight-fitting

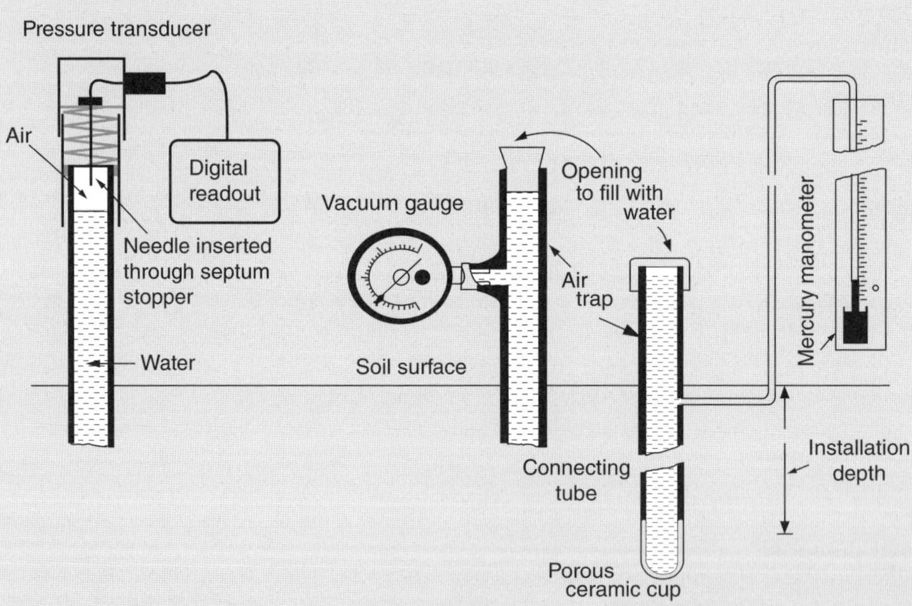

Fig. B5.1. Body of a tensiometer, either with a vacuum gauge (middle) or combined with a mercury manometer (right) (after S.J. Richards, 1965), or (left) with a pressure transducer temporarily connected via a hollow needle (based on Or, 2001).

Continued

Box 5.1. Continued

hole to the appropriate depth. After insertion, the tube and, more importantly, the cup have to be in close contact with the soil matrix.

Tensiometers have a range of measurement that is restricted to potentials above about –800 cm (–0.8 bar). Significantly, extremely small volumes of air get trapped on the surfaces of the ceramic of the cup and within the body of the instrument during the filling process. As the subatmospheric pressure within the tensiometer decreases, any existing bubbles of air expand but others will be released from the internal surfaces (Take and Bolton, 2003). Bubbles expand because, in contrast to liquid water, they have no tensile strength. Finally they interrupt the water column and prevent the correct transmission of pressure from the cup to the manometer. In addition, under large negative pressure, bubbles of water vapour can form with the same instrument failure resulting. As the tensiometer cup is permeable to both water and solutes, the osmotic potential of the soil cannot be determined. A mercury type manometer allows more precise measurement than a commercial vacuum gauge (Fig. B5.1) but there is always the danger of environmental contamination by unintentional mercury spillage. Nowadays, pressure transducers – connected to an electronic recording device – are available (Or, 2001), which are easier to maintain. These pressure transducers do not need a permanent connection with the tensiometer body but can be joined to the tensiometer for just the short period required to make the measurement. For that purpose a hollow needle is pushed through the cap on top of the connecting tube, which includes a self sealing septum (Fig. B5.1 left). By inserting the needle the hydraulic connection between tensiometer and pressure transducer is established, ideally without any change of pressure.

For field measurement of soil water potential in the high-tension range, i.e. less than –800 cm potential, ordinary water-filled tensiometers are unreliable. To overcome this restriction 'high capacity tensiometers' have been developed. These consist of a porous ceramic disk just 5–10 mm in diameter set into a cylindrical unit sealed behind the disk to a strain gauge pressure sensor (Ridley and Burland, 1993, 1999; Guan and Fredlund, 1997). The volume of the water-filled space between the disk and the strain gauge is also small (3–70 mm^3). The maximum pore size within the ceramic disk is such that they do not drain even as the soil dries to water potentials close to the PWP. To reduce the problems of air bubbles the tensiometer unit has to be filled with degassed water under vacuum and then be subjected to positive pressure before being inserted into the soil. Tensiometers of similar structural design, but filled with an osmotically active synthetic polymer rather than water, have the benefit that the pressure within the body of the tensiometer is positive. This allows measurements of tension to values beyond the PWP (Bakker *et al.*, 2007; van der Ploeg *et al.*, 2008) but as the osmotic potential of the polymer is a function of temperature, a further calibration is required to determine the soil water potential. The applicability of these two techniques still needs to be confirmed for use in the field.

Another method for measuring soil moisture tension is the electrical resistance block. These instruments were invented in the middle of the last century (Bouyoucos and Mick, 1940). The blocks are made of gypsum, nylon or fibreglass, all of them in porous form. Buried in the soil such a unit will adjust its water content according to the water tension of the soil in contact with the block. The blocks contain a pair of plate electrodes to measure the electrical resistance. The resistance is related to the water tension of the block material, and after tension equilibration between block and soil that tension is indicative of soil moisture tension.

The 'porous-matrix sensor' (Whalley *et al.*, 2007) has been developed more recently from the more voluminous resistance blocks. Like an ordinary tensiometer the instrument can be inserted into the soil after drilling a tight-fitting access hole to the desired depth of measurement. The zone of contact with the soil, the cup, is made from porous ceramic

Continued

Box 5.1. Continued

material. Within the cup the electrical resistance or – more recently – the dielectric constant are measured (Whalley *et al.*, 2013). These properties change with the water tension of the surrounding soil. This tension is inferred from a calibration function. The measurement range is limited to about 3 bar.

An inexpensive method with simple materials provides the *filter paper method* (Hamblin, 1981). A standard filter paper of certified quality is brought into contact with the soil, and is allowed to equilibrate with the soil water tension. After reaching equilibrium, which can take more than a week in relatively dry soil, the water content of the paper is determined. Ahead of time a calibration function between water content and tension must have been established for the paper. With this known 'paper moisture characteristic' the water tension of the soil is derived from the registered water content of the paper. The possible tension range extends from 0.01 bar to 100 bar.

Soil water content

The *water content* of the soil can be measured easily and precisely (when necessary to some metres depth) by taking soil samples with an *auger or sampling tube*. The method is more difficult in soils containing gravel and stones. After weighing the moist samples, separated according to depth, they are dried at 105°C and again weighed. The gravimetric water content is expressed as a percentage of the *dry* soil, not of the wet soil. For obtaining the volumetric water content, the gravimetric content has to be multiplied by the corresponding bulk density of the soil. Expressing the water content in volume per cent is numerically identical to the storage of water in the soil profile expressed by the unit in millimetres (mm) of water per decimetre of soil depth (see Section 5.1).

Over a period of time, soil augering will disrupt the soil profile at a study site. With the aid of radioactive probes the water content can be determined without excessive disturbance. Aluminium or PVC access tubes have to be inserted into the soil for guiding the probes, the insertion depth being dependent on the intended measuring depth. After calibration at the site, a *neutron-scattering moisture meter* will show directly the water content on a volumetric basis. Unlike the neutron probe with one tube, the *gamma ray transmission technique* requires two exactly parallel access tubes for receiving a double probe. This double probe consists of the gamma ray source, separated from the detector by some horizontal distance. The depth resolution of the gamma ray technique is more accurate than that of the neutron probe, which emits and detects the neutrons in a sphere of soil that changes in volume according to the water content. For the determination of the volumetric water content with the aid of gamma rays, the bulk density must be determined separately.

A more recent technique for measuring soil water content is *time-domain reflectometry* (TDR) developed by Annan, Davis and Topp (Topp *et al.*, 1980, 2003). The TDR technique makes use of the large differences in the dielectric constant between air (1), dry soil (3–7) and water (80) (Zegelin *et al.*, 1992). This constant characterizes the tendency of molecules to orient themselves in an electrostatic force field (Hillel, 1998). Because of the water molecule's polarity (Section 4.1), the dielectric constant of water is comparatively large. For the determination of the soil water content a unit something like a tuning fork with parallel rods is inserted into the soil, though two, three or more rods arranged equidistant from a central one are also in common use (Jones *et al.*, 2002). An electromagnetic pulse is propagated along the rods, which serve as a wave-guide. The soil volume between and around the rods provides the dielectric medium, which changes its dielectric constant with water content. An oscilloscope can be used to show when the step voltage pulse leaves the source and when

Continued

Box 5.1. Continued

it returns after reflection at the free end of the rods. This is the time of travel. Travel speed is calculated from the length of the rods and the travel time. It varies inversely with the dielectric constant in a known manner. Then the volumetric water content can be derived from the dielectric constant using an empirical calibration (Jury *et al.*, 1991). The water content is an average value for the soil over the length of the rods, which may be inserted vertically or horizontally or in any other direction into the soil (Kutilek and Nielsen, 1994). For measuring steep changes in water content, as in a soil profile where roots are extracting water, it is advisable to create a permanent installation for at least a season, having several units that can be inserted horizontally at different depths.

By placing samples of undisturbed soil in a *pressure-plate apparatus* (Richards, 1941) the moisture content at predetermined values of water tension can be determined in the laboratory. Before placing the core samples in the pressure-plate apparatus, they are normally saturated with water by *capillary wetting*. On the other hand, sample saturation by extracting air from the soil pore space under vacuum causes the complete filling of all the pores present, leaving no air entrapped in dead-end pores and holes. Such a technique of *vacuum wetting* is preferable for determining the soil's pore size distribution. But one has to be aware that vacuum wetting can change porosity especially in clayey soils. After wetting, samples are equilibrated at various pressures during desorption. The pressure applied can be much greater than the suction that can be determined by normal water-filled tensiometers (Fig. B5.1). As a first approximation the results can be used to estimate water tensions from water content values measured in the field for the range where the data from tensiometers are unreliable during dry periods.

fluxes. The calculated change of state is stored and these data will serve as initial conditions for the succeeding flow event, considered again over a short time and a small distance. The non-linear, partial differential equation (Eqn 5.4) is difficult to solve analytically. Therefore it is taken as a starting point for applying a huge number of difference equations, which attain reliability and accuracy compared with the actual events by choosing quite restricted, 'finite differences': Eqn 5.4 is being solved using *numerical methods* (Kutilek and Nielsen, 1994).

5.2 Evaporation

Evaporation is one of the components of the water balance equation (Eqn 3.1). It is an output quantity, which causes the soil water content to decrease. Quite often the word evaporation is supplemented by the term 'unproductive', as *unproductive evaporation* is a form of water use that is, in contrast to transpiration from a leaf, not directly linked to net primary production. Due to evaporation, bare soils that are wetted-up by winter rains will dry from the surface in spring. Such a soil condition is highly appreciated by some farmers, as it supports cultivation for seedbed preparation and seeding. On the other hand, evaporation causes a significant *water loss* in regions of dryland farming. Often the annual rainfall is insufficient to support annual crop growth and yield. Under such conditions farmers may find themselves compelled to introduce a break between cropping years with a year of fallow so that the rainfall of 2 years can be used for just one cropping season. This dryland farming technique is not very effective, mainly because water loss results mostly from evaporation rather than seepage. Often the 'water storage efficiency' does not amount to more than 20–30% of the annual precipitation in the fallow year. We will return to this problem and its solution in later sections.

During evaporation, water that leaves wet soil changes from the liquid to the vapour phase at or near the soil surface. This change is an energy consuming process, the largest proportion being derived from radiant energy. The water molecules will diffuse away from the soil surface into the air, according to the difference in vapour pressure. In addition to diffusion, the removal of water vapour will be markedly accelerated under conditions of mass flow, also called convection. Convective vapour movement can originate from local heating of the air near the soil surface or can be caused by wind. All three *pre-conditions* for evaporation – energy supply, saturation deficit and convection – are dependent on climatic factors that determine the *evaporative demand* or the 'potential evaporation'. Evaporation from moist soil is influenced to a much lesser degree by the nature of the soil surface (i.e. by colour and albedo, aggregation and roughness, presence of mulch as a barrier to water vapour) than it is by atmospheric conditions.

The earth receives a minute part of the radiation that the sun emits (about the two thousand millionth part, which is about 5×10^{-8}%). Outside the earth's atmosphere and normal to the solar beam the energy supply is about 1.36 kJ m^{-2} s^{-1} or 1.36 kW m^{-2} or something less than 2 cal cm^{-2} min^{-1} (the *solar constant*). The units given explain that the rate of solar radiant energy transmitted to the earth is expressed as a flux density. Solar radiation is *short-wave radiation*, the wavelength varying between 300 and 3000 nm (Fig. 5.5). When passing through the atmosphere, part of the radiation is absorbed and reflected. Thus over the northern hemisphere, on average about 47% of the solar constant strikes the soil surface in the form of direct and diffuse radiation (Fig. 5.5). In arid regions with sparse amounts of cloud, the percentage will be higher, up to 70% or so. That part arriving at the ground is called *global* or *total radiation* (R_T). The radiation density of 1 cal cm^{-2} (4.19 J cm^{-2}) was known as a langley (ly) before the general adoption of SI units. During a day with 12 h of sunshine, the energy reaching the outer atmosphere can accumulate to 1440 ly (2 ly min^{-2} × 60 min × 12 h) (6.03 kJ cm^{-2}). Based on the calculated radiant energy of 47% of the solar constant, the global radiation striking the ground per day in the northern hemisphere is around 677 ly (2.83 kJ cm^{-2} or 28.34 MJ m^{-2}). In actuality of course, R_T depends on the season and the time of day, on latitude, altitude, cloudiness, atmospheric turbidity, slope and aspect.

The global radiation R_T supplies the energy for the evaporation of water. The equation of the radiation balance reads:

$$R_T\,(1-r)-R_L=R_N=G+H+LE+A \tag{5.5}$$

The unit of all the energy flux terms in Eqn 5.5 is J cm^{-2} day^{-1}. The term r stands for the reflectivity coefficient, also called the *albedo*. The albedo, a material-specific property, marks the fraction of the incoming short-wave radiant energy being reflected. A dark, wet, bare soil with a rough surface may have an albedo of 0.1–0.2 (10–20%), whereas a dry, light-coloured soil can attain an albedo of 0.3–0.35.

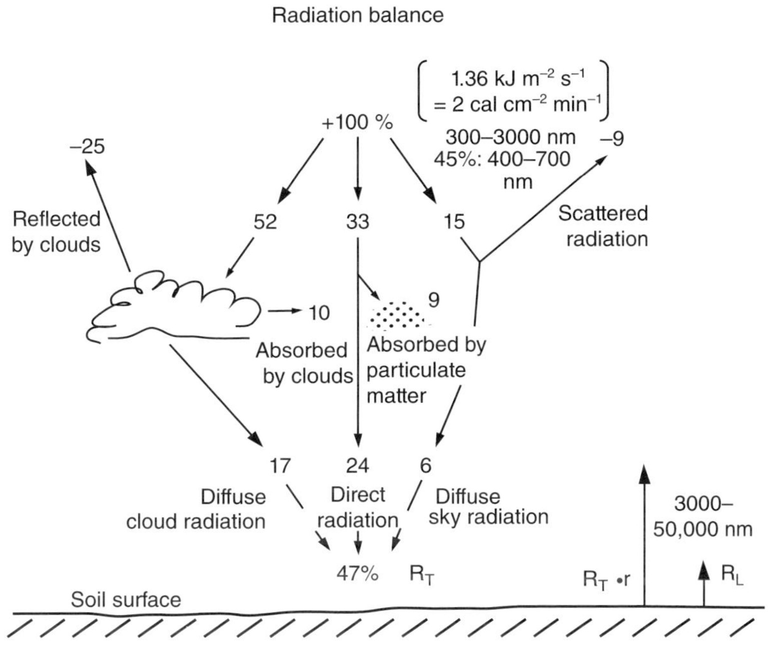

Fig. 5.5. The radiation balance in the northern hemisphere. Explanations are given in text (after Larcher, 1973).

As r increases, less energy remains for evaporation. The radiation not reflected, i.e. $R_T(1-r)$, will be *absorbed* by the bare soil, usually in the range of 60–90%. Part of the absorbed radiation will be *back radiated* from the ground to the atmosphere in the form of *long-wave radiation*, R_L, the wavelength varying between 3000 and 50,000 nm (Fig. 5.5). A very small part of the outgoing long-wave radiation is backscattered by the clouds. Therefore, R_L is the net long-wave radiation at the ground.

What is left as a balance of incoming and outgoing radiant energy fluxes at the ground is the *net radiation*, R_N. The net radiation dissipates in a number of ways, including as an energy flux to heat the soil (G) and the air (H), the heated air being perceptible (*sensible heat flux*). Quite a small part of the energy flux is fixed during assimilation in the form of chemical compounds (A), when plants cover the ground, and the rest of the energy flux is consumed in the evaporation of water (LE). In LE the L stands for the latent heat of vaporization (585 cal g^{-1} or 2.45 kJ g^{-1} at 20 °C) and E is the evaporation rate (g cm^{-2} day^{-1}). The heat of vaporization is used for changing water from the liquid to the gas phase. Therefore the energy flux expended for evaporation is latent and not perceptible, unless some energy is taken from the evaporating body itself, which will cool down. The term LE is also named the *latent heat flux*.

The *actual soil evaporation rate* at a site will normally not exceed the *potential rate of evaporation*, unless there is some additional energy inflow from other sites on top of the radiant energy flux. At such sites, air might have been heated up under conditions of short water supply, which restricts the energy consumption by evaporation (Eqn 5.5). Heated air, created at drier sites, can be transferred to other sites by wind, thereby increasing the evaporative demand on that site. The energy transfer in form of heated air is called *advection*, and has a significant impact in arid regions on local water use, for example at irrigated sites. Of course advection can also affect the water balance in more humid regions.

The *potential evaporation*, set by atmospheric conditions, can be estimated using special devices. One of them, small and useful in remote terrain, is the Piche evaporimeter, often used by plant ecologists. Another device, much bigger and not transportable, is the well-known 'Class A pan evaporimeter' (Fig. 5.6), constructed by the US Weather Bureau. It is a circular pan, 121 cm in diameter and 25.5 cm deep (Hillel, 1998). Now it is used in many countries. Besides estimation from direct evaporation measurements, the potential evaporation can be calculated on the basis of the radiation or energy balance equation (Eqn 5.5), but it may be more reasonable from the standpoint of the technology of measurement to apply a combination of energy balance and mass transfer (or aerodynamic) methods (Taylor and Ashcroft, 1972). Penman (1948) developed such a combination method (Box 8.1). The *Penman equation* for the daily evaporation from a free water surface is given by:

$$LE_p = \frac{(\Delta/\gamma)(R_N - G) + LE_a}{\Delta/\gamma + 1} \tag{5.6}$$

In Eqn 5.6: L is the latent heat of vaporization (J g^{-1}), E_p is the potential evaporation rate (g cm^{-2} day^{-1}), Δ is the slope of the function of saturated vapour pressure versus temperature (mbar °C^{-1}), γ is the psychrometric constant (mbar °C^{-1}), and

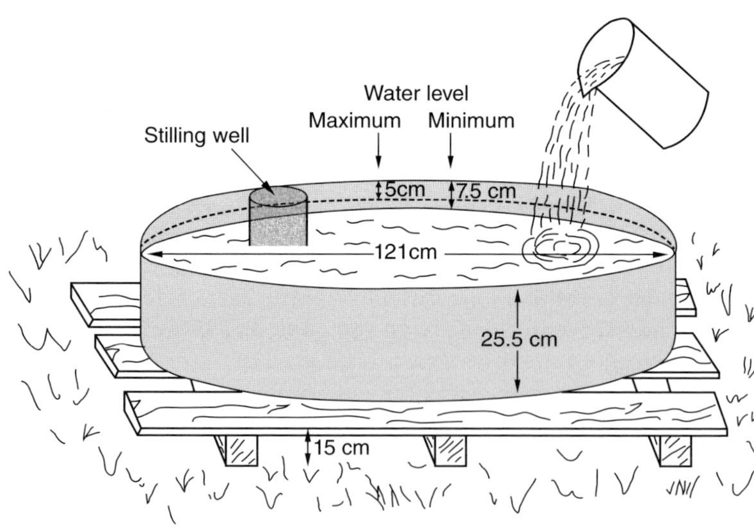

Fig. 5.6. The Class A pan evaporimeter, originating from the USA. Water loss by evaporation is estimated daily in the early morning by levelling the drop of the water surface in the stilling well. In replacing the evaporated water, the level has to be maintained within 5–7.5 cm from the rim (based on Doorenbos and Pruitt, 1977; Allen *et al.*, 1998; Hillel, 1998).

R_N and G are the energy flux of net radiation and of heating the soil (J cm^{-2} day^{-1}), respectively. The expression E_a is a ventilation–humidity term, which takes into consideration the influence of wind speed and the vapour pressure deficit of the air on evaporation. Equation 5.7 expands on this term:

$$E_a = 0.263(0.5 + 0.00622\,U)(e_s - e) \qquad (5.7)$$

In the equation, U is the daily wind speed (i.e. the 'run of wind' over the day) and $e_s - e$ is the mean saturation deficit of the air during that day. This deficit is the difference between the saturated vapour pressure of the air (e_s), which depends on air temperature (Fig. 8.3), and the actual vapour pressure (e). The coefficients given in Eqn 5.7 require that U has the unit km day^{-1} and $e_s - e$ is in hPa or mbar (Table 7.1). As a result, the numerical value of E_a obtained has the unit mm day^{-1}, which is a desirable unit. But in Eqn 5.6, LE_p, R_N and G are all in J cm^{-2} day^{-1}. To obtain that unit for E_a, the numerical value calculated by use of Eqn 5.7 has first to be divided by 10, thus converting mm day^{-1} into g cm^{-2} day^{-1}. Secondly, that value has to be multiplied by L (J g^{-1}).

The term LE_p of Eqn 5.6 is calculated in the unit J cm^{-2} day^{-1}. To obtain the potential evaporation rate (E_p) directly in mm day^{-1}, we have to divide LE_p by L (2450 J g^{-1}). We obtain the unit g cm^{-2} day^{-1}. Assuming the density of water to be 1 g cm^{-3}, we obtain the unit cm day^{-1}. We end up with the rule: for conversion of LE_p (J cm^{-2} day^{-1}) into E_p (mm day^{-1}), divide the numerical value of LE_p by 245. About 250 J cm^{-2} (or 60 ly) correspond to 1 mm H_2O.

To calculate the potential evaporation by use of the Penman equation, the following weather variables must be measured in the field: (i) *net radiation*; (ii) *air temperature*; (iii) *wind run*; and (iv) *relative humidity*. Temperature and relative humidity determine the saturation deficit of the air. Therefore, all the measured variables completely satisfy the three pre-conditions for evaporation mentioned above. In most cases soil heating (G in Eqn 5.6) is ignored, something that strictly speaking is incorrect. Warming the soil will consume radiant energy, especially during spring in temperate regions when it can absorb as much as 30% of R_N (Abdul-Ghani, 1979). As the soil cools down towards the colder part of the year, it releases energy and adds to the heat supply from radiation. (However, when calculating potential 'evapotranspiration', see Section 8.2, it is reasonable to assume that there is no net heating of the soil during the main period of crop growth.) Where net radiation is not measured directly, it can be approximated from measured *hours of sunshine* (Withers and Vipond, 1974). The great advantage of the combination method like that of Penman is that all the climatic variables can be measured at only one height (2 m), whereas aerodynamic methods require measurements to be made at two or more heights. The measurement of gradients is much more difficult and subject to error than is the recording of relevant data at only one height.

The actual evaporation rate from a bare soil will account for about 90% of the calculated Penman evaporation (Hillel, 1980; Hartge and Horn, 1991). But that is only true when the soil is wet and in the first stage of evaporation (Fig. 5.7). Altogether there are three distinct stages of evaporation (Idso *et al.*, 1974). The first stage is a stage having a constant evaporation rate. The daily rate depends solely on external, meteorological conditions, the potential evaporation (E_p).

The smaller the evaporative demand of the atmosphere and the greater the initial water content, the longer this stage of constant evaporation rate lasts (Fig. 5.7).

When the soil dries near the evaporating surface during the *first stage*, the decrease in the unsaturated hydraulic conductivity is compensated for by an increase of the hydraulic gradient (Darcy equation, Eqn 5.1), keeping the evaporative water flux at a constant level. But when the soil surface layer has reached air-dryness (i.e. the minimum achievable water content through exchange with the air) the gradient will not increase any more. Rather the gradient will decrease with increasing depth of the dry soil layer, and the first stage of evaporation will have come to an end. In a sandy soil the conductivity will decrease during drying much more sharply than in a clayey soil (Fig. 5.3). That is the reason why, in the sandy soil, the first stage lasts only a comparatively short time (Fig. 5.8). The same argument can be employed for the comparison between a soil layer with

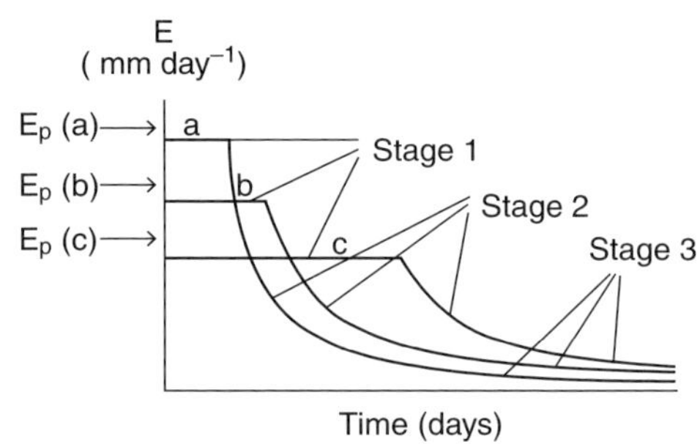

Fig. 5.7. Daily rates of soil evaporation indicating the three consecutive stages of evaporation, namely a constant rate stage, a falling rate stage and finally another nearly constant rate stage. The actual daily soil evaporation during these stages is modified by three levels of potential evaporation. At a low rate of potential evaporation E_p(c), the first stage (c) will last longer than the first stage (a) under the condition of a higher potential evaporation E_p(a). The medium rate of potential evaporation is E_p(b). The evaporation process starts at a soil wetness that is close to field capacity (based on Hillel, 1971, 1980).

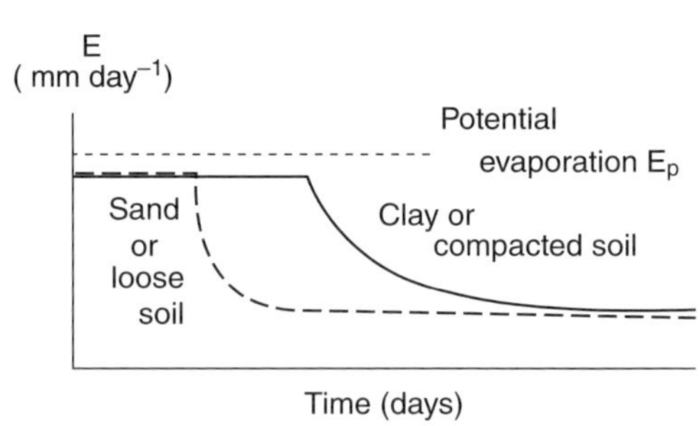

Fig. 5.8. Daily rates of evaporation from soils close to field capacity, as influenced by soil texture and structure. In a sandy soil the stage 2 evaporation will be attained earlier than in a clayey soil. The same is true when comparing a soil with a loose aggregated structure and one with a massive (compacted) structure.

loose aggregated structure and one with a massive structure. Loosening the soil by tillage will interrupt the 'capillary conduction' (Fig. 5.8) by decreasing the unsaturated conductivity at tensions > pF 1.8–2 (Fig. 5.3), thus shortening the time period of the first constant rate stage (see Section 17.3).

During stage 1 (*a constant rate stage*) the level of the daily evaporation rate is fixed above all by the level of E_p during that day (Fig. 5.7). This means that the actual rate is controlled mainly by *external meteorological conditions*, among other things by the energy supply. Also the duration of the first stage depends on E_p (Fig. 5.7), but is also influenced by soil texture and structure (Fig. 5.8).

The *second stage* is a *falling rate stage*, and the decreasing rates of soil evaporation are always smaller than E_p. The actual level of the daily rates is determined during this period of soil drying by the ability of the soil to conduct the water to its surface. During this stage the actual evaporation is controlled by *internal soil conditions*, which are governed by the soil hydraulic properties. This ability depends on the conductivity function (Fig. 5.3) and the build-up of hydraulic gradients. The ability is much more developed in coherent clayey soils than in single-grained sandy soils. Field observations in spring or after the end of the rainy season demonstrate that sandy soils with a deep groundwater level will dry from the surface much more rapidly than neighbouring clayey soils. In an arable field, the compressed traffic lanes will often remain wet and dark coloured over a longer time period than in the seedbed between the lanes, which retains an uncompacted coarse granular structure. The seedbed with the friable structure appears to dry and brighten up from the top quicker than the lanes, as the water flow to the surface is reduced earlier and declines faster.

From these considerations we can draw the following conclusion: rapid top layer drying is a sign of early water saving. But very often the opposite is believed, namely that the early drying off from the surface of a soil of good tilth is a sign of greater water loss, whereas the wet surface of compacted tracks is considered to signify only a limited evaporative loss.

The coarser a soil is in texture or aggregation, the earlier the onset of *soil-based regulation* of evaporation, which results in the reduction of water loss during dry periods. But there is an exception to the rule: when aggregates exposed at the soil surface are too coarse, even where a seedbed has been prepared by secondary tillage, wind can force air through the soil. This can enhance the loss of soil moisture. The wind brings drier air in contact with the surfaces of exposed aggregates, increasing evaporation and mass flow from coarse secondary pores and cracks within the topsoil. Sometimes there are situations when such a forced rapid drying down to seeding depth is desirable.

During the *third stage* of evaporation almost *constant rates* are again attained, but at a much lower level (Figs 5.7 and 5.8). At this stage the drying front will have penetrated the soil to a depth of some centimetres. In the dry surface layer, the soil equilibrates with the humidity of the air just above the soil surface. A soil that has dried down to PWP (pF 4.2), is still surprisingly moist, relative to the humidity of the outside air. At PWP the relative humidity (RH) is as high as 98.85% (Fig. 5.9). A RH of 50% corresponds to pF 6. When the soil reaches such an extreme state of dryness, only very thin water films remain at the surfaces of clay minerals, just a few molecular layers thick. The flow of liquid water is then practically zero (Fig. 5.3), and most of the

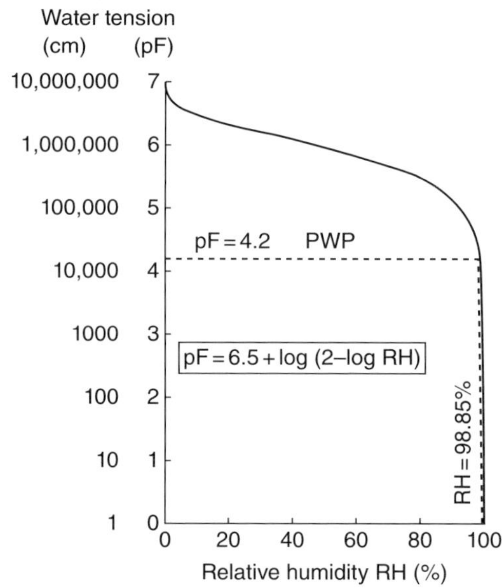

Fig. 5.9. The relationship between relative humidity of the air (RH) and soil water tension. The water tension is given in cm and in pF values. The permanent wilting point (PWP) corresponds to pF 4.2 and a relative humidity of 98.85%. The relationship shown holds for 20 °C (based on Schofield, 1935).

water movement is in the vapour phase. The rate of vapour diffusion is relatively small when there is only a small concentration gradient to drive water vapour transport.

It is not too difficult to determine soil evaporation in the absence of plants. Water tension and water content have to be measured as a function of soil depth over time. The water tension, recorded by tensiometers, indicates the depth of the *hydraulic water divide*. This is a *plane of zero flux* within the soil profile. At this depth the hydraulic gradient is zero (Section 5.1, Example 2, Case A). As the soil dries after being thoroughly wetted by rain, the zero flux plane will migrate from the soil surface downwards in the profile. Above the zero flux plane, the daily changes in volumetric soil water content, integrated over depth, correspond to the daily evaporation rate (Ehlers, 1978). Some of our field investigations were carried out on a Luvisol derived from loess near Göttingen. In Fig. 5.10, the relative evaporation rate (i.e. the actual rate divided by the potential rate) is plotted as a function of the water tension, measured at a depth of 10 cm. When the tension approached the range of FC, the soil began to restrict evaporation. At 1000 cm water tension the evaporation rate was near zero (Fig. 5.10). At that stage the topsoil layer (0–2 cm) had become air-dry.

5.3 Infiltration and Water Transport

A soil profile, where much of the soil water has been used, can be replenished by means of water infiltration. Infiltration is the process of water entry into the soil. The water, coming from rain or irrigation, has to enter the soil surface. The less

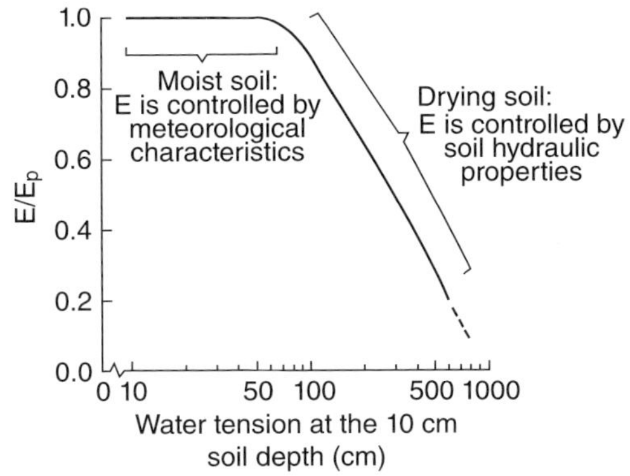

Fig. 5.10. The relative evaporation rate (E/E_p) as a function of the water tension at 10 cm depth in a fallow loess-derived Luvisol. E is the actual and E_p the potential evaporation rate. This function is soil-specific and has to be re-established according to texture (sandy to clayey) and structure (after Müller and Ehlers, 1987).

water the soil can absorb at a high rate, the sooner water accumulates at the soil surface. This *surface water* can cause *soil erosion* on sloping terrain as it runs off downhill.

Both infiltration and evaporation processes are governed by *external* climatic variables as well as by *internal* soil properties. In a uniform soil, the actual infiltration rate can never be greater than the precipitation rate. The maximum intake rate that is possible under conditions of prolonged heavy rainfall is a potential infiltration rate. This is an intrinsic soil property, also called the *infiltrability* (Hillel, 1971) – an expression that is adopted here. The infiltrability is determined by texture and structure of the soil. But it is also highly dependent on the soil water content. As long as the precipitation rate does not exceed the infiltrability, surface water cannot accumulate. Rather all the rainwater will infiltrate into the soil. This *pre-ponding period* is characterized by non-ponded conditions. Yet when the rain continues at a constant and sufficiently high rate some moment will occur called incipient ponding (Hillel, 1971), when *non-ponded infiltration* ceases and surface water will become evident. The reason for the puddle formation during the *ponding period* is the steady decline of the infiltrability with increasing water storage in the soil profile. At the moment when surface water becomes apparent, *ponded infiltration* has started (Fig. 5.11).

When rainwater infiltrates into a homogeneous, dry soil, the soil profile more or less becomes saturated, and always from the top downwards, at a rate depending on the level of the rainfall rate (q_r). This infiltration process represents an extreme case of the non-stationary or transient flow condition (Fig. 5.4), causing changes in water content, water tension and hydraulic conductivity. For the vertical flow process the Richards equation had been introduced (Eqn 5.4):

$$\frac{d\theta}{dt}=\frac{d(\text{flux})}{dz}=\frac{d}{dz}\left(K\frac{d\phi}{dz}\right)=\frac{d}{dz}\left[K\frac{d\Psi+dZ}{dz}\right]=\frac{d}{dz}\left[K\frac{d\Psi}{dz}+K\right] \tag{5.8}$$

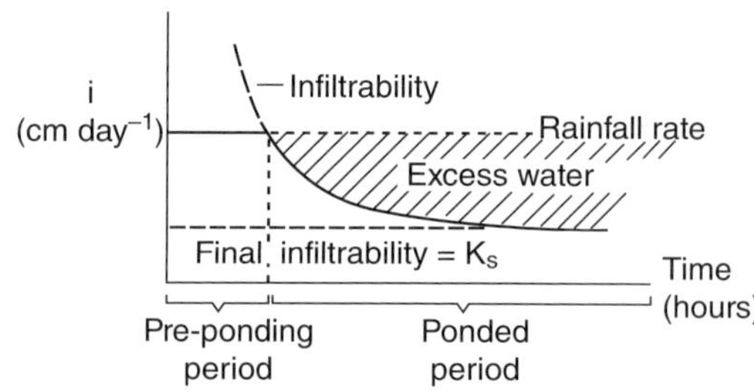

Fig. 5.11. The actual infiltration rate (i) at constant rainfall rate as a function of time. During the pre-ponding period the infiltrability, a variable soil property, is greater than i. During that period i corresponds to the rainfall rate. But once the soil is ponded, i corresponds to the infiltrability, and both are smaller than the rainfall rate, causing an excess of surface water. After a long time the final infiltrability will be reached, which corresponds to the field-measured saturated hydraulic conductivity (K_s) (based on Hillel, 1971, 1980, 1998).

With the aid of a computer the equation can be used to calculate the infiltration process. This requires that the differential equation be replaced by a huge number of difference equations. A difference equation portrays the differential form of that equation for finite but very small steps in time and distance. For instance, the difference form of Eqn 5.8 describes the situation of water flow during infiltration by the wording: when an ingoing water flux diverges from an outgoing water flux over a minute soil depth, the volumetric water content in a minute layer of that depth will change over a very short time interval. The computed outcome of a discrete water flow event over a small increment of time and depth is stored by the computer as an interim result. And there are large quantities of interim results to be stored that serve as the initial conditions for the next step of an ongoing flow process, again taking place over small increments in time and depth. For the computer simulation, the moisture characteristic and the conductivity function have to be known. From the input data, the computer will not just report the change in water content according to depth and time, but will also indicate the variation in tension and conductivity.

From the foregoing description it is evident that for the execution of the computer program a *boundary condition* must be known, namely the rainfall rate during the period of the simulation. Only in the pre-ponding time will the rainfall rate correspond to the flux through the soil surface. Afterwards, during the ponded time, the flux gets smaller. Finally, to get the program started, an *initial condition* has to be established – the water content distribution within the soil profile at the beginning of the simulation period. The water content distribution fixes the variation in water tension within the profile and hence the hydraulic gradients between soil layers, the initial hydraulic conductivities, and the initial fluxes between layers. To obtain stable solutions, the program internally adjusts the step width in time and depth.

Now the physical background of the infiltration process can be explained. A homogeneous soil will be considered with identical physical properties in the various soil layers. When water infiltrates into the soil at constant rate i, eventually the upper horizon, subdivided into layers or compartments, will be filled with water to some percentage. At such an advanced stage of infiltration, the term $d\theta/dt$ of Eqn 5.8 is zero in the compartments. Then the ingoing and outgoing fluxes of the upper compartments will be identical: the water flux passing through these compartments will coincide with i, and will correspond to q_r, as long as the infiltrability does not limit infiltration (Fig. 5.12). In the example q_r is taken to be relatively small.

As depicted in the figure, at the moment t_1, the uppermost layer (compartment 1) had not refilled to a water content of 30 vol.%, the maximum attainable value under this small q_r. It is assumed that the unsaturated conductivity was relatively small and that the flux through the compartment, set by q_r, could be achieved only in the presence of a relatively steep hydraulic gradient (Darcy equation). The steep gradient is depicted in Fig. 5.12 by a sharply confined wetting zone (for explanation see Fig. 5.13). The sharp boundary at this zone is characterized by a steep change in water content over a short increment of depth. At time t_3, compartment 1 was filled with water up to 30%. At that moment the conductivity was greater than at t_1, the gradient is less steep and the wetting zone spreads.

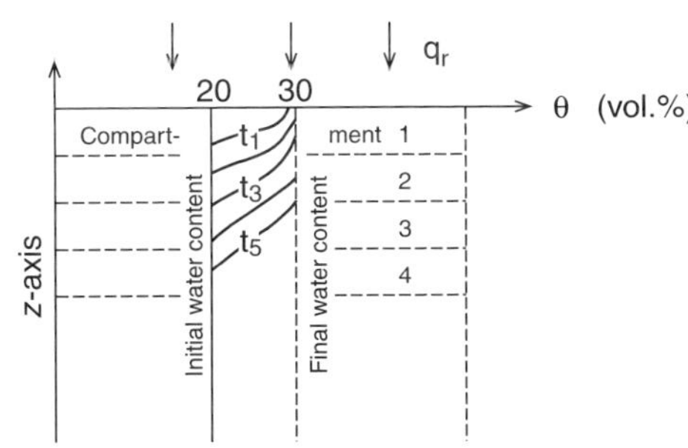

Fig. 5.12. Water content profiles during infiltration into a homogeneous soil at different times (t_i) at low rainfall rate (q_r). The initial water content is 20 vol.%, the final water content is 30 vol.%.

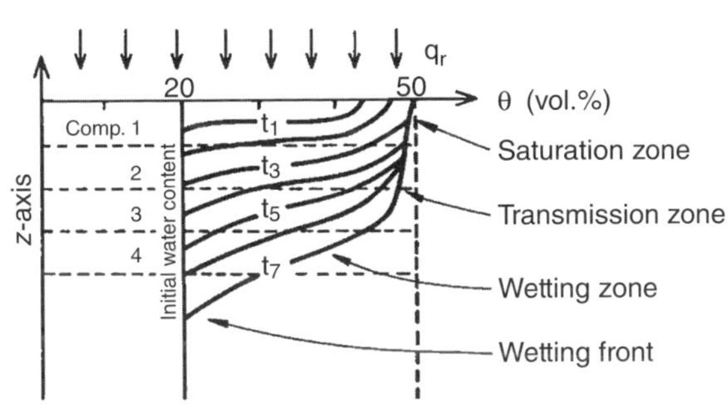

Fig. 5.13. Water content profiles during infiltration into a homogeneous soil at different times (t_i) at a large rainfall rate (q_r). In the saturation zone the pore volume of the soil is almost completely filled with water. In the transmission zone the air content in blocked pore spaces is slightly higher. In the saturation and transmission zones the hydraulic gradient is unity. The infiltration rate is equal to the saturated conductivity (K_s) (based on Hillel, 1971, 1980).

We shall have a look again at the Darcy equation:

$$\begin{aligned} q &= -K(d\phi/dz) \\ &= -K(d\Psi/dz) - K(dZ/dz) \\ &= -K(d\Psi/dz) - K \end{aligned} \quad (5.9)$$

After a prolonged period of infiltration (t_5) the gradient in matric potential $d\Psi/dz$ attains the value of zero in the uppermost compartment 1. At that stage the hydraulic gradient is reduced to the value of the gradient in the gravitational potential, that gradient being one. The water content in compartment 1 will not increase any more, and stays fixed at 30 vol.%. The fluxes in and out of the compartment are the same. We come to the realization that under the *steady state* condition in the upper compartment, the flow rate, the flux q, attains the quantity of the unsaturated conductivity K (Eqn 5.9). Under these conditions q is numerically equal to K and has the same value as the infiltration rate i and rainfall rate q_r.

At greater rainfall rates (Fig. 5.13) the water content profile will be shifted to larger values than when it is raining at slow rates (Fig. 5.12). The maximum pore filling may correspond to the total pore volume less the part occupied by pockets of soil air that cannot escape. Under such conditions, i attains the value of the saturated conductivity (K_s) but only after a long time (Fig. 5.11).

We must conclude that the final infiltrability can never be greater than K_s (Fig. 5.11). Faster infiltration rates or larger infiltrabilities in dry soils are always the consequence of steeper hydraulic gradients. The gradients will be reduced in the course of time and will tend to a value of one. That is the most important physical reason why *infiltrability declines with time*. But there are other reasons for the decline, including: (i) disruption of soil aggregates (distinctive in loamy soils); (ii) swelling (clayey soils); (iii) blocking of pores by fine material (fine sands); (iv) clogging by silting-up (silty soils); and (v) compression of trapped air at the wetting front.

Where shrinkage cracks are present, the initial infiltration rates in clay soils can be very large (cf. Fig. 5.3; for a more general assessment of water flow along favoured pathways see Box 5.2 on pp. 57–58). As the soil wets, values can drop off quite rapidly due to swelling and the decrease of K_s as cracks close. Reduction in the infiltrability of silty soils is caused by poor aggregate stability and surface sealing. Generally the infiltrability of soils accords with texture and is in the sequence: sand > loam > clay, corresponding to the sequence of K_s (Fig. 5.3). We can conclude that loamy-silty as well as clayey soils can run into problems during irrigation.

During ponded infiltration, the infiltration rate also decreases over time (Fig. 5.11). The relationship between infiltration rate (i) and time (t) was formulated theoretically by Philip (1957). His two-term *infiltration equation* reads:

$$i(t) = 1/2\,S\,t^{-1/2} + K_s \quad (5.10)$$

The S stands for sorptivity (cm day$^{-1/2}$) and K_s for the hydraulic conductivity of the saturated soil near the surface (Youngs, 1995). Sorptivity is determined by

Box 5.2. Soil structure and preferential flow

The relation between soil water content and water tension is the *soil moisture characteristic* (Fig. 5.1). It varies according to the texture (Fig. 3.2), organic matter and structure of the soil (Fig. 17.1A). *Soil structure* is the spatial arrangement of solid particles comprised of minerals and organic matter. The solids serve as a framework for the voids, the spaces between the particles. Particles and voids determine the number and volume of pores available for the storage (Fig. 5.1) and the transmission (Fig. 5.3) of water. In a soil with sand, silt and clay particles aggregated into discrete units, structure refers to the shape of aggregates as well as to the pore system within and between them (Section 17.2). The shape of the moisture characteristic curve is determined by the volume of water removed from the pores as the tension increases. From capillary rise (Eqn 4.2) we know that each value of water tension corresponds to an 'effective pore diameter' (Figs 5.1, 5.3). Therefore, the water released is also equivalent to the volume of pores with the range of effective diameters given by the lower and higher tensions. Calculations for discrete changes in tension along the moisture characteristic allow the *pore size distribution* of the soil to be determined. Taking values of water content measured in the field, together with the corresponding values of water tension, allows the '*field moisture characteristic*' to be established. Data are often highly variable. Although there are many reasons for the variability in field data, one cause is *hysteresis*. Because of this phenomenon a single water content cannot be ascribed to a specific water tension (or vice versa). In many instances the water content is greater during desorption when the soil is drying, than during sorption when the pore space of the dried soil is rewetted. The phenomenon of hysteresis can be explained by the distinct effect of pore diameter on the retention of water in a pore that varies in width. According to this '*ink bottle*' effect, during desorption the smallest diameter of the pore will be effective in preventing pore water from draining, whereas during sorption the larger diameter determines when the pore completely refills with water. This means that during desorption the same pore water content is maintained at higher tensions than during sorption, when this same water content is only attained after the suction has decreased sufficiently to allow capillary rise into the section of the pore with the largest diameter (Marshall *et al*., 1996; Hillel, 1998).

Pore size distribution, together with the effective diameter of individual pores are important for water transport. However, the transport of water in soils with strongly aggregated structure, or with large and continuous pores, is not well described by models based on the Richards equation (Eqn 5.4) because preferential flow occurs (Beven and Germann, 1982; Wagenet, 1990). *Preferential flow* is the process whereby the movement of water and its constituents is focused along favoured pathways through a porous medium. It means that part of the *soil matrix*, which surrounds these pathways and contains water under tension, is effectively *bypassed* by tension-free or near tension-free water. The term preferential flow does not itself convey a mechanism for the process (Helling and Gish, 1991), whereas the term that is often used – *macropore flow* – implies transport through relatively large pores, channels, fissures, or other semi-continuous voids within the soil. Although there is no standardized definition for macropores, some pore classification has been proposed. Luxmoore (1981) suggested classes of micro-, meso and *macropores* defined by effective pore diameters of less than 10 μm, 10–1000 μm and *more than 1000 μm*, respectively. Skopp (1981) defined macroporosity as 'that pore space which provides preferential paths of flow so that mixing and transfer between such pores and [the] remaining pore space is limited'. Helling and Gish (1991) have summarized some other classifications of the size of soil pores and their functions with respect to water movement or root penetration.

As flow through tubes at a given hydraulic gradient is proportional to the fourth power of their radii (Eqn 7.12), drainage will be much more rapid from large continuous macropores than from pores of smaller diameter. Initial soil moisture content, rainfall intensity and

Continued

Box 5.2. Continued

duration affect the distribution of water movement among small and large pores (Jardine *et al.*, 1990). A small rain event (5–10 mm) on relatively dry fields may move water into the soil matrix, thereby increasing the conductivity of the finer pore system as more of them become water-filled (Fig. 5.3). Subsequent rainfall may then be moved entirely through the soil matrix. However, during periods of high rainfall intensity, or when water application exceeds the momentary infiltrability attributed to soil matrix flow, or under conditions of saturated flow, preferential flow can be initiated.

Macropores may be developed in soil by *physical* or *biological processes*. Physical processes are swell–shrink and freeze–thaw cycles, as well as mechanical disturbance by tillage. Biological activities include burrowing by earthworms, insects and other soil fauna, or the growth of roots. Mouldboard ploughing may destroy the continuity of pores between the plough layer and the subsoil on an annual basis. Long-term no-tillage plots, on the other hand, often develop a high density of continuous, relatively large vertical channels (Goss *et al.*, 1984a). Continuous macropores can be formed in soil due to the activity of soil macro-fauna, especially earthworms (Ehlers, 1975b). Manure application may encourage the activity of earthworms, which may result in a greater continuity of macropores. In soils with significant swell–shrink behaviour, cracking may be important in the development of a preferential flow domain (Fig. 5.3), and the extent of crack development to depth is generally related to water extraction by roots. The channels created by roots as they extend through the soil can also dominate the transport process once the original roots have decayed (Barley, 1954). Freeze–thaw cycles may also result in fractures. The installation of tile drains provides some continuous porosity between the soil surface and the drains. These macropores therefore constitute a part of the rapid conduit between the field and the surface water body into which the tile drains discharge. Macropore flow is important for the functioning of mole drains that are drawn in clay soils (Goss *et al.*, 1983).

The essential feature of preferential flow is that percolating water can bypass a large fraction of the soil matrix, thus moving deeper and with less displacement of the initial soil solution than would have been predicted by piston displacement (Bouma, 1981; Beven and Germann, 1982; Quisenberry *et al.*, 1993). Watson and Luxmore (1986) found that under ponded conditions, 73% of the flux through the loamy to fine loamy surface layer of a forest soil was conducted through macropores (pore diameter >1000 μm). Furthermore, they estimated that 96% of the water was transmitted through only 0.32% of the soil volume. Ponding allows water to enter very large pores at the soil surface, and therefore promote preferential flow. It is possible that between 70% and 90% of chemicals applied to the soil may be moving preferentially through macropores (Ahuja *et al.*, 1993). Hence, there may be a faster breakthrough of water and associated contaminants than would otherwise be predicted (Munyankusi *et al.*, 1994).

Preferential flow occurs even in coarse-textured soils that are considered to be homogeneous (Andreini *et al.*, 1990; Beven and Germann, 2013). Quisenberry and Phillips (1976) reported that macropore flow commenced at the tilled–untilled boundary in a tilled Maury silt loam. Similar results were observed on a Cecil sandy clay loam in the Piedmont of South Carolina (Hatfield, 1988). Importantly, flow that does not conform to Eqn 5.4 is commonplace but we still lack a physically based theoretical framework that accounts for all water movement in soil (Beven and Germann, 2013).

In conclusion, preferential flow can reduce the amount of water that is stored in the soil and available to plants, or at least move water deeper into the profile. Such flow can help to reduce soil erosion by water and to improve the aeration status of the topsoil. The phenomenon can be particularly important in allowing contaminants to move very much more quickly through the soil than expected. On the other hand, preferential flow can slow down the rate of leaching of chemicals held in the matrix that is bypassed by the water.

the difference in initial and final water content during infiltration. It is a soil property that determines the rate of water intake into a soil column set in a horizontal position. After a long time of infiltration the term $1/2\ S\ t^{-1/2}$ approaches zero. At that time the water entry into the horizontal column stops. The capacity of the soil for water uptake has been satisfied by complete saturation of the capillary forces. However, when the water is infiltrating vertically through the soil profile, the infiltration rate i will approximate the saturated conductivity K_s after a 'long' time (Figs 5.11 and 5.13).

The final infiltrability or K_s can be ascertained in field soils by measuring infiltration rates when ponded conditions are maintained for some time (4–8 h, or even longer). The measured infiltration rates are plotted against the inverse time axis $1/\sqrt{t}$. The slope of the line is 0.5 S. The intercept of the line with the ordinate, obtained by extrapolation, represents the final infiltrability or K_s. The achievement of stationary flow conditions – a steady state – can be checked by measuring matric potential and water content at different depths of the soil profile. At steady state flow, both these quantities are constant over time.

6 The Root – the Plant's Organ for Water Uptake

6.1 The Role of the Root in the Plant

Even in the embryo of seeds the root primordium can be identified. During germination either a primary root will develop, as happens in many dicotyledons, or several radicles emerge from the seed. Many species of the grass family (*Poaceae*) form three to five of these *seminal roots*. The roots grow into the soil, anchoring the plant and nourishing the growing shoot by providing *water* and mineral *nutrients*. Water is transported from soil to root by *mass flow*, driven by a difference in total water potential between that at the root surface and that in the surrounding soil. Strictly speaking it is the hydraulic gradient that generates the convective flow of water towards the root surface. Those plant nutrients that are dissolved in the soil solution are inevitably drawn to the root with the convective flow of water. Roots may not take up some solutes as quickly as they arrive with the water. Consequently the soil in the vicinity of the roots will become enriched with nutrients such as calcium (Ca) (Goss, 1991) and possibly magnesium (Mg) (Jungk, 2002).

Besides mass flow, nutrients will be transported to the root surface by another mechanism: the process of *diffusion*. Transport by diffusion is triggered by nutrient uptake into the plant itself, which lowers the nutrient concentration at or near the root surface. Diffusion is the more important transport process for those nutrients like phosphorus (P) and potassium (K), which are present at only small concentrations in the soil solution (Jungk and Claassen, 1989). But even with nitrate, diffusion can be the major transport mechanism towards roots (Strebel and Duynisveld, 1989).

Roots are not only effective in removing those nutrients from soil solution that are in an available form, but it is significant that they are also able to secrete protons, organic acids and chelating agents. Using these materials they can modify the availability of certain nutrients for themselves. These *root exudates* stimulate the release of ions from soil minerals, and therefore the bioavailability of macronutrients, such as P, and some micronutrients that normally are only sparingly soluble (Jungk, 2002).

Most of the material secreted by plant roots is in the form of mucilage, much of it originating in the root cap cells. Root mucilage has a very large water-holding capacity and is 99.9% water. The water potential of the gel, when fully hydrated, is about −70 to −100 cm H_2O. Even at a potential of −500 cm H_2O it still consists of 99% water (Carminati *et al.*, 2010). The large water content of the exuded mucilage provides a greater hydraulic conductivity around the root tip than in the bulk soil, thereby allowing a smaller hydraulic gradient (Eqns 4.4 and 5.1) to act as the driving force for water to be taken up by the roots (Carminati *et al.*, 2011). It seems that the plant actively changes the hydraulic properties of the root–soil interface by the secretion of material that supports the flow of water to its roots. In addition, the mucilage provides an attractive environment for microorganisms that use some components as a substrate for their growth, and in doing so will release other mucilaginous material. The mixture of plant and microbial mucilages is called *mucigel*. Mucigel has been shown to link different clay particles, thereby increasing the cohesion of soil materials and leading to the formation of microaggregates. The clay–mucigel combination will hold more water than the clay alone.

The population density of microbes in the *rhizosphere*, the soil that is directly influenced by root activity, may be 10–200 times greater than that in the bulk soil. This increased concentration of soil microorganisms can be either beneficial or detrimental to the plant, depending on the species that dominate. Microbial activity may increase availability of nutrients by mineralization of organic forms or by increasing the solubility of mineral forms. There are P-solubilizing fungi (e.g. *Penicillium balaji* strains) some of which have been exploited commercially.

The most important microbes for P are mycorrhizal fungi that form symbiotic associations with plant roots. The majority of plants establish such an association with certain types of soil fungi; this association is known as a *mycorrhiza* ('fungus-root'). Mycorrhizas are generally mutualistic symbioses. Carbohydrate is passed from the plant to the fungus, and in return the fungus facilitates increased nutrient uptake, particularly of P, from the soil to the plant. On one hand, mycorrhizal fungi can also increase the availability of other ions of limited mobility in soil, such as zinc (Zn) to the root. On the other hand, the rapid colonization of roots by mycorrhizal fungi can protect the plant from the toxic effects of some metals and metalloids, such as manganese (Mn), by restricting their uptake (Brito *et al.*, 2014).

Interest in these symbioses has increased dramatically in recent years because of their potential benefit in agriculture, forestry and revegetation of damaged ecosystems. Some plants cannot become established or grow normally without an appropriate fungal partner. Even when plants can survive without mycorrhizas, those with 'fungus-roots' grow better on infertile soils and areas needing revegetation. It is hyphae external to the root that are important for nutrient acquisition. The mechanism by which this occurs is a combination of increased surface area for absorption from the soil solution and inward translocation of P from beyond zones of depletion around roots. P moves more quickly in hyphae than it can diffuse in soil. Without mycorrhizas, the depletion zone of P around a root is ~1 mm, but mycorrhizas can extend for 100 mm, thereby greatly increasing the zone from which P is absorbed. In addition to the larger volume of soil explored, the kinetics of nutrient adsorption may also be enhanced.

The higher plant experiences a significant cost associated with supplying carbon (C) to the fungus, although this is not as great as the total C used by the fungus since much of the C used by the latter would normally be released into the rhizosphere as root exudate.

Mycorrhizas have been shown to increase the *drought tolerance* of plants (Safir *et al.*, 1972). Possible explanations include changing the content of growth control substances such as abscisic acid (Danneberg *et al.*, 1992) and cytokinins (Drüge and Schönbeck, 1992), increasing the rate of transport to the root in very dry soil because of better soil structure (Fitter, 1985) and improved P nutrition.

Mycorrhizal fungi have also been shown to transport nitrogenous compounds between plants (e.g. Mårtensson *et al.*, 1998), but the most effective microbes for providing higher plants with nitrogen (N) are *rhizobia*. These form *symbioses* with *leguminous plants*. The rhizobia enter the host plant through a root hair, and travel into the cortex via an infection thread. The bacterium undergoes a transformation to produce many bacteroids, and the growth and division of infected root cells leads to the formation of root nodules. Under ideal conditions, the symbiosis between soybean and *Bradyrhizobium japonicum* can fix about 500 kg N ha^{-1} $year^{-1}$. In the temperate climate of north-west Europe the N fixed by grain legumes like field pea (*Pisum sativum*) and faba bean (*Vicia faba*) is usually less, but with 100–250 kg N ha^{-1} $year^{-1}$ it is still considerable.

While there is little evidence to suggest that rhizobia are sensitive to water stress, either from drought or salinity, the symbiosis, such as that between faba bean and *Rhizobium leguminosarum*, is usually vulnerable to such stresses (Serraj *et al.*, 1999). Prolonged exposure to water stress adversely affects the functioning of nodules, leading to them being shed.

Most of the water required by nodules is supplied from that taken up by the supporting root rather than by direct absorption from the soil. However, some of the water used comes from the phloem and accompanies the photosynthates transported into the nodules. Nodules can withstand water loss equivalent to 20% of their fresh weight and still recover their N_2-fixation capacity on rehydration (Lie, 1981). The large energy requirement for N_2 fixation means that any shortage of photosynthates induced by drought will reduce the effectiveness of the symbiosis in this regard. Reduced respiration by the rhizobium bacteroids is one factor that restricts N_2 fixation under drought conditions (Swaraj *et al.*, 1986). Accumulation of ureides in the plant can also occur under drought stress, and this causes negative feedback regulation of N_2 fixation.

A second major series of symbioses is affected by *actinomycetes* in the family *Frankia*, which also form nodules on plants such as alder (*Alnus* spp.) and *Casuarina* species. Because actinomycetes are able to regenerate in greatly disturbed sites, such as mining spoil heaps, these N_2-fixing symbioses are very important in non-agricultural soils.

In addition to symbiotic N_2 fixation, there are also free-living organisms that are able to reduce atmospheric N_2. *Cyanobacteria* (also known as blue-green algae) are autotrophic and use energy generated in photosynthesis to reduce N_2 gas. Witty *et al.* (1979) suggested that up to 28 kg N ha^{-1} $year^{-1}$ might be contributed to arable soils in temperate regions by this process. Other groups of N_2-fixing bacteria are heterotrophic, using organic C molecules to generate the reducing power

Table 6.1. Development of shoot–root ratio of winter wheat, grown in Göttingen, Germany, 1971. Shoot and root mass in g dry matter m^{-2}. The decimal code for growth stages is that of Zadoks *et al.* (1974). Unpublished data.

Date	15 April	6 May	1 June	22 June	14 July	3 August
Growth stage	Beginning of tillering	Middle of tillering	Stem elongation	Milk development	Dough development	Ripening
Decimal code	21	25	39	71	83	91
Shoot	6.5	25	257	618	976	1110
Root	4.7	9.9	39.4	56.5	56.2	55.6
Shoot–root ratio	1.3	2.5	6.5	10.9	17.4	20.0

required. However, their contribution is small and likely to be less than 1 kg N ha^{-1} $year^{-1}$ (Harris, 1988).

The root is the organ that, together with its attendant microorganisms, provides the shoot with nutrients and water. In return, the shoot supplies the root with assimilates. Root and shoot depend on each other, each representing a sink for the materials supplied by the other. A major part of the products of photosynthesis will be retained in the shoot for the generation of shoot tissue, as long as there are buds developing. Under those circumstances, only a minor fraction of assimilates will be diverted to the root. But when the root cannot supply sufficient of the materials it normally provides (because the soil dries or because the supply of available nutrients is inadequate for uptake), shoot growth slows. Under such conditions of deficiency, few of the assimilates formed by the existing leaf canopy can be used in the shoot. They will then be redirected to the root, encouraging its growth and thus improving the impaired root functions. In this way *shoot* and *root growth* are functionally related. The regulation is fine-tuned by plant hormones. These hormones originate both in shoot and in root tissue and are transported upwards and downwards in the vascular tissue, which is composed of xylem and phloem.

Immediately after germination (i.e. during the juvenile phase of growth) root growth and activity are of relatively greater importance for plant establishment than is shoot formation, so the *shoot–root ratio* tends to be small. However, with the transition to a complete autotrophic mode of life, growth of the plant shoot is much more significant and the shoot–root ratio will accordingly be enlarged. Cereals will usually end the juvenile phase when tillering begins. The shoot–root ratio of winter wheat (*Triticum aestivum*), based on the dry matter content, is illustrated in Table 6.1 starting at the end of the juvenile phase.

6.2 Structure of the Root Tip

Whatever method of root investigation is applied (Box 6.1), the aim is to gain an overall view of the *whole root system*. The root system of a plant is the total below-ground organ that has developed from the primary root or roots of the seedling plant, and results from longitudinal growth of individual roots and by root branching. *Root tips*, the growing ends of individual root branches, have a striking appearance. They look fresh and white, rather like the colour of blanched asparagus (*Asparagus officinalis*). At some distance from the tip, the colour of the branch changes to light brown. This is the older part of the branched root system, where the root epidermis has been suberized. When the root tissue starts to decay, the colour turns to dark brown.

Depending on the species, the root tip may extend to several centimetres in length. The tip can be subdivided into four sections: (i) the root cap; (ii) the meristematic zone; (iii) the zone of elongation; and (iv) the root hair zone (Fig. 6.1).

A growing root pushes the root apex ahead through the soil. In the apex new cells are constantly being formed. The part of the tip that is elongating rapidly is that part nearer to the apex than the root hair zone. The zone behind the root hair zone, that is further from the apex, gets older and cells of the endodermis become suberized. Therefore the root tip with its various sections is not an organ that remains static and unchanged. The tip is always being regenerated, the tissue changing appearance and function when passing through the stages of development. The starting point of growth is the meristematic zone, which produces new cells by division. The meristem is bowl-shaped, with the bottom pointing in the direction of root growth. The cells here form the *root cap*. The cap is the protecting shield of the meristem. It excretes mucilage, lowering considerably the friction between the pushed 'spearhead' and the surrounding soil particles. The slimy material protects the tip from desiccation and favours plant water uptake (Section 6.1). It supports the colonization of the immediate surroundings of the tip by microorganisms, forming the *rhizosphere* (Section 6.1). The cells of the cap perceive gravity, thus enabling geotropic orientation to root growth.

On the other hand, the cells of the meristematic zone that form the rest of the bowl generate embryonic cells backwards – so to speak. Filling the bowl and encapsulated within the two zones of dividing cells is the quiescent centre. Thus the root cap is moved forwards through the soil as new cells that will form the epidermis, cortex and stele are initiated and then expand. It is in the *zone of elongation* that these cells expand, particularly growing in length, thereby allowing the root system to grow to depth. Thereafter the enlarged cells turn into the maturation zone, also named the *root hair zone*. Here the cells become specialized into tissues of different function. The epidermis and beneath it the cortex are formed, and inside is the stele, the vascular tissue (Fig. 6.1).

In this region of differentiation the most conspicuous external features are the *root hairs*. They turn out to be protrusions of epidermal cells, forming the so-called *rhizodermis*. In general, root hairs are shorter than 1000 μm (Table 6.2) but may extend to some millimetres. Their diameter varies between 10 μm and 20 μm. Root hairs increase the diameter as well as the surface area of roots (Table 6.2), thus contributing essentially to the water and nutrient uptake. They also anchor the root tip in the soil, a prerequisite for roots to be effective in pushing the root apex through the soil.

The life of root hairs only lasts a few days. Then they will disappear together with the cells of the rhizodermis. They will be replaced by a new epidermatic tissue, the *exodermis* or *hypodermis*. Suberin and cutin, which are highly polymeric esters of fatty acids, are deposited within the cell walls of the exodermis. These waxes make the cells relatively impermeable to water. But some of the cells are spared from suberization. They provide the remaining

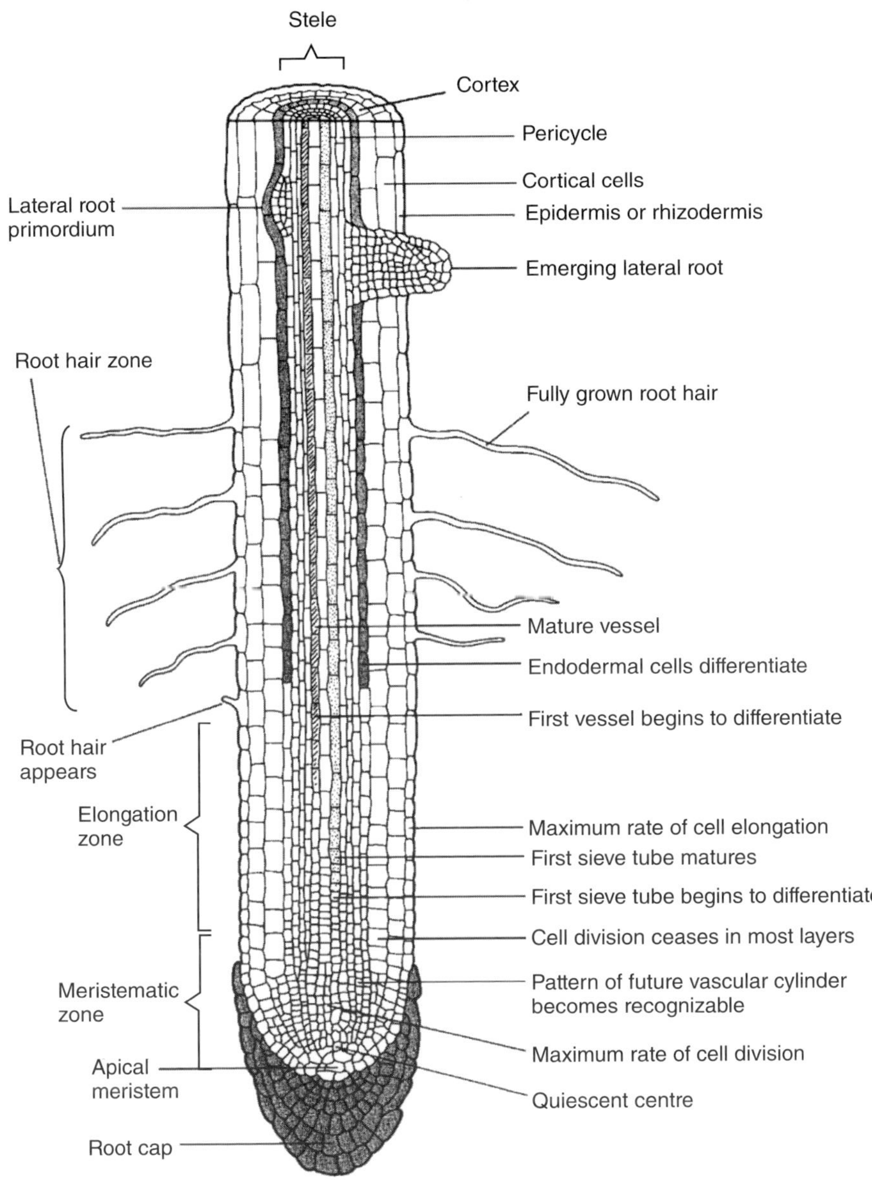

Fig. 6.1. Structure of a root tip. Longitudinal section, semi-schematic sketch (based on Taiz and Zeiger, 1991).

gates of water and nutrient entry into the older root sections. They are called *passage cells*.

Beneath the exodermis is the *cortex* (Fig. 6.1). The clear cells of the cortical tissue are in the main loosely arranged, leaving intercellular spaces between them. In the older parts of the root these cells may store assimilates. The innermost cell layer of the cortex is the *endodermis*. The cells of the endodermis present a special anatomical feature. The radial walls are lined with suberin.

The hydrophobic wax tends to rule out the passage of water and nutrients within cell walls and through intercellular spaces into the stele. The barrier is the so-called *Casparian strip* (Fig. 7.4).

The *stele* contains the root's *vascular tissue*. Whereas in the shoot of dicotyledons the vascular bundles are arranged in concentric rings, in the root they radiate from the centre. The *phloem* is located between strips of *xylem* (Fig. 7.4). According to the number of xylem strips, one can separate diarch, tetrarch, pentarch and polyarch roots. The root of the sugarbeet, for instance, is diarch and that of cereals is tetrarch. The outer cell layer of the stele is the *pericycle*. It adjoins the endodermis. Lateral roots originate in the pericycle (Fig. 6.1). In dicotyledonous plants, a *cambium* is formed between xylem and phloem. The cambium layer produces a secondary phloem to the outside and a secondary xylem to the inside, thus causing the secondary thickening of the root. In this way tap and storage roots are formed.

Table 6.2. Morphological features of root tips of several crop plants (after Föhse *et al.*, 1991).

		Root hairs			
Crop	Root radius (μm)	Length (μm)	Number per mm root length	Length (mm) per mm root length	Surface area (mm^2) per mm^2 root surface area
Onion	229	50	1	0.05	0.007
Common bean	145	200	49	9.8	0.4
Winter wheat	77	330	46	15.2	1.2
Oilseed rape	73	310	44	13.6	1.3

Box 6.1. Methods of studying roots

Compared to the knowledge we have about shoot growth, our information on root growth is rather meagre. This lack of knowledge has to do with the difficulties in observing roots. Roots grow largely undetected and represent the hidden half (Waisel *et al.*, 2002) of the entire plant. A key point is that the observation of roots requires that they be separated or somehow distinguished from the medium in which they are growing, normally the soil. A comprehensive description of root studying methods was given in the classic book by Böhm (1979). A more recent methodological handbook has been published by Smit *et al.* (2000).

Essentially four basic, direct field methods are available for quantitative root research in crop stands. The first is the *profile-wall method*, where roots are counted in the field after soil removal. In this technique a pit has to be dug, normally at least 1 m wide and of similar breadth, so that one person can work in the pit. The width and depth have to be adjusted according to planting density and rooting depth, respectively. The face of the pit is usually lined up at a right angle to the plant rows (Fig. B6.1A) and serves as the profile wall for root observations and measurements. The wall is smoothed and any protruding roots are severed before a vertical soil layer of – let's say – 5 mm thickness is carefully washed away under a fine spray of pressurized water and with the help of a scraper to loosen the soil. The newly revealed root segments are located and counted, often using a 5 cm × 5 cm grid

Continued

Box 6.1. Continued

positioned at the wall. Once the thickness of the soil layer has been removed, the number and length of the exposed root segments as well as the grid dimensions provide the basis for calculating the *root length density* (in short: the rooting density) in various soil layers. Often the unit chosen is cm root length cm^{-3} soil. Accumulating rooting density in soil layers over depth results in the *total root length*, related to a unit of land surface area, i.e. in cm cm^{-2} (or km m^{-2}). The advantage of the method is that the rooting data can be obtained directly in the field, arranged according to soil width and depth. The data can be assigned to individual plants if they are cultivated in spacious rows, and related to features such as earthworm channels in the soil, which can promote root penetration, or to some other soil structural feature that can modify root growth (Fig. B6.1A).

With the *monolith method* the soil, containing the roots, is taken to the laboratory. Rectangular blocks of soil (often 20 cm width × 10 cm × 10 cm) are dug from a pit, pre-soaked in a bucket containing water, and the roots are washed out over sieves that allow soil particles to wash through. The roots can be quantified according to mass or length. Length measurements on washed roots are tedious. Therefore the line-intersect method of Newman (1966) as modified by Tennant (1975) is often applied. A subsample of the roots is spread over a grid and the number of intersects between roots and lines is counted. This method is much more convenient than measuring the length of each individual piece of root directly. Total root length of the sample and the number of intersects are correlated, and that relation can easily be calibrated beforehand by using measured lengths of woollen threads or fine wire. With the aid of computer-assisted data acquisition systems based on image analysis, counting and analysing can be performed automatically.

Like the profile-wall method the monolith method is also quite destructive. Large areas of soil destruction may not be acceptable, especially in long-term field trials with limited plot

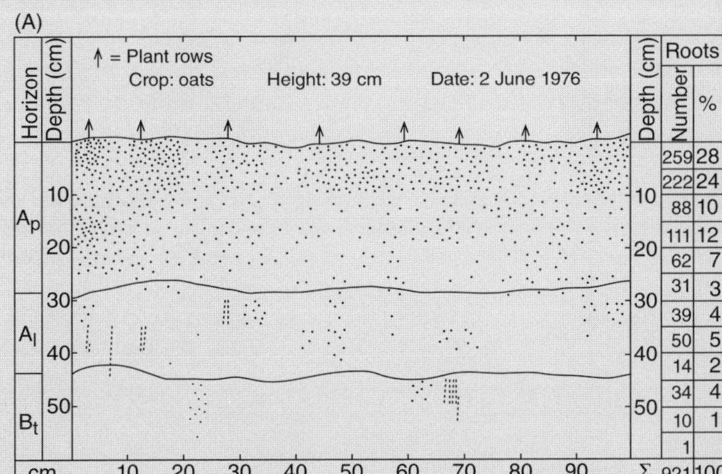

Fig. B6.1A. Rooting profile of oat, determined by the profile-wall method at the beginning of June, 5 weeks after seeding. Roots grown in earthworm channels are presented by broken lines. Each dot or dash corresponds to a unit of root length. Such a unit is a root segment of 5 mm length. The arrows mark the plant rows. The soil is a silt loam, a lessivè or Luvisol derived from loess (after Böhm and Köpke, 1977).

Continued

Box 6.1. Continued

size. The *auger method* avoids the digging of pits. Soil cores, often 5–10 cm in diameter, are augered out from various depths. Some augers are specially designed for this purpose, named root augers. Augers can be driven by hand or motor. Whatever procedure for core sampling is followed, it should ensure that definite relationships between root length and soil depth and sample volume are obtained. As cores are much smaller than monoliths, coefficients of variability may be higher, demanding a larger number of replicate samples. Approaches to controlling spatial variability include confining sampling to well-defined positions within and between plant rows. For auger samples, which are relatively small in volume compared with a monolith, a semi-automatic root separation procedure has been developed – the 'hydropneumatic elutriation system' (Smucker *et al.*, 1982). Again roots can be quantified in terms of mass or length.

The fourth method, *mini-rhizotrons*, does not require the separation of roots from soil. The method, like the others, is suited for the field and avoids digging pits. Insertion of a transparent wall allows continuous root observation *in situ*, whereas all the other methods of root assessment inevitably change in space, from date to date. One approach was proposed more than 70 years ago but was not recognized until it was re-discovered by Böhm (1974). Holes are augered into the soil either vertically or at an angle and glass or Plexiglas tubes are fitted into the holes (Fig. B6.1B). Angled insertion offers two advantages, it helps reduce shrinkage

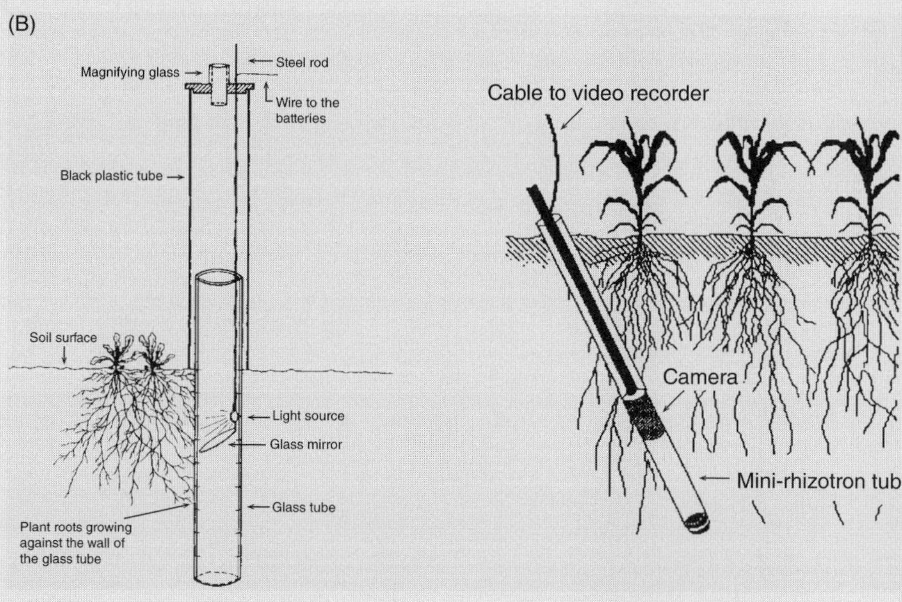

Fig. B6.1B. The development of the glass-tube technique for studying roots: on the left, the original mini-rhizotron with a vertical tube, using a magnifying glass, a light bulb and a mirror for identifying and observing the roots crossing the grid marked on the tube (Böhm, 1974). On the right, in a modern version, the roots are recorded using a video camera (Nickel *et al.*, 1995). The positioning of the tube at an angle to the surface stops the roots from growing vertically downwards, preferentially along gaps and planes of weakness close to the tube, and experiencing less mechanical resistance than in the bulk soil.

Continued

Box 6.1. Continued

cracks from opening an air gap around the tubes and reduces the likelihood of roots tracking along the length of the tube. Horgan *et al.* (1993) showed that the angle of insertion that would be least sensitive to the angle that roots were growing was 54° to the vertical. When dealing with problem soils, such as swelling clay or gravel soils, it may be preferable to replace the tube by a modified one with special features. The tube is smaller in diameter than the hole, drilled with an auger. It is enclosed with a plastic bag. After insertion the bag around the tube is pushed against the wall of the drill hole by a constant pressure of compressed air (Merrill *et al.*, 2005). The required depth of insertion depends on plant species, the stage of plant development, soil type and the maximum rooting depth. During the course of the growing season roots intersect with the mini-rhizotrons and grow over a grid, engraved on the tube wall. The number of contact points or intersections with the grid lines, their circular orientation and depth are recorded either manually (Fig. B6.1B, left) or by use of a video camera (right). Buckland *et al.* (1993) reported that for fibrous root systems the number of intersections with grid lines gave the closest agreement with the auger method, whereas the number of first contact points was better for sparcely rooted plants, such as fruit trees. Not only the formation, but also the decay of the roots can be determined. Data stored on video tape can be evaluated and subjected to further studies, using special computer software for image analysis.

6.3 Root Systems

Lateral roots are formed by branching, the growth originating in the meristematic pericycle of a main root, for instance that of the primary root (Fig. 6.1). The lateral branches are always formed opposite to the primary xylem strands. For sugarbeet, having a diarch root, there are two lines of lateral roots formed. These primary lateral roots also form further orders of lateral branch roots, as a result of which a coherent root system is created (Fig. 6.2). The formation of lateral branch primordial is controlled by the plant hormone indole-3-acetic acid (IAA) (Laskowski, 2013). IAA not only determines the number produced per unit length of root but also where they emerge. For example, when a root axis curves around an object, such as a soil aggregate, laterals only form on the outer or convex side of the curve, where cells are in tension, and not on the inner, concave, side, where they are in compression (Goss and Russell, 1980; Richter *et al.*, 2009). After the location of lateral root initials has been established their development and emergence appear to depend on a further supply of IAA from the root meristem (Babé *et al.*, 2012) and an additional supply from the shoot (Bhalerao *et al.*, 2002). Babé *et al.* (2012) showed that subjecting the region undergoing cell elongation of barley and maize to water stress for only 8 h reduced the levels of IAA and suppressed the development of lateral roots even though their formation had been initiated prior to the treatment.

There are two types of root systems. With the *taproot system* (sometimes called 'allorhizious' rooting), the primary root is the initial organ to form lateral roots. This primary root corresponds to the original primary seminal root, the *radicle*. Taproot systems are characteristic of gymnosperms and dicotyledons (Figs 6.2 and 6.3A: sugarbeet and oilseed rape, *Brassica napus* var. *oleifera*). In monocotyledons like grasses the radicle is short lived. Of much greater importance are the *seminal roots*, originating from an internal node in the seed. The number of seminal roots varies according to species. In addition to seminal roots, the shoot forms *adventitious roots* from basal nodes just below or above the soil surface. Just like seminal roots the adventitious roots, also called nodal or crown roots, produce lateral branches. Seminal and adventitious roots thus form a *fibrous root system* ('homorhizious' rooting) (for example wheat, Fig. 6.3A). Both the adventitious roots and the seminal roots can stay alive until the plant dies. They all contribute to water and nutrient uptake throughout the life cycle, a characteristic feature of small-grained cereals. Maize (*Zea mays*) seems to be different. Here the seminal roots are only effective in early support. They die relatively early in the season (Gardner *et al.*, 1985).

Another distinguishing feature of root systems is the thickening of the primary root. In many dicotyledons a *stout taproot* is formed by the cambium, as happens in oilseed rape (Fig. 6.3A), lucerne (*Medicago sativa*) and red clover (*Trifolium pratense*) (Fig. 6.3B). The storage root of sugarbeet (Fig. 6.3A) is formed from the primary root by the activity of not just one, but of several 'secondary' cambium rings, a special feature of some species within the genus *Beta*. The monocotyledons lack a cambium for secondary thickening. They form a more or less homogeneous *fibrous root* system, the main roots being of similar fine diameter (wheat, maize; Fig. 6.3A). Also some dicotyledons do not form a taproot, and have fibrous roots, like tansy phacelia (*Phacelia tanacetifolia*), potato (*Solanum tuberosum*) and sunflower (*Helianthus annuus*) (Fig. 6.3B).

According to the intensity of branching and the attainable depth of rooting, plants can be subdivided in forming an *extensive* or *intensive type of root system*. Plants with an extensive rooting system penetrate the soil deeply but with little branching.

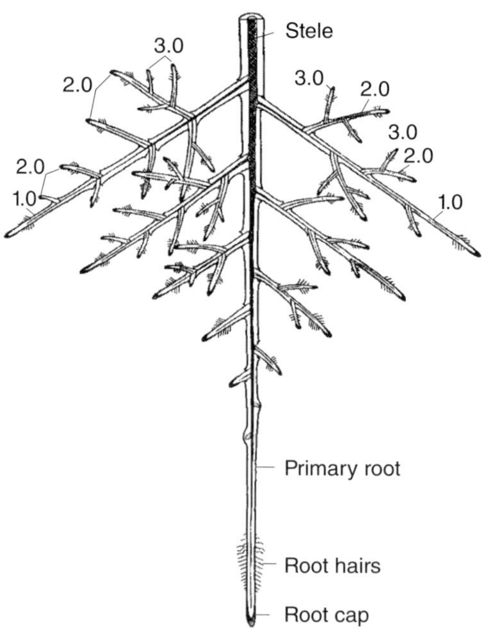

Fig. 6.2. Structure of a taproot system. Beside the lateral roots formed on the primary roots, the various orders of lateral roots are shown. Branching starts at the base of the primary root and of the first order lateral roots and continues in the direction of the root tip (after Rauh, 1941).

Such species may have a great advantage in competition with other plants growing on oligotrophic and dry sites (lucerne; Fig. 6.3B). On the other hand, plants with a shallower, but densely branched system can benefit from more favourable site conditions. They are intensively rooted (red clover; Fig. 6.3B). The distinction may also be applicable within one species. On the one hand, there is the old extensively rooted, undemanding native variety and on the other, the modern, intensively rooted top-yielding variety, but with it comes a greater fertilizer requirement.

The *rooting depth* within a species and even within a cultivar is modified by soil type, soil structure and moisture, and growing conditions. That is also true for how densely roots explore a given soil layer. But some common rules for root development can be identified. For example, in Germany (Vetter and Scharafat, 1964) fodder grasses and potato are regarded as shallow rooted (50–90 cm). Roots of pea and field bean extend deeper in the soil (80–100 cm), followed by small-grained spring cereals (120–140 cm) and winter cereals (140–180 cm). Maize, sugarbeet and sunflower will reach 180 cm depth or even deeper, and lucerne can extend its roots down to a depth of several metres. These values will apply to optimum rooting conditions as occur in loamy soils of medium moisture. In sandy and clayey soils the rooting depth is usually shallower. Constraint is common due to soil dryness or waterlogging. Diminished air-filled pore space in wet soil is harmful to root growth, as aerobic root respiration is restricted. Thus, the above values for rooting depth are likely to be the maximum attained over the course of the season. The maximum rooting depth is usually reached at the beginning (cereals) or the end (legumes) of flowering with the onset of the reproductive phase (cf. Figs 6.4 and 6.5).

Water and nutrient uptake are essentially dependent on the distribution of the roots within the soil profile. For uptake processes, root distribution based on root length is much more critical than is the variation in root mass. Therefore, the more recent literature has concentrated on rooting profiles based on root length. The intensity of rooting is measured by the term *root length density*, or rooting density in short. This is the length of roots (cm) per unit of soil volume (cm^3). Figure 6.4 shows that field or faba bean (*Vicia faba*) has a less dense rooting system than oat (*Avena sativa*), and that root development in oat is deeper than in field bean. The maximum rate that oat roots extend to in depth is about 2 cm day^{-1} (Fig. 6.4). Ellis and Barnes (1980) reported average values of about 5 mm day^{-1} for root penetration by winter wheat over the whole growing season.

The *driving process* for cell expansion and *root growth* is the *intake of water*, which results in an expansion of the protoplast and causes it to push harder against the cell wall. The reaction of the cell wall to deformation increases the hydrostatic pressure – the *turgor pressure* or pressure potential (P) within the cell (see Section 4.2). In a non-growing cell, the increase in P increases the total water potential (ϕ) of the cell (Eqn 4.7), such that the difference between ϕ inside the cell and ϕ outside the cell falls to zero. As the original influx of water into the cell resulted from the difference in ϕ, water uptake will now stop.

In the zone of elongation for a root (Fig. 6.1), the potential gradient is maintained by the *biochemical loosening* of the cell wall, which then stretches irreversibly as water enters the cell and P increases. The intake of water dilutes those cell contents that lower the osmotic potential (O) (Eqn 4.7) within the cell. So continuing growth depends on further *accumulation of solutes*. Growth therefore depends on solute accumulation, water uptake and cell wall loosening. It has been

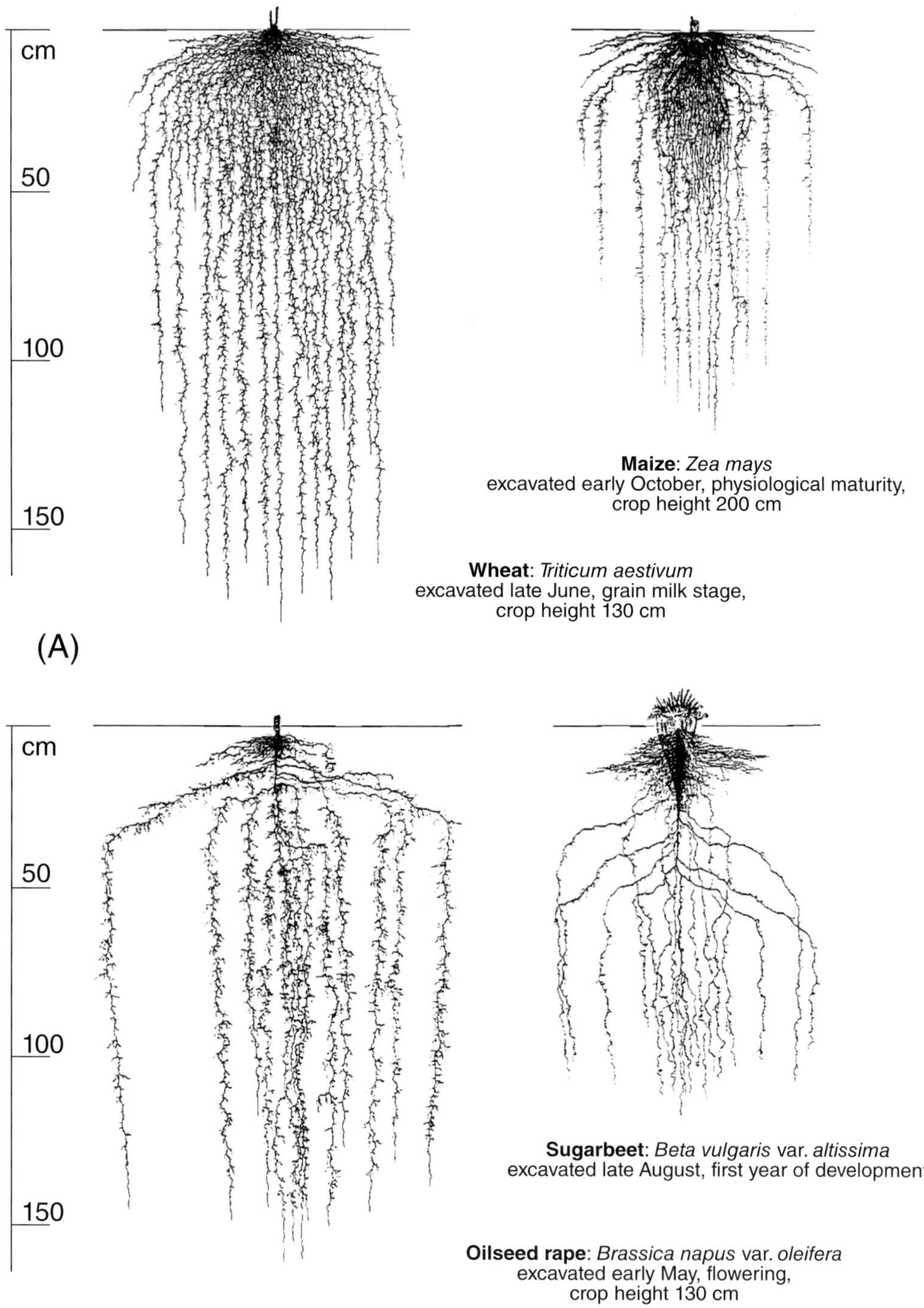

Fig. 6.3. Architecture of the root systems of various crops (after Kutschera, 1960).

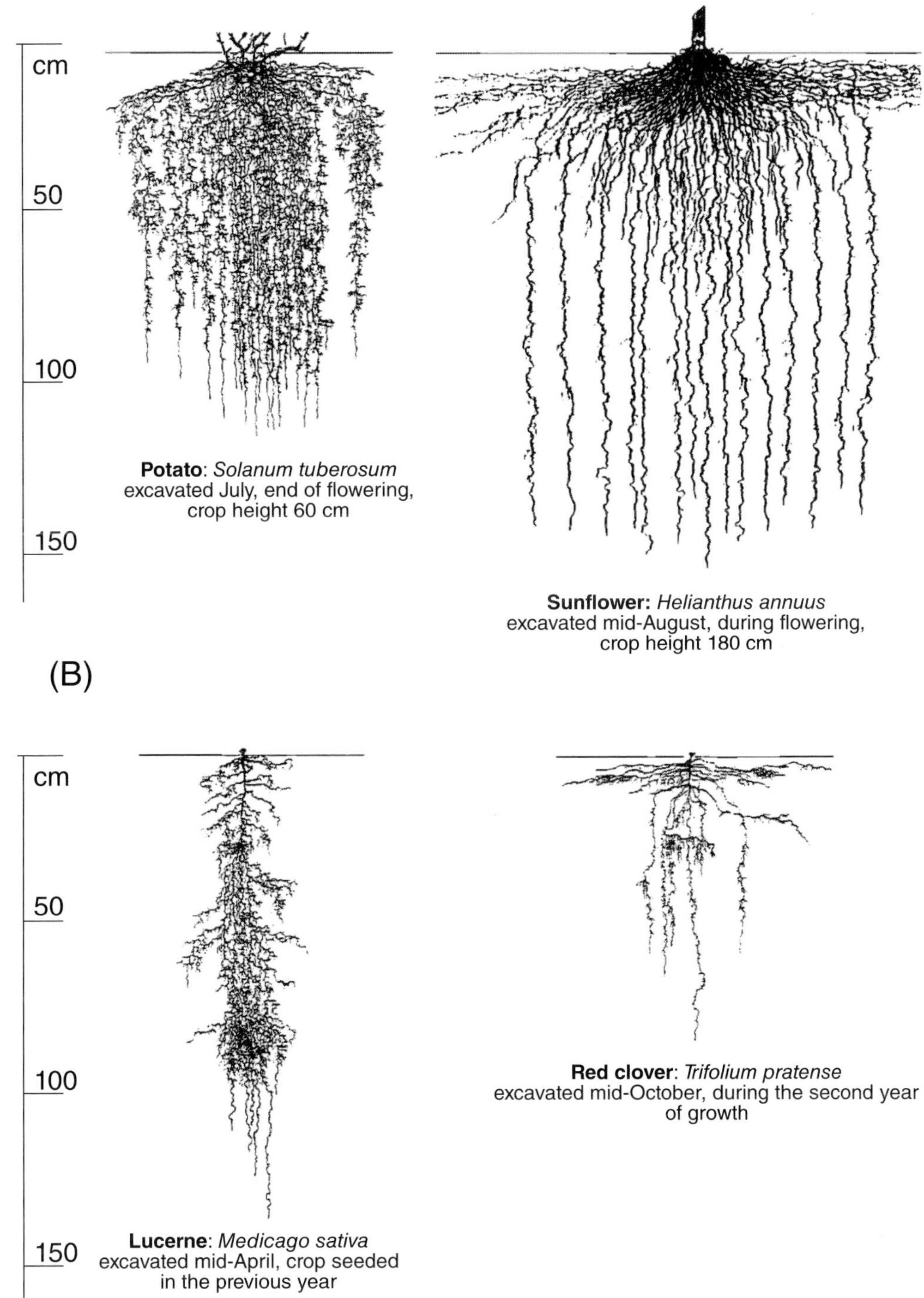

Fig. 6.3. Continued.

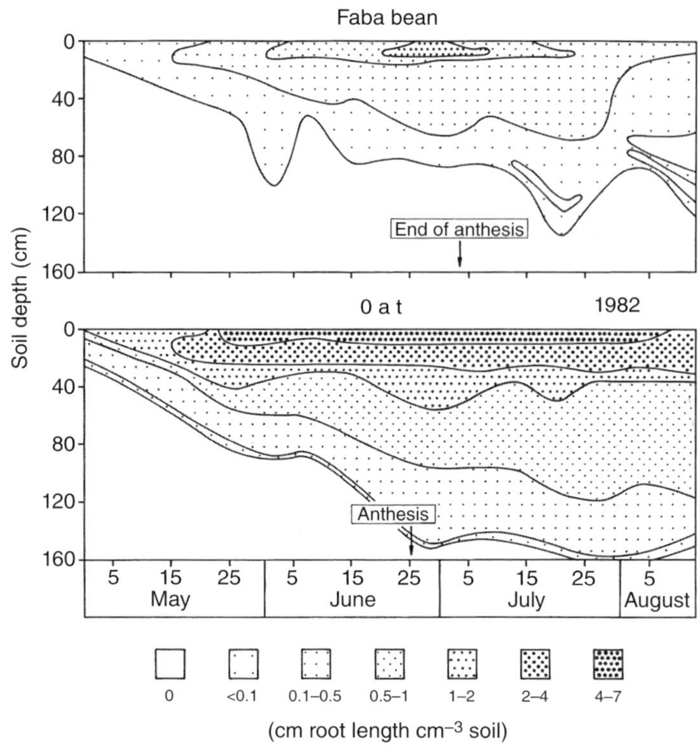

Fig. 6.4. Root length density of faba bean and oat as a function of soil depth and time. The soil is a silty Luvisol derived from loess (from Müller and Ehlers, 1986).

found empirically that P needs to be maintained in cells, or growth will cease. In soil, growth not only involves extension of cell walls, but also overcoming any *resistance to deformation* of the soil material. To describe root elongation in soil, Greacen and Oh (1972) and Veen (1982) adapted an equation for cell expansion (see Section 7.1), developed by Lockhart (1965). It relates the change of root length per unit time (dL/dt) to unit length of root (L):

$$1/L \cdot dL/dt = \varepsilon(P - Y - M) \quad (6.1)$$

Length (L) is given in cm and time (t) in days. ε is the cell wall extensibility (day^{-1} MPa^{-1}), P is the turgor pressure to be expanded by the root cells to overcome the restraint (MPa), Y is the minimum turgor pressure for tissue expansion (MPa), and M is the pressure of the soil (MPa), as it resists deformation by the moving root tip (Greacen and Oh, 1972). As a rule the greatest density in root length is found in the upper layers of the soil. With increasing depth the density declines. According to Gerwitz and Page (1974) the rate of change in *rooting density per depth* increment is proportional to the density of root length in the considered layer. Their empirical equation reads:

$$dL_v/dz = -a \cdot L_v \quad (6.2)$$

The root length density is L_v and a is the constant of proportionality. The factor is valid for more or less homogeneous soils and depends on soil texture, soil type and

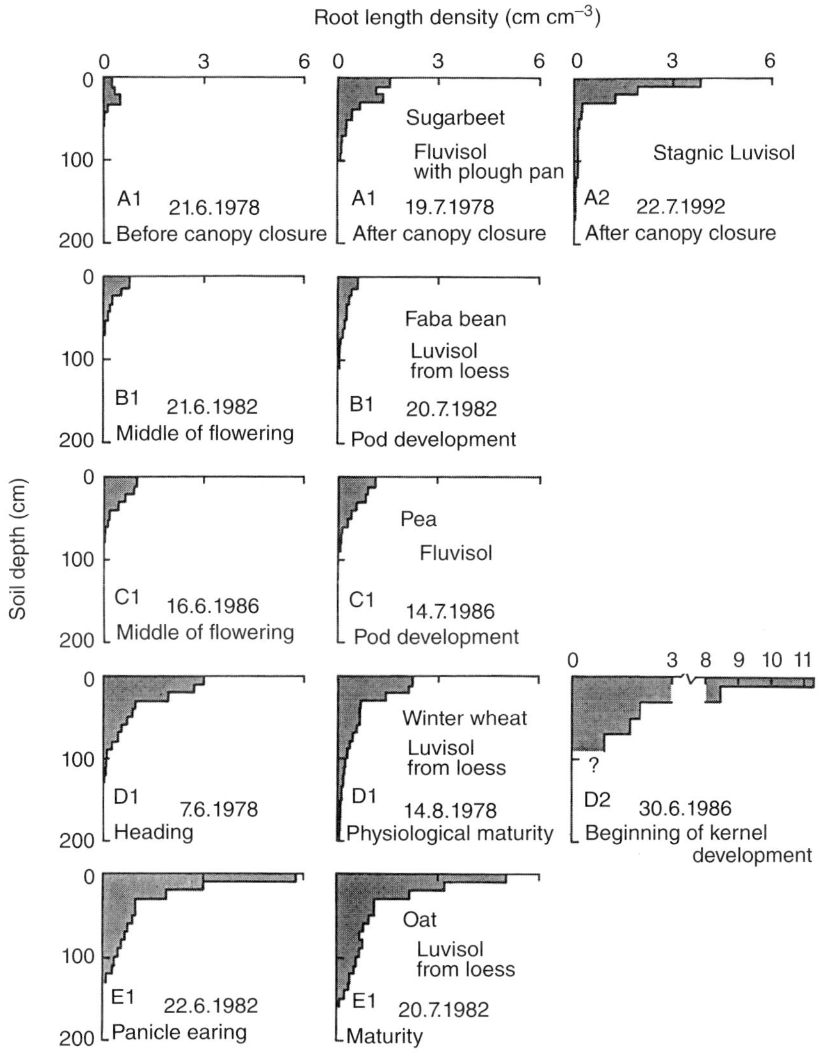

Fig. 6.5. Profiles of root length density (L_v) of five crops at different growth stages. References: A1: Schmidt (1979), profile-wall, L_v calculated using correction factor 2.5 (Schmid, 1991). A2: Windt and Märländer (1994), monolith. B1: Müller (1984), profile-wall, correction factor 1.0. C1: Könings (1987), profile-wall, correction factor 2.5. D1: Böhm (1978), profile-wall, correction factor 2.0. D2: Müller (1988), auger. E1: Müller (1984), profile-wall, correction factor 2.2. The correction factor makes allowance for the underestimation of L_v by the profile-wall method.

plant species. Moreover, constant a will change during the season. Integration of Eqn 6.2 with respect to depth z results in:

$$L_v = L_{v|0} \cdot e^{-az} \quad (6.3)$$

The term $L_{v|0}$ stands for the rooting density at 0 cm depth – in other words, at the soil surface. Plotting ln L_v against z results in a straight line with the point of intersection being $L_{v|0}$ and the slope –a.

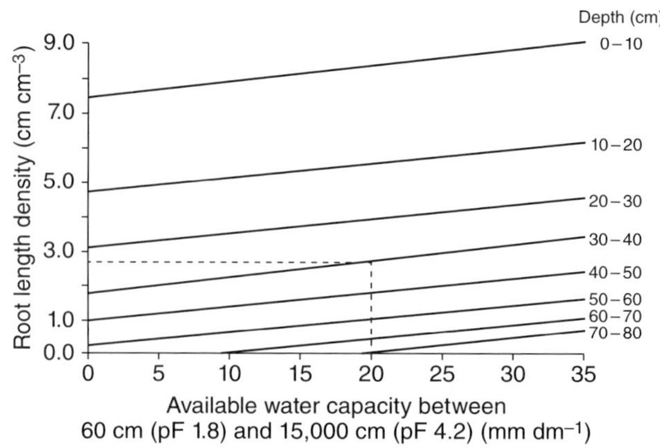

Fig. 6.6. Root length density of cereals at the growth stage of ear emergence as a function of the available water capacity effective in the field (cf. Section 9.4) in various 10-cm soil layers. The example shows, for a capacity of 20 mm dm^{-1} soil depth in the 30–40-cm layer, a root length density of 2.7 cm cm^{-3} of soil (Gäth *et al.*, 1989).

Figure 6.5 shows the *measured root length distribution* in the soil for various crops. Generally L_v decreases with depth exponentionally, as suggested by Eqn 6.3. But there may also be some deviations from this typical distribution, when root growth is influenced by specific soil characteristics that vary with depth. In the case of sugarbeet (Fig. 6.5A1), a plough pan in the 30–35-cm layer limited deep rooting. As a consequence, roots were concentrated in the 20–30-cm layer above. The maximum rooting depth attained after row closing in July 1978 was about 100 cm. In another soil without a pan the sugarbeet crop had rooted down to almost 200 cm depth, when the rows had closed in July 1992 (Fig. 6.5A2). During the season, rooting depth of grain legumes and cereals usually increased until the beginning of grain fill. At that time the older part of the root system in the upper soil layers was probably progressively decaying, as can be seen by the decline in root length density (Fig. 6.5B1–E1).

Generally, rooting data show a high variability in space. But results for a given crop also vary between years, sites, and investigators, especially if different methods of investigation are used (Fig. 6.5, cf. A1 with A2 and D1 with D2).

So far it is obvious that a quantitative assessment of roots is costly in time and labour and can prevent an area of land from being used for other studies. Therefore it is not surprising that researchers have tried to produce *estimates of root distribution* as a function of depth and time under various conditions. Such data are needed for instance as input variables in some types of computer models to simulate water use and nutrient uptake of crops. One empirical approach was developed in Germany for small cereals by Gäth *et al.* (1989). Actual rooting data from field experiments conducted on 25 sites were evaluated for prediction purposes. The starting point was the estimation of L_v at the growth stage of ear emergence in 10-cm soil layers down to 80 cm depth as a function of the available water capacity (Fig. 6.6). It is apparent that both L_v and rooting depth increase with available water (cf. Fig. 3.2). Rooting data for growth stages before or after earing were obtained by additions to or reductions of L_v due to net growth or decay. The climatic water balance (Section 5.1) of a site also influenced L_v, the necessary changes being different on sandy or loamy-silty soils (not shown).

7 The Water Balance of the Plant

7.1 Water Potentials in Plant Cells

During the first stage of nutrient uptake by plants, ions pass from the soil solution into the root. Inside the root, nutrients are routed to the endodermis, either through the 'free space' within cell walls, or from cell to cell of the rhizodermis and cortex via the cytoplasm (see below). Within a cell, the concentration of many solutes is greater than in the solution outside the root. Hence these nutrients have to be transported across the plasmalemma of the cells against an existing concentration gradient. In some cases, ions are selectively excluded from cells. That is to say, *nutrient uptake* is an active and energy-consuming process. The energy required is generated by cell metabolism (Mengel and Kirkby, 1982). On the other hand, the *uptake of water* by roots and the conduction within the plant does not rely on expenditure of energy. These processes are passive events. They follow the rule: *water flows from sites of higher potential to sites of lower potential.* Water flow within the plant, from the rhizodermis through the cortex to the xylem in the central stele, and from the stele to various organs of the plant, and finally from the leaves to the atmosphere, is caused by differences in total water potential, just as in the case of water flow in the soil.

To develop an understanding of the water potential in plants, that is to say how the total water potential in the plant develops, how it differs within the various parts of the plant, and how it brings about water movement within the plant, the *water balance of a single cell* will be outlined first. In a turgid plant cell, the cellular content (the protoplasm) is filled with water. The *protoplasm* contains the *organelles* (i.e. the nucleus, mitochondria, chloroplasts and the vacuole). In a water-saturated cell the total water potential is zero. At saturation the cell cannot take in any more water and is unable to induce water movement from cell to cell towards itself. A turgid cell, together with turgid neighbouring cells, causes the tautness of the tissue, which in turn determines the form of herbaceous plants.

Water-filled cells have a maximum *turgor pressure* – 'full' turgor. Turgor originates in the protoplasm, saturated with water. Water, being incompressible, forces the *plasmalemma* against the *cell wall*, the wall being slightly, but elastically, stretched. Without this spatial limitation established by the rigid cell wall and the surrounding turgid tissue, the cell volume would steadily increase because of continuing water entry and would possibly end with the bursting of the cell. But in reality the cell wall encloses the cellular content, and its reaction to deformation is to pressurize the cell contents. This idea of a cell is like that of a balloon, where a stretched rubber skin encloses the filling gas. In other words we can say that the hydrostatic pressure inside the cell (the turgor) is developed by the restricting cell wall.

Turgor is an excess pressure, being higher than the surrounding atmospheric pressure. The turgor has a strong impact on cell elongation and vegetative growth, including the growth of leaves or roots (Eqn 6.1).

Similar to the equation describing root length growth (Eqn 6.1), the equation for *cell enlargement* reads:

$$1/V \cdot dV/dt = G_r = \varepsilon\,(P - Y) \qquad (7.1)$$

The V stands for the cell volume (cm^3), t for the time (day), G_r for the relative rate of the three-dimensional cell enlargement (day^{-1}), ε for the volumetric cell wall extension coefficient (day^{-1} MPa^{-1}), P for the turgor (MPa) and Y for the minimum turgor for tissue expansion (MPa). For cell growth, the minimum turgor has to be reached and surpassed. Then P, causing a tension on the cell wall, will induce a cell wall elongation, which is now no longer an elastic stretching but an irreversible extension. Wall thickening and stabilization takes place through cellulose accumulation. Even a minimal extension of the cell wall reduces the turgor considerably, as water is almost incompressible (in contrast to air). Due to the turgor reduction, water will start to move again into the slightly enlarged cell. The turgor is re-established, and once again water absorption comes to an end.

The positive turgor pressure raises the energy content of the water. The water gains the ability to carry out some work, for instance for moving out of the cell or for enlarging the cell volume. Relating the turgor to unit volume of water results in the expression of a component potential of water. That is the pressure potential, P, with the unit of a pressure (MPa, bar). The pressure potential is positive. Its reference level is that of water at atmospheric pressure, the normal pressure surrounding us. In a fully turgid cell that momentarily does not expand, the pressure potential is at maximum (Fig. 7.1). On the other hand, it was already mentioned that the total potential (ϕ) is zero in the state of full turgor. Therefore the positive pressure potential must be counterbalanced by a potential, numerically of the same magnitude, but with negative sign. This is the osmotic potential, O (Fig. 7.1).

The osmotic potential is caused by soluble substances or *solutes*, like salts or some organic compounds, for instance sucrose. These solutes are *osmotically effective* and lower the water potential within the plant cell. Consequently, the reference level for the osmotic potential is pure water without any solutes. The osmotic effectiveness of solutes relies on the presence of *cell membranes* like the *plasmalemma*, which separates the protoplasm from the external medium, or the *tonoplast*

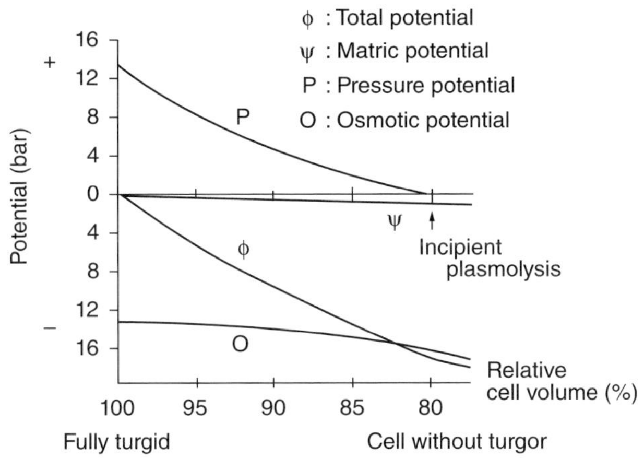

Fig. 7.1. Total potential and component potentials of water as a function of the cell volume, which changes elastically with the state of water filling. The cell volume is at maximum when the cell is fully turgid. With the complete loss of turgor plasmolysis will start, a state when the protoplasm separates from the wall (based on Kramer, 1983 and Larcher, 1994).

between the protoplasm and the vacuole. These cell membranes are, on first impression, impermeable for solutes but permeable to water, allowing the build-up of a higher solute concentration inside than outside the cell. Because of the concentration difference, water will enter into the cell by passing through the '*semipermeable*' membrane, tending to lower the solute concentration inside the cell.

Before coming back to a discussion of the component potentials, a contemporary and more detailed consideration is presented of the structure and particular *permeability of cell membranes* for solutes and water. A biological membrane consists of a phospholipid double layer into which some glycoproteins are inserted. These transmembrane proteins represent *passage pores*, permitting a catalysed transport of specific ions like calcium or potassium across the membrane, which is quite often an energy-consuming process. These pores permit a selective ion transfer across the membrane, which otherwise would be impermeable to these ions. Membranes are *selectively permeable* for special solutes, as long as the pores exist and are open. Membranes prevent salts or sugars from free diffusion, but membrane pores will allow a controlled diffusion process or a catalysed 'carrier' transport into or out of the cell cytoplasm.

As the solute concentration is normally greater inside than outside the cell, the solvent water tends to diffuse through the membrane into the protoplasm and the vacuole. Here, owing to the greater solute concentration, the concentration of water molecules is less. A *diffusion* process between sites is always triggered by differences in specific, *partial concentrations* of molecules, whether those are ions, soluble compounds or water. The diffusion across membranes of a solvent such as water is called *osmosis*.

As described, water will pass the membrane by osmosis in the direction of the lower osmotic potential (with greater solute and smaller water concentrations) in the cell. Rapid transport of water molecules across the membrane is facilitated by a special type of *transmembrane pore*, which was identified in the early 1990s. The pores are built up from proteins and amino acid strings and are called *water channels* or *aquaporins*. The pores are quite narrow, 0.3–0.4 nm in diameter, and not much larger than the water molecules (diameter 0.28 nm). Water channels let the water molecules pass in a highly selective manner from side to side, one after the other. The channel length can accommodate about 20 molecules end to end across the membrane during water passage. Water channels can open and close very effectively, allowing or interrupting the passage of water into and out of cells. Key features of the control are the distribution and frequency of pores, and their rapid disintegration and reformation within hours (Blanke, 1998).

Now we have to come back to the component potentials of the cell, shown in Fig. 7.1. In a flaccid cell, only partly filled with water, the negative osmotic potential O causes the water to move across the plasmalemma and the tonoplast inwards into the protoplasm and the vacuole, a process we call *osmosis*. Filling the cell with water causes the cell volume to increase. While the osmotic potential increases moderately (to less negative values), the pressure potential P increases more intensively (Fig. 7.1) and the cell becomes taut. If the cell becomes fully turgid, the movement of water molecules by diffusion across the membranes comes to an end. The net movement stops, although there is still a larger concentration of solutes and a smaller concentration of water molecules inside the cell compared with the outside. The reason for the cessation is that the osmotic potential of water, causing *diffusion* to the inside, is at this point counterbalanced by the pressure potential (Fig. 7.1), which encourages *mass flow* or bulk flow of water to the outside. It follows that osmosis is driven by the combination of the two potentials, the osmotic and the pressure potential.

The equation for describing the potential of water in plant cells has already been introduced (Eqn 4.7):

$$\phi = O + P \qquad (7.2)$$

Within the cell the osmotic potential O is always negative and the pressure potential P always positive (Fig. 7.1).

The introduction of the *potential concept* has a real benefit, even a bit of luck, for those dealing with soil–plant–water relations. The concept is concise and therefore helps to understand the movement of water in the soil, from the soil to the root, in the plant from root to shoot and leaves and finally to the atmosphere. The concept eliminates the linguistic confusion in this field and helps to bridge the gap between the approaches developed in life sciences and in soil science. The concept should bring together those interested in plants and in soils in a common approach to grasp what is commonly but vaguely defined as 'water dynamics'.

In older literature, Eqn 7.2 has commonly been presented in the form:

$$S = \pi - W \qquad (7.3)$$

The S describes the 'suction force' of the cell, π is the '(potential) osmotic pressure' and W is the 'cell wall pressure' or the turgor pressure. With this formulation, two opposing pressures, π and W, are defined, both bearing a positive sign. Even the suction force is a positive term. However, this representation of cell water status is not as desirable as Eqn 7.2.

When plant tissue and therefore individual cells lose water, cell volume decreases (Fig. 7.1). The reduction in water content causes the osmotic compounds to become concentrated. Coming back to Eqn 7.2, it is clear that the negative osmotic potential O decreases. But it will not decrease to the same extent as the positive pressure potential P declines. For that reason the total potential ϕ declines together with the pressure potential, but the change per unit of cell volume is more pronounced (Fig. 7.1).

With the loss of water in a shrinking cell some structural elements of the cell wall become drained. Water will be left in small pores, created for instance by the rigid network of cellulose microfibrils of the wall. A water meniscus indicates that the pore water is under tension. Such an argument may serve to explain the presence of a component potential in the cell (Fig. 7.1) that is well known as a principal component of the soil water potential, the *matric potential* Ψ. Therefore we expand Eqn 7.2 for the plant cell:

$$\phi = \Psi + O + P \tag{7.4}$$

As a cell dries, at some time the state of complete loss of turgor or pressure potential will be attained. This is the state of *incipient plasmolysis* (Fig. 7.1). The plasmalemma will pull away from the cell wall, and the cell will die.

Water potentials in cells and plants are most frequently reported in the units of pressure, like bar (Fig. 7.1) or MPa. In the soil the unit preferred is often that of a length, the head of a water column (Section 4.2). The head can be presented in cm (Fig. 5.1). The possible units are drawn together in Table 7.1.

Based on the table we can judge that the component potentials in the cell fluctuate over an additional order of magnitude (Fig. 7.1) than the potentials in a moist soil (Fig. 5.1). Depending on the degree of turgidity, the potentials in plant cells (beside Ψ) are larger in number than those in the soil supplying the water. They are similar in magnitude to the potential associated with the soil's permanent wilting point. We can also conclude that in a cell with 90% of the volume of a turgid cell (Fig. 7.1), the total potential ϕ is much lower (–10 bar) than ϕ in moist soil (–0.1 to –1 bar, see Fig. 5.1, where actually water tension is shown as an expression of Ψ). As a consequence of this potential drop, the water will tend to move spontaneously from the soil towards the plant cell.

Table 7.1. Some of the common units of water potential, related to unit weight (length) or unit volume of water (pressure).

Potential based on:	Unit weight	Unit volume			
Unit of potential:	(cm)	(bar)	(mbar = hPa)	(Pa)	(MPa)
	0	0	0	0	0
	–10.2	–0.01	–10	–1,000	–0.001
	–102	–0.1	–100	–10,000	–0.01
	–1,020	–1.0	–1000	–100,000	–0.1
	–15,300	–15.0	–15,000	–1,500,000	–1.5
	–102,000	–100.0	–100,000	–10,000,000	–10.0

7.2 Water Uptake by Roots

From the previous considerations it should be obvious that plant roots are able to extract water from soil, as the water potential is usually lower in the roots than in the surrounding soil (Box 7.1). How much water can be absorbed by a

Box 7.1. Early experiments for determining water suction and water pressure of roots

Stephen Hales (1677–1761) was an English clergyman, living in Farringdon, Hampshire and in Teddington, Middlesex. As well as religious affairs, he was interested in the mysterious workings of nature. He admired 'the beauty and harmony of the scene of things,

Fig. B7.1. Hales' experiment for determining the 'imbibing force' of tree roots (Hales, 1727; from Bettex, 1960).

Continued

Box 7.1. Continued

created by the divine Architect'. Especially he was interested in the circulation of saps and fluids in plants and animals. In this context the Bible speaks of a tree at a watercourse, staying green and fresh, even in blazing heat, as it stretches out its roots towards the brook (Jeremiah 17: 8). To explore the significance of roots in water uptake from the soil, the pastor investigated in 1723 the 'imbibing power' – the *suction force* – of an elderly pear tree (*Pyrus communis*) growing in the garden of his vicarage (Fig. B7.1). In his experiment he cut a root branch of ~1.2 cm diameter at point i (Fig. B7.1), which he had excavated at a profile wall (Hales, 1727). He put the bare root stump, still connected to the tree, into the top of a glass cylinder. Then he made an airtight seal at the upper filling hole of the cylinder (point r) using a sheep skin and wet pig bladders, which he covered with beeswax and turpentine and then tied tight. Into the bottom of the glass cylinder he inserted a glass tube, which he made airtight to the cylinder (point d) in the same way. Turning the tube upwards he filled tube and cylinder with water. In this upward position he closed the filled tube with his finger. Directing the tube downwards he immersed it in a cistern filled with mercury, where he removed his finger. He found out that the suction force of the root was large enough to draw the mercury upwards into the tube to point z (20 cm height) within 6 min (Fig. B7.1).

Hales recognized that the mercury rose faster and higher the more strongly the sun was shining. Towards evening the mercury column fell, but the next day it rose again. The priest found that the power of the roots to soak up moisture was linked to the perspiration of the leaves, and that the perspiration was controlled by the warmth of the sun. The absorption of moisture was sustained by the capillary attraction of the sap vessels, which were replacing the loss of perspiration water by a steady attraction of fresh supplies (Hales, 1727).

The restless researcher experimented also on grapevine (*Vitis vinifera*). During the period of sapflow (in spring before leaves appear) he measured the level of the (positive) *root pressure*. It is a striking phenomenon, observing this climber losing quantities of sap after injuring the shoot. Hales tightly connected single glass tubes, 0.64 cm in diameter, one after the other to a total height of 11.4 m (38 feet), after having fastened the lowest pipe to the cut shoot. He found that the sap ascended in the tube system to 30 cm or even 7.5 m height, depending on the plant's vigour during the season of sap rise. He compared these results with findings he got from experiments with various domestic animals. He measured the blood pressure in the great left crural artery, where it first enters the thigh. Similar in concept to the vine experiments, he fixed a glass tube more than 3 m long and 0.32 cm diameter in bore to the opened artery. The experimentalist finally considered the force of the vine root to be five times stronger than the force of the blood in a horse, seven times higher than in a dog and eight times greater than in a female fallow deer (Hales, 1727; Guerlac, 1972).

Hales undertook a large number of investigations, which he initiated after insightful contemplation and performed with great skill. He read his reports at several meetings before the Royal Society. But he was far ahead of his time, and his achievements later fell into oblivion.

root system over a day will depend on various factors. On the one hand, meteorological quantities determine almost exclusively the evaporative demand of the atmosphere. The demand sets the upper level for transpiration by the leaves and, consequently, for water uptake by the root system. On the other hand, uptake will depend on plant factors like the size of the canopy, which governs whether it covers the soil surface only sparsely or more completely. Finally the soil, in its function as a medium for storing and conducting water, will influence the uptake by roots. Important soil properties are water content, water potential and hydraulic conductivity.

It is certainly not an easy task to understand and to formulate mathematically the contribution of the various factors in root water uptake. A starting point to understand the process may be the '*single root model*' of Gardner (1960). Imagine a root system in the soil that is composed of single root axes in parallel arrangement. Water will flow from the soil to a single root (Fig. 7.2). The geometry of the flow region is characterized by cylindrical coordinates, like the flow region around a drainage tube. Gardner made it his task to find a solution for the flow problem in such a system.

The root is considered to be a long, hollow cylinder. At the outer wall (the root surface) the matric potential is considered to be less than the matric potential in the surrounding soil. Hence the water is conducted from the soil to the root. The equation of continuity for the cylindrical system reads:

$$d\theta / dt = 1 / r \cdot d / dr(r \cdot K(\Psi) \cdot d\Psi / dr) \qquad (7.5)$$

The contribution of the gravitational potential to the difference in total potential, $\Delta\phi$, is assumed to be very small. Therefore only Ψ but not ϕ is considered in Eqn 7.5. The term θ is the soil's volumetric water content, t is time, r is the radial distance to the root centre and K is the unsaturated hydraulic conductivity of the soil.

The cylindrical soil volume around the single root is bounded by a neighbouring cylinder centred on the next single root. Increasing root length density narrows the mean distance between single roots. The distance between two adjoining single roots is taken to be 2 r_2, where r_2 is the radius of the soil cylinder from which the single root is extracting water (Fig. 7.2). Then the following relationship holds:

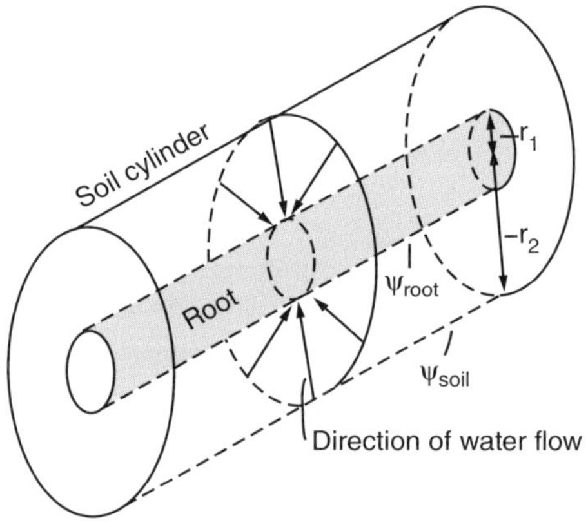

Fig. 7.2. Water flows in a geometrical system with cylindrical coordinates from the soil to the root surface. The single root is a long, hollow cylinder with the radius r_1. The root cylinder is surrounded by a soil cylinder of radius r_2 (the root cylinder inclusive). The difference in matric potential ψ as a driving force for the water flow is $\psi_{soil} - \psi_{root\ surface}$ ($\psi_s - \psi_r$). A possible difference in the gravitational potential is left out of consideration.

$$r_2 = 1/\sqrt{(\pi \cdot L_V)} \tag{7.6}$$

The term L_v stands for root length density. When r_2 is taken in cm, then the unit of L_v is cm (root length) per cm^3 (of soil).

Equation 7.5 can be solved for a limited soil volume around the single root. For the solution, a 'boundary condition' has to be introduced. This condition says: there is no water content change in the soil during water uptake. This *steady state condition* is intellectually obeyed by supposing, for instance, a water supply at the circumference of the soil cylinder in the distance r_2 around the single root. Under these conditions, the solution of Eqn 7.5 is given by:

$$UR = 2\pi K(\Psi_s - \Psi_r)/\ln(r_2 - r_1) \tag{7.7}$$

The term UR presents the water uptake rate of the single root, also named the '*specific root water uptake rate*'. The unit is cm^3 water per cm root per day (cm^3 water cm^{-1} root day^{-1}). The K is the mean effective hydraulic conductivity of the soil in the cylinder; Ψ_s is the matric potential in the soil and Ψ_r the potential at the root surface. The radius of the root is r_1. Introducing Eqn 7.6 into Eqn 7.7 results in:

$$UR = -4\pi K(\Psi_s - \Psi_r)/\left[\ln(\pi \cdot L_v) + 2\ln r_1\right] \tag{7.8}$$

The larger the value of L_v (Eqn 7.8) and the smaller r_2 (Eqn 7.7) becomes, the greater is UR, as the hydraulic gradient rises for a given potential difference. Such a situation might be assumed under fairly dry soil conditions with a sparsely rooting crop, when the water flow to the single root is limited (Kage and Ehlers, 1996). In contrast, for a crop growing in moist soil with non-limiting flow conditions, the opposite might be expected (i.e. the value of UR will decline with increasing L_v) (Ehlers *et al.*, 1991). Thus when water uptake by the whole root system is governed by a given evaporative demand, increasing L_v means that there will be a reduction in the flow rate to the individual root.

The matric potential must decline from a position in the bulk soil to the root surface (Fig. 7.2) to drive the water towards the root. The greater the potential difference driving the flow of water, the larger is UR (Eqn 7.7). Until the development of technologies aided by the microchip, neither the potential drop near the root nor the decline in water content were measurable. But in 1986 the Australian soil scientists, Hainsworth and Aylmore (1986), were the first to report measurements of the water content distribution around a single root by computer-aided tomography (CAT). Figure 7.3 depicts the result for a single root of radish (*Raphanus sativus*). The plant was 9 days old. Measurements were taken 8 cm below soil surface.

Further advances in imaging techniques mean that nowadays magnetic resonance imaging (Segal *et al.*, 2008) or neutron radiography (Carminati *et al.*, 2010) permit the quantification of water content of a single root and the adjacent rhizosphere. Even tiny and pliable-tipped micro-tensiometers have been developed and used to measure directly the drop of matric potential as a function of the distance from the root surface and of time (Segal *et al.*, 2008).

Now we will follow the pathway of water from the root surface into the root. First, it should be clearly stated that water can enter the root principally anywhere

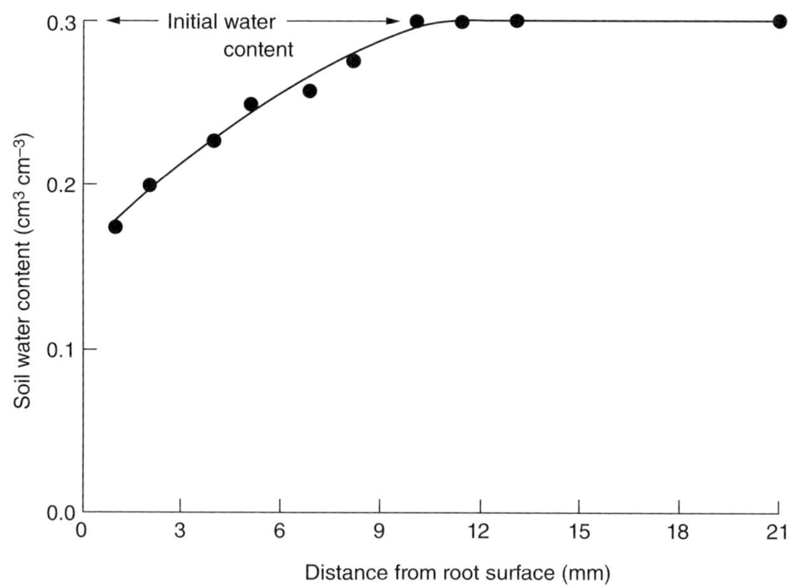

Fig. 7.3. Water content distribution around a single root of radish after 10 h of water uptake (after Hainsworth and Aylmore, 1986).

along the root axis. Water is taken up at the root tip, especially in the root hair zone (Fig. 6.1), as well as in older portions. Here the *rhizodermis* of the root tip has been replaced by a suberized epidermal tissue, the *exodermis* or *hypodermis*. Because of the suberin, the exodermis is largely non-permeable to water. But even so, water entry is made possible by special *passage cells*, set into the exodermis. Anyway, the specific water uptake rate (UR) will be greater in the hair zone than in the exodermis, all other factors, especially soil moisture, being equal. Reasons for the preferential water uptake in this zone are: (i) cells are not suberized in the hair zone; (ii) root hairs increase the absorbing surface of the rhizodermis cells; (iii) root hairs extend into the surrounding soil cylinder of water uptake, thus shortening the flow distance from soil to root surface (Fig. 7.2); and (iv) root hairs ensure an intimate contact with the surrounding soil. Furthermore, as outlined in Section 6.1, the mucilage released from root cap cells enhances water flow towards the tip region of a single root, when the soil is drying because of water uptake.

A root surface, bare of any hairs and contact mucilage, may lose *hydraulic connection* with the soil. Either the root shrinks when the plant suffers water shortage or the soil contracts and separates from the root due to drying, a feature particularly associated with clayey soils.

But also without any shrinkage of root or soil the hydraulic contact will decrease, as the soil dries owing to root water uptake, because the *wetted contact area* between root and soil diminishes. The wetted contact area is, so to speak, the bridgehead for the passage of water from soil particles to root cell surfaces. Root hairs and mucilage (see Section 6.1) are important for bridging the gap. In the language of hydraulics: root hairs and mucilage lower the *hydraulic resistance*.

We can re-write the Darcy equation (Eqn 5.1) in the following format:

$$q = -1/[\Delta z / K(\Psi)] \cdot \Delta\phi \quad (7.9)$$

In this form the equation is like that of Ohm's law (Eqn 4.3). The term $\Delta z/K(\Psi)$ is the hydraulic resistance. The resistance increases with the length of the flow path Δz, the flow path being marked by the potential drop $\Delta\phi$. When the unsaturated hydraulic conductivity $K(\Psi)$ becomes reduced in a drying soil, the resistance increases likewise. At the root surface this increase will be sharper, the more the wetted contact area retreats from the interface.

It is presumed that water can enter the root interior through either of two pathways, and that it can be conducted in the cortex parenchyma radially towards the stele, utilizing three routes (Fig. 7.4). Water can be absorbed firstly by passing into root hairs of the rhizodermis cells or by entering into passage cells of the exodermis. In both cases, water is penetrating into the protoplasm of these cells. Secondly, water can move into the root without immediately entering the cells. As a bulk flow, water can be conducted between cells, along neighbouring cell walls or through intercellular air spaces. This route is named the *free space pathway* or the *apoplast pathway* (Fig. 7.4).

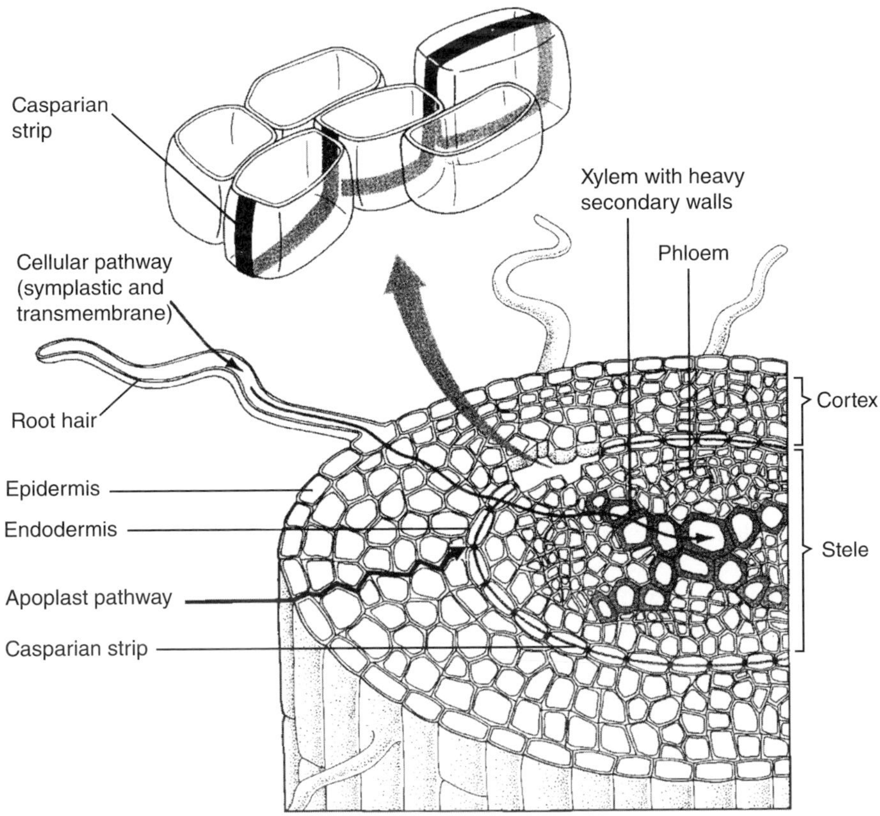

Fig. 7.4. Possible pathways of water conduction in the root tip (after Taiz and Zeiger, 1991).

After entering the protoplasm the water will be conducted from cell to cell. The transport can take place by diffusion through cytoplasmic connecting strands, the plasmodesmata. This is the *cytoplasm pathway* or *symplast pathway*. The final possibility is that the water will move osmotically by diffusion and mass flow from one cell to the next, passing through plasmalemma, cell wall, then plasmalemma of the neighbouring cell, protoplasm, possibly through tonoplast and vacuole, and so on in continuous repetition. This route is the *transmembrane pathway*.

It is largely unknown which pathway prevails under which circumstances. But it is surmised that plants are able to adapt to different situations by changing the routes. Whatever pathway through the cortex is chosen, the water arriving at the innermost concentric cellular layer of the cortex has finally to enter the cell protoplasm. Here at the *endodermis* the apoplast pathway is obstructed by suberization of the lateral cell walls, the barrier being named the *Casparian strip* (Fig. 7.4). At the endodermis layer, the plant has its last chance to modify the composition of solutes coming in with the water before they are transported from root to shoot. Nutrients in abundance and non-essential or even toxic ions can be excluded from passing onwards to the stele containing the vascular system.

In the section above it was proposed that there was a hydraulic resistance to the movement of water as it passed through the contact interface between soil particles and cells of external root surfaces, such as those of the rhizodermis or exodermis. Despite the different pathways for radial water movement within the root, a mean hydraulic resistance for the cortex parenchyma can similarly be derived. This *resistance of the cortex* will be very much influenced by the permeability of the cell membranes. And the permeability depends to a large extent on cell respiration. Respiration again relies upon temperature and oxygen supply. These interrelations explain the phenomenon that plants may wilt in poorly drained or *waterlogged soils*. The plants show signs of water deficiency although the soil's water supply is plentiful.

For maintenance of *root tissue permeability*, most of the crop plants need an *oxygen supply* to the roots via the pore system of the soil and an adequate *soil temperature*. In the soil, oxygen is transported from the soil surface to the oxygen-respiring roots by diffusion. As long as some soil pores are drained, oxygen *diffusion* through the aerated pore system may satisfy *oxygen requirements*. But in a soil with excessive water, oxygen diffusion is greatly impaired. When the soil pores are largely filled with water, oxygen diffuses through water-filled pores. However, the diffusion coefficient of oxygen in water is 10^4 times smaller than in air (Table 7.3). As a result, the oxygen supply rate to the roots is very much reduced. Not only does the lack of oxygen reduce the root tissue permeability, but so does the cool soil temperature. Wet soils stay cool because of the large heat requirement for warming the water. From research and experience, a practical rule has been derived for describing a soil with *sufficient aeration*. Such a soil has a minimum percentage of large pores >30 μm in diameter, which drain at 100 cm water tension (Fig. 5.1). The percentage should be 8–10 vol.% or greater. The volume percentage of large pores that drain at field capacity (Fig. 5.1) is called *air capacity*.

Plants living under submerged conditions are able to adapt to the limited oxygen supply through the pore system of the soil. In these plants the parenchyma cells in the cortex of the shoot and root are only sparsely packed, leaving air-filled channels in between. This special tissue is called *aerenchyma* and supports internal longitudinal transport of oxygen from shoot to roots by diffusion. Rice (*Oryza sativa*) is a crop with an effective aerating tissue. Wheat is better adapted to conditions of waterlogging than barley, as wheat can develop more porous root tissue in the event of flooding (Marschner, 1993).

Aerenchyma can also form in roots under conditions of limited water supply or nutrient availability. The radial transfer of water and ions, such as phosphate, nitrate and potassium, only takes place via the few transverse 'bridges' of cortical cells that separate one large air-filled channel from the next. In this case the aerenchyma does not so much sustain aeration but facilitates greater exploration for nutrients by the root system through growth that requires fewer structural carbohydrates from photosynthesis (Postma and Lynch, 2011).

7.3 Transpiration by Leaves

As with the process of soil evaporation, three conditions have to be met for *transpiration* through the leaf. Within the leaf, water changes from the liquid to the vapour phase. This change is an energy-consuming process, the energy being supplied by radiant energy. Hence radiation is the first condition. Secondly, a drop in vapour pressure is necessary to start the diffusion of the water molecules in the vapour phase from the intercellular spaces through the stomatal pores and out of the leaf. And thirdly, the water vapour must be removed from the leaf surface to the atmosphere. That also occurs by diffusion across a thin boundary layer, but the removal is greatly increased by mass flow, caused by the wind.

The water is supplied to the leaf in liquid form through *vascular bundles*, which end in the *mesophyll* in the form of a finely ramified network (Fig. 7.5).

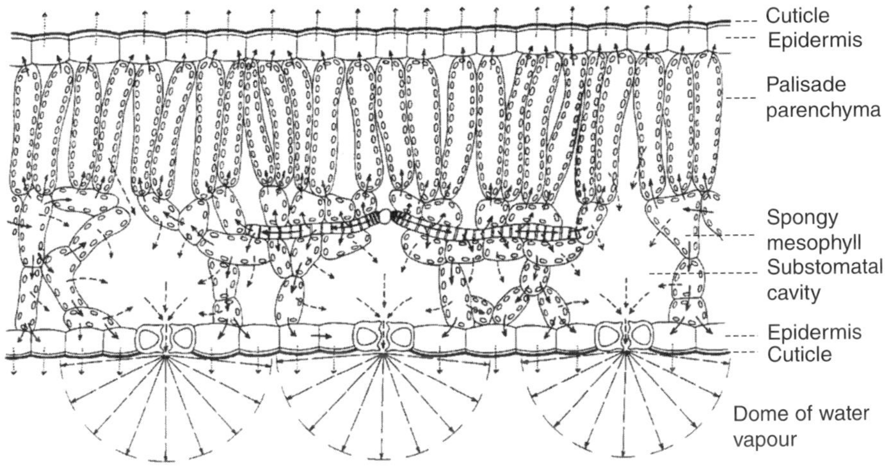

Fig. 7.5. Transpiration of a foliage leaf with abaxial stomates. Arrows indicate the pathway of liquid water and of water vapour. The molecules in vapour diffuse from the substomatal cavities through the stomates to the outside of the leaf (after Nultsch, 1991).

The bundles contain two conducting systems, the phloem with the *sieve tubes* and the xylem with the *vessel elements*. Leaving the terminal vessels, water enters the mesophyll cells. Part of the water is conducted in liquid form through the cells of the *palisade parenchyma* or the *spongy mesophyll* to the *epidermal cells*. Another, and by far the largest part, escapes from the mesophyll cells into the *intercellular spaces* in vapour form (Fig. 7.5). The largest of these spaces is called the *substomatal cavity*. From here the water vapour is lost from the leaves, using the diffusion path from the intercellular system through the *stomatal pores*. A much smaller part is released from the epidermal cells through the waxy cuticle. The *cuticular transpiration* of mesophytes comes to no more than 3–5% of the total water loss, which mainly comprises the *stomatal transpiration*. The stomates represent the prime passageways for the escaping water vapour. At the same time the stomates are the entrance openings for CO_2. By closing their stomates, plants are in a position to cut down the water loss very effectively when supplies run short, though closure has a harmful consequence for photosynthesis as the CO_2 supply is also cut.

Stomatal pores are formed by guard cells, which are present in the epidermis (Fig. 7.5). In many crop plants the openings are placed not only on one side (Fig. 7.5) but on both sides of the leaf (Table 7.2). The *density of stomates* varies between less than 20 apertures mm^{-2} to more than 300 (Table 7.2). In many trees, however, like apple (*Malus pumila*), cherry (*Prunus avium*), olive (*Olea europaea*) or common oak (*Quercus robur*), the upper or adaxial leaf side is completely free of stomates. All the openings are located on the lower or abaxial side (Fig. 7.5). The density changes between 250 (*Prunus*) and 550 stomates (*Olea*) mm^{-2} (Flindt, 2000). Interestingly, aquatic plants like the water lily (*Nymphea* spp.) have concentrated their openings on the adaxial side for obvious reasons. Despite the high stomatal density the *pore area* is rather small, covering no more than 2% of the one-sided leaf area.

Two guard cells are always required to form a single pore, the cells being shaped quite diversely among plant species. The guard cells of many dicotyledons

Table 7.2. Density of stomates on upper (adaxial) and lower (abaxial) leaf surface and length of stomatal pores for various crop plants (after Flindt, 2000).

		Density of stomates (no. mm^{-2})		
Species	Latin name	Adaxial	Abaxial	Pore length (μm)
Bush bean	*Phaseolus vulgaris*	40	280	nd[b]
Faba bean[a]	*Vicia faba*	44	65	33
Field pea	*Pisum sativum*	100	200	nd
Lucerne	*Medicago sativa*	170	140	nd
Sunflower	*Helianthus annuus*	175	325	22
Potato	*Solanum tuberosum*	50	160	nd
Oat[a]	*Avena sativa*	57	56	31
Wheat	*Triticum aestivum*	33	14	38
Maize	*Zea mays*	52	68	19

[a]From Müller *et al.* (1986). [b]nd, not determined.

and of some monocotyledons appear *bean-shaped* (Fig. 7.6A). The cell wall is thickened and strengthened at the ventral side, whereas the wall at the dorsal side is thinner and more elastic.

When fully turgid, the elastic wall is pressed outwards into the neighbouring cells, being either epidermal cells or subsidiary cells. By this dorsal extension the rigid ventral wall is concavely bent inwards into the guard cell, forming an open pore with the concave wall of the opposite guard cell. When the plant loses water and consequently turgor, the dorsal walls will shrink, and the ventral walls will relax, straighten and move close together. As a result, the pore will shut again and only the external atrium will still be visible (Fig. 7.6A).

In grasses the guard cells are *dumb-bell-shaped* with bulbous swellings at the end (Fig. 7.6B). The protoplasm is concentrated in the swollen areas, and here the cell walls are elastic, whereas in the middle section of the cells the walls are thickened. For this type of stomate the guard cells are always accompanied by subsidiary cells (Fig. 7.6B). The maximum opening is less than in the bean-shaped type.

In spite of the small pore area, plants can transpire water through the stomates per unit leaf area in an order of magnitude that is near the rate of water loss per unit area from an open water surface. This phenomenon can be explained

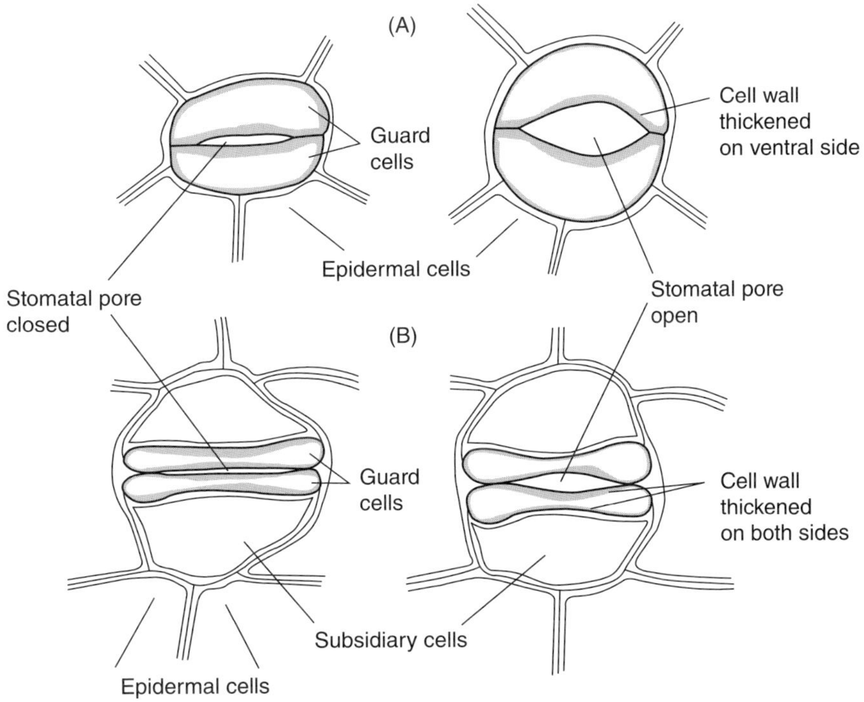

Fig. 7.6. Bean-shaped stomatal guard cells (A) of dicotyledons and some monocotyledons. The dumb-bell-shaped guard cells (B) are found in grasses. For (A) and (B) the left-hand side shows closure of the pores. On the right-hand side the pores are open. At closure just an external atrium is visible, below which the closed pore is located.

by the geometry around a stomate of the hemisphere that is taken up by water vapour during diffusion (Fig. 7.5). Within an open pore a water molecule diffuses in only one direction to the outside of the leaf. But after having reached the leaf surface, the diffusion can proceed in three directions. The three-dimensional movement causes the water molecules to form a *dome of water vapour* above a stomate. By deflection of the molecules at the boundary of a stomate, the *field of diffusion* is largely extended. This extension accelerates the rate of diffusion through the pore, compared to an 'endless' pore without any edge effect.

Immediately near the transpiring leaf the air is rather humid. Water molecules have to diffuse through this humid air layer, called the *boundary layer*, before reaching the atmosphere. As the gradient in water vapour, the driving force for diffusion, is only small across the layer, the boundary layer can be regarded as a diffusion barrier.

The barrier is just a few millimetres thick, and wind will shrink the layer or even remove the barrier intermittently, which explains why transpiration is accelerated in windy conditions compared with calm air (Fig. 7.7).

7.4 The Action of Stomatal Guard Cells

Opening and closing of stomates is controlled by the *turgor of the guard cells*. As osmotically active substances accumulate in the guard cells, the total potential of water will be lowered compared with that in surrounding cells. This potential drop causes an inflow of water by *osmosis*, a diffusion process. The gain of water raises the turgor in the guard cells, and the pore between two guard cells is widened. Conversely, a loss of osmotic substances will cause the turgor to decrease, when the guard cells release water to the neighbouring tissue. The active change of solute concentration in the guard cells is called *osmoregulation*. By osmoregulation the plant maintains an active control of turgor and stomatal aperture. The osmotic components regulating inflow and outflow of water in guard cells are mainly *potassium ions*. These ions are transported actively from surrounding epidermal or subsidiary cells into the guard cells or vice versa. For the transport, the ions have to pass the biological membranes of the cells involved. The passage is managed by special passage pores, the *potassium channels* (Schröder *et al.*, 1987). The catalysed ion transport is activated by plant hormones, of which the best known is abscisic acid.

The opening and closing action of guard cells depends on a number of environmental factors (Raschke, 1975). A prominent factor is the water supply. When plants are losing water from the tissue, the turgor will decrease. Either after attaining a critical threshold or by a gradual decline of the turgor, the guard cells will close the stomates more or less abruptly. This *H_2O control circuit* protects the plant from early and deadly wilting. The feedback mechanism is directly supported by the water *vapour pressure deficit* of the air. The deficit will normally increase towards noon with increasing temperature and decreasing relative humidity in the atmosphere. The deficit sensed near a transpiring leaf can cause the plants actively to reduce the stomatal openings during the time of the day when, without control, the transpirational water loss would be greatest (cf. Fig. 9.14).

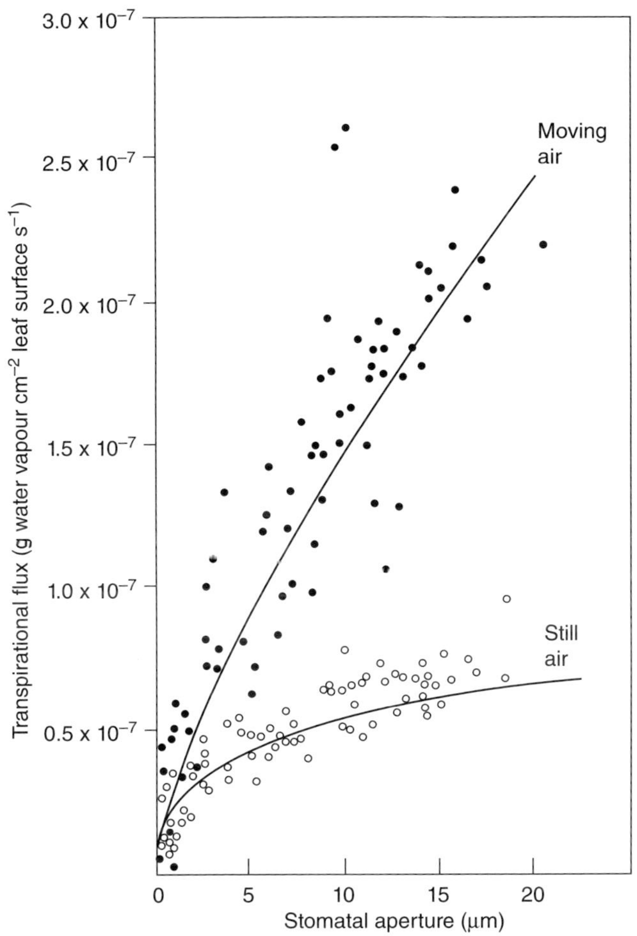

Fig. 7.7. Transpiration rate of *Zebrina pendula*, a common house plant, as a function of stomatal aperture in still and moving air. Wind accelerates transpiration, but in calm air the larger boundary layer slows down the water loss. In that case the stomatal aperture has less control on transpiration than in wind (after Bange, 1953; from Taiz and Zeiger, 1991).

With a more balanced water supply, another feedback system takes effect, the *CO_2 control circuit*. In the morning hours, when illumination starts along with photosynthesis, the CO_2 concentration within the intercellular spaces of the mesophyll decreases. This causes an opening of the stomatal pores in C_3 and C_4 plants during the day. That is to say, the demand is regulating the supply. The same mechanism is operating in crassulacean acid metabolism (CAM) plants, which keep the stomates open during the night. During the night the plants with CAM metabolism bind the CO_2 in the form of malic acid (Section 1.2). Beside the indirect CO_2 mechanism, the guard cells of C_3 and C_4 plants seem to respond directly to *light* by opening the stomates. Within the 'normal' range at a site, the stomatal aperture is hardly affected by the *temperature*. High temperatures above 30–35°C, however, may cause stomatal closure, depending on the species. The closure may be caused by high internal CO_2 levels, owing to increased respiration, or may be the consequence of insufficient water supply at high evaporative demand.

7.5 Water Transport within the Plant

Liquid water can be conducted in the cortex of the root towards the xylem in the stele by passing from cell to cell, using cytoplasmic strands called plasmodesmata (Section 7.2). Taking this *symplast pathway*, water is driven by *diffusion*, and the driving force is a gradient in water concentration. With the transmembrane pathway, water is moving by *osmosis*, and the driving force is a gradient in total water potential (Section 7.1).

In the leaf, water molecules in vapour form move from the substomatal cavity through the stomates and the boundary layer by diffusion. The diffusion rate or flux is given by *Fick's first law of diffusion*, which reads:

$$J = -D\Delta c / \Delta x \tag{7.10}$$

In the equation J is the flux and is defined as the quantity (either mass or volume) of a substance that is transported by diffusion per unit cross-sectional area and per unit time (the unit is, for instance, $cm^3\ cm^{-2}\ s^{-1}$). The letter D ($cm^2\ s^{-1}$) stands for the *diffusion coefficient*, which is defined by how fast a substance will diffuse through a medium when the concentration gradient is 1. D is characteristic of the substance, but depends much more on the 'packing density' of the molecules that make up the medium for diffusion (Table 7.3). The density of molecules within a liquid is much higher than within a gas. Therefore the diffusion of a substance through a liquid is comparatively slow. The concentration gradient in Eqn 7.10 is constructed from the concentration difference (Δc) and the diffusion distance Δx. The gradient is defined to be negative in the direction of net particle movement (i.e. from larger to smaller concentration). The minus sign in Eqn 7.10 then causes the diffusive flux to be positive in the direction of the (positive) x-axis (compare Section 5.1, Example 2, Case C and Eqn 5.1).

We imagine that right before the beginning of a diffusion process through a medium a diffusible substance is concentrated at a location, and that there is a neighbouring location, which is free of the substance. That is to say, there exists a large gradient in concentration between the two locations at starting time zero. After the diffusion has started, the gradient steadily declines, as the concentration

Table 7.3. Diffusion coefficients at 25°C of carbon dioxide and oxygen in air and liquid water, and for water vapour in air (after Grable, 1966).

	Diffusing gas ($cm^2\ s^{-1}$)		
Medium	CO_2	H_2O	O_2
Air	1.81×10^{-1}	2.57×10^{-1}	2.26×10^{-1}
Water	2.04×10^{-5}	–	2.60×10^{-5}

of the substance gets smaller at the starting location and increases at the destination location. With the decline of the gradient the diffusive flux slows down. But a 'long' time will pass before the concentrations at the two locations will have completely equalized.

The time that passes for the concentration at a certain distance Δx to reach half the initial value at the originating location is called $t_{0.5c}$, which is:

$$t_{0.5c} = (\Delta x)^2 / D \cdot K' \quad (7.11)$$

In the equation, D again stands for the diffusion coefficient and K′ is a constant. The smaller the diffusion coefficient, the more time will pass for a certain concentration to be obtained in a particular distance. The necessary time increases with the square of the diffusion distance.

Consider a cell of 50 μm length. A substance has to move over this distance by diffusion. How much time will be necessary for the transport process? For convenience we will take a value of 1 for the constant K′. The diffusion coefficient will be taken to be 2×10^{-5} cm^2 s^{-1} (cf. Table 7.3). The time $t_{0.5c}$ is:

$$\begin{aligned} t_{0.5c} &= (50 \times 10^{-4}\ \text{cm})^2 / (2 \times 10^{-5}\ \text{cm}^2\ \text{s}^{-1}) \\ &= 1.25\ \text{s} \end{aligned}$$

The calculation shows that the diffusion over short distances in aqueous media from cell to cell may be quite rapid, unless selectively permeable cell membranes are restricting the diffusive transport. If the medium through which the diffusion is taking place is not water but air, then the diffusion is still more rapid. How much time will it take for H_2O molecules to diffuse over a distance of 1 mm from the leaf epidermis into the boundary layer? The D is taken to be 2.5×10^{-1} cm^2 s^{-1} (Table 7.3):

$$\begin{aligned} t_{0.5c} &= (10^{-1}\ \text{cm})^2 / (2.5 \times 10^{-1}\ \text{cm}^2\ \text{s}^{-1}) \\ &= 0.04\ \text{s} \end{aligned}$$

Now we will assume that water has to be transported by diffusion from the root to the top of a tree, a distance of 10 m. How much time would then elapse?

$$\begin{aligned} t_{0.5c} &= (1000\ \text{cm})^2 / \left(2 \times 10^{-5}\text{cm}^2\text{s}^{-1}\right) \\ &= 50 \times 10^9\ \text{s} \\ &= 578{,}704\ \text{days} \\ &= 1585\ \text{years} \end{aligned}$$

That is a lengthy period of time, and very few tree species ever reach such an age.

What then is the mode of water transport within the plant, if not from cell to cell by diffusion or osmosis?

After passing through the endodermis of the root, the water enters the *stele*, where it is conducted through the cell tissue by osmosis, finally arriving at the xylem strands. The xylem of the angiosperms is composed of a few living parenchyma cells and cells of large diameter that have lost their protoplasts. One of these types of cells is called the *trachea* or the *vessel member*. Strung together these members form a *vessel* (Fig. 7.8). The vessels represent the 'hydraulic pipelines' (Sperry, 2000) for long-distance transport of water within the plant. They permit a rapid conduction from the root through the stem axis to the leaf. The vessel members are joined together at their ends by open perforations or *perforation plates* (Fig. 7.8), restricting to some degree the vertical flow path by transverse constrictions. *Bordered pits* are inserted into the strengthened, lignified longitudinal walls of the vessels, which are passage openings in a horizontal direction (Fig. 7.8). They allow the transverse transport of water into neighbouring cells, but may seal off the vessel by closing membranes, when by accident the vessel has dried out.

In the vessels, water is conducted by *mass flow*, like water through water mains in the street and through water pipes in the house, driven by a difference in pressure, more precisely by a difference in pressure potential. However, in the plant tissue it is normally not a difference in excess pressure, but in low (negative) pressure, which causes the water to flow through the vessels (Box 7.1).

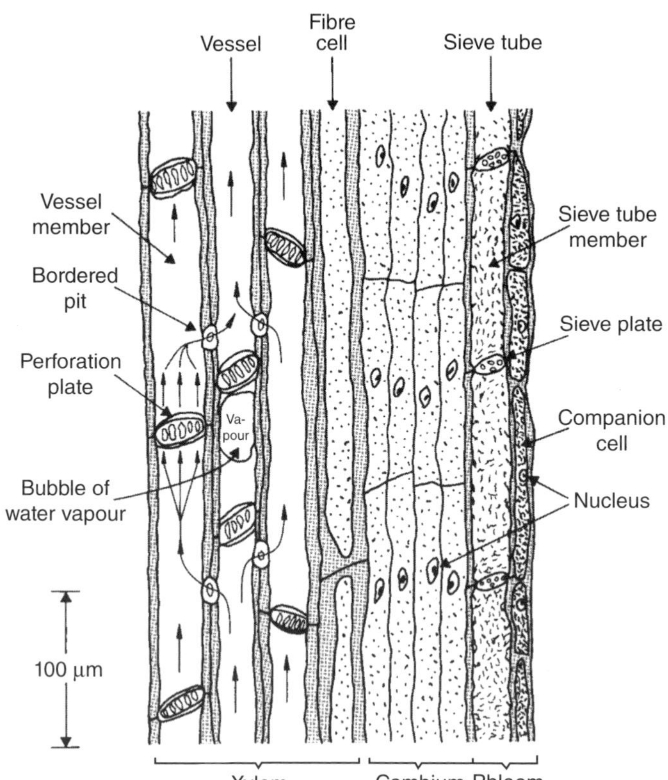

Fig. 7.8. The vascular tissue of plants (angiosperms) is composed of vessel tubes and of sieve tubes. The sieve tubes serve for transport of organic compounds like assimilates, whereas the vessel tubes convey water and mineral nutrients as well as organic compounds, metabolized in the roots. Arrows within the vessels indicate the direction of sap flow. The pit density is in reality many times the number shown in the diagram (based on Nobel, 1974 and various other sources).

The long-distance transport is carried out at high speed. With deciduous trees the velocity ranges from about 1 m h^{-1} to 40 m h^{-1} (0.03–1.1 cm s^{-1}) and with herbaceous plants from about 10 m h^{-1} to 60 m h^{-1} (0.3–1.7 cm s^{-1}) (Larcher, 1973). In the following we calculate the *pressure potential gradient* necessary to attain the flow velocity at the scale mentioned. *Hagen–Poiseuille*'s *law* relates volume flow rate Q of water to the gradient and the tube radius.

$$Q = -\pi r^4 / 8\eta \cdot \Delta P / \Delta x \qquad (7.12)$$

The Q stands for the volume of water, flowing per unit time through a tube of radius r, driven by a difference in pressure potential ΔP, which exists across the length Δx of the tube. The symbol η signifies the viscosity of water, which is 1×10^{-2} poise (1 poise = 1 g cm^{-1} s^{-1} = 10^{-1} Pa s) at 20°C. The minus sign means that the volume flow within the tube is positive in the direction from higher to lower pressure potential.

Q has the unit of cm^3 s^{-1}, but we are dealing with velocities in cm s^{-1}. Hence we must relate the volume flow rate Q to the unit cross-sectional area A (cm^2) of the tube, thus obtaining the water flux q (cm^3 cm^{-2} s^{-1}), with the apparent unit of velocity (cm s^{-1}):

$$q = Q / A = -r^2 / 8\eta \cdot \Delta P / \Delta x \qquad (7.13)$$

We assume that the radius of the vessel is 50 μm and the flow velocity is 0.3 cm s^{-1}. How large will the pressure potential gradient need to be under these conditions?

According to Eqn 7.13 the gradient can be calculated as:

$$\begin{aligned} \Delta P / \Delta x &= -8 q \eta r^{-2} \\ &= -8 \times 0.3 \text{ cm s}^{-1} \times 10^{-3} \text{ Pa s } \times \left(50 \times 10^{-4} \text{ cm}\right)^{-2} \\ &= -0.96 \times 10^{2} \text{Pa cm}^{-1} \\ &\cong -1 \times 10^{4} \text{ Pa m}^{-1} \cong -0.1 \text{bar m}^{-1} \end{aligned}$$

At the suggested flow velocity of 0.3 cm s^{-1}, the gradient in the vessels of 100 μm diameter comes to –0.1 bar m^{-1} of conducting path. For each metre the pressure potential declines by 0.1 bar. The pressure gradient is valid for an ideal tube with quite smooth walls. But with the vessels one has to expect a higher *flow resistance*, caused by the roughness of vessel walls and presence of perforation plates. Therefore the gradient has to be slightly greater for maintenance of the flow velocity, coming up probably somewhere between −0.1 and −0.12 bar m^{-1}. Please note that this is the gradient required for water to flow through a horizontal tube with the same diameter as a vessel.

With plants growing upright, the gravitational potential also has to be considered when calculating the gradient necessary for the suggested water flux in the vertical direction. The *gradient for the gravitational potential* is +0.1 bar m^{-1} of plant height. This is the amount by which the pressure potential gradient has to decrease m^{-1} height, thereby ensuring that the given flux of ascending water is not just overcoming the physical resistance to flow in the tube, but also overcomes the earth's gravitational pull. Based on these considerations (allowing for the gain in gravitational potential of the ascending water) we estimate that the gradient in pressure potential has to be near to −0.21 bar m^{-1} plant height. For a 10 m high tree we calculate a potential difference that is easily −2.1 bar between soil surface and canopy, and for a coastal sequoia (*Sequoia sempervirens*) from California of 100 m height or a karri tree (*Eucalyptus diversicolor*) of the same size from the south coast of Western Australia a difference of at least −21 bar. How can these pressure potential differences be generated?

7.6 Water Potentials in Plants

For quite a long time it was thought that the pressure potential difference in the vessels was caused by excess pressure, which was supposed to be generated in the roots. Today we know that the maximum difference in pressure potential within the bounds given above (i.e. up to 21 bar or even higher; Sperry, 2000), is not achieved by *pressure*, but by pull. That is pull by a subatmospheric or negative pressure. Speaking in terms of the potential concept (Section 4.2) we have to state that it is a difference in the total water potential (negative) that causes the water to flow at a defined velocity. The *difference in total potential* between the root and top of the plant has to achieve a value that enables the water to overcome not only the flow resistance in the xylem but also the gravitational pull of the earth. Furthermore, the potential difference has to ensure an adequate flow velocity, which will change during the day in accordance with the course of the evaporative demand. From the root to the leaf the total potential will be composed of the four component potentials (Eqn 4.5), each varying in level.

In real terms we can imagine that the total potential ϕ reaches to −12.3 bar in the finely ramified vascular bundles of the leaves in the canopy of a tree of 100 m height (Table 7.4). Here in the xylem the osmotic potential O has a small negative value, just −1 bar. The matric potential Ψ is assumed to be −0.3 bar, but the strongly negative pressure potential P is, let's say, around −21 bar. The *gravitational potential* Z has to be added, and that is +10 bar with reference to the soil surface. All the four component potentials add up to −12.3 bar for ϕ (Table 7.4). From the vascular bundles the water is transmitted into the cells of the mesophyll. Here ϕ may amount to −14 bar, composed of O = −22 bar, Ψ = −4 bar and P = +2 bar, with Z again being +10 bar. In the cell wall of a mesophyll cell, bordering the substomatal cavity, the potential allocation might be: O = −3 bar, Ψ = −23 bar, P = 0 bar and Z = +10 bar. The highly negative value of Ψ can be explained by the assumption that the wall is losing water in the form of vapour to the intercellular spaces, as a result of which the fine spaces, formed by the fibrillas in the wall, tend to dry. Here ϕ adds up to −16 bar. Supposing that the relative humidity in the intercellular spaces is 98% at 20°C, the water potential of the vapour is −27.3 bar (cf. Fig. 7.9). As Z is again +10 bar, ϕ results in −17.3 bar.

These are 'made-up' numbers, but nevertheless they are in the correct order of magnitude. They explain that there is a drop in the total potential ϕ from the water-conducting xylem along the cells of the mesophyll into the intercellular spaces. From here the potential drop continues to the outside air.

Table 7.4. Total water potential ϕ and component potentials of water (O, osmotic potential; Ψ, matric potential; P, pressure potential; and Z, gravitational potential) at various locations along the water transport system of a giant tree, 100 m in height.

Location within the transport path	Height[a] (m)	Total potential and component potentials (bar)				
		ϕ	O	Ψ	P	Z
Atmosphere outside the leaves (relative humidity 75%)	100	–378	–388[b]		0	+10
Air in intercellular spaces (relative humidity 98%)	100	–17.3	–27.3[b]		0	+10
Cell wall in mesophyll	100	–16	–3	–23	0	+10
Cell of mesophyll	100	–14	–22	–4	+2	+10
Xylem of leaf	100	–12.3	–1	–0.3	–21	+10
Xylem at base of trunk	0	–5.3	–1	–0.3	–4	0
Root stele	–1	–5.1	–9	–1	+5	–0.1
Root cortex	–1	–3.1	–7	–1	+5	–0.1
Soil	–1	–1.1	0	–1	0	–0.1

[a]Reference level: soil surface.

[b]The potential of water vapour is equal to the sum of the component potentials O and Ψ under isothermal equilibrium conditions, when Z is neglected (Hillel, 1971).

Table 7.4 exemplifies total and component potentials from the soil to the atmosphere in a system with a very tall tree.

The large *subatmospheric pressure within the xylem sap*, the negative pressure potential, is caused by the evaporation of liquid water from the mesophyll cells into the intercellular spaces, including the substomatal cavity (Fig. 7.5). Each of the evaporating water molecules is instantly replaced by a new molecule, moving up through the liquid phase. The pull or suction is propagated by a decrease in the absolute concentration or partial pressure of the water vapour along the transport path from the inner spaces of the leaf through the stomates, through the boundary layer at the outer leaf surface up into the atmosphere. Depending on absolute vapour pressure and air temperature, which both determine the relative humidity of the air (explained in Section 8.2), the water potential in the air can be lowered drastically (Table 7.5).

Pulling the 'string of water' through the xylem upwards causes a negative pressure in the vessels. Because of wall thickening (Fig. 7.8) the vessel walls will not collapse, but can withstand the pressure difference between inside and outside. Normal tissue cells with more elastic and unreinforced walls would break down more easily with the application of such pressures.

There is another problem with water transport through the plant. Air, dissolved in water, is released under the conditions of increasing negative (subatmospheric) pressure. The gas forms bubbles, a process called *cavitation*. The bubbles cause *embolism*, by which the water transport through the vessel is interrupted (Fig. 7.8). Water molecules in the liquid state are linked together by high intermolecular binding forces (Section 4.1). The cohesion gives water a high *tensile strength*. A pulling force of more than 300 bar is necessary to separate water molecules from each

Table 7.5. Relation between relative humidity (RH) at 20°C and the water potential (Hillel, 1980; Taiz and Zeiger, 1991). The formula for the calculation of the water potential (not including gravitational potential) at 20°C is: $O + \Psi = 1350 \cdot \ln(RH)$. The RH is taken as a fraction of 1 (100% ≡ 1). The result is indicated in negative values of bar. Compare with the formula in Fig. 5.9, which achieves an almost identical result.

Relative humidity (%)	Water potential[a] (bar)
100	0
99.99	–0.135
99.9	–1.35
99	–13.57
95	–69.25
75	–388.37
50	–935.75
20	–2172.74

[a]Compare note b of Table 7.4.

other. Even in the highest trees the cohesion between water molecules should be sufficiently high to ensure the intactness of the water threads within the vessels. But in reality the linkage can be broken by cavitation. Air bubbles have no tensile strength. The formation of 'gas seeds' breaks the water threads immediately (Box 7.2).

Plants will protect themselves from disastrous consequences of cavitation by partial breaks of the conducting vessel tubes. These take the form of open perforations or perforation plates (Fig. 7.8) with elliptical holes cut in the cross-sectional area, where the vessel members or tracheas border. But plants cannot avert the onset of cavitation. As soon as the water thread is broken in a trachea, the water flow will be diverted into neighbouring vessels by the bordered pits of a trachea below and above the occluded vessel member (Fig. 7.8). At the same time the expansion of the bubble from one vessel member to the next is limited by the specific vessel structure. The holes at the perforations and the pits at the walls are good for liquid, but not for gas transport. Because of the surface tension of water at the liquid–vapour interface, the holes will stop the air bubbles from escaping or steadily enlarging. During the night, when the transpiration rate falls to zero and the tension eases, the air bubbles can be dissolved again. Thus the damage is repaired and the diversion will be closed.

In recent literature it is suggested that cavitation is not always harmful and can, in fact, be beneficial. A '*controlled*' *cavitation* can be beneficial for plant life, especially under conditions of short water supply. Controlled cavitation supports stomates in their action to constrain transpiration, resulting in a more gradual use of soil water (Sperry, 2000).

The everyday occurrence of cavitation in trees and in maize has been demonstrated by researchers using acoustic detectors. When the water threads break, high frequency sound waves are created. The shock waves spread over the plant and the signals can be recorded as a 'click' (Tyree and Dixon, 1983).

Box 7.2. Searching for the cause of sap ascent

For quite a long time the cause of the transpirational flow in plants remained a mystery. But from the 1850s it was definite that the fundamental driving force for water flow in plants did not originate from root pressure, but had to be attributed to the transpirational suction. But why could the water rise up to the crown of even the highest trees? If plants simply acted like a vacuum pump in raising the water upwards into the canopy, the water column in the xylem would not exceed a height of 10 m and probably even less. The 10-m height corresponds to the water head of normal pressure. Some attempts at an explanation were based on the phenomenon of capillarity, on osmotic sucking up and on unknown 'plasmatic' effects, stemming from living cells. Schwendener (1829–1919), a professor of botany in Berlin, wrote in the year 1886:

> I absolutely stand by the fact that the vital activity of cells is somehow intervening in sap motion. The lift of water up to heights of 150 to 200 feet and more, is simply impossible without this intervention. And all the endeavours to break through existing barriers by uncertain physical concepts, are not much more than seeking the philosopher's stone. As often we investigate the processes in living organisms more closely, we will again and again encounter, in addition to the effects of physically understood concepts, that unknown something. That is the vital activity of the protoplasm, the mechanics of which we are at present completely in the dark
>
> (Schwendener cited in Böhm, 1893)

Josef Böhm (1833–1893) was a professor of botany at the 'Hochschule für Bodenkultur' (University of Agricultural Sciences) in Vienna. With an almost unbelievable enthusiasm for his work, he pursued his aim of investigating the true explanation for the high ascent of the sap flow in plants. For the purpose he used an apparatus that could measure the negative pressure in the water-conveying system of the plant to a first approximation (Fig. B7.2). He fixed twigs of crack willow (*Salix fragilis*) and of thuja (*Thuja* sp.) into this equipment. The rise of the mercury column in the right-hand arm indicated the suction force of the transpirational flow. The value had to be corrected for the hanging water column in the left-hand arm.

But for most of his experiments a minor mishap tended to occur, putting the effectiveness of his effort into question. After a noticeable rise of the mercury column a 'gas seed' formed in the system, which is marked with a 't' in Fig. B7.2, indicating a 'Torricelli space'. With the development of an air bubble, formed in a subatmospheric system by release of dissolved air from water, the rise of the mercury was immediately interrupted. Böhm tried to remove the air by boiling the water for hours, but this was apparently futile. During his lecture-free vacations in the years from 1882 to 1889 he undertook about 400 experiments of this kind, but almost none was successful, in spite of every painstaking effort. Nevertheless, at least three of his experiments were on target. During one day, when the barometer was pointing to 74.5 cm, the mercury column rose to 76.5 cm, and with thuja even to 90.6 cm! But then the gas seed he had been fearing formed again, and the mercury column instantly dropped below the barometer reading.

Böhm explained the 'apparently paradoxical fact' that the mercury rise was higher than the surrounding pressure by the cohesion of water. According to Böhm's insight the cohesion between the water molecules becomes effective due to the adhesion between water molecules and the walls of the water-conducting vessels. Based on his paper 'Capillarität und Saftsteigen' ('Capillarity and the rise of sap'), published in 1893 (the year of his death), Josef Böhm is regarded as the 'father of the cohesion theory' (Huber, 1956). The articles of the research scientist Askenasy (1845–1903) from Heidelberg (Askenasy, 1895) and of the Irishmen Dixon and Joly on the same topic (Dixon and Joly, 1895) were published only 2 years later.

Continued

Box 7.2. Continued

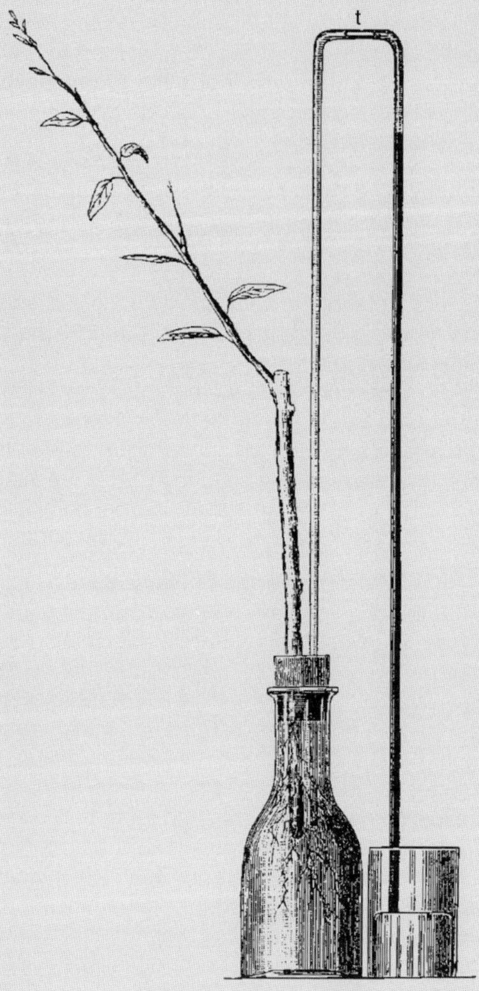

Fig. B7.2. Böhm's experimental set-up for the determination of the highest possible sap ascent, using twigs of the crack willow. At the beginning of the experiment the system is freed from all visible air bubbles. Moreover, the dissolved air, which is invisible, has to be removed from the water by boiling. Transpiration generates a negative pressure in the closed system. The subatmospheric pressure is indicated by rise of the mercury column in the right-hand arm of the apparatus. The cohesion of the water molecules, the tensile strength of the water column, extending from the left-hand arm to the vessels in the twig, supports the build-up of a high negative pressure. The (negative) value may surpass the (positive) value of the atmospheric pressure of around 1 bar. As soon as a 'gas seed' is formed, the progress of the mercury rise is interrupted, just as in a tensiometer (Box 5.1). The gas seed will increase in volume, forming an air bubble, depicted in the figure at 't' (Böhm, 1889).

It is not always a difference in negative pressure that moves the water through the plant. Occasionally the water in the xylem vessels of the plant is under positive pressure, sometimes called the *root pressure*. Root pressure may arise in periods of low transpiration, for instance during the night or at times of high relative humidity. Under such conditions, mineral nutrients absorbed by the root from the soil may accumulate in the xylem, lowering the osmotic potential of the xylem sap. As a result, the total potential declines, thereby increasing the hydraulic gradient between root xylem and surrounding soil. The gradient causes the water to flow from the soil into the root. The water flux induces a positive pressure within the xylem, squeezing out water from special outlet openings, the *hydrathodes*, present in the leaf. The outflow of liquid water from the leaf is called *guttation*. It may be observed in cereal crops on spring mornings in the early hours. Droplets of sparkling water cling to the apex of the leaf lamina, and often they are confused with dew. The exerted positive pressure exceeds 1 bar only in exceptional cases (Box 7.1). Probably it is the root pressure that helps plants to recover from cavitation (Sperry *et al.*, 2003).

8 The Plant as a Link between Soil and Atmosphere: an Overview

8.1 The Soil–Plant–Atmosphere Continuum (SPAC)

In the soil, the water potential may be above −1 bar or may fall below that value as far as the permanent wilting point (PWP), which is taken to be −15 bar. But the potential may even drop below the wilting point, when soil drying is advanced (Fig. 5.1). In the atmosphere, which surrounds the shoot and leaves, the potential can reach values during the day that are normally much lower than those in the soil. Depending on the actual water content or the relative humidity of the air, respectively, the potential will easily reach values beyond −15 bar, falling to −50 or −1000 bar (Table 7.5). Generally, there will be a potential difference between soil and atmosphere, with the plant acting as a connecting water conductor. As a crude simplification, the plant can be thought of as being much like the wick of an oil lamp. The lower end of such a wick soaks up the paraffin from the supply reservoir, and delivers it to the zone at the upper end of the wick, where it is burned. The burning causes a suction that controls uptake and delivery to meet the demand.

The plant differs from a wick in so far as it can actively maintain water uptake by extending its roots into previously untapped regions of the soil reservoir. Moreover, the plant can restrict water loss, but if uptake does not keep pace with the loss, there can be some development of a water deficit within the tissue. Plants can adapt to periods of water shortage by anatomical modifications, such as reducing the number of leaves, producing smaller leaves, leaf rolling, increased rooting depth, and by physiological adaptations, including the active lowering of the osmotic potential by accumulating solutes in the cells. But just as in the example of a wick and oil, the transport of water in the *soil–plant–atmosphere continuum* (SPAC) (Philip, 1966) is also a passive process.

Within SPAC, water moves down a potential gradient (i.e. from places of higher to places of lower potential). The water is conducted either in the liquid or vapour state. The movement is either by mass flow or by diffusion. In the course of such transfer, water has to overcome *hydraulic* or *diffusion resistances*. Under otherwise identical conditions, the larger the resistance, the longer the time required for the transfer. Let us go back to Eqn 7.9, which is a modified form of the Darcy equation:

$$q = -1/[\Delta z/K(\Psi)] \cdot \Delta\phi = -1/R \cdot \Delta\phi \quad (8.1)$$

Measuring q, the water flux, in $cm^3\ cm^{-2}\ day^{-1}$, and the total potential ϕ in cm, then the hydraulic resistance R will attain the dimension of time. In this example, the unit is day. If instead of cm we take the unit of bar for ϕ, the unit of R changes into bar day cm^{-1}.

We will also re-define Eqn 7.10, which is Fick's first law for the diffusive gas transport:

$$\begin{aligned} J &= -D\Delta c/\Delta x = -D\beta^{-1} \cdot \Delta p/\Delta x \\ &= -(1/\beta) \cdot 1/(\Delta x/D) \cdot \Delta p \\ &= -(1/R^*) \cdot \Delta p \end{aligned} \quad (8.2)$$

J is the diffusive flux of water vapour, which we will define in units of $cm^3\ cm^{-2}\ day^{-1}$. The D stands for the diffusion coefficient ($cm^2\ day^{-1}$), c is the concentration, here expressed in the unit $cm^3\ cm^{-3}$, and Δx is the distance in the direction of the *x*-axis (cm) that has to be crossed by the diffusing water vapour molecules. We may replace Δc by Δp, which is the difference in the partial pressure. One possible unit of p is the bar. To use that unit the factor β has to be introduced, which is the ratio of partial pressure to concentration (bar/($cm^3\ cm^{-3}$)). Consequently, the diffusion resistance R^* ($= \beta \cdot \Delta x/D$) has the units bar day cm^{-1}, which is the unit of the hydraulic resistance.

From Eqns 8.1 and 8.2 we can deduce that both resistances, hydraulic and diffusive, increase in proportion to the transport distance, and decrease in proportion to the size of the transport parameters $K(\Psi)$ and D. For water flowing from soil to root, the drier the soil and smaller the value of K, the greater is the hydraulic resistance. But plants can modify the hydraulic resistance in the soil. By increasing root length density, plants reduce the transport distance from a position in the bulk soil to the root surface (Eqn 7.6), and thereby decrease the resistance. Variable resistances along the flow and diffusion path of water within SPAC are marked by an arrow in Fig. 8.1.

The *contact resistance* in Fig. 8.1 changes with the soil water content, with the length and density of root hairs, with the development and extent of the mucilage layer and with the diameter of a shrinking or swelling root – all of them influencing the wetted contact area at the transition from soil to root (see Section 7.2). The *radial resistance* of the pathway from the root surface through the cortex, including the endodermis, to the stele is considered to be much larger than the *axial resistance* in the root xylem. Also, the resistance in the xylem of the shoot is comparatively small. For vapour phase movement there is a considerable resistance at the *cuticle* and variable resistance in the *stomates*. As already shown, the *boundary layer resistance* changes with the speed of air movement outside the leaf (Fig. 7.7).

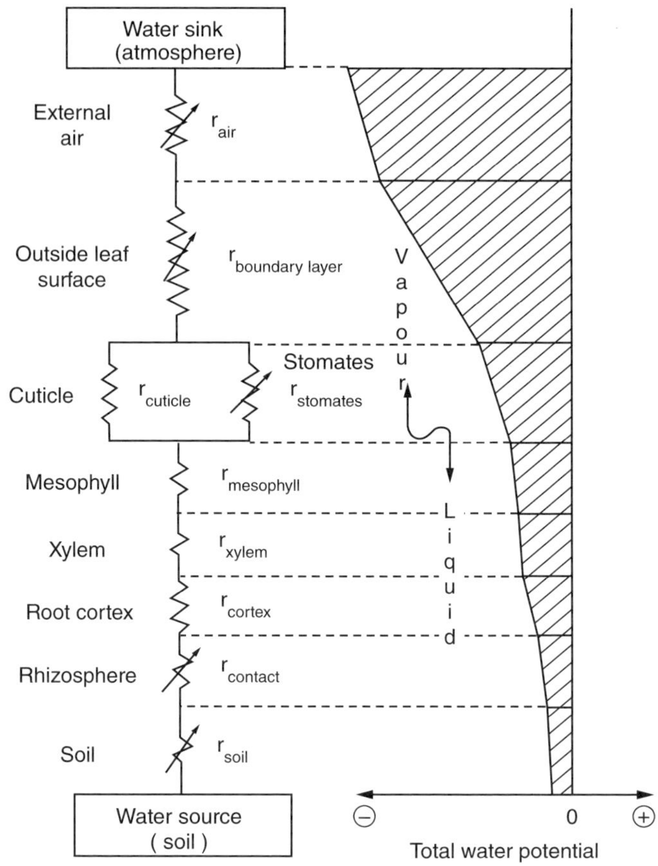

Fig. 8.1. Hydraulic and diffusion resistances in the soil–plant–atmosphere continuum (SPAC). The resistances are connected one after the other in series. Only in the leaf are the resistances of the cuticle and the stomates arranged in parallel. The height of the potential drop depends on the magnitude of the partial resistance (based on Hillel, 1980).

The outside air resistance will depend on weather variables like the vertical temperature gradient, relative humidity and wind (Fig. 8.1).

As indicated in Fig. 8.1, the resistance of water transport in the liquid phase is generally greater in the plant than in the soil (Newman, 1969). Especially in the root cortex, notably at the endodermis with its Casparian strip (Fig. 7.4), the hydraulic resistance to liquid water flow seems to be comparatively large relative to others along the path from the root surface to the leaf. Accordingly, there is quite a large potential drop from the cortex to the stele, in which the xylem is located (Fig. 8.1). Only when the soil has dried close to the PWP, as a result of water uptake by the plant, will the *flow resistance* in the soil reach the same order of magnitude as the total resistance to flow in the plant (Reicosky and Ritchie, 1976).

The reduction in water potential within the SPAC is shown in Fig. 8.2 for four hypothetical situations: case 1 depicts the potential drop when the soil is moist (high

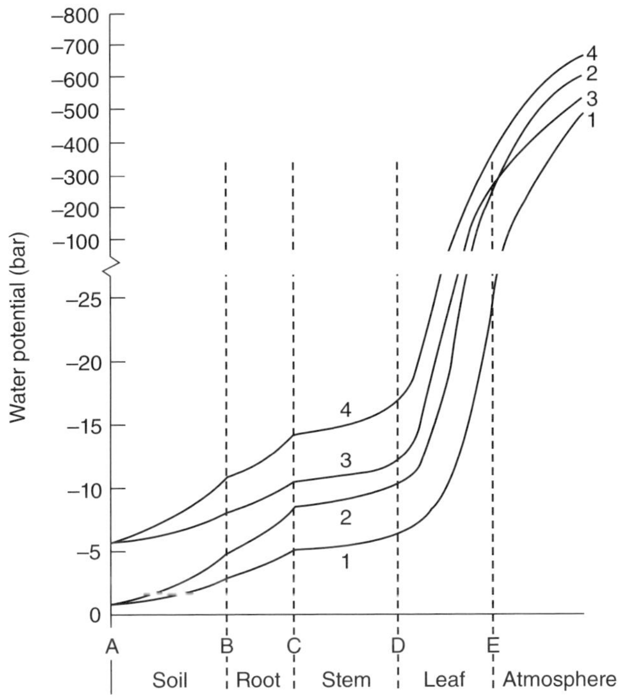

Fig. 8.2. The potential decline in the soil–plant–atmosphere continuum (SPAC) at two soil moisture levels and two transpiration rates. Explanations in text (after Hillel, 1980).

potential). Within the mesophyll cells (DE) of the leaf, the water potential stays much above a critical limit of, let's say, –20 bar, below which the plant will start wilting. In case 2 the transpiration rate is greater at an identical soil water potential, so that the critical potential within the leaf is almost attained. There is a similar situation just before wilting in case 3 when the available soil moisture content has been greatly depleted, the leaf water potential is close to –20 bar in some of the mesophyll cells, and the transpiration rate is also small. But finally, if the transpiration rate increases when soil moisture is in short supply, which is the situation in case 4, the leaf potential will fall below the critical level and the plant wilts.

8.2 Potential Evapotranspiration

The potential evapotranspiration demand determines the loss of water of single plants and crop stands to the atmosphere at the maximum possible rate (i.e. the potential rate). Like potential evaporation (Section 5.2; Box 8.1), potential evapotranspiration can be explained in terms of net radiation and other meteorological parameters. But in contrast to the site of evaporation in the soil, the transpiring surface of the plants is not represented by a single uniform layer. It is structured over several leaf storeys, varying in the angle of inclination and insertion height above the soil surface. As a result, water vapour molecules are transferred from the transpiring leaf surfaces to the air more effectively, especially when wind is

Box 8.1. Potential evaporation

By the middle of the 18th century, the components of the water balance were known but there was insufficient quantification to explain the origin of springs. The work of Stephen Hales (Box 7.1) had established the importance of evaporation in the water balance. Edmond Halley (1656–1742) had shown that evaporation varied seasonally and described the loss from an open pan of water, suggesting that it provided a measure of the maximum rate of evaporation from the land. It was left to Dalton (1802a) to quantify the water balance of England and Wales with sufficient precision that precipitation fully accounted for the flow of water in rivers and springs. Dalton (1802b) also presented an equation to describe evaporation from a water surface, E, as a function of the temperature of the water and the vapour pressure of the atmosphere:

$$E = c(e^* - e)$$

where c is a factor depending on wind speed, e* is the saturated vapour pressure at the temperature of the water and e is vapour pressure of the air during the period of observation.

In 1870, Lawes and Gilbert constructed a series of drain gauges, each having a cross-sectional area of approximately 4 m^2, filled with soil that had been maintained in its original state as it was contained. The surface of the soil was kept bare and the average annual evaporation for the first 10 years was 57% of precipitation (Lawes *et al.*, 1881). Later estimates showed values closer to 50% (Russell, 1907; Keen, 1936). Thus, although the water balance had been quantified, when Howard Penman began his research programme in the Soil Physics Department at Rothamsted during 1937, the processes involved in the water loss by evaporation from the drain gauges had not been studied.

Penman had investigated high frequency acoustic impedance of porous bodies during his postgraduate studies, so there was a natural progression in his choice of topic for his first work at Rothamsted: the movement of gases and vapours, including CO_2, in soil (Penman, 1940a,b). According to Monteith (1986a), Penman's start on his key study of evaporation was purely fortuitous and prompted by a refurbishment of the department, which temporarily prevented laboratories being available. Penman used the time to analyse the autographic records from the drain gauges together with rainfall records and calculate the rate at which water evaporated from bare soil (Penman, 1940c). This work was interrupted in the early part of the Second World War but was reactivated as part of a broad effort to provide advice to military commanders on land trafficability following the allied landings in France and continued after the end of hostilities (Monteith, 1986a). In 1948 it resulted in the classic paper that included the derivation of what we now refer to as the Penman equation. During his research Penman had found that the rate of evaporation was proportional to the driving force, the 'saturated vapour pressure deficit' and inversely related to the resistance provided by a thin non-turbulent layer of air next to the surface (boundary layer), ~1–3 mm thick, which vapour crosses by diffusion. When bare soil formed the surface there was another resistance caused by the presence of dry pores. Initially, the rate of evaporation from a saturated soil was very similar to that from an open water surface at temperatures around 22°C but as the cumulative loss reached ~20 mm the rate of evaporation from soil declined. At warmer temperatures about 35°C, the rate of loss from soil could exceed that of water but, once the cumulative evaporation amounted to a few mm, the rate declined rapidly. This was the first demonstration of the self-mulching effect of soils (Section 17.3), which is particularly relevant to soils in arid climates. Penman was also able to establish that more prolonged evaporation from soil at the same rate as from an open water source occurred when the temperatures of the soil surface and the air above the boundary layer were closely similar.

Continued

Box 8.1. Continued

For the case with greater loss from soil than from open water, the temperature of the soil surface exceeded that of the air. However, as accurate measurement of surface temperature was difficult with the equipment available, this was a barrier to greater understanding of the exchanges of heat with the air and hence the prediction of evaporation rates. Penman, therefore, chose to combine thermodynamic and aerodynamic equations for heat and mass transfer so as to eliminate the need to measure surface temperature (Eqn 5.6). In this equation, the parameter Δ is the change in saturation vapour pressure with temperature. Penman (1948) argued that in most systems involving natural evaporation, Δ could be treated as a constant over the temperature range involved. The constant g is the psychrometric constant, which relates the partial pressure of water in air and the temperature and allows the vapour pressure to be determined from paired readings of dry and wet bulb thermometers.

Penman tested his equation using a set of concrete-lined tanks, ~1.8 m deep and 0.76 m diameter, containing a sandy soil and sunk into the ground until the rim was just above the surface. Each tank was linked via an outlet at the bottom to a paired, covered tank containing water. The evaporation from the soil was determined from the volume of water required to maintain the water level in the paired tank. Soil was either kept bare of vegetation or was turfed. One tank was water filled and a second, shallower tank, containing water to a depth of only 53.3 cm below the soil surface, was added in the second year of measurement. With the exception of the shallow water-filled tank, all the tanks were arranged around a brick-lined pit to allow access to the bottom of the tanks. He found good agreement between the measured values of evaporation from the shallower open water tank and those predicted by the aerodynamic or the thermodynamic equations and hence also with the combination equation. However, the value measured from the deeper tank tended to exceed that predicted by the thermodynamic equation. The evidence suggested that the water in the deeper tank was warmed more because of the presence of the open pit on one side. Throughout the year the evaporation from the bare soil that was maintained with a wet surface was close to 90% of that from the shallow water-filled tank and the turfed surface was 75%. In addition, Penman also reported good agreement between measured values of evaporation with those estimated using his equation for: (i) open water located in Boston, USA, Molato, Italy, Sevan Lake, Armenia and San Juan, Puerto Rico; (ii) for bare soil located at Pusa, India; and (iii) for the difference between annual rainfall and runoff at six catchments in the UK. In January 1948 the American climatologist C.W. Thornthwaite published his classic paper dealing with evaporation from cropped land. He coined the word 'evapotranspiration' to describe the combination of water lost directly from the soil and through the transpiration stream of plants. He argued that 'potential evapotranspiration', defined as the water use that was determined only by the climate, could be used in the logical classification of climates (Thornthwaite, 1948). His approach to determining potential evapotranspiration was on a statistical rather than physical basis. Penman disliked the word evapotranspiration and always referred to 'potential transpiration' defined as the 'loss of water from short, green vegetation, completely shading the ground and never lacking water' (Monteith, 1986a).

As Penman's equation allowed potential evaporation to be estimated, it had obvious implications for the irrigation of agricultural and horticultural crops but it was also important to catchment hydrologists. Important aspects to both areas included corrections related to the height and structure of the plant canopy (Penman and Long, 1960). For the main growing season with height exceeding that of short turf and crops plentifully supplied with water, the ratio between actual and potential evaporation, determined by the Penman equation, was found to be 1.15 for winter wheat, spring barley, faba bean and potato and 1.3 for spring

Continued

Box 8.1. Continued

wheat, sugarbeet and kale (*Brassica oleracea*) (French *et al.*, 1973a,b,c; Long and French, 1967). The evaporation of water intercepted by the canopy of trees can be very fast and hence drainage can be overestimated unless the aerodynamic term in the Penman equation is suitably amended (Thom and Oliver, 1977). Penman considered the deviation of evaporation from the potential value as the soil dried. He determined the soil water deficit at which the measured water use by a crop departed from the potential evaporation. Penman called this soil water deficit the 'root constant'. Results from a number of locations suggested a soil water deficit of between 70 mm and 80 mm for the root constant of short turf (Penman, 1950).

Penman became the Head of the Physics Department at Rothamsted in 1954, the same year that John Monteith was appointed to the staff. Monteith had been working on dew formation for his PhD thesis at the suggestion of Penman (Monteith, 1957). He developed new and very sensitive micrometeorological instruments for his investigations and adopted a balance approach to incoming and outgoing energy to quantify dew formation at ground level. He was to use this approach together with the analogy of Ohm's law in a number of other studies. Monteith's classic paper published in 1965 built on the foundation of the Penman equation. In the article he integrated developments in knowledge and understanding of meteorology and plant physiology (Monteith, 1965). This meant enhancing the components in the Penman equation that were related to transport through stomates and the aerodynamic properties of leaf surfaces (Eqn 8.8). This new formulation rapidly became known as the Penman–Monteith equation. Adoption of this equation by the FAO (Section 8.2) also resulted in the formulation of potential evaporation to include greater specificity of the reference crop characteristics to meet the data requirements for solving the formula.

Howard Penman was born on Tyneside in the north-east of England and experienced the deprivations of the period of the Great Depression. He developed into an austere person with a strong sense of right and wrong. He constantly challenged fellow scientists with his penetrating questions about their research. One of his passions was choral singing, though he enjoyed entertaining with songs from Tyneside (Monteith, 1986a).

John Monteith was born in Fairlie on the west coast of Scotland but moved with his family to Edinburgh at the age of 11. Quietly spoken and friendly, he had a great sense of humour. At one workshop on thermocouple psychrometry led by his friend Gaylon Campbell, Monteith circulated a spoof menu for the dinner, offering Campbell's soup, welded joints à la Wescor with pomme de Peltier and hot junction leeks, a sweet of chromel caramel with cooling currants all washed down with Sauterne Chateau Spanner, vintage 1951.

To extend Monteith's summation of Penman's work (Monteith, 1986a): by showing how classical physics can be used to solve environmental problems, both Penman and Monteith made unique contributions to agriculture, horticulture, forestry, hydrology and natural ecosystems. Their contribution is brilliantly encapsulated in a formula that is used worldwide to improve the efficient use of a scarce resource and make the supply of food more plentiful and reliable.

blowing into a transpiring plant stand, when there is a rapid removal of the vapour. In using the phrase 'more effectively', we mean that *per unit ground surface* more molecules are transferred to the atmosphere per unit time by a freely transpiring canopy than by a wet soil surface. Therefore the potential evapotranspiration rate is usually slightly greater than the potential evaporation rate.

Originally the *potential evapotranspiration rate* was considered by Penman (1948) to be the evaporation from a green plant stand, well supplied with water and cropped short, as a well-trimmed lawn. Since Penman used this concept, some scientists have sought to broaden the idea to include taller plant stands and even forests (Lu *et al.*, 2005). Taking such an approach emphasizes the fact that potential evapotranspiration is not only influenced by meteorological conditions but also by properties of the plant community itself.

The *Penman equation* has already been introduced in a form for calculating the evaporation (LE_p) from a free water surface (Eqn 5.6; see also Box 8.1). The term L was the latent heat of evaporation and E_p the potential evaporation. The equation is also valid for calculating the potential evapotranspiration of a short grass stand ($L\ ET_p$), but with one modification: here the ventilation–humidity term E_a (see Eqn 5.7) is of the form:

$$E_a = 0.263(1+0.00622U)(e_s-e) \tag{8.3}$$

In Eqn 8.3 the term U is the wind speed (km day^{-1}), e_s is the saturated vapour pressure of the air, and e is the actual vapour pressure (hPa = mbar, see Table 7.1).

Another equation was derived by van Bavel in 1966. The *van Bavel equation* has the same form as shown by Eqn 5.6, and is used for calculating $L\ ET_p$. But van Bavel presented a new approximation for E_a:

$$E_a = \frac{\rho \cdot \varepsilon \cdot k^2}{P_a} \cdot \frac{U(e_s - e)}{[\ln z / z_0]^2} \tag{8.4}$$

The terms in Eqn 8.4 are: ρ the density of air (1.205×10^{-3} g cm^{-3} at 20°C and standard air pressure), ε the ratio of the molecular weights of water (18 g) to air (28.9 g) = 0.622, P_a the ambient air pressure, and k the von-Karman constant (0.41). All these terms contained in the first fraction on the left-hand side of Eqn 8.4 result in a factor of 1.24×10^{-4} g cm^{-3} bar^{-1} at 20°C and the standard atmosphere air pressure of 1.013 bar. In the second fraction the term U stands for wind speed, the daily run of wind in km day^{-1}. The difference $e_s - e$ is the saturation deficit (mbar) of water vapour in the air. The height, z, is the vertical position where the run of wind is measured (200 cm above crop surface). The term z_0 stands for the 'roughness length'. Depending on the height h of a standing crop, a wind blowing over the field will be retarded to a greater or lesser extent. The value z_0 is the height within a crop stand above ground where the wind will come to a standstill. From the height of the stand h, the roughness length z_0 can be calculated approximately:

$$z_0 = 0.132\,h \quad (z_0 \text{ and h in cm}) \tag{8.5}$$

From Eqn 8.4 it can easily be seen that a maize crop of 2 m height will achieve at a greater E_a and ET_p than a short cut lawn of about 5 cm height under the same conditions. And the ET_p of the lawn will be greater than the E_p of a wet, bare soil.

A critical aspect of water use is the *advection* of energy into a crop stand from the surrounding area. Advection is the horizontal supply of energy in the form of sensible heat. It is the stream of energy entering a field from the surroundings that is additional to the input coming from solar radiation directed on the field. We can imagine a narrow field of winter barley in the spring, only 20 m in width. This strip of barley is surrounded by large fields of sugarbeet. The beet plants are still small and at an early stage of development, having only two foliage leaves. The plants hardly cover any of the soil surface. In these fields radiation and wind have dried the soil surface and a layer of some mm thickness below. The dry spell has also dried the soil with barley, but here the plants with elongating stems extract water from deeper layers, explored by their roots. In the sugarbeet fields, however, the evaporation has been drastically reduced by self-regulation (Section 5.2), and LE of Eqn 5.5 is near zero. Thus the energy flux for heating the soil (G) and the air (H) has been increased (Eqn 5.5). Wind will carry the heated air into the barley field, where part of the sensible heat is intercepted by the standing crop. The supply of sensible heat will increase the transpiration of the barley, without increasing the crop's net assimilation. The additional transpiration will account for much of the transferred energy flux. Increased transpiration due to advection is called the '*oasis effect*'. A special case of advective energy supply to a canopy, such as the barley strip, that is taller than the surrounding plants, is called the '*clothesline effect*' (Tanner, 1957). The heated air will not be blown over a dense crop stand but will spread between row plants or between bushes or trees of plantations, causing increased transpiration from all leaf storeys.

The complete van Bavel equation is now presented for comparison with the Penman equation (Eqn 5.6). It will not be expressed in terms of $L \cdot ET_p$, but as ET_p:

$$ET_p = \{[(\Delta/\gamma)(R_N - G)/L] + E_a\}/(\Delta/\gamma + 1) \tag{8.6}$$

As for E_p (Eqn 5.6), to calculate ET_p the following quantities have to be measured: net radiation, air temperature, daily wind run and the *saturation deficit* of water vapour in the air. The saturation deficit can be derived from the recorded *relative humidity* (RH) and air temperature. The RH is defined as:

$$RH = (e/e_s) \times 100 \tag{8.7}$$

The value of the saturated vapour pressure (e_s) is strongly dependent on air temperature. It is reported in handbooks of meteorology and can be determined from Fig. 8.3. With e_s and the measured RH the actual vapour pressure (e) can be calculated by use of Eqn 8.7. The saturation deficit is then easily obtained from the difference $e_s - e$ (see example calculation in Fig. 8.3).

Net radiation can directly be measured in the field by use of a net pyranometer. Such an instrument records in- and outgoing radiation, the radiation balance. A solar pyranometer, however, measures only the incoming radiation and not the balance, but from daily values the net radiation can be estimated. Net radiation can even be derived from the duration of bright sunshine, which is easily measured by use of a Campbell-Stokes heliograph. It consists of a spherical lens, burning a recording strip during the hours of sunshine. Air temperature and actual vapour pressure are needed as well as an estimate of the albedo. The albedo of crop plants with complete canopy cover varies mostly between 0.15 and 0.25. Published tables exist that provide the necessary basic data on radiation constants that will vary with season and latitude (Withers and Vipond, 1974; Allen *et al.*, 1998). The energy flux of heating the soil, G, is assumed to be near zero during the main period of crop growth (Section 5.2).

Finally for the calculation of ET_p by Eqn 8.6, the plant height h (Eqn 8.5) has to be measured. In many cases ET_p may be calculated on a daily basis. For that

State	T	e_s	e	RH	τ	$e_s - e$
	°C	.... mbar		%	°C	mbar
a	25.0	31.2	25.0	80	21.4	6.2
b	21.4	25.0	25.0	100	21.4	0.0
c	12.1	14.0	14.0	100	12.1	0.0
d	25.0	31.2	14.0	45	12.1	17.2

$\overline{a' - a}$ = Saturation deficit at a

$(\overline{a - a''} / \overline{a' - a''}) \times 100$ = RH at a

$\overline{c' - c}$ = H_2O condensation by cooling b → c

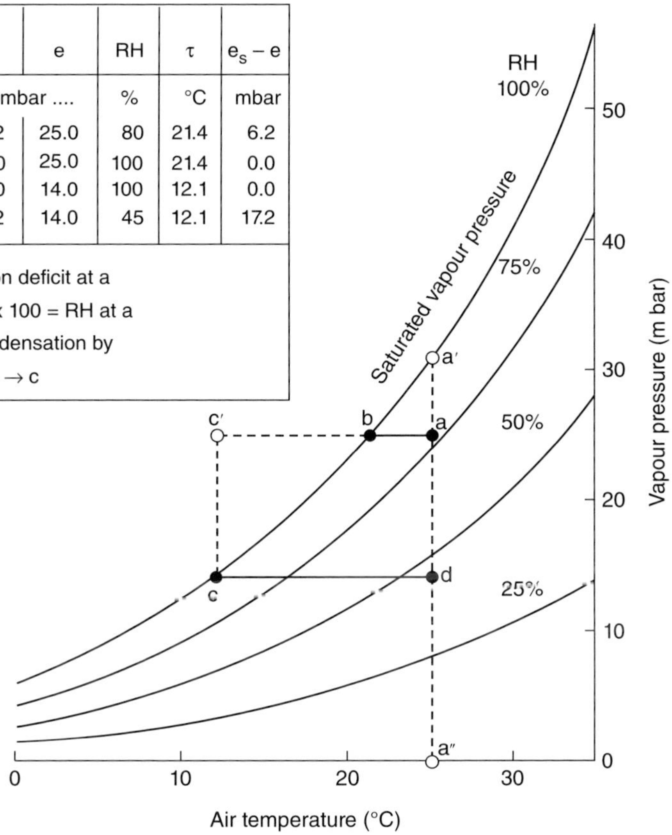

Fig. 8.3. The saturated vapour pressure (e_s) of the air is highly dependent on air temperature. Consider an actual vapour pressure (e) that is smaller than e_s. When e remains constant, but the temperature falls, the saturation deficit, $e_s - e$, steadily decreases. When the dew point (τ) is reached, e will equal saturated vapour pressure. Then $e_s - e$ will be zero, the relative humidity will be 100% and the water vapour becomes liquid. Dew and fog are formed. The water vapour, e (mbar), can be transformed into absolute humidity, ρ (g water m^{-3} air), by applying the formula: $\rho = 217\ e/T^*$, where T^* (K) is the absolute air temperature (after van Eimern and Häckel, 1979).

purpose the corresponding values of daily average or daily sum have to be entered into Eqns 8.4 and 8.6.

Another equation has been introduced by Monteith (1981). It is an extension of the Penman equation, incorporating a canopy resistance that takes into account the stomatal control of water loss (Hatfield, 1990; Ritchie and Johnson, 1990). The so-called *Penman–Monteith equation* (Hillel, 1998) is employed for calculating actual evapotranspiration under both limited and non-limited conditions of water supply. In the latter case of a non-limited water supply, the canopy resistance is small, and the evapotranspiration rate is equivalent to the potential rate. But the uncertainty in fixing the canopy resistance under various circumstances pertaining to soil, plant and meteorological conditions has hindered the broad application of this scientifically based equation.

It is not surprising that the results of ET_p calculations differ depending on which equation is used and the particular meteorological factors required in the calculation. We mentioned above that, for example, crop height or canopy resistance can be of crucial importance to the calculated value of ET_p. The same is true for the formulation of the ventilation–humidity term. There are many more empirical equations for calculating ET_p than listed in Section 8.2. Many of the approaches are only of local importance. This diversity causes problems in irrigation advice, as the magnitude of ET_p predicted will influence the timing of irrigation and the quantity of water to be applied. Allen and coauthors (1998), all of them experts in irrigation and working on behalf of the Food and Agriculture Organization of the United Nations (FAO), tackled this problem. They recommended the use of a 'globally valid standard for crop water requirement calculations'. This standard has to be independent of crop type, development stage and agronomic practices. They proposed the 'FAO Penman–Monteith equation' that only requires meteorological data as input. Allen *et al.* (1998) called this calculated evapotranspiration the reference crop evapotranspiration rate ET_o. This 'reference' ET applies to a 'hypothetical crop, 0.12 m in height, a fixed canopy resistance of 70 s m^{-1}, and an albedo of 0.23. The reference surface closely resembles an extensive surface of green grass of uniform height, actively growing, completely shading the ground and with adequate water' (Allen *et al.*, 1998). This formulation is reminiscent of the original approach of Penman (1948). The FAO Penman–Monteith equation reads:

$$ET_o = \frac{0.408\Delta(R_N - G) + \gamma \frac{900}{T+273} U(e_s - e)}{\Delta + \gamma(1 + 0.34U)} \tag{8.8}$$

where all parameters, other than T, have been explained before. T is the mean daily air temperature (°C) at 2 m height. Equation 8.8 relies on daily weather data (Ventura *et al.*, 1999). We will come back to potential and reference evapotranspiration when defining the crop water requirement as the basis for irrigation (Section 19.3).

8.3 Relations between Potential Evapotranspiration, Soil Water and Transpiration

Under a given evaporative demand of the atmosphere, the magnitude of water potentials within the sequence of SPAC is governed by the level of the soil water potential (Fig. 8.2). Because of water uptake by roots, the content as well as the matric potential of soil water will decrease. The change in soil moisture is more pronounced in periods without rain and when the extraction of water by plant roots is restricted to a limited soil volume.

This idea of uniform water extraction from a defined soil volume, confined to a plant pot of a few litres, is the basis of the next figure (Fig. 8.4). Along with the removal of water from the pot the matric potential decreases steadily from day to day. At noon on these sunny days, the *evaporative demand* is greatest and correspondingly the water potential within the leaf clearly declines. The reason is that the high rate of transpiration at midday is not counterbalanced completely by the

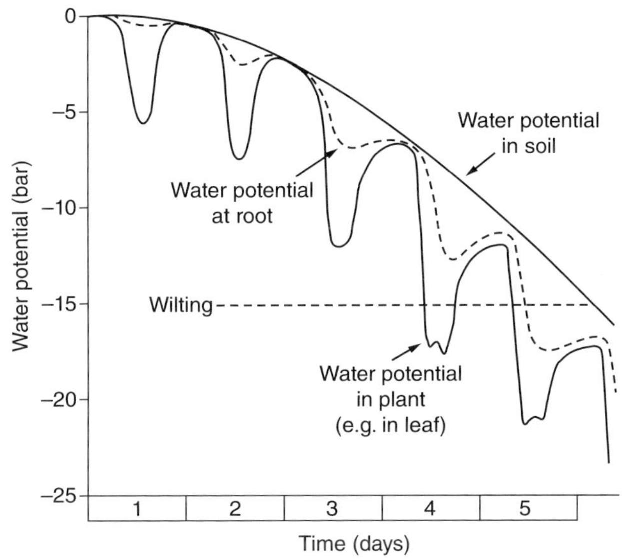

Fig. 8.4. Diagram of temporal change of the water potentials in the soil, at the root surface and in the leaf during a desiccation period of 5 days (after Slatyer, 1968).

water uptake rate of the root system. This deficit causes a *water loss* from the plant tissue, which may be very small. But nevertheless the loss induces the plant water potential to decline (Fig. 7.1). That again increases the hydraulic gradient $\Delta\phi/\Delta z$ (Eqn 8.1), causing a tendency to accelerate the water flux. The fall in water potential in the leaf is continued throughout the hydraulic pathway in the plant down to the root surface. Even here, between root and soil, the gradient will increase at noon (Eqn 7.7), but not as much as between leaf and root surface in the plant.

In the afternoon the evaporative demand gradually declines. At that point the period of the plant's water loss comes to an end: more water enters the plant through the roots than is transpired by the leaves. The tissue again becomes filled with water, and the water potential in the plant and at the root surface increases. At the end of the night an almost complete balance will be achieved between the potentials in the plant and in the soil. This phase of a complete recovery and equilibration is attained after the first, second and third day, but is no longer achieved on the fourth and the fifth day in the example given (Fig. 8.4).

When the water potential in the plant declines at noon, the pressure potential, the turgor, is also reduced (Fig. 7.1). That will affect the rate of *cell enlargement* (Eqn 7.1), which may decline noticeably in periods of a restricted water budget. Only in the evening after relaxation will the rates of cell enlargement and growth increase again (Fig. 8.5). This period of growth by cell elongation can be continued until late at night. Possibly the growth at night is greater than during the day.

During the course of water extraction and soil drying in the pot (Fig. 8.4), the daily amplitudes of the water potentials in leaf and root become more pronounced. By the fourth day, the critical water potential is reached, where *wilting* occurs. On the fifth day, the plant water potential does not rise above the wilting point, and the system 'runs out of control'. The plant cells undergo *plasmolysis*

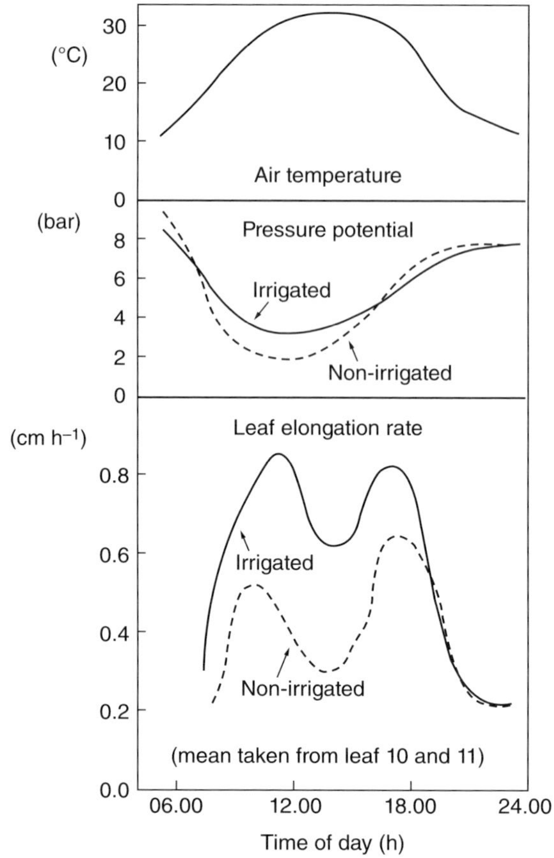

Fig. 8.5. Growth rate of two maize leaves during a day of fine weather in Davis, California, 37 days after seeding. The maize was irrigated or was rainfed. The longitudinal growth of the leaves is influenced by temperature (and radiation) as well as by the pressure potential, the turgor of the leaves (after Acevedo Hinojosa, 1975).

(Fig. 7.1) and the potted plant wilts irreversibly. Finally the plant dies. Admittedly, stomatal control can slow down water loss and can cushion the daily potential drop, but the ultimate end point will just be delayed, not prevented.

Under otherwise identical conditions, for increasing values of potential evapotranspiration, the daily amplitudes of water potential are likely to become more intense, thereby allowing the rate of transpiration to increase, but the onset of wilting will take place earlier in time. It is not the complete story to say that because of a large water use the supply of water from the soil will be rapidly exhausted. What has to be considered is the dynamic aspect of water supply (i.e. the capability of the soil to conduct the water at a sufficient rate towards the root system). What does 'at a sufficient rate' mean? On the cloudy and humid day, the soil water content could drop to only 23% before the transpiration rate of maize plants, growing in containers, declined (Fig. 8.6). On the warm and dry day, much more water was transpired per unit time because of the higher evaporative demand. However, under these conditions the transpiration rate was already reduced at a water content of 35%. Apparently the hydraulic gradient and the unsaturated conductivity of the

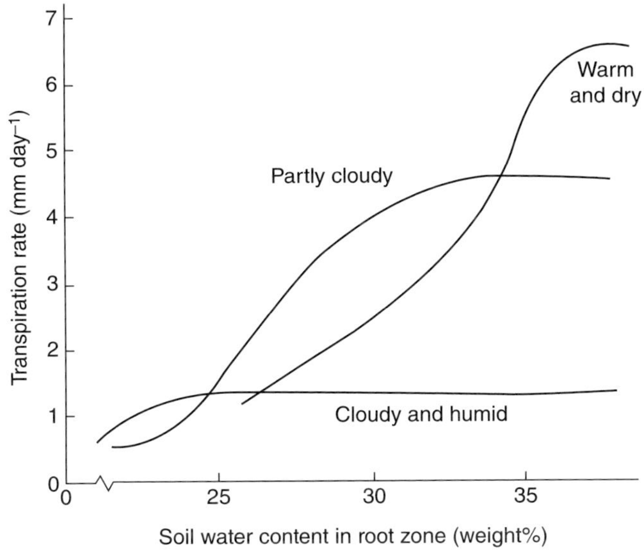

Fig. 8.6. The actual transpiration rate is limited at different soil water content, depending on the prevailing meteorological conditions (after Denmead and Shaw, 1962).

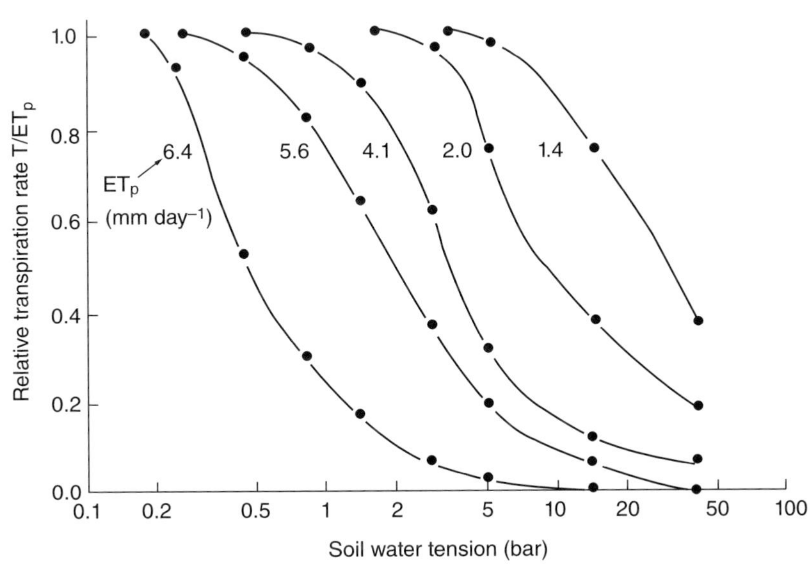

Fig. 8.7. The greater the potential evapotranspiration rate (ET_p), the smaller is the soil moisture tension (and the larger is the soil water content) at which the relative transpiration rate (T/ET_p) declines as the actual transpiration rate (T) drops below ET_p (after Denmead and Shaw, 1962).

soil were not sufficient to supply enough water to meet the potential evapotranspiration, even in a wet soil. Presumably the maize plants closed their stomates, thus avoiding irreversible damage from excessive water loss. For the partly cloudy day, the conditions were intermediate between these two extremes (Fig. 8.6). We

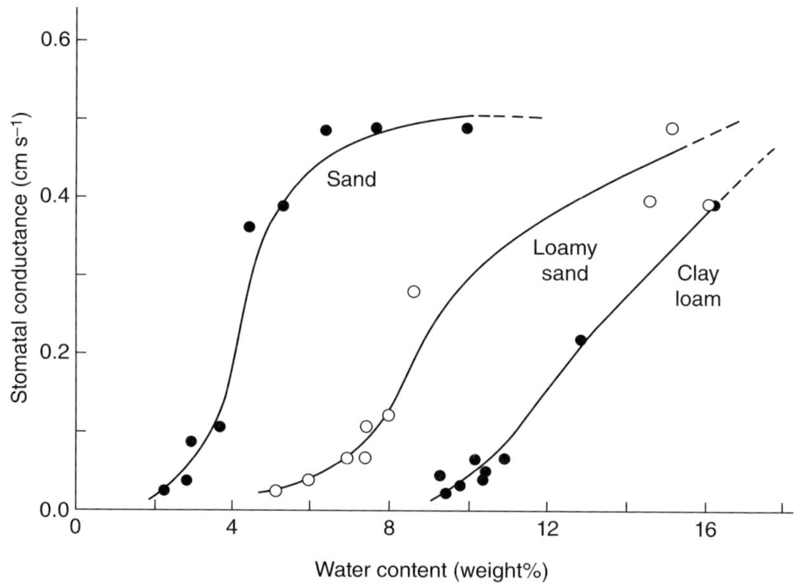

Fig. 8.8. Change in stomatal conductance of onion (*Allium cepa*) at varying water content in three soils of different textural classes (Millar *et al.*, 1971; from Baver *et al.*, 1972). The conductance was measured using a porometer dating from the initial phase of development (Kanemasu *et al.*, 1969).

conclude that it is the magnitude of the potential evapotranspiration that will determine the critical soil water potential at which transpiration will fall below the potential evapotranspiration rate (Fig. 8.7).

For crops experiencing similar potential evapotranspiration, the rate of water uptake and thus the rate of transpiration will decrease more sharply when grown in soil where there is a greater reduction in soil hydraulic conductivity with each decrease in soil water content. Figure 8.8 depicts how *stomatal conductance*, a measure of stomatal aperture, declines with soil drying. In the clay loam and the loamy sand it gradually declines with decreasing soil water content. In contrast, stomatal conductance in plants from the sandy soil falls sharply. Thus on sandy soils, water supply turns out to be problematic. This is not only because of the potentially smaller amount of available water (Figs 3.2 and 5.1) and the likelihood of relatively shallow rooting (Fig. 6.6); it is also because of the sharp transition between abundant supply and insufficient provision (Fig. 8.8), which is a consequence of the severe drop of the hydraulic conductivity with increasing water tension (Fig. 5.3).

9 Water Use by Crops

9.1 Growth of Roots and Leaves

We have already commented that the shoot–root ratio of crop plants gets bigger during the course of development. The ratio is shifted in favour of the shoot, as shown for winter wheat in Table 6.1. During the period of grain filling it reaches 10 or even 20:1. If the mass of transpiring leaves is considered rather than the total mass of the shoot, the ratio narrows and comes close to a value of 1:1 (Table 9.1). But the ratio of water uptake by roots to water loss by leaves is not so much a matter of mass but rather a function of area. Expressing the areas of transpiring leaf and of absorbing root system for cereals, oat in this example, as a ratio, results in a value of about 0.5:1, which is nearly constant during the vegetative period (Table 9.1). However, it has to be recognized that in calculating the ratio, the total root surface area was used but not the most active area in water extraction, the region of the root tip, since that is unknown. On the other hand, the effect of root hairs on increasing the area of absorbing surface at the root tip has also been neglected.

The *leaf area index* (LAI) best characterizes the size of the canopy with respect to transpiration and net assimilation. LAI is the ratio of the green leaf surface area (one side only) of a crop to the surface area of the soil that is covered by the standing crop. The time course for LAI expansion of a small-grain cereal and a leguminous crop during the season is shown in Fig. 9.1. Although both summer crops were seeded in mid-March, leaf development in oat occurred earlier than in faba bean.

Compared with cereals, grain legumes like faba bean have a relatively small root length density (Figs 6.4 and 6.5) and total root length (Fig. 9.2). But the roots of grain legumes are thick. The diameter varies somewhere between 400 μm and 800 μm. In cereals, however, it varies between 150 μm and 300 μm. Therefore the crops differ less in root surface area than in root length.

Table 9.1. Development of the leaf:root ratio of oat, averaged over 2 years (1982 and 1983). The ratio is presented either in terms of dry mass or surface area (unpublished data from Göttingen, Germany).

Part of plant	Unit	15 May Tillering	5 June Stem elongation	5 July Anthesis
Leaf blade	g m^{-2}	15	130	140
Root	g m^{-2}	50	110	130
Leaf:root ratio	g m^{-2}:g m^{-2}	0.3	1.2	1.1
LAI[a]	m^2 m^{-2}	1.0	3.4	3.1
Two-sided LAI[b]	m^2 m^{-2}	2.0	6.8	6.2
Total root length	km m^{-2}	5.5	10.5	14.0
Total root surface area[c]	m^2 m^{-2}	5.2	10.0	13.2
Leaf:root ratio	m^2 m^{-2}:m^2 m^{-2}	0.4	0.7	0.5

[a]Leaf area index.
[b]Stomates on upper and lower leaf side.
[c]Root radius assumed to be 150 μm.

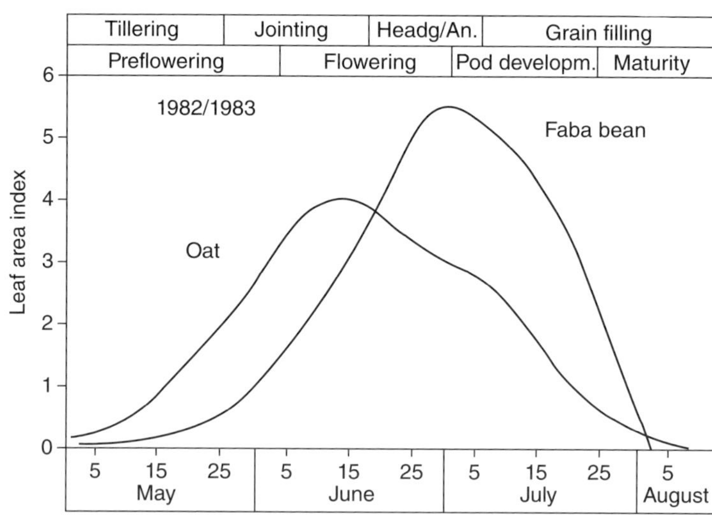

Fig. 9.1. Development of leaf area index (LAI) of oat and faba bean, averaged over 2 years. Jointing is stem elongation; Headg, heading; An., anthesis; developm., development (from Ehlers, 1986).

9.2 Leaf Area Index and Transpiration

The actual transpiration rate for a crop stand is determined by the potential evapotranspiration rate and the soil water content (Figs 8.6 and 8.7). But the rate of water loss will also be influenced by characteristic plant properties, as long as the water supply to roots is non-limiting. Here we will start the discussion by dealing with the role of LAI on transpiration. First, the leaf is the principal organ

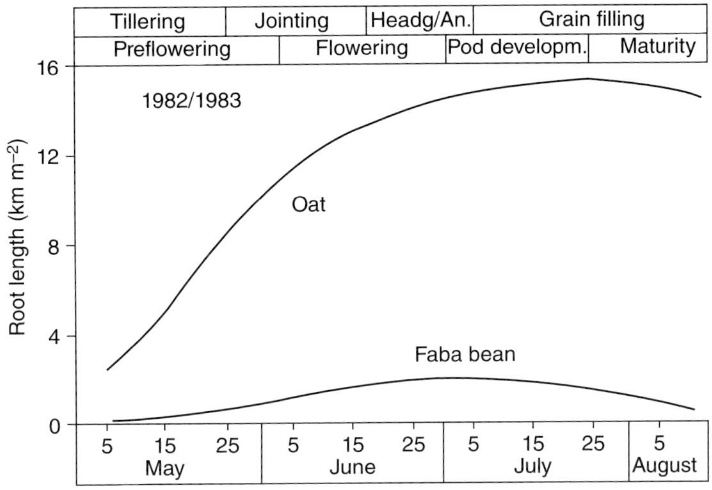

Fig. 9.2. Development of total root length of oat and faba bean, averaged over 2 years. Headg, Heading; An., anthesis; developm., development (based on Müller *et al.*, 1985).

of water loss. Therefore it seems reasonable to assume that the actual transpiration rate per unit soil surface area will increase with the green leaf area per unit soil area. Earlier (Section 9.1), we called this latter ratio the LAI. Secondly, the loss of water vapour is based on the supply of radiant energy. With increasing LAI the energy supply is spread over a larger canopy and leaves are mutually shading. Therefore, when LAI increases, the increment in transpiration rate will be less per unit of additional LAI. Thirdly, based on the foregoing, we could expect that after attaining a 'high' LAI, the transpiration rate will be near its maximum. A further increase in LAI will only contribute an insignificant addition to water loss. The reason is that the soil surface is now completely covered and shaded by the canopy, and the radiation interception by leaves reaches a maximum. In that case the energy supply solely limits the actual rate by which water is converted from the liquid to the vapour phase. In other words, after attainment of a certain 'threshold' LAI, the transpiration rate per unit soil surface is no longer increased to any great extent by the addition of more foliage.

Based on these considerations a *saturation curve* can be predicted between LAI and transpiration rate. Such a function is presented in Fig. 9.3 for three crops. As in Fig. 8.7 the actual transpiration rate (T) has been related to the potential evapotranspiration rate; the ratio is called *relative transpiration*. At an LAI of 3.5–4.5 the relative transpiration approaches 90% of the maximum. However, even when the LAI is only 1, then the relative transpiration is already 40% of the maximum. All three response curves are similar in shape, although the growing conditions were different. *Grain sorghum* (*Sorghum bicolor*) was grown as a row crop in a sub-humid climate of Texas, irrigated *cotton* (*Gossypium hirsutum*) as a row crop in a semi-arid continental climate of New Mexico, and *oat*, closely sown in 15-cm rows, in a temperate, humid climate of Lower Saxony in the northern part of Germany.

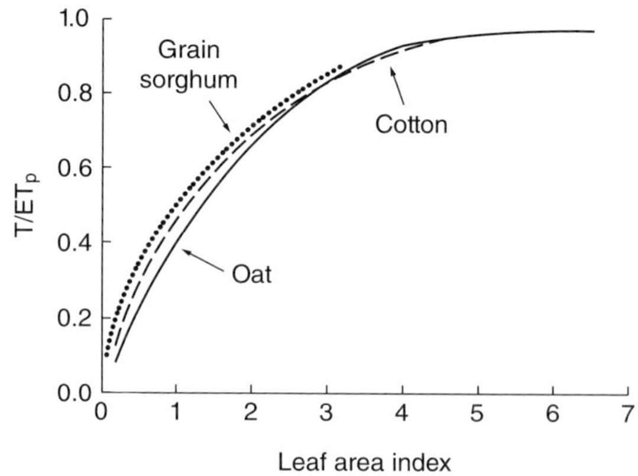

Fig. 9.3. The ratio of actual transpiration rate (T) to potential evapotranspiration rate (ET_p) as a function of leaf area index (LAI) for three crops: grain sorghum (after Ritchie and Burnett, 1971); cotton (after Al-Khafaf *et al.*, 1978); and oat (after Ehlers, 1991).

Under conditions of ample water supply the seasonal transpiration of a crop will be determined by the development of the LAI over time. Greatest transpiration rates can be achieved after attainment of a LAI that is roughly between 3 and 4. Such a LAI will permit nearly 90% of the potential evapotranspiration (Fig. 9.3). As a result of its earlier canopy development (Fig. 9.1) the relative transpiration rate in oat reaches the maximum no later than the end of May or beginning of June, but for faba bean this does not occur until the end of June (Figs 9.4A and B).

In a temperate climate (like that of Germany) the *potential evapotranspiration rate* (ET_p) varies between 4 mm day^{-1} and 6 mm day^{-1} during the main growing season from May to August (Fig. 9.4A). During stem elongation (jointing) the *actual transpiration rate* (T) of oat markedly surpasses the level of ET_p (Fig. 9.4B). Such a phenomenon has been described by a number of authors for cereals. The 'oasis and the clothesline effects' may be the primary explanations. Additional reasons include the increased surface roughness of the crop and the sudden exposure and uplifting of non-lignified tissue during this growth stage, which allows a comparatively high cuticular transpiration in response to a stream of advective energy.

By the time spring-sown crops are planted, winter crops are usually well established with roots and foliage because of autumn seeding and their development during early winter. This is the reason why the water use of *winter cereals* is greater than that of *spring cereals*, especially in spring. Figure 9.5 presents an example from England with winter wheat compared to spring barley. In this and the next example the actual evapotranspiration rate (ET) is considered instead of the actual transpiration rate (T). The reason for doing so was given earlier (Section 3.2). The separation of E from T is troublesome and requires information on conditions that are sometimes not reported. Just as in the case of T, ET can be considered in relation to ET_p, through the *relative evapotranspiration* ratio, ET/ET_p.

Another example of seasonal evapotranspiration is presented for *sugarbeet* in Fig. 9.6. In Germany the sugarbeet crop is planted in rows, 45 cm apart, and canopies will not close before the end of June. By this time at the latest, a sufficient

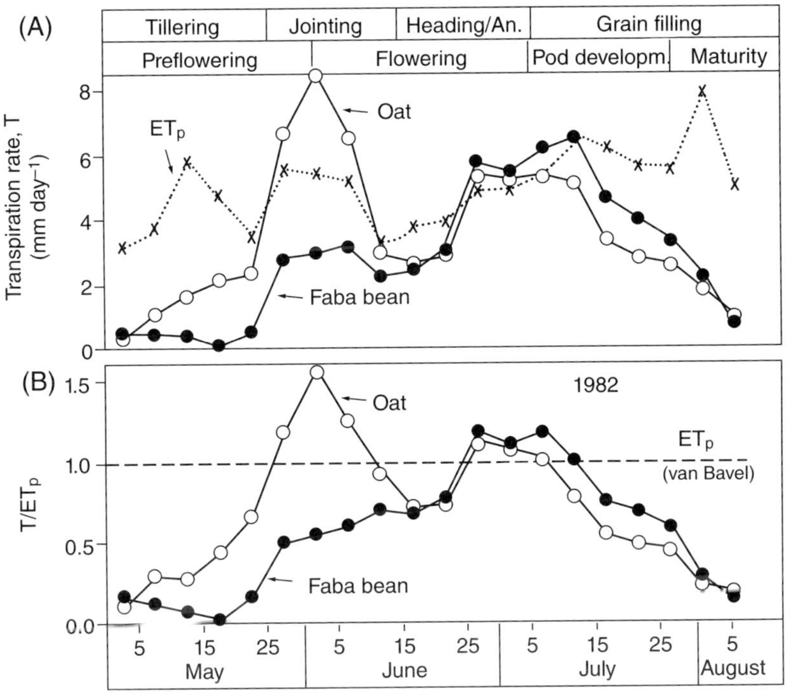

Fig. 9.4. Actual transpiration rate (T) of oat and faba bean as well as potential evapotranspiration rate (ET_p) as a function of time (A). An., anthesis; developm., development. (B) The relative transpiration rate (T/ET_p) during the growing season for oat and faba bean (based on Müller *et al.*, 1985).

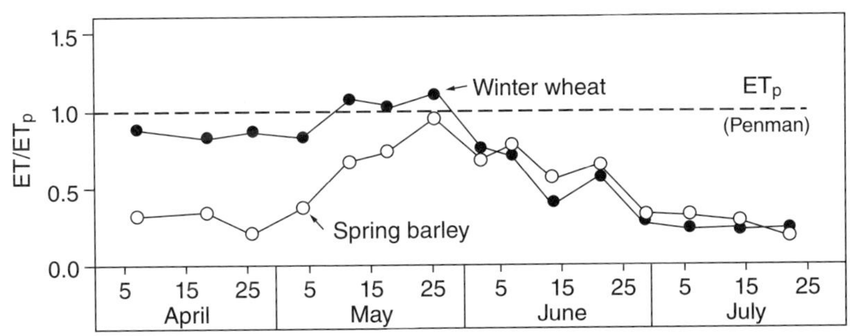

Fig. 9.5. The rate of actual evapotranspiration (ET) related to potential evapotranspiration (ET_p), called the relative evapotranspiration, as a function of time for winter wheat and spring barley in England (after McGowan and Williams, 1980; from Wild, 1988).

LAI is achieved that permits high rates of transpiration. Correspondingly, the relative evapotranspiration (i.e. the ratio of actual evapotranspiration ET to ET_p) is at maximum between the middle of June to the middle of August (Fig. 9.6). With depletion of soil water and the yellowing of the leaves due to ageing, the relative evapotranspiration is reduced in autumn, until it drops to near zero at harvest time.

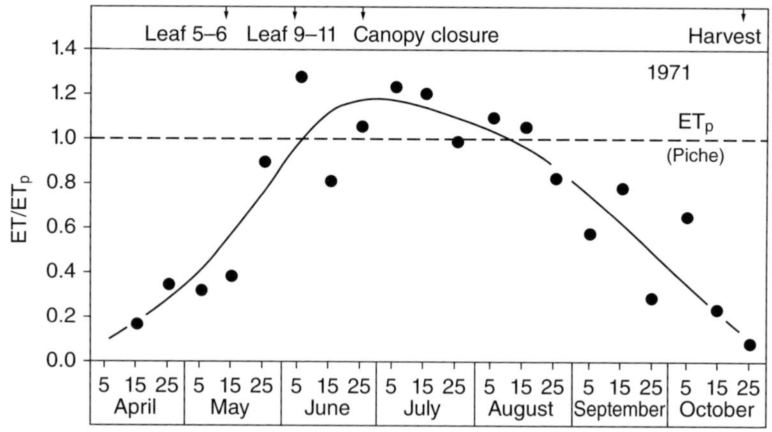

Fig. 9.6. Time course for the relative evapotranspiration in sugarbeet. ET is the actual evapotranspiration and ET_p the potential evapotranspiration rate (Göttingen, Germany, 1971, unpublished data).

9.3 Root System Development and Water Uptake

Plants can transpire water via the canopy into the atmosphere at roughly the same rate per unit soil surface area and time, as can be taken up by the total root system from the various layers of soil containing roots. The synchrony between water loss and water uptake is explained by the fact that mesophytes like our crop plants can endure only a slight loss of the water stored in their tissue. That makes them fast responders in replacing water loss by water extraction from a soil, when water is available to roots.

What kinds of factors determine the water uptake rate of roots? Again ET_p and LAI have to be regarded as the primary factors. Nevertheless, the rate of water uptake is also influenced and restricted by soil properties like *water content, matric potential* and *hydraulic conductivity*. When a soil dries, there is a decrease in all three properties that characterize soil water availability, so the water uptake by roots declines. Near the permanent wilting point (PWP) the uptake rate is minute (Fig. 9.7) and the hydraulic conductivity is of the order of 10^{-6} cm day^{-1} (cf. Fig. 5.3). Such a limiting conductivity in drying soil has been reported in the literature on several occasions (Newman, 1969; Taylor and Klepper, 1975; Reicosky and Ritchie, 1976).

When a soil is wet, uptake rates can be impaired too (Fig. 9.7). Reduced uptake can be caused by *oxygen deficiency* in roots due to the small percentage of soil pores that are air-filled (Section 7.2). Another apparent reason for differences in uptake rates by roots is that water uptake from a re-wetted, rooted soil profile will be confined principally to the upper soil layer, which is normally quite densely rooted (Fig. 6.5). Here the uptake rate can be sufficient to meet the evaporative demand of the atmosphere. Under such circumstances the uptake from deeper soil layers remains small, although these layers are for the most part the wetter zones of the soil profile. The optimum soil matric potential for root water uptake in a

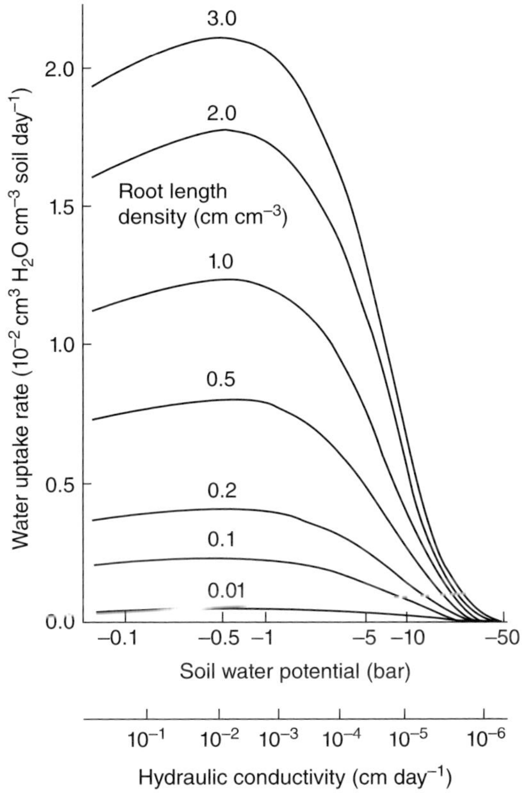

Fig. 9.7. Water uptake rate per unit volume of soil (WU) as a function of soil matric potential, Ψ, or hydraulic conductivity, respectively, and of root length density. Oat was grown on a loess-derived Luvisol, a silt loam. The equation reads: $WU = 1.13\ T^{0.073}\ e^{-0.14\,T} \ln (2.39\ L_v + 1.01)$, $r^2 = 0.42^{**}$; $n = 299$ observations on WU, T and L_v from all rooted soil layers during the season. T is water tension, which is the positive value of soil matric potential Ψ. The equation calculates WU in units of 10^{-2} cm^3 H_2O cm^{-3} soil day^{-1} (after Ehlers *et al*., 1980a).

loess-derived silt loam ranges between −0.5 and −1 bar (Fig. 9.7). In a sandy soil, however, the optimum was found at a higher potential near −0.1 bar (Herkelrath *et al*., 1977). The reason for the difference is that, for a given soil water potential, the contact resistance in the rhizosphere of a coarse-textured sandy soil will be greater than that in a finer-textured silt loam derived from loess. Feddes *et al*. (1978) also hypothesized that there is an optimum rate of water uptake in relation to soil water potential (Fig. 9.7). They proposed that a 'reduction function' operates to moderate root water uptake. This function forces water uptake to zero not only at low potentials (Fig. 9.7) but also at very high water potentials, such as in waterlogged soil. It is worth noting that in the loess soil (Fig. 9.7) waterlogging never occurred.

The water uptake rate depends also on the rooting density (Figs 9.7 and 9.8). The greater the *root length density* (L_v) the smaller is the radius of the soil cylinder

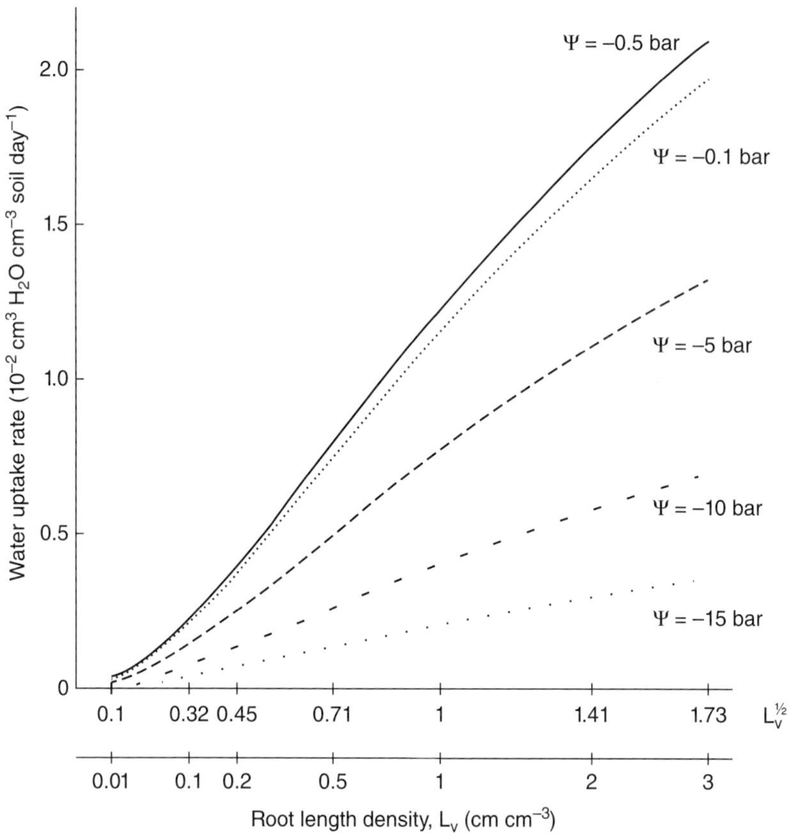

Fig. 9.8. Water uptake rate of oat per unit volume of soil as a function of root length density at various soil matric potentials. Data are derived from Fig. 9.7 and can be calculated with the equation given in the caption of Fig. 9.7.

of water extraction around the single root (Eqn 7.6). Increasing L_v can increase or decrease the *specific root water uptake rate* (UR), depending on limiting or non-limiting flow conditions in the soil (Section 7.2). In any case, with increasing L_v the *uptake rate per unit soil volume per unit time* (WU, for instance in cm^3 water cm^{-3} soil day^{-1}) will increase to a greater or lesser extent (Ehlers *et al.*, 1991) as WU is UR × L_v. In other words, an increase in L_v lowers the distance of flow through the soil to the root surface (Eqn 7.6) and consequently reduces the hydraulic resistance of the flow path (Eqn 7.9).

The way that rooting density L_v affects the relationship between water uptake rate (WU) and soil matric potential (Ψ) shown in Figs 9.7 and 9.8 is the result of a non-linear regression analysis. For the regression, field measured variables (Ψ and L_v) and computed data (WU) were obtained for the main growing season of oat. At a given Ψ the WU increases more or less in proportion with the square root of L_v (Fig. 9.8). Such an approach, however, cannot adequately reflect the actual field situation since it does not fully account for the interaction of the different parts of the growing root system in withdrawing water from the various soil layers. When the uptake rate (WU) declines in an upper layer of the soil because of the diminishing water content, WU will increase in deeper layers of the profile that have, by that time, been explored by roots. As one logical consequence of the phased water extraction from a moist soil profile, a rule can be suggested that the sum of water uptake rates from successive soil layers will be roughly similar to the potential rate of evapotranspiration (Sections 9.2 and 9.8).

Examples of *successive water withdrawal* (Fig. 9.9) can be identified readily in the field, particularly for periods without rain, if one employs suitable equipment, the correct analysis and evaluation. Largest uptake rates are measured in the topsoil layer (here 0–20 cm) early in the season, when the LAI of oat is around 3 or so (Fig. 9.1), and when the top layer is densely rooted (Fig. 6.4). As the top layer dries, the uptake rate increases in the next deepest layer (20–40 cm). Pursuing such a control strategy results in the greatest uptake rate achievable in the profile being shifted from one soil layer to the next lower layer in a temporal sequence (Fig. 9.9). Normally the uptake is largest in the top layer with intensive rooting. In lower layers that have a smaller root length density (Fig. 6.4) uptake rates are usually smaller (Fig. 9.9).

The stepwise progression of water uptake from soil layer to soil layer is not an abrupt one but occurs gradually, so that the individual time periods of water extraction from adjoining soil layers are overlapping (Fig. 9.9). Water extraction

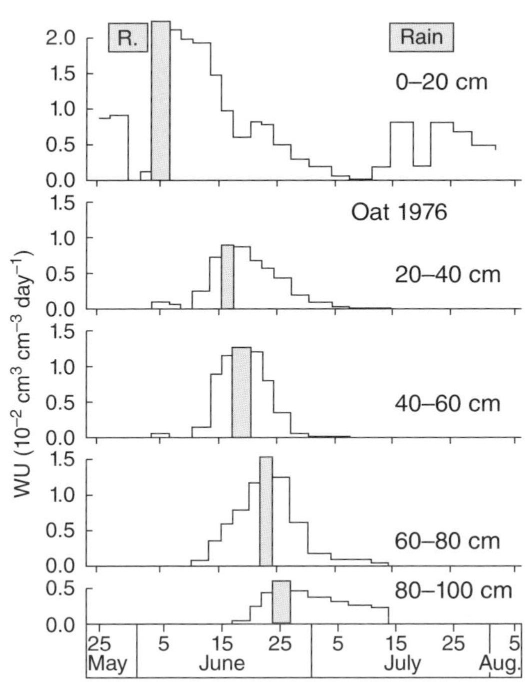

Fig. 9.9. Water uptake rate (WU) of oat in separate 20-cm soil layers during the season. The individual maximum rates, represented by shaded columns, were determined in periods without rain. During rainy periods (their duration is indicated) water uptake is small and the uptake calculations are more difficult (see Section 9.7) (Göttingen-Rosdorf, Germany, 1976, unpublished).

from a layer can only start when the root system of a crop stand has reached that layer with the '*rooting front*' (Fig. 6.4). Thereafter the uptake rate will depend strongly on the value of root length density (L_v) in that layer (Figs 9.7 and 9.8), the increase in L_v being a function of time (Figs 6.4 and 6.5).

From what we have seen so far it has become clear that water uptake is governed by meteorological conditions influencing the transpiration rate, particularly when a critical LAI of 3 has been attained and exceeded. The evaporative demand is met by root water uptake from the various rooted layers of the soil profile. Within the various layers uptake is dependent on soil water characteristics and root length density (Figs 9.7 and 9.8). We have learned, and it is obvious, that the total uptake of water from the profile cannot exceed evaporative demand. That means that the uptake rate within a soil layer will be influenced by the uptake pattern in the other rooted layers, which in turn is related to root and soil properties. It is a challenge to understand the influence of root system parameters, such as root length density, total root length or rooting depth, on water uptake pattern in a specific soil layer, when conditions and properties are steadily changing throughout the rooted profile as a consequence of multiple dynamic processes. Therefore we will focus on the situation, when in a specific soil layer optimum conditions have developed for root water uptake.

Under such optimum conditions the *maximum water uptake rate* (WU_{max}) will be attained, first in the top layer and subsequently in the various layers of the profile below (Fig. 9.9). The LAI is >3 (Figs 9.1 and 9.3), but usually root length density has not yet reached its peak value in the specific layer of interest and the root system is still expanding (Fig. 6.4). Therefore one may hypothesize that at the moment when WU_{max} occurs (Fig. 9.9), the roots, still fresh and non-suberized, are highly permeable to water and physiologically younger than in later stages of growth. At the same time soil conditions are ideal for water uptake, not too wet and not too dry (Fig. 9.7). Such a situation with optimum water uptake will occur only once in a soil layer during the vegetation period.

Under these optimum conditions the value of WU_{max} in a layer will be influenced by the potential evapotranspiration ET_p. WU_{max} will also depend on how much water is being taken up from the remaining soil layers, which is a function of the distribution of roots within the profile (Fig. 6.5). To make our approach easier, we define a maximum water uptake rate, which now is not related to a unit volume of soil but to a soil layer of defined depth, let's say 20 cm. This is the *maximum water uptake rate in a soil layer* (LWU_{max}, mm day^{-1}). It depends on ET_p. Then it follows that the ratio LWU_{max}/ET_p, a relative quantity without units, is not only influenced by the *root length in* this *20-cm layer* (RL, cm root length cm^{-2} soil surface area) but also by the *total root length* in the soil profile, which is the sum of root length (SRL) of the individual 20-cm soil layers (cm cm^{-2}). Such a relation between LWU_{max}/ET_p and RL/SRL is shown for three crops in Fig. 9.10.

From the relationship shown in Fig. 9.10, we arrive at the equation:

$$LWU_{max} = b \cdot ET_p \cdot (RL/SRL)^{1/2} \quad (9.1)$$

In this equation b is the regression coefficient. As a first approximation the small intercept on the *x*-axis is neglected. The equation can be converted into an

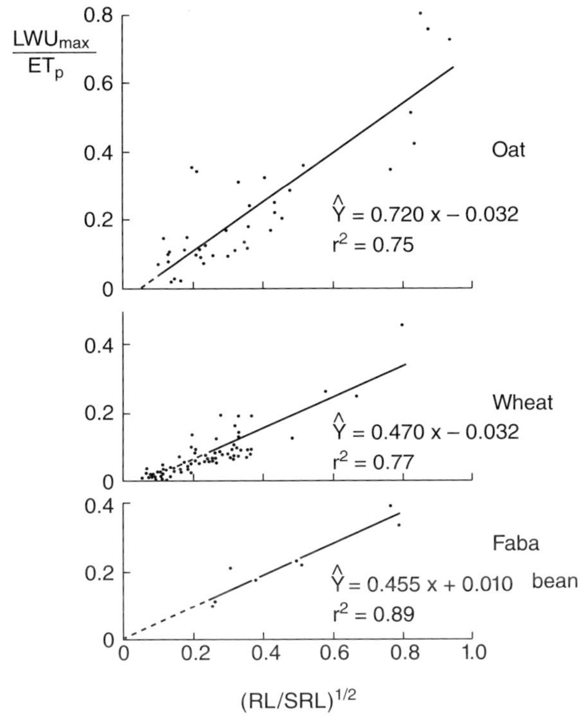

Fig. 9.10. The maximum water uptake rate in a 20-cm soil layer, normalized by the potential evapotranspiration rate (LWU_{max}/ET_p) as a function of normalized root length in the corresponding layer (RL/SRL). Oat and faba bean were grown on loess-derived Luvisol near Göttingen, Germany, with a temperate, humid climate. Spring wheat was cultivated on an Arenosol (uniform loamy sand) and two 'duplex' soils in the wheat belt of Western Australia with a Mediterranean climate (after Ehlers *et al.*, 1991).

equation valid for the maximum water uptake rate (WU_{max}) per unit volume of soil (cm^3 H_2O cm^{-3} soil day^{-1}). This equation reads:

$$WU_{max} = b^* \cdot ET_p \cdot L_v^{1/2} \cdot SRL^{-1/2} \tag{9.2}$$

The b^* is a constant factor and L_v is the root length density. The equation indicates that within a soil layer the water uptake rate per unit volume of soil is linearly related to ET_p, when the hydraulic conditions of the soil and the physiological state of the root for water uptake are most favourable in that layer. Under such favourable conditions WU is WU_{max}. But WU_{max} also increases with L_v, in this case in a non-linear manner (Fig. 9.7). It is a type of square root relationship, very similar to the relation depicted in Fig. 9.8 for WU and L_v under optimum and non-optimum conditions. In addition we can say: under the condition of increasing SRL, more roots are contained in other layers, the number of which increases with increasing rooting depth. Root proliferation causes WU_{max} to decrease (Eqn 9.2) because the roots extract more water from deeper layers and less water from the layer concerned. The conclusion from these considerations is that the rate of water extraction in a soil layer under optimum conditions is not only governed by the rooting density in that layer but also by the *relative distribution of the roots* in the soil profile.

In a soil profile, free of impeding pans and allowing deep rooting, a relationship is to be expected between total root length (SRL) and *rooting depth* (z_r). Such a relationship is shown in Fig. 9.11, where z_r includes 99% of SRL. Because of the close relation, Eqn 9.2 can now be re-formulated:

$$WU_{max} = b^* \cdot d \cdot ET_p \cdot L_v^{1/2} \cdot z_r^{-1} \tag{9.3}$$

where d is the regression coefficient of the equations in Fig. 9.11.

Equation 9.3 exemplifies the fact that under non-limiting conditions of water flow in the soil and of water extraction in the root tissue, the water uptake in a soil layer will be less the deeper the roots grow (Ehlers *et al.*, 1991). Shallow rooting crops with a small rooting density, such as faba bean (Figs 6.4 and 6.5), may therefore exploit the water in the soil as heavily as crops with deeper root systems and greater rooting density, such as oat (Figs 6.5 and 9.12) and wheat (Figs 6.5 and 9.11). The difference between deep-rooted and shallow-rooted crops is that the potentially extractable amount of soil water increases with the rooting depth of a crop (Fig. 9.12, Section 9.4).

When a soil layer dries, water uptake by roots declines (Figs 9.7 and 9.8). In the case that the soil layer is re-wetted by rain, roots are in a position to start the water withdrawal again at a high rate (Fig. 9.9, 0–20 cm layer). This ability to recover from inactivity is linked to the *regeneration of the root system*. Remoistening the soil stimulates the root system into branching and growing in length again (Asseng *et al.*, 1998). In this way plants are able to adapt to particular situations of soil water supply.

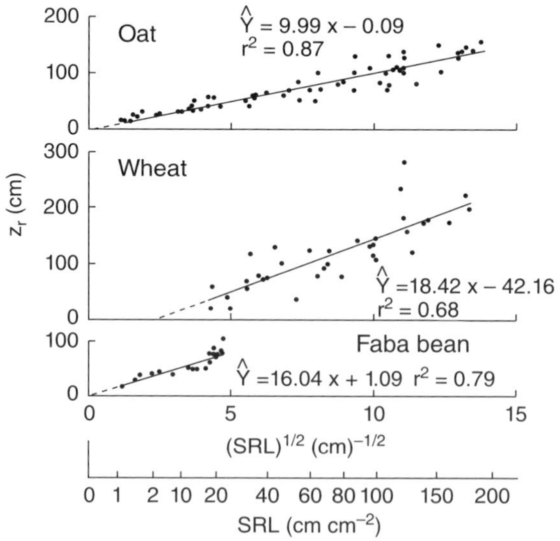

Fig. 9.11. The relationship between rooting depth (z_r) and total root length (SRL). Both quantities vary during the season. Crops and soils as in Fig. 9.10 (after Ehlers *et al.*, 1991).

Water stored in a shallow soil layer is taken up by plants earlier and in a larger amount than is water in deeper layers (Fig. 9.9). This is because of the time taken for root exploration of a particular soil layer (Fig. 6.4), and the root length density that is attained in a given layer (Figs 9.7 and 9.8) in relation to the total root length in the soil profile (Fig. 9.10). But that is just one part of the explanation of why roots start to extract water from the top layer, even when the soil is homogeneous in texture and structure and evenly moist. Evidently the water stored in the topsoil is in principle easier to withdraw than water in the layers of the subsoil. What follows is based on the statement made earlier (Section 7.5, Fig. 8.1) that the water, flowing through the vessels of the xylem, has to overcome axial flow resistances. Depending on the flux through the vessels a more or less steep gradient is necessary for water to flow. The gradient can attain a value in the order of 0.1 bar per metre plant height (Section 7.6). Also the gradient in the gravitational potential has to be taken into account for an erect plant, which is 0.1 bar m^{-1}. Let us now consider a plant with a rooting depth of 2 m. We assume that at the 1.5 m soil depth the total potential ϕ in the root system is exactly −3 bar. It follows that 1 m above, at the 0.5 m soil depth, ϕ within the root system has to be lower, let us say around −3.2 bar, to keep the water moving upwards in the root xylem. This is the main reason why the gradient between a water-absorbing single root and the surrounding soil has to be larger at shallow depths than in deeper positions within the root system, assuming that the water content in the homogeneous soil is the same or that the soil has drained to field capacity (FC). In any case, the larger the potential gradient, the faster the flow rate to a 'single root' (Eqn 7.7) at shallow depth, which means that water extraction starts in the topsoil.

It is this physical principle that results in plants conserving the water stored deep in the soil. These water supplies will be at the plant's disposal later in the season, when survival has to be ensured during the reproductive phase, even when there is little rain. Crops with a deep root system will show the greater yield stability within a region, where the period of yield formation (i.e. of grain filling and ripening) is accompanied by drought on a more or less regular basis. These outlines are only true for the case that once a year prior to the growing period, either during winter or during a rainy season, the soils are replenished with water to depth.

How the configuration of the root system determines the uptake pattern of crops in the soil profile is shown in Fig. 9.13. In this figure, the uptake rate for each individual soil layer is cumulated over the growing season. Oat, with roots extending to 150 cm depth (Figs 6.4 and 9.12), extracts 70 mm of soil water from the profile below 80 cm depth. That amount corresponds to 22% of total water uptake. Roots of faba bean, however, reach no more than 100–120 cm depth (Figs 6.4 and 9.12), and they scoop just 30 mm or 12% from the subsoil below 80 cm (Fig. 9.13). But in spite of the very much smaller root length density the shallower-rooted faba bean extracts 95 mm (36%) of water from the middle layer, 30–80 cm deep, whereas the deeper-rooted oat extracts less water from this layer, just 82 mm (26%). The figure also demonstrates that about 50% of total water is extracted by roots from the topsoil (0–30 cm), the cultivated tilth. The topsoil contributed a comparable amount to the water supply of sugarbeet and winter wheat in the sub-oceanic or temperate humid climate of Göttingen (Ehlers, 1976). Under faba bean,

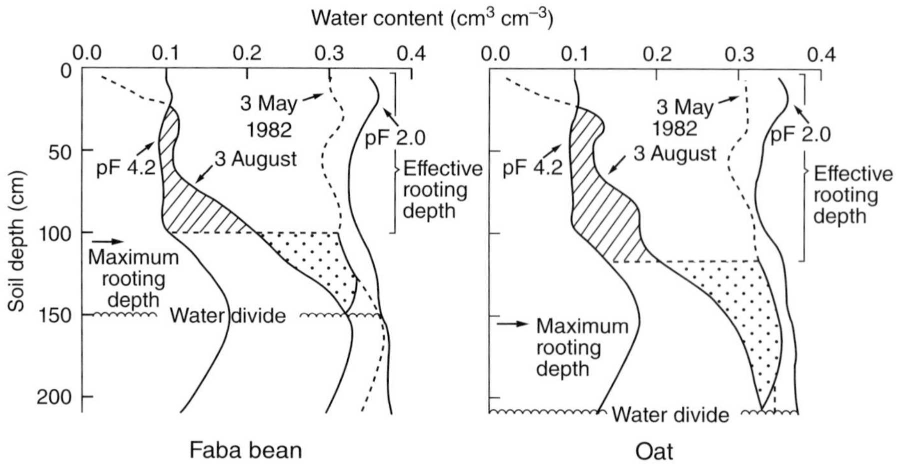

Fig. 9.12. Water content profiles in a loess-derived Luvisol under two crops early in the season (3 May) and immediately after harvest (3 August) in a year (1982) low in rainfall. Beside the two lines for field-measured water content two additional lines are shown, representing laboratory-measured water content at field capacity (pF 2.0) and permanent wilting point (PWP) (pF 4.2). Also the deepest position of the hydraulic water divide is indicated. This plane of zero flux separates upward water flux from downward water flux through the soil matrix. An arrow marks the maximum depth of rooting, observed in the field. The hatched areas as well as the 'effective' rooting depth are explained in Section 9.4 (Göttingen, Germany, 1982, unpublished).

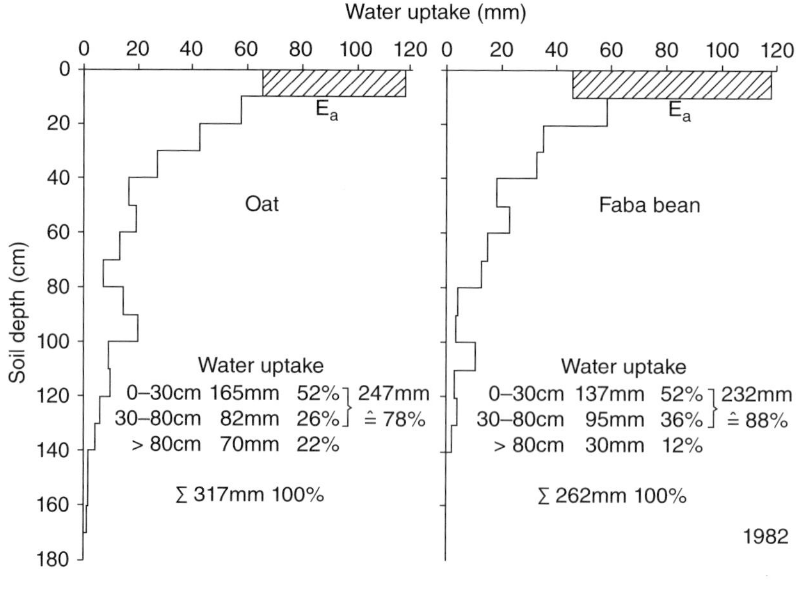

Fig. 9.13. Cumulative water uptake from individual soil layers by roots of oat and faba bean. The crops were grown on a loess-derived Luvisol in 1982. The actual soil evaporation (E_a), determined in the presence of the covering crops, is indicated by the hatched area (Müller and Ehlers, 1986).

which develops later and canopy closure is also slower (Fig. 9.1), the actual soil evaporation (E_a) is greater (70 mm) compared with oat with 50 mm (Fig. 9.13).

Oat reduced E_a more than faba bean because of its rapid *canopy development* (Fig. 9.1), which favoured transpiration during the period from May until the middle of June (Fig. 9.4). It is also true that root development of oat was faster and more intense than in the leguminous plant (Fig. 6.4). As a result, the cereal achieved a greater cumulative water uptake from the 0–10 cm layer (Fig. 9.13). These relationships hint at another *feedback system* that favours growing plants, and results because root water uptake dries the upper soil layer first. The hydraulic conductivity and the hydraulic gradient, associated with water flow in an upward direction, become reduced. Compared with bare soil, the reduction in hydraulic properties (Fig. 5.10) will occur earlier and will be more severe in the presence of plant roots. The following statement summarizes the logical outcome: *soil evaporation is reduced by root water extraction*. In other words, root activity decreases evaporation in favour of transpiration. The plant canopy, shading the soil, reduces the evaporation during the first stage of soil drying (Section 9.7). The plant root system, in extracting water, reduces the length of this period and the duration of the following period – the stage of falling evaporation rates.

9.4 How Much of the Soil Water is Extractable by Plant Roots?

The amount of water that potentially can be extracted by plant roots is held in the soil in an 'available' form. Plants can only acquire water from the available pool. Furthermore, the actual volume of water acquired from the soil profile depends on the depth of the root system. But not all of the water in the rooting zone will actually be taken up by roots. Availability of stored water is defined by limits set by soil physics (Section 5.1). The volume of available water, stored in a unit volume of soil, is approximated by taking the difference between the water content at FC and at PWP (Figs 3.2 and 5.1). To calculate the amount of water that is stored in the rooted profile and available for water extraction, *rooting depth* has to be considered (Section 6.3). However, determination of the rooting depth is difficult and the use of such an approach to determine the quantity of extractable water is somewhat questionable. Nevertheless, it is absolutely essential to have an estimate of the quantity of water in the rooting zone that is available for extraction so that we can assess the quality of an arable soil with respect to its productivity and for irrigation scheduling.

Commonly, the root system extracts less water from deeper soil layers than from shallower ones (Figs 9.9 and 9.13). One factor influencing water uptake is the distribution of root length density (L_v) throughout the profile (Figs 6.5, 9.7, 9.8 and Eqn 9.2). But L_v is seldom known. L_v also changes as a function of soil depth (Eqn 6.3) and time (Figs 6.4 and 6.5). Therefore it is highly desirable to find a field method for fixing the depth that frames the profile from which water is being taken up by roots during the season. This soil depth has an *imaginary lower boundary*, marking the extent of the profile that supplies water to the roots *within the physical limits of availability* (i.e. FC and PWP). This lower boundary of an 'effective root zone' (McKeague *et al.*, 1984) is the '*effective rooting depth*' ($z_{r\ eff}$)

of Renger and Strebel (1980). In addition to the physical limits of availability (FC and PWP), two other parameters, the '*drained upper limit*' and the '*lower limit*', need to be determined in the field so that actual water uptake from the soil profile can be calculated (Ritchie, 1981). These are measurable soil water contents that help define water extraction under field conditions.

For a meaningful evaluation of $z_{r\ eff}$, the lower limit within the soil profile has to be determined in a 'dry' year with inadequate rainfall (Renger and Strebel, 1980), when crops will have suffered from drought towards the end of the season (Ritchie, 1981). This approach, which we will now outline, actually identifies the *quantity of extractable water* (Ritchie, 1981) in the profile without requiring the root distribution to be determined.

We will focus on Fig. 9.12, which shows how $z_{r\ eff}$ was determined for faba bean and oat in a loess-derived Luvisol during 1982 – a dry year in Germany. The soil was free-draining, and the groundwater table was far below the rooting zone. The lower section of the profile contains a *hydraulic water divide*, a plane of zero flux (Section 5.2). The lowest position of the divide is indicated, which was attained near the end of the growing season. The depth of the divide, where the gradient of water potential is zero, was indicated from tensiometer readings (Section 5.1, Example 2, Case A).

Below the lowest position of the divide, *seepage water* is draining from the profile at all times during the year. Above the lowest position, water lost by drainage is assumed to be small in these loess-derived soils. Here water is extracted by plant roots, either because the roots are taking up the water directly from the supply in the rooted layer or because water flows upwards through the soil matrix from non-rooted into rooted layers, where it is extracted. An idea of the scale of this '*capillary rise*', by which non-rooted layers contribute to plant water use, is obtained by comparing the depth of the divide and the *maximum rooting depth*, that is the depth that includes 99% of all the roots identified (compare Fig. 9.12 with Figs 6.4 and 6.5).

The lines for FC and PWP, determined at a suction of 100 cm (pF 2.0) and 15,000 cm (pF 4.2), respectively, in the laboratory, are also included in Fig. 9.12. In addition, lines corresponding to the water content measured in the field on two occasions are shown, one taken early in the spring (3 May) when the plants were still small (Fig. 9.1), and one taken at harvest (3 August). The gravimetric water content for these lines was obtained from auger samples, and was converted to volumetric water content by multiplying by the appropriate bulk density (Box 5.1). The actual field water content in spring was less than FC at pF 2.0 (cf. Fig. 5.2). This line for the spring water content is the *drained upper limit*. It seems that the field-measured *lower limit* does not correspond at all well to the PWP – the closest fit was in the 20–60 cm layer. In the top 0–20 cm layer the soil dried below the PWP, presumably because water was lost by evaporation as well as by root water uptake. The lower limit increases steadily between the 60-cm depth and the water divide. This steady increase in the remaining water content indicates that the contribution of individual soil layers to plant water supply decreases down to the zero-flux plane.

The positioning of the *effective rooting depth* ($z_{r\ eff}$) at a definite distance below the soil surface is based on the following idea. In a dry year, the lower limit indicates the *maximum of water extraction* by roots from the profile down to the water divide. For that reason the depth of $z_{r\,eff}$ is fixed in such a way that the actual amount of water extracted below $z_{r\,eff}$ (dotted area in Fig. 9.12) corresponds to the amount remaining above $z_{r\,eff}$ (hatched area), which has not been extracted but is still potentially available to the crop. Completing calculations in this way is consistent with the idea that the total amount of water actually extracted from the soil by plant roots is taken from a profile, which is bounded at depth by $z_{r\,eff}$.

The quantity of *extractable water* (W_{extr}) in a rooted soil profile can easily be calculated by the equation:

$$W_{extr} = FC_{av} \cdot z_{r\,eff} \tag{9.4}$$

where FC_{av} is the '*available field capacity*' or water available at FC. It is calculated as the difference in soil water content between the drained upper limit in spring or after irrigation and PWP. W_{extr} was calculated to be 165 mm for faba bean, and 190 mm for oat because of the greater $z_{r\,eff}$ (Fig. 9.12). Table 9.2 contains data for $z_{r\,eff}$, FC_{av} and W_{extr} for different soil textural classes, surveyed for various soils in Germany.

In semi-arid and arid regions it may well be that after rainfall or irrigation and water redistribution within the soil profile there will be no plane of zero flux, because downward water flow has ceased within the profile. Tensiometer readings then indicate that water flow is solely in an upward direction. Upward flow from deeper and non-rooted layers can contribute to water uptake, and has therefore to be calculated, for example using the approach demonstrated in Section 5.1, Example 3. The method includes some uncertainties, linked to soil heterogeneity and to possible errors, when identifying the correct conductivity

Table 9.2. Effective rooting depth ($z_{r\,eff}$), available field capacity (FC_{av}) and amount of extractable water (W_{extr}) in the rooted profile, evaluated for sugarbeet and winter wheat grown on soils of various texture. The results are given for soils with small or medium bulk density, which varies typically with the texture. In compacted soils with higher bulk density, FC_{av} as well as $z_{r\,eff}$ are smaller. The $z_{r\,eff}$ will be smaller by 10–20 cm for spring cereals and grassland (Renger and Strebel, 1980).

Textural class	Clay (< 2 μm) (%)	Silt (2–60 μm) (%)	$Z_{r\,eff}$ (cm)	FC_{av} ($cm^3\ cm^{-3}$)	W_{extr} (mm)
Coarse sand	< 5	< 10	50	0.06	30
Medium sand	< 5	< 10	60	0.09	54
Fine sand	< 5	< 10	70	0.12	84
Loamy sand	8–18	10–40	75	0.16	120
Silty sand	< 8	10–40	80	0.18	144
Silt	< 8	80–100	100	0.25	250
Loamy silt	8–18	65–90	110	0.21	231
Sandy loam	18–25	15–40	100	0.17	170
Silty loam	18–30	50–70	110	0.19	209
Clay loam	25–45	30–50	110	0.15	165
Loamy clay	45–65	15–35	100	0.14	140
Silty clay	45–65	35–55	100	0.14	140

function. Therefore it might be advantageous to follow the soil water content to considerable depth, where the difference between the upper and lower limit is considered to be small (Ritchie, 1981).

In the presence of a shallow groundwater table, capillary rise can make a major contribution to the water use of crops. Either the water moves upwards by capillary rise to the rooting front or the roots directly access the water table. In these cases it may be difficult to quantify extractable water in the soil profile.

9.5 Stomatal Control of Water Vapour Loss

The time delay between the loss of water vapour through leaves into the atmosphere and water uptake by the root system means that plant tissues undergo water deficit. The deficit causes a decrease in the total water potential (ϕ), in the pressure potential (P) (Fig. 8.5), in the osmotic potential (O) and in the matric potential (Ψ) (Fig. 7.1). Plants experience this imbalance in their water budget on a daily and a seasonal basis. A daily course of some relevant water budget parameters is shown for two crops in June 1983 (Fig. 9.14). The course of *photosynthetically active radiation* (PAR) characterizes 23 June as a fully sunny day, whereas in the afternoon of 22 June clouds occasionally covered the sun. The *saturation deficit* of the air (Section 8.2) is greater during the sunny day than on the day with partial cloud cover. On both days after sunrise, ϕ declines more rapidly and to a greater extent in oat than in faba bean. At *noon* and during the *afternoon*, ϕ in the leguminous plant attains a *minimum value* of −1.2 to −1.5 MPa (−12 to −15 bar), but ϕ in the cereal reaches −2.0 to −2.2 MPa.

Figure 9.14 also shows the diurnal variation in *stomatal conductance* to water vapour, a measure of *stomatal opening*. Generally, conductance is much greater for oat than for faba bean. Apparently the stomates open wider in the cereal crop than in the legume, so that the much larger decrease in ϕ for oat is understandable. It also indicates that the stomates of the bean plants are able to do some *fine tuning*. The legume responds to the evaporative demand by a dynamic rapid opening and closing of stomates, such that ϕ remains at a 'medium' level. The faba bean quite 'sensitively' restricts the water loss through leaves. Oat reacts very differently. These plants only start to close their stomates shortly before noon, though apparently even then the level of opening was greater than that of faba bean. Despite incomplete or even only slight closure, the potential 'recovers' a little during the early hours of the afternoon. The oat plant responds by re-opening of the stomates later in the afternoon, as indicated by a more or less distinct second peak in stomatal conductance between 16.00 h and 18.00 h. Thereafter, when radiation is declining, the saturation deficit and the evaporative demand fall. In late afternoon and evening the imbalance in the water budget of both crops gradually declines. The stomates close progressively until sunset, and ϕ rises again.

When the twilight ends towards 22.00 h and night falls, the stomatal conductance drops to very small values (Fig. 9.14). But the conductance is not zero, probably because of a small amount of *cuticular transpiration*. At sunrise around 06.00 h the stomates of both crops open again rapidly and widely. At this early

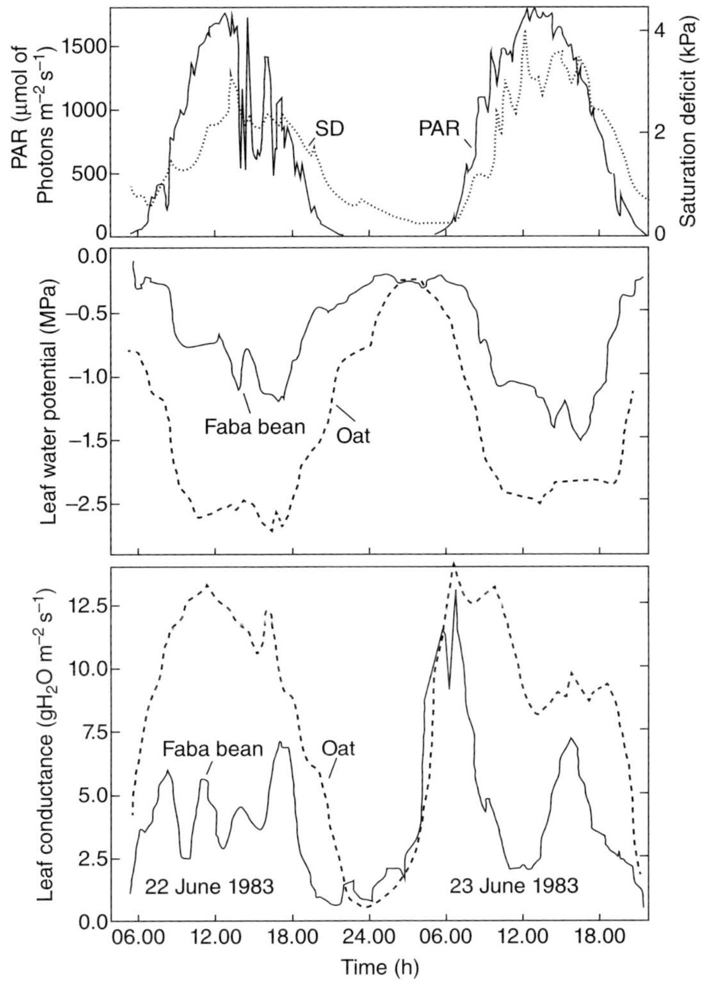

Fig. 9.14. Daily course of total water potential in leaves and of stomatal conductance for faba bean and oat in June 1983. The conductance was determined using a steady state porometer. The course of photosynthetically active radiation (PAR) and of the saturation deficit of the air (SD) is also indicated (after Müller *et al.*, 1986).

time on the sunny day (23 June) the saturation deficit is least. But once again stomatal control takes effect more obviously in faba bean than in oat.

Faba bean, with a sparse and shallower root system, has an effective mechanism for controlling water vapour loss. With that control the faba bean is able to maintain the daily *minimum value of* ϕ at a relatively high level, not falling below −15 bar or −1.5 MPa (Figs 9.14 and 9.15). Other grain legumes like soybean, cowpea (*Vigna unguiculata*), green gram (*Vigna radiata*), black gram (*Vigna mungo*) and snap bean (*Phaseolus vulgaris*) are well known for keeping the minimum ϕ at a comparatively high level and behave similarly. It seems that the pulses mentioned belong to the group of *water savers* (Fig. 1.1).

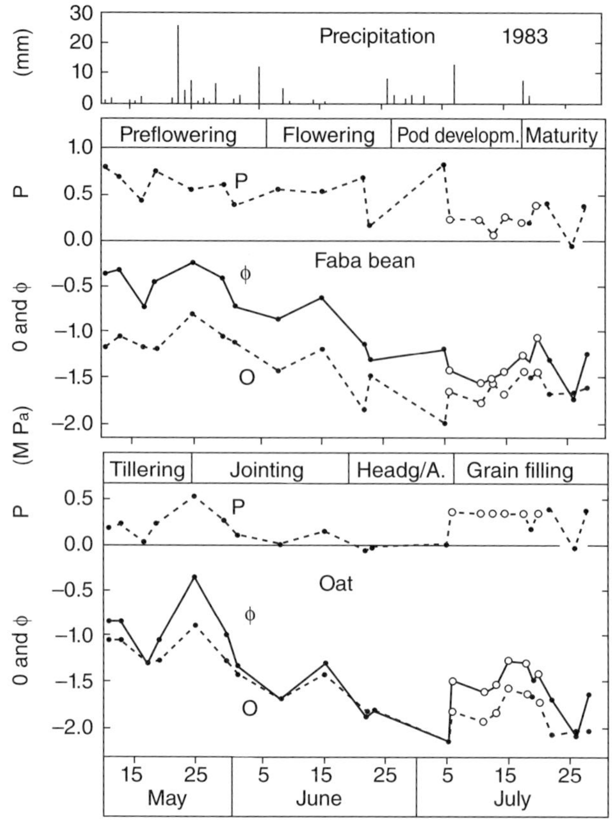

Fig. 9.15. Seasonal course of total water potential (ϕ) and of the partial potentials, the osmotic potential (O) and the pressure potential (P), for faba bean and oat. Dots represent 14.00–15.00 h, circles 09.00–10.00 h measurements. Developm., development; Headg, Heading; A., anthesis (after Müller *et al.*, 1986).

Oat allows ϕ to fall to lower values, around −21 bar (Figs 9.14 and 9.15). Along with the dense and deeper root system, oat more likely belongs to the group of *water spenders*. Accordingly, the transpiration rate per unit leaf surface area is greater for oat than faba bean (Fig. 9.16). Such a result seems reasonable for July, when the LAI and therefore the transpiring leaf area of oat is smaller than that of faba bean (Fig. 9.1). But the transpiration of oat, adjusted for leaf area, is even greater in June (Fig. 9.16), although until mid-June the LAI is larger, not smaller than that of the legume. Comparatively low values of ϕ measured in oat were also observed with other field-grown cereals like wheat, barley, sorghum and maize (references in Müller *et al.*, 1986).

By delicate fine tuning, ϕ in faba bean is kept at a higher level than in oat during the daytime (Fig. 9.14) and over the season (Fig. 9.15). On the other hand, differences in the *osmotic potential* (O) between the two crops are only minor (Fig. 9.15). O drops similarly in both crops during the growing period, roughly following the same time trend as the total potential (ϕ). A less negative

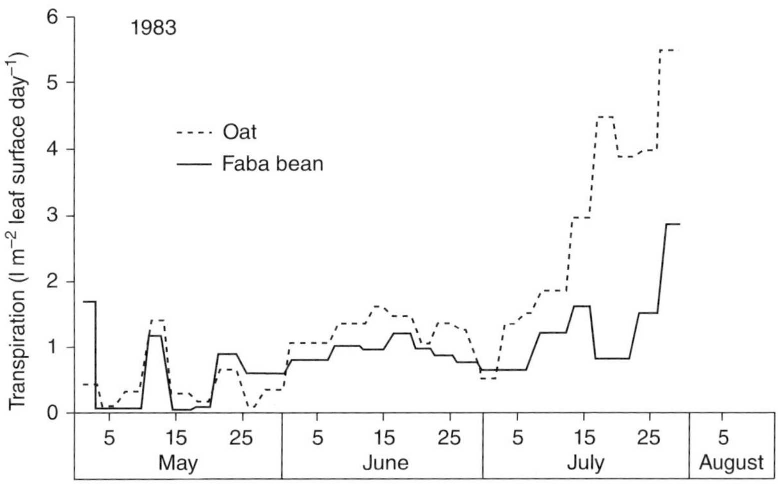

Fig. 9.16. Seasonal course of transpiration for faba bean and oat. The transpiration rate, evaluated by the soil water budget approach, is related to one-sided leaf area (after Ehlers *et al.*, 1986b).

ϕ at a given O means a higher *pressure potential* (P) (Fig. 7.1, Eqn 7.2). The higher P or turgor in bean relative to oat existed almost throughout the season (Fig. 9.15).

In contrast to oat, a water spender, faba bean is a water saver with several plant features that seem to complement each other. The sparse and shallow root system to some extent limits water uptake (Fig. 9.13) and hence requires control of the plant water economy, achieved by delicate stomatal control of transpirational water loss. Effective control seems to be necessary to protect the plant tissue from drying out. Such protection seems to be all the more important since the plant apparently requires a comparatively high turgor for maintaining vital processes like growth.

9.6 Water Use Throughout the Growing Season

Weather conditions and meteorological factors determine the evaporative demand and influence the crop's total water use during the growing season. The water use will be greater in years with high net radiation, air temperature, saturation deficit and wind than in years with less demanding atmospheric conditions. This concept, which applies to one location, can be extended to regions with different *climates*. A crop's water use will be greater in warmer climates than in cooler climates. But there is one proviso that we must mention. Our statement on water use and climatic conditions is only true when the supply of *extractable soil water* is sufficient for crop transpiration to meet the evaporative demand through most of the season. The quantity of extractable water (Section 9.4) is modified by soil texture, organic matter content, and by the position of the groundwater table. Moreover, for the assessment of the extractable water supply, the amount of *precipitation* in a region is of prime importance, as it will determine whether or not the extractable water will be replenished. The approach of extractable water becomes meaningless when insufficient rainfall prevents the soil water from being resupplied in the rooting zone prior to the cropping season, unless refilling is achieved artificially by irrigation. Although not meaningless, it is difficult to apply the approach in regions with a highly *variable rainfall pattern* between years, a characteristic of arid, semi-arid and sub-humid regions of the world, where crop production is mostly limited by water use.

Whenever precipitation is sufficient to refill the rooting zone of the soil by the start of the growing season, the *root system* of a crop becomes important, as it influences the amount of extractable water and hence the potential water use of a crop. For the cumulative water use during the vegetative period, it is the depth of the root system rather than its density that appears to be decisive (Section 9.3). Another plant property influencing water use during the season is the LAI (Section 9.2). The transpirational water use will depend heavily on LAI, as long as it is smaller than about 3 m^2 leaf surface area m^{-2} soil surface (Fig. 9.3). When a LAI of 3 or so has been attained, the water use of a crop will be more or less at its maximum (i.e. it will remain independent of LAI when this further increases above 3). What is important for seasonal water use is the time period over which a given LAI is maintained during the growing season. The integral of LAI with time is named the *leaf area duration* (LAD).

$$\mathrm{LAD} = \int \mathrm{LAI} \cdot \Delta t \qquad (9.5)$$

According to the pre-set unit of the time t, the LAD has the unit of hour, day or week. The LAD of a crop stand is assessed by determining the LAI several times during the season. Usually the above-ground parts are harvested per unit area, for instance per m^2. The leaf area (normally restricted to the leaf blade) is determined by use of predetermined area-to-mass relations, which change with the age of the stand. This method requires a great deal of work. Quite neatly the area of harvested leaves can be determined directly by use of a planimeter or *leaf-area meter*. The device integrates the leaf area with a laser beam. A third non-destructive method is to use a device with 'fish-eye' optics. The sensor measures radiation above and below the canopy, which allows foliage gaps and LAI to be derived (Welles and Norman, 1991).

To use Eqn 9.5 for evaluating the role of LAD on cumulative water use, strictly speaking the integration has only to consider leaf area indices that are restricted to less than 3 or so, since above this level of LAI the transpiration rate approaches the maximum. Nevertheless, one may expect the water use of crops to increase with the length of the vegetation period (Table 9.3).

In 1971 the cumulative *evapotranspiration* (ET) of sugarbeet was greater than the ET of winter wheat, as the growing period lasted longer. In 1982 leaf area development (Fig. 9.1) and root growth (Fig. 6.4) of oat occurred earlier, which resulted in more *transpiration* (T) than in faba bean. Because of retarded foliage cover and rooting density, soil *evaporation* (E) and seepage losses due to subsoil *drainage* (D) were increased under beans compared with oat. 1977 was a wet and cool year. Under these climatic conditions the ET of oat was smaller than during the dry and warm season of 1982.

Table 9.3. Evapotranspiration (ET), soil evaporation (E), crop transpiration (T) and subsoil drainage (D) measured under some crops and in fallow soil from spring to harvest in Göttingen, Lower Saxony, Germany. The soil is a loess-derived Luvisol.

Year	Crop	Period of measurement	ET	E	T	D
			(mm)			
1971	Winter wheat	5 April–18 August	343	–	–	53
	Sugarbeet	5 April–10 November	492	–	–	82
	Fallow	5 April–10 November	–	228	0	142
1977	Oat[a]	27 April–15 August	309	45	264	?
1982	Oat[b]	1 May–9 August	398	52	346	6
	Faba bean	1 May–9 August	366	74	292	30

[a]Wet and cool season.
[b]Dry and warm season.

The *fallow* soil in 1971 reached almost half the ET of sugarbeet, just by evaporation (E) within the same time period from April to November. This was quite a wasteful and unproductive use of water, and was further accentuated by the increased amount of drainage losses (Table 9.3). Water extraction by roots cuts down seepage and evaporation under sugarbeet and winter wheat.

In traditional *dryland farming* a fallow year is introduced as a break between two adjoining cropping seasons for additional rainwater collection for the benefit of the next crop. The low storage efficiency of rainwater during the fallow period makes the enterprise risky, not least because fallowing can result in more erosion by wind or water.

9.7 How to Determine the Components of the Field Water Balance

Water that has been released by crops to the atmosphere under natural field conditions is *no longer available for* quantitative *analysis*. This contrasts sharply with the situation for mineral nutrients, which after being taken up by roots are stored, at least for some time, within the various parts of the plant. Therefore, as long as there is no appreciable release of minerals from the plant or any decay of tissue, the harvested plant dry matter can provide the basis of analysis for nutrient content and uptake. The question of the quantity of nutrients extracted from the unit soil area by a crop stand is much more easily answered than that of water use by a crop. The term '*water use*' has to be specified at this point, and poses the question of how one can determine evapotranspiration, evaporation, transpiration and water uptake by roots from single soil layers *in the field*.

The basic principle is to measure *water content* and *water tension* of the soil as a function of depth and time (Box 5.1). Where possible, measurements should be replicated in the field to cope with the spatial variability of the site under investigation. At the Institute of Agronomy and Plant Breeding in Göttingen the principal soil type investigated has been a silty loam soil derived from loess, a Luvisol, which fortunately is more uniform in its properties on a spatial scale than many other soil types. The *gravimetric* water content was determined by augering the silty, stone-free soil down to 200 cm depth two or three times a week. After taking a sample from a given depth, it was immediately cut into 10-cm sections, which were stored in plastic boxes for weighing, drying (105 °C) and weighing again. The 200-cm depth was chosen to ensure that the maximum rooting depth was exceeded. Volumetric core samples were taken in spring, when the soil was slightly drier than FC, and were used to determine the soil bulk density (BD). The appropriate BD values were then used to derive the *volumetric* water content. A bank of tensiometers down to 210 cm depth recorded water tension. The installation depths of the porous cups down the profile were 10, 20, 30, 40, 60, 80, 100, 120, 150, 180 and 210 cm. Readings were taken daily or every second day. The hydraulic *conductivity functions* were also established for principal soil horizons, either for bare soil in the field (Ehlers and van der Ploeg, 1976) or in the laboratory (Ehlers, 1977; Wendroth *et al.*, 1993). The field method was outlined in detail and graphically illustrated by Hillel *et al.* (1972).

The investigation of the water content gives information on the daily *changes in water supply* of the *total profile* and of *individual soil layers*. The changes can be provoked either solely by 'capillary' water flow through the soil matrix, which is predominantly a flow in the unsaturated state, or they are caused by an (additional) water withdrawal by plant roots. The roots absorb the water like a sink, from which the water disappears (Section 7.2). In the literature, the phrase '*sink term*' is used to denote this quantity of water 'loss' from the soil to the roots.

As well as these soil variables, *rainfall* also has to be recorded. Quite often meteorological variables, essential for calculating the *potential evapotranspiration*, are measured in the field at the same time. These include net radiation, run of wind, relative humidity and air temperature (Section 8.2). These variables are also needed to assess the actual soil evaporation in a crop stand. For that approach, the LAI must also be measured.

With all the measured values and functional relationships mentioned, it is possible to calculate the quantities needed to determine the components of the field water balance. We shall explain the mode of calculation on the basis of daily rates.

Evapotranspiration

The water balance equation (Eqn 3.1) is formulated for a cropped soil that is flat and where there is no runoff. For a given period of time the equation is:

$$\mathrm{P}-\mathrm{E}-\mathrm{T}-\mathrm{D}=\Delta\mathrm{S} \tag{9.6}$$

The P stands for precipitation, E for evaporation of the soil below the leaf canopy, T for transpiration, D for subsoil drainage and ΔS for change in profile water storage. Here in this formulation the transpiration includes the evaporation of water intercepted by the canopy. The unit for all the quantities is the same (i.e. unit water depth per unit time). We may select mm per day or cm per day. For convenience, we start here with cm per day.

Let us now imagine a soil profile of 100 cm depth (depth L = 100 cm). The position of L is below the rooting zone (Fig. 9.17A). We can rewrite Eqn 9.6 in the notation:

$$P - E - T - D = \int_{L}^{0} \frac{d\theta}{dt} \cdot dz \tag{9.7}$$

where θ is the volumetric water content ($cm^3\ cm^{-3}$), t is the time (day) and z is the soil depth (cm). Now Eqn (9.7) is rearranged:

$$-E - T = \int_{L}^{0} \frac{d\theta}{dt} \cdot dz + D - P \tag{9.8}$$

We obtain E and T (together ET) as a negative outcome of the change in profile water content. That interim result has to be corrected for seepage, as deep drainage contributes to a change in water content over and above that due to ET. Precipitation adds to the soil water content, and has therefore to be subtracted. We obtain the drainage term from the measured hydraulic gradient below the rooting zone and from the corresponding conductivity function $K(\Psi)$ (Eqn 5.1). The hydraulic gradient also indicates when water uptake from the rooting zone changes the direction of the water flux below the rooting zone from downward to upward direction. That is, when the drainage water efflux changes to an influx by capillary rise. For the lysimeter approach see Box 9.1.

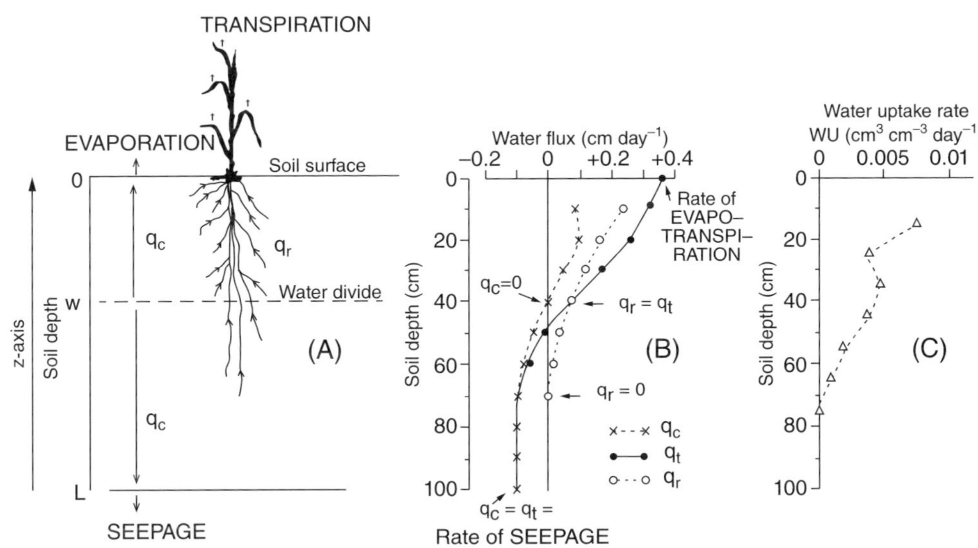

Fig. 9.17. (A) In a soil profile, water flows through the 'capillary' pore system of the soil matrix, represented by the capillary flux (q_c). Above the water divide (which is a plane of zero flux) q_c is directed upwards; below the divide it is downwards. There is also an upward water flux through roots (q_r). (B) The total water flux (q_t) is shared in a 'capillary' water flux of the soil (q_c) and a water flux through roots (q_r). (C) The water uptake rate of roots (WU) within a 10-cm soil layer is obtained by differentiation of q_r with depth (after Ehlers, 1975a).

Evaporation

It is exceptionally difficult to measure the soil evaporation (E) in the presence of transpiring plants under field conditions. Therefore we applied a calculation procedure that is outlined below. First of all the potential daily soil evaporation in the presence of the crop (E_p^*) is needed. E_p^* is related to the net radiation (R_N^*) below the canopy, that reaches the soil surface. R_N^* is determined by the LAI:

$$R_N^* / R_N = e^{-0.398\,\mathrm{LAI}} \tag{9.9}$$

In the equation R_N^* is presented as a fraction of the net radiation above the crop stand, R_N (Ritchie, 1972). Equation 9.9 is illustrated in Fig. 9.18. Ritchie (1972) assumed that within a crop stand the wind speed was so slowed that its contribution and that of the saturation deficit term in E_p could be ignored. Under these simplified conditions the following relation holds:

$$E_P^* = (\Delta / (\Delta + \gamma)) \cdot R_N^* / L \tag{9.10}$$

In Eqn 9.10, Δ is the slope of the function relating the saturated water vapour pressure of the air to temperature. This function is depicted in Fig. 8.3. The γ stands for the psychrometric constant and L is the latent heat of vaporization (cf. Eqn 5.6). By combining Eqns 9.9 and 9.10, E_p^* can be calculated from:

$$E_P^* = (\Delta / (\Delta + \gamma)) \cdot R_N / L \cdot e^{-0.398\,\mathrm{LAI}} \tag{9.11}$$

Equation 9.11 states that the potential soil evaporation (E_p^*) can be obtained from measurements of air temperature, net radiation (R_N) and LAI. The equation underestimates E_p^* when the soil is sparsely covered by plants (Duynisveld and Strebel, 1983). In consequence, these authors suggested that Eqn 9.11 should be expanded with a ventilation and vapour pressure deficit term when LAI is less than 1. The modified equation reads:

$$\begin{aligned} E_P^* = {} & (\Delta / (\Delta + \gamma)) \cdot R_N / L \cdot e^{-0.398\,\mathrm{LAI}} \\ & + (1 - \mathrm{LAI}) \cdot (\gamma / (\Delta + \gamma)) \cdot f(h) U^{0.75} \cdot \Delta e \end{aligned} \tag{9.12}$$

in which $f(h)U^{0.75}$ is a wind function that is dependent on height of the crop stand (h) and daily wind run (U) measured at 2 m above ground (Ehlers, 1989). Specifications of $f(h)$ are listed in Rijtema and Aboukhaled (1975).

To calculate E under the crop canopy, E_p^* has to be adjusted. To make this adjustment, the relationship between the actual soil evaporation (E_f) and the potential soil evaporation (E_{pf}) of the fallow loess soil, shown in Fig. 5.10, has been used. To find the correct ratio E_f/E_{pf}, the water tension at 10 cm soil depth under the crop has to be measured. Then E is calculated according to the equation:

$$E = E_f / E_{pf} \cdot E_p^* \tag{9.13}$$

E is the actual evaporation of the soil covered by crops (Müller and Ehlers, 1987), which is less than E_p^* because of soil drying.

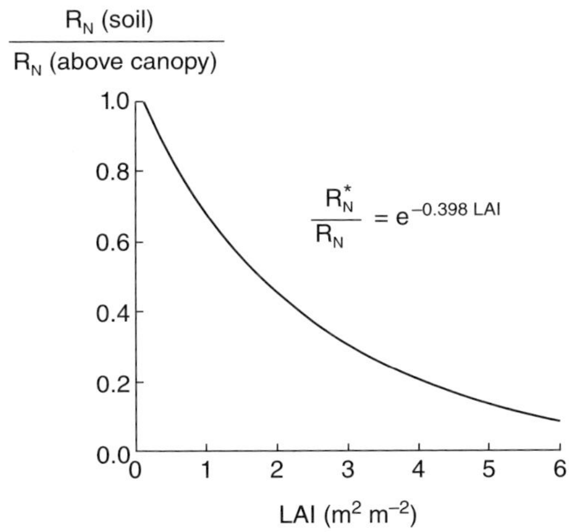

Fig. 9.18. In a crop stand, radiation arriving at the soil surface is reduced by the canopy. Only that part of radiation not intercepted by the canopy supplies energy for evaporation: the fraction of net radiation at the soil surface (R_N^*) and net radiation above the canopy (R_N) as a function of the leaf area index (LAI) (based on Ritchie, 1972).

Transpiration

Transpiration is quite easily calculated from the difference between ET and E. Actually, it is the total water uptake by the roots that has been determined by the water balance. The real transpiration may deviate slightly from total water uptake, as the water content in the plant tissue can change a little, although this effect is considered to be negligible. Another approach for transpiration assessment is the measurement of sap flow (Box 9.2).

Water uptake by roots

We come back to the continuity equation (Eqn 5.2):

$$dq/dz = d\theta/dt \tag{9.14}$$

The equation presents the relationship between the change in water flux over a certain depth, the flux divergence (dq/dz) and the temporal change in water content ($d\theta/dt$) in the soil layer considered. The thickness of the soil layer (dz) is defined as unit depth, for instance 1 cm. In the case of a soil layer with roots, a *sink term* (see above) has to be introduced. In the literature a sink term is often symbolized by S. Following Eqn 9.2 we call the term WU, the *water uptake rate* (cm^3 H_2O cm^{-3} soil day^{-1}).

The following serves to introduce WU in Eqn 9.14. In rooted soil there is a divergence of the *total water flux* (q_t) when passing the unit depth (dz). The divergence is dq_t/dz. The divergence of q_t and not just of the capillary flux through the soil matrix causes the change in soil water content in the rooted profile. The total water flux q_t (cm day^{-1}) comprises the capillary water flux and the water flux through roots. The *capillary flux* is denoted here by q_c, and not just by q as we used before. The *flux through roots* is q_r, which is directed upwards through the root system towards the soil surface. In contrast to q_r, q_c can be directed upwards during evaporation or downwards during seepage conditions. These capillary fluxes, opposite in direction, are separated by a hydraulic water divide (Section 5.1, Example 2, Case A), the plane of zero flux (Section 5.2). At the depth of the divide (Fig. 9.17A), where q_c is zero, it follows that q_r is equal to q_t (Fig. 9.17B).

Now Eqn 9.14 is expanded by the sink term WU:

$$\begin{aligned} dq_t/dz &= dq_c/dz + dq_r/dz \\ &= dq_c/dz + WU = d\theta/dt \end{aligned} \tag{9.15}$$

All the terms have the unit of day^{-1}. The equation states that the divergence of the total water flux is created by divergence of capillary water flux and a divergence of water flux through roots. The divergence of root water flux is caused by root water uptake. The rate of water withdrawal corresponds to the sink term WU. Equation 9.15 is an enlarged form of Eqn 9.14, expanded by the sink term.

For practical purposes it might be easier to calculate root water uptake on the basis of fluxes through the soil (Fig. 9.17B) than by use of the differential Eqn 9.15. Therefore Eqn 9.15 is integrated with respect to depth. We integrate from depth L, which is below the rooting zone, to any depth L + Δz above L. In so doing Eqn 9.15 is reformulated as:

$$\int_L^{L+\Delta z} (d\theta/dt)dz = q_{t|z=L} - q_{t|z=L+\Delta z} \tag{9.16}$$

At depth z = L the term $q_{r|z=L}$ is zero, as there are no roots. At that depth q_c is q_t (Fig. 9.17B). Therefore we can modify Eqn 9.16:

$$\int_L^{L+\Delta z} (d\theta/dt)dz = q_{c|z=L} - q_{t|z=L+\Delta z} \tag{9.17}$$

Equation 9.17 can be rearranged:

$$q_{t|z=L+\Delta z} = q_{c|z=L} - \int_L^{L+\Delta z} (d\theta/dt)dz \tag{9.18}$$

Now we see that q_t can be balanced in any depth above L (z = L + Δz) when the capillary seepage flux at depth L (free of roots) and the daily change in volumetric water content in the profile section above L are known. Figure 9.17B shows diagrammatically the quantity of q_t for the depths z = L + Δz. The quantity of Δz is preset in 10-cm distances. From tensiometer readings and corresponding conductivity functions, q_c is calculated for the various layers of the profile (Fig. 9.17B). Knowing q_t and q_c, we can easily calculate q_r:

$$q_{r|z=L+\Delta z} = q_{t|z=L+\Delta z} - q_{c|z=L+\Delta z} \tag{9.19}$$

So in summary, the procedure for obtaining q_r requires that first we have to determine the level of q_c at depth z = L. In the example presented, q_c is a seepage flux. In exactly the same way q_c may represent capillary rise, by which roots in the profile section above L are supplied with water. Starting with q_c at depth z = L, we calculate q_t in the layer z = L + Δz by considering the daily water content changes in the various soil layers (Eqn 9.18). For all the depths z = L + Δz the capillary fluxes (q_c) must be known. Subtract from q_t the value of q_c for all the depths z = L + Δz and obtain q_r (Eqn 9.19). All the fluxes have units of cm day^{-1}. The estimation of q based on values of q_t and q_c is demonstrated in Fig. 9.17B.

Once q_r is known, the water uptake rate by roots in the various layers of the rooted profile can easily be calculated. Values are determined by taking the difference between the values of q_r from one 10-cm layer to the next (i.e. from z = L + Δz to z = (L + Δz) + 10 cm). This way the water uptake rate within a 10-cm layer is obtained. That is the LWU (Section 9.3). In contrast to WU of Eqn 9.15 (day^{-1}), LWU (cm day^{-1}) is related to a soil layer of given depth, here a depth of 10 cm:

$$LWU = q_{t|z=L+\Delta z} - q_{r|z=(L+\Delta z)+10\ cm} \tag{9.20}$$

In the uppermost layer from 0 cm to 10 cm depth, q_c is unknown, as a tensiometer at the soil surface is missing. Therefore q_r cannot be calculated either (Eqn 9.19). For the top layer, bordering the atmosphere, we have to write:

$$LWU = q_{c|z=-10\ cm} - \int_{-10\ cm}^{0\ cm} (d\theta/dt)\cdot 10\ cm + P - E \tag{9.21}$$

The 'capillary flux at the soil surface' can be estimated by calculating the actual soil evaporation (E) within the crop stand (compare with information in Evaporation, see p. 144), as a result of which Eqn 9.21 is solvable. In rainy periods, P has to be accounted for in the calculation of LWU for the 0–10 cm depth (Eqn 9.21). In the case that water from precipitation infiltrates deeper into the soil than the corresponding thickness of the surface layer, the estimation of q_r (Eqn 9.19) and of LWU (Eqn 9.20) becomes difficult below the top layer. During heavy rain periods it is therefore advisable to restrict the calculation to total ET (Eqn 9.8).

Once LWU (cm day^{-1}) for the various 10-cm soil layers has been evaluated by use of Eqns 9.20 and 9.21, WU (day^{-1}) is obtained by dividing LWU by 10 cm, the preset layer thickness. By doing so, water uptake is reduced to unit depth (i.e. to 1 cm). The determination first of LWU according to Eqn 9.20 by taking the difference of q_r (Fig. 9.17B) from layer to layer, and secondly the conversion into WU can be understood and worked out with the aid of Fig. 9.17C.

9.8 Numerical Simulation

So far we have presented an experimental method to determine the components of the water balance and crop water use in the field. This approach may appear quite cumbersome and elaborate. Computer techniques and the application of numerical procedures allow the water balance to be determined in another way. The backbone of such a mathematical approach is a *model of the water budget*. Such a model is made up of a series of equations. One of them is the Richards equation, which is extended by a sink term. The starting point for the mathematical treatment is the initial distribution of the water tension and the water content within the soil profile. Based on this initial condition the capillary water flux taking place from layer to layer is calculated. As a consequence of divergent water fluxes, the water content in the layers of the profile changes (Richards equation, Eqn 5.4). But the soil water content also changes due to root water uptake (Eqn 9.14), which is evaluated by taking into consideration the distribution of roots and of the water tension (Fig. 9.7) in the soil profile. The water budget models consider important and necessary boundary conditions like the evaporative demand of the atmosphere or rainfall events, the latter causing water infiltration into the soil or interception by the canopy.

Crop transpiration is then obtained by integrating the water uptake over the rooted depth of soil. The calculated transpiration cannot be greater than the *'potential' transpiration*. The potential transpiration is defined as the potential evapotranspiration (Eqn 8.6) less the potential evaporation of a soil covered by a crop canopy (Eqn 9.11). The potential transpiration is the unlimited transpiration at a given evaporative demand. This feedback control within the model aims to constrain the calculated rate of water uptake to an upper limit, which is set by atmospheric conditions.

The result of a *simulation study* can be evaluated (i.e. to establish whether it portrays the real world more or less exactly) by relatively simple field measurements that are much easier than a complete set of water budget measurements. For instance, the following procedure is suggested. The calculated tensions for a few selected depths of the soil profile are read out from the computer for various dates of the study period. The simulated data are then compared with corresponding readings from field-installed tensiometers. In the case of a major discrepancy between model outcome and measured results, some of the model parameters have to be adjusted. The goal of these efforts is to reconcile simulated values with measured results. The fine tuning of the model leads to its *validation*, an indication that the model is suitable for replicating the field situation in the case that was studied. The validation does not imply that the model will truly portray the water budget when field conditions like weather, soil and crop are changed. The method of feeding back simulated tensions for comparison with measured tensions was chosen by Benecke and van der Ploeg (1976) among others. The outcome of such a 'controlled' simulation study is presented in Fig. 3.3.

There is an extensive literature, giving hints on the structure of simulation models for determining the soil water balance (Leenhardt *et al*., 1995) and on the application of numerical methods for studying root water uptake (Hillel *et al*., 1975, 1976; Luo *et al*., 2003), evapotranspiration of crops (van der Ploeg *et al*., 1978; Ragab *et al*., 1990) and of crop sequences (Maraux and Lafolie, 1998). Some of the models can separate soil evaporation and plant transpiration (Lascano, 2000). Others deal with the dynamics of root growth and with the corresponding effects on water uptake pattern (Wang and Smith, 2004). There are two introductory textbooks on modelling processes in soils, one by Campbell (1985) and one by Richter (1987). Root water uptake and transpiration greatly influence plant growth, development and yield (see following chapters). Therefore it is not surprising that models have been extended to soil properties and plant responses. There is a comprehensive book on simulation studies at an advanced level published by the American Society of Agronomy (Hanks and Ritchie, 1991). This book is not just restricted to various physical and chemical processes in the soil, but it extends the understanding to soil–crop interactions and the enhancement of crop growth models. More recently the American Society of Agronomy published a collection of papers outlining current science and modeling approaches associated with predicting the response of crop plants to limited water supply under various management scenarios. The articles deal with the impacts on physiological and growth processes in crops subject to stress from water shortage as well as plant traits that might help to offset adverse effects (Ahuja *et al.*, 2008).

Box 9.1. How lysimeters work

Lysimeters are measuring devices for solving the water balance equation (Eqn 3.1) in the sense that the individual components can be determined. In particular, lysimeters are helpful in separating seepage from evapotranspiration (Goss and Ehlers, 2009). Concentrations of solutes and suspended material, such as nutrients, pesticides and pathogenic micro-organisms (Goss *et al.*, 2010) can also be measured in the drainage water. With the flux of seepage water known, the load of various substances that are leached in this way is relatively easy to calculate. Otherwise, the quantitative assessment of the discharge in the field may prove to be a difficult operation. As lysimeters are limited in space, surface runoff and lateral flow do not occur. Therefore the determination of the evapotranspiration (ET) turns out to be rather simple.

In the very basic designs, lysimeters are square or round boxes or containers, which are filled with soil and which are usually lowered into the ground. The soil surface within the lysimeter is set level with the surface of the surrounding soil (Fig. B9.1A). Lysimeters are bordered at the side by vertical walls. The bottom of the lysimeter soil is separated from the underlying soil by a metal sheet. Above this sheet, tension-free drainage water may occur during wet seasons, which can empty through a pipe (not shown in the figure).

To avoid microclimate edge effects, lysimeters have to be cropped like the surrounding soil. The plant stand within the lysimeter is adjusted to correspond to the field crop outside with respect to planting pattern (stand density, row spacing), so that there is no difference in temporal growth stage development. This ‘fitting in’ into the surrounding area is to avoid abnormal radiation conditions and the oasis effect, which is an excessive transpiration caused by advective energy.

For determining evapotranspiration, in addition to precipitation and seepage, the temporal change of the stored soil water supply has also to be measured. That is done in the main by weighing the lysimeter. In the past, small lysimeters were used that could be moved on to a balance (Fig. B9.1A).

Nowadays, however, they are more likely to be weighed in place on a frequent or continuous basis (Fig. B9.1B). For this purpose platform scales fitted with ultra-sensitive strain gauges have been employed successfully (Marek *et al.*, 2006), although electronic ‘compensation scales’ are preferred (Howell *et al.*, 1995). The accuracy of this latter type of scale is based on the fact that the heaviest part of the total lysimeter weight is balanced by

Continued

Box 9.1. Continued

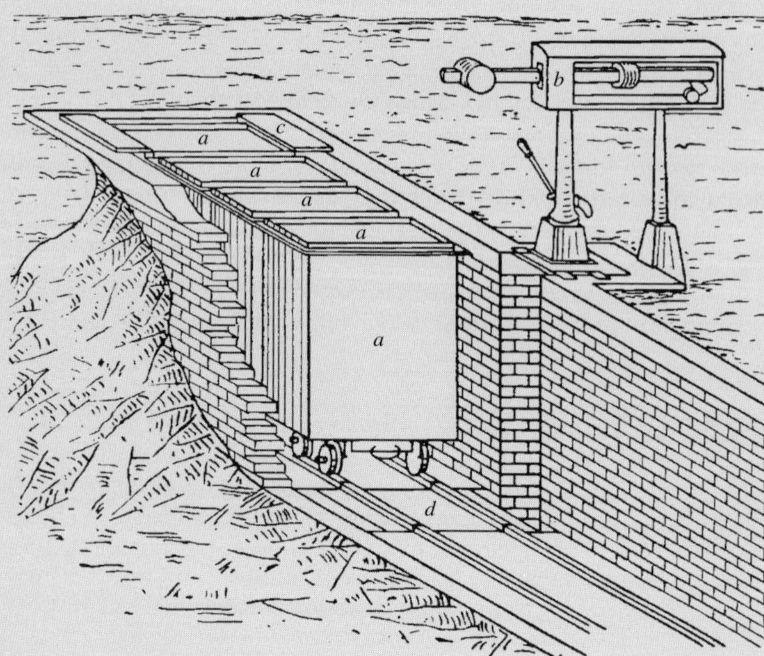

Fig. B9.1A. Lysimeter facilities in Göttingen at the time around 1900. The lysimeters can be rolled to the scales on rails. a, 'Box carriage' (the pipe for the draining water is not shown); b, 'scale beam', the checking accuracy is 20 g; c, 'covering planks' protect the lysimeters from large temperature deviations relative to the surrounding soil; d, 'scale table' to support the lysimeters (from von Seelhorst, 1902).

counterweights. As a result, only the very fine changes in weight 'beyond' the basic load are recorded. In the case of a lysimeter not equipped with scales, the water content changes have to be measured by different means, for instance by use of measuring techniques like time domain reflectrometry, neutron scattering or gamma ray adsorption. If there is no provision for such techniques either, the one remaining method is to sum up the components of the balance for periods when the changes in soil water supply are small, so that a cumulative value is obtained. That may be a suitable method for estimating the balance from one spring or from the end of one rainy season to the next, when the soils will have re-attained field capacity (FC) after whatever water has been used by crops.

A modern lysimeter has to meet still more requirements. But it has to be borne in mind that with increased capability the costs may rise heavily. The lysimeter area should not be smaller than a minimum of 2–4 m^2. Moreover, the container should be sufficiently deep (2–3 m) to accomodate the root system. The soil in the lysimeter has to correspond to the natural soil not only in layering but also in structural characteristics. For that reason, lysimeters are often built by encasing an undisturbed soil core. And finally, the water needs to flow within the lysimeter soil under the same conditions as in the outside field soil (Palmer *et al.*, 2011). However, at the bottom of the lysimeter, the hydraulic continuity is interrupted by the sealing plate. Water may just leave the lysimeter bottom on top of the metal sheet via a

Continued

Box 9.1. Continued

drainpipe, when the water tension is zero or when even a slight positive pressure potential has been built up. The prerequisite for drainage water flow is the build-up of 'free' water without tension. Therefore, unlike the surrounding soil, in a lysimeter sealed at the bottom, conditions will prevail that are too wet, at least during part of the time. As a consequence, the water regime is shifted in favour of evapotranspiration rather than drainage.

Controlling the water tension at the bottom of the soil cylinder can prevent such a shortcoming. The control is made possible by use of a porous plate, put underneath the bottom plane of the soil, or by use of some porous cups inserted into the bottom soil layer. The tension is generated by a suction device, creating a low pressure (Fig. B9.1B). The pressure is adjusted according to the tension, recorded by tensiometers at the corresponding depth in the soil outside the lysimeter. The seepage water, removed by the suction device, can be analysed for solutes and pollutants for a quantitative assessment of the leaching losses per unit area and time.

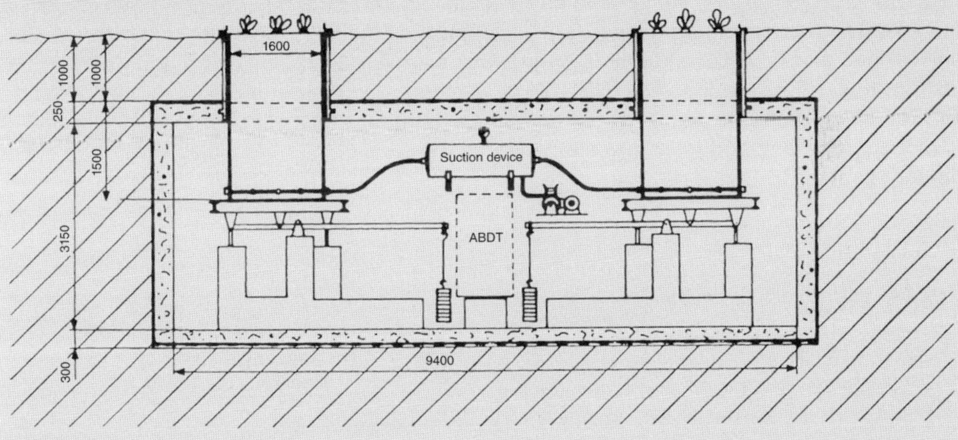

Fig. B9.1B. Modern facilities with two cylindrical, weighable lysimeters. The tension at the lysimeter bottom can be monitored with the aid of a suction device according to the soil water tension in that depth outside in the field. The lysimeters contain the undisturbed soil with its natural layering, a deep loess-derived chernozem. The automatic balance and data transmission device (ABDT) is situated below the suction device. The dimensions are given in mm (after Roth *et al.*, 1994).

Box 9.2. Measurement of water flow through plants

We have considered the flow of water through the plant in terms of movement through a crop stand, although we have identified the key processes taking place within individual leaves. As noted earlier, the water use of a crop can be calculated from the field water balance, where we first obtain a value of evapotranspiration that combines evaporation from the soil and the transpiration through the plant. Determination of transpiration then requires subtraction of the amount of soil evaporation (Section 9.7). Our estimate of daily evapotranspiration from the field water balance is usually an average determined for a period of a few days, but more elaborate micrometeorological methods, such as Bowen ratio or

Continued

Box 9.2. Continued

eddy covariance, allow values to be obtained over much shorter time periods. However, the micrometeorological approach is much more complex, needs more expensive equipment and requires relatively large areas of flat uniformly cropped land (Smith and Allen, 1996). An alternative approach to the determination of transpiration is the direct measurement of sap flow in an individual plant. For annual field crops the only available technique is the *stem heat balance method*, but for woody horticultural crops like kiwifruit (*Actinidia deliciosa*), the *heat-pulse method* is also available.

The principle of the *stem heat balance method*, first used by Vieweg and Ziegler (1960), is to heat a small section of the plant stem, and then determine how much heat is moved away from the region by the sap under steady state conditions. By measuring the specific heat capacity and density of the sap, it is then possible to determine the volume flow of sap.

Commercial stem heat balance gauges are available that consist of a flexible heater and pairs of thermocouple junctions; the latter being used to measure temperature changes. The heater is wrapped around the stem to be investigated, and is then surrounded with insulating material. The thermocouple junctions are placed against the stem, on either side of the heater, and aligned axially. A second pair is arranged in a similar alignment, but staggered with the first set (Fig. B9.2). Gauges are available to fit stems with diameters ranging from 2 mm to 125 mm. The larger the diameter, the greater the number of sets of thermocouples required to measure the temperature differences between the pairs of thermocouples either side of the heater.

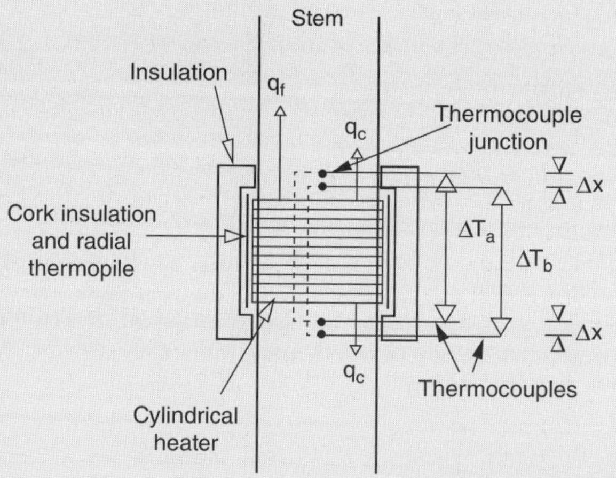

Fig. B9.2. Schematic diagram of the working of a stem flow heat balance gauge. A cylindrical heater is carefully wrapped around a stem of the plant. This heater warms the stem tissue, and some heat, q_c, is dissipated by conduction away from the heated area. The insulation surrounding the heater is aimed at minimizing radial losses, but the radially mounted thermopile allows the heat dissipated in this way to be determined. Two pairs of thermocouple junctions are placed on the stem, with one junction from each pair above the heated section and one below to determine the temperature differences, ΔT_a and ΔT_b. Because of upward flow, the temperature of the sap will be greater above the heater than below. See text for further details.

Continued

Box 9.2. Continued

The heater generates heat (q_h, cal day^{-1}), which is dissipated by conduction through the stem tissue (q_c) and by mass heat flow in the xylem sap (q_f) such that:

$$q_h = q_c + q_f \tag{1}$$

The heat dissipated by conduction (q_c) is calculated, assuming one-dimensional heat flow, from the difference between the two temperature differentials (°C) ΔT_b and ΔT_a measured across the heated stem segment (Smith and Allen, 1996). q_c is then obtained from Eqn 2:

$$q_c = A \cdot k_s \frac{(\Delta T_b - \Delta T_a)}{(\Delta X)} \tag{2}$$

A is the cross-sectional area of the heated stem area (cm^2) and k_s is the thermal conductivity of the stem tissue (cal °C^{-1} cm^{-1} day^{-1}). ΔX is the vertical separation of the two thermocouple junctions above and below the heater (Fig. B9.2).

By subtracting the calculated value of q_c from the heat output (q_h) the value of q_f can be determined.

The volume flow of sap, f_v (cm^3 day^{-1}), is then given by:

$$f_v = \frac{q_f}{c_{xs} \cdot \rho_{xs} \cdot (\Delta T_a + \Delta T_b)/2} \tag{3}$$

where c_{xs} is the specific heat capacity of the xylem sap (cal g^{-1} °C^{-1}), and r_{xs} (g cm^{-3}) is the density of the sap. $(\Delta T_a + \Delta T_b)/2$ is the increase in temperature of the flowing sap, as it is warmed when passing through the heated part of the stem in an upward direction.

These equations assume ideal conditions, and in reality account must be taken of the radial conduction of heat. The radial heat flow through the tissue and surrounding insulation is determined during periods of no sap flow (usually at night) from the temperature gradient measured with the thermopile, and again by subtracting the value of q_c from q_h. There can be a further error present if heat storage in the plant tissue is neglected, especially for stems with larger diameters (Swanson, 1994). Ishida *et al.* (1991) described an improved gauge design that reduced the errors in stem flow values by controlling stem temperature. Peressotti and Ham (1996) developed a dual heater gauge that also greatly improved the accuracy of the method, and had the added advantage that determination of the radial flow of heat was unnecessary with their design.

The stem heat balance method has been used to demonstrate that the transpiration of soybean plants with a lower frequency of stomates was reduced relative to that of plants with higher stomatal frequency (Tan and Buttery, 1995). Senock *et al.* (1996) used this approach to determine the water use of wheat grown under irrigation in Arizona. Cumulative water use of plants grown under a CO_2 concentration of 550 µmol mol^{-1} in a free-air carbon dioxide enrichment (FACE) facility used between 7% and 23% less water than those grown under 370 µmol mol^{-1}.

The *heat-pulse method*, originally proposed by Huber (1932), is suitable only for woody stems, and in this case the approach is to determine the movement of a heat pulse of 1 s or 2 s duration, rather than establishing steady state conditions. In this case, holes have to be drilled into the sapwood of the stem, and these accommodate the heat probe units. Each unit consists of a heater probe and two sensor probes, the latter bearing thermistors for temperature measurement. The three probes are inserted into holes that are arranged axially, with the lower sensor being closer to the heater probe than is the upper sensor. As a result of the short heat pulse created by the heater, the temperature at the lower sensor

Continued

Box 9.2. Continued

rises by conduction through the wood. The temperature at the upper sensor rises as heat is moved with the sap flow until the temperature at the two sensors becomes equal at time t_{eq}. For the distance between the lower sensor and the heater being x_l (mm) and the separation for the upper sensor being x_u (mm), the velocity (V) of the heat pulse is given by:

$$V = \frac{(x_u + x_l)}{2t_{eq}} \quad (4)$$

Marshall (1958) showed that if the heat flow is homogeneous within the tissue the sap flux density (J) can be calculated using the relationship:

$$J = d_w(0.33 + M) \cdot V \quad (5)$$

where d_w is the density of the sapwood tissue and M is its moisture content. Cohen *et al.* (1981) introduced an alternative approach, based on the heat-pulse approach, that uses the time taken (t_m) for a single temperature probe, located 15 mm above the heater probe, to record a temperature maximum from a single heat pulse. In this case the velocity of the heat wave, V_m, is given by:

$$V_m = \sqrt{\frac{x_u^2 - 4\kappa t_m}{t_m}} \quad (6)$$

where κ, the thermal diffusivity of the live wood, is equal to $x_u^2/4t_m$, when there is no sap flow taking place. Sap flow usually stops at night under conditions of very small vapour pressure deficit, when stomates are closed and so transpiration ceases.

The insertion of the heat probe and sensors for these heat-pulse methods causes some heterogeneity in the thermal properties of the sapwood, and it also stimulates a wound reaction resulting in the blockage of some conducting strands. As a result, values obtained from Eqns 4 and 6 underestimate sap flow by 50% or more (Cohen *et al.*, 1981; Green *et al.*, 2003). Swanson and Whitfield (1981) derived correction factors that could be used to improve the values obtained with the above relationships. These have been further refined by Green *et al.* (2003). However, the analytical approach to the problem of heterogeneity cannot always be used. Green and Clothier (1988) reported that in vines such as kiwifruit and grapes, the water is conducted through large diameter vessels that are often distributed within a matrix of smaller woody tissue. This introduces additional heterogeneity in the thermal regime that further contravenes Marshall's assumption of homogeneity. Empirical calibration of the heat pulse method is required under these conditions.

10 Radiation and Dry Matter Production

10.1 Radiation and Net Photosynthesis of Single Leaves

The sun's radiation is the driving force for evaporation (Section 5.2) and transpiration (Section 8.2) and hence of the water balance of soils and plants. Moreover, radiation is also the prerequisite for *photosynthesis* (Box 1.1) and hence of dry matter production by plants. In this way radiation, water use and dry matter production are intimately linked with each other. It has been mentioned before that only a part of the sun's radiation reaches the earth's surface. This is the 'global radiation'. In Section 5.2 we calculated that on average the quantity of global radiation is about 28.3 MJ m^{-2} striking the northern hemisphere during 1 day. The wavelength of global radiation varies between 300 nm and 3000 nm (Fig. 5.5).

Plants do not absorb radiant energy with equal effectiveness across this wide range of wavelengths. Rather there are two *absorption maxima*, one positioned at around 430–450 nm (blue) and the other at around 640–670 nm (red). Pigments like *chlorophyll a* and *chlorophyll b* are especially effective in capturing radiant energy in both bands, whereas *β-carotene* is only active in the 420–480 nm range. The two maxima mentioned for the blue and red spectrum correspond roughly to the wavelength limits of visible light (390–760 nm). Within the medium waveband of visible light (i.e. within the green part of the spectrum) plants absorb comparatively less radiant energy. A larger part of the energy within the green spectrum is reflected and transmitted by leaves, thus producing the characteristic '*leaf-green*' colouring. Hence, for photosynthesis plants only use effectively that part of global radiation which is within the wavelength spectrum of visible light. This part is called the '*photosynthetically active radiation*' – PAR. PAR can be measured by special sensors. Normally the wavelength of PAR sensors is set between 400 nm and 700 nm. Within this range PAR accounts for about 44–50%

of global radiation reaching the earth's surface. Solar infrared from 700 nm to 3000 nm is mostly reflected or transmitted by leaves.

Provided that no unidentified environmental factor is limiting plant metabolic processes, increasing radiation will enhance the *CO_2 exchange rate* (CER) at the leaf surface. Consequently a larger quantity of CO_2 is assimilated per unit leaf area and time (Fig. 10.1). The CER is comparable to the *net CO_2 assimilation rate*, also called the *apparent* CO_2 assimilation rate, which is the result of *gross CO_2 assimilation rate* minus *respiration rate* (Fig. 1.4).

At the *light compensation point*, gross assimilation is balanced by respiration and so the net assimilation is zero. Increasing radiant energy causes CER to increase in the form of a saturation curve, the *light response curve* (Fig. 10.1). C_3 plants reach the level of *light saturation* at lower levels of radiation than do C_4 plants, where light saturation may never be attained even when radiation is high. Nevertheless, leaf surfaces of C_3 as well as C_4 plants use the unit light more effectively for CO_2 assimilation at lower than at higher radiation, as indicated by the slope of the curves. The slope also indicates that C_4 plants use light more efficiently than C_3 plants. The difference in efficiency between plant species shows itself best at high radiation. To put it another way, high levels of radiation are a prerequisite to demonstrate the superiority of C_4 plants in CO_2 assimilation.

The response to increasing light intensity is a decrease in CER per unit of light intercepted. This result indicates that, as radiation increases, the CO_2 concentration of the air limits net assimilation of leaves to an ever-greater extent. Eventually light saturation is approached. As the demand for CO_2 in the chloroplasts rises

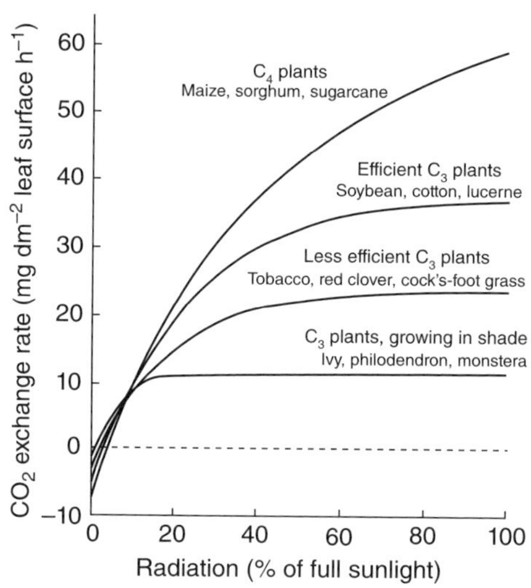

Fig. 10.1. Idealized light response curves for leaves of different plant species. The broken horizontal line marks the light compensation, where the CO_2 uptake rate due to photosynthesis is the same as the CO_2 release rate from respiration. The radiation level, where light compensation is reached, differs between species. Plants with light compensation at higher radiation are usually more efficient in CO_2 assimilation. That is true for C_4 plants relative to C_3 plants, which again show large differences between species (after Gardner *et al.*, 1985, modified).

with light intensity, it is the transport rate, more exactly the diffusion rate, that increasingly causes the supply not to keep pace with the demand. For any given conditions the diffusion rate is dependent on the concentration gradient (Eqn 7.10), which is limited by the CO_2 concentration of the air. At present under typical field conditions, CO_2 gas is available for uptake by the stomates at a concentration of about 400 ppm. Although at this concentration the diffusion flux is generally small and consequently CO_2 is only marginally available for uptake, not less than 85–92% of plant dry matter is produced from CO_2 that is fixed by photosynthesis.

The *superiority of the C_4 plants for net CO_2 assimilation* depends not only on performance at higher levels of radiation but also on air temperature. The optimum temperature range is higher (30–40°C) for many C_4 plants originating from tropical and hot climates, than for many of the C_3 plants (15–30°C). C_3 plants show a somewhat wider range of optimum temperatures, but these are generally lower on the scale (Larcher, 1994). Reasons for the temperature difference between C_3 and C_4 plants are the strong affinity of the phosphoenolpyruvate carboxylase for CO_2 and the lack of photorespiration, which are both outstanding characteristics of C_4 plants. In contrast, the CO_2 affinity of ribulose-1,5-bisphosphate carboxylase/oxygenase (RuBisCO) in *C_3 plants* is not only generally lower but moreover it declines as the temperature rises. Furthermore, the rise in temperature accelerates the *photorespiration* in C_3 plants. Photorespiration is the release of fixed CO_2, a process catalysed by the same enzyme that alternatively is responsible for linking CO_2 to ribulose-1,5-bisphosphate (RuBP). In addition to having a low affinity for CO_2, this double function of RuBisCO further reduces the photosynthetic efficiency of C_3 plants compared with C_4 plants.

10.2 Radiation Interception and Dry Matter Accumulation in Crop Stands

As well as atmospheric conditions, the input of global radiation at a site depends on the latitude, aspect and inclination of the location. Season and time of day are also important variables. Of course CO_2 assimilation and dry matter production of a crop stand are directly influenced only by that part of incoming radiation that is intercepted by the crop's canopy. Under given radiation conditions, the greater the *leaf area*, the more light a crop stand will intercept. Similar to the relation of leaf area index (LAI) and transpiration (Fig. 9.3), LAI and intercepted radiation are related to each other in the form of a saturation curve (Fig. 10.2).

Direct solar and diffuse cloud and sky radiation (Fig. 5.5), intercepted by an upper leaf, is either *absorbed* by the leaf pigments, or is *transmitted* or *reflected* by the leaf. Therefore only part of the incoming radiant energy arrives at lower storeys of leaves. The deeper the radiation penetrates into the crop stand, the more of its energy is attenuated. The energy attenuation is a function of number of the leaf storeys penetrated by the radiation. And the number of leaf storeys is related to LAI. The larger the LAI, the less energy arrives at the soil surface (Fig. 9.18). The less energy reaching the soil surface, the more that soil evaporation is reduced (Section 9.7). The larger the LAI, the more of the radiant energy remains within the crop stand to be used in photosynthesis.

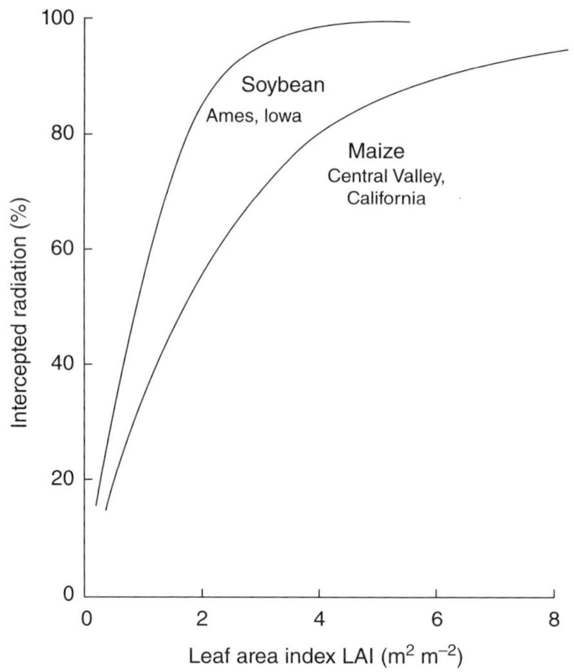

Fig. 10.2. Relationship between leaf area index (LAI) and radiation intercepted by the canopy of two crops. The leaves of soybean are more horizontally spread (planophile leaf inclination); with maize, the leaves are in a more upright (erectophile) position (data from Shibles and Weber, 1965; Williams *et al*., 1965a).

Depending on *leaf inclination* (Fig. 10.2), a larger part of incoming radiation is either absorbed within the upper leaf storeys with predominantly horizontal leaves (*planophile* leaf inclination) or penetrates into deeper storeys, which will be the case when the leaves are in a more upright position (*erectophile* leaf inclination). In the case of high radiation intensity, an erectophile canopy has a beneficial effect on *light utilization*, as the supplied radiant energy is more uniformly distributed over the total leaf area at all heights of leaf insertion. When more leaf storeys share in interception of radiation, the energy input per unit leaf area will admittedly be lowered. But this is more than balanced by the fact that the gain in CO_2 assimilation per unit of radiant energy is greater at low radiation than at high radiation (Fig. 10.1). As a result, deeper light penetration can increase the crop's capacity to fix CO_2 per unit of ground surface.

Summing up, it turns out that at high global radiation erect leaves can improve the photosynthetic rate of the whole canopy per unit ground surface. That will support the *crop growth rate* (CGR, Eqn 1.1). However, when radiation levels are low, leaves in lower storeys may not receive enough radiation to arrive at the light compensation point (Fig. 10.1). In such a case, respiration will outweigh gross photosynthesis and the CO_2 balance will become negative. After some time the leaves located low down in the canopy will turn yellow and die. Therefore, under conditions of low radiation, a planophile canopy with fewer leaf storeys and a smaller LAI may be beneficial for light utilization. A planophile orientation can also be advantageous in interspecific competition, particularly with respect to weed control. As the ground is covered earlier with a planophile orientation, weeds can be shaded out and soil water can be saved for the crop, both because of the exclusion of competition and because there will be less soil evaporation. Some of the recent maize cultivars change leaf inclination during the season, an adaptation for early competitiveness and best possible radiation use throughout the season.

Whichever way the canopy is orientated, the LAI grows during the early season (Fig. 9.1). And an increasing LAI promotes the mutual shading of the leaves. Because of the greater shading, the average net assimilation rate (NAR) of the leaves of a crop will decrease (Fig. 10.3).

Although on average for a crop stand, the NAR of the leaves (per unit leaf area) decreases, the net primary production of the crop (per unit ground area), the crop growth rate (CGR), will not be reduced. There will be a much greater increase in the CGR as LAI gets larger (Eqn 1.1). More of the incoming radiation is intercepted and the CGR tends to a maximum. The maximum is realized over a wide range of LAI encountered in the field. During the early growing season (the period of crop establishment) the maximum CGR is attained when the LAI is between 3 and 5 (Figs 10.4 and 10.5). After reaching this 'critical' LAI, roughly 95% of the global radiation is intercepted (Fig. 10.4), allowing for nearly maximum production. At any given location the particular maximum CGR is attained only under non-limiting conditions (i.e. when the only growth factor restricting leaf and plant growth is radiation). Under such conditions the production at a site is limited solely by the capture of available light by the canopy at the greatest possible rate per unit

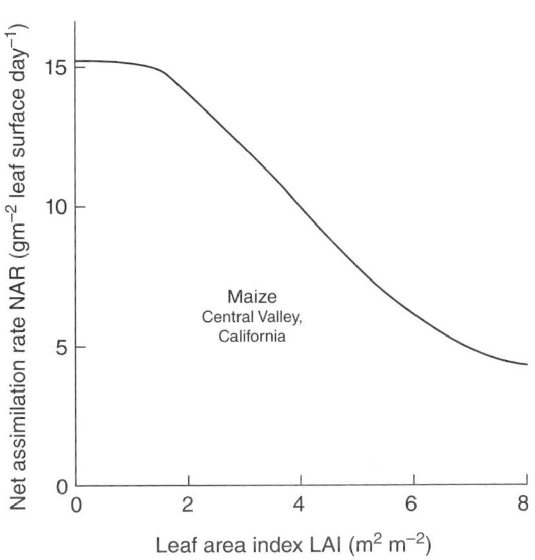

Fig. 10.3. The average net assimilation rate (NAR) of maize leaves as related to the leaf area index (LAI) of the crop stand (based on Williams *et al.*, 1965b).

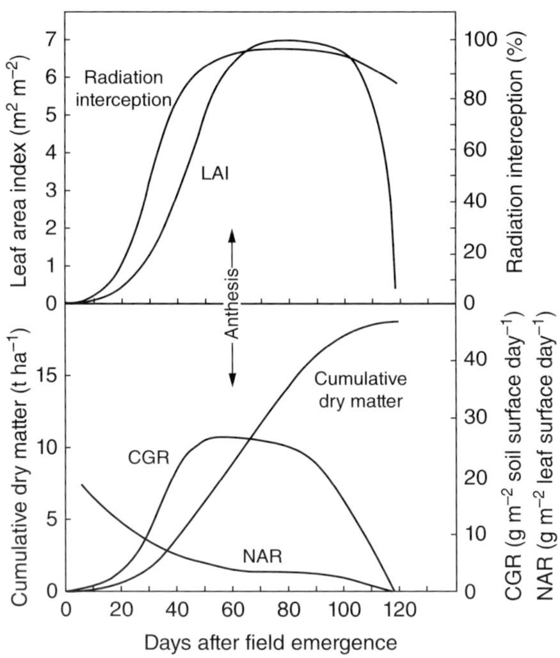

Fig. 10.4. The temporal development of factors that determine total yield during the growing season for a spring cereal. The radiation interception is ahead in time compared with the leaf area index (LAI) because at low LAI, light interception and shading increase heavily per unit of LAI increase (Fig. 10.2). Because shading increases in time, the net assimilation rate (NAR) decreases (Fig. 10.3). NAR is also reduced by rising temperature during the growing period, accelerating respiration and ageing of the photosynthetic apparatus. The crop growth rate (CGR) is the product of LAI and NAR (Eqn 1.1). The cumulative dry matter is the integral of CGR over time (after Gardner *et al.*, 1985). The NAR of a winter cereal (not shown) will probably start at a lower level after emergence than with the spring cereal because of limited global radiation and temperature during the cool season.

soil surface area. In many regions of the world the climate is warm and dry, and water and plant nutrients can be deficient. Under such limiting conditions the LAI remains smaller than 3, and the maximum CGR or the *potential yield increase* will never be attained.

The relationship between the LAI of a crop and the quantity of intercepted radiation (Fig. 10.2), as well as that with the CGR (Fig. 10.5) takes the form of a saturation function. Because these functional relationships are of the same kind, a more or less linear relationship between CGR and intercepted radiation is expected (Fig. 10.6). The more global radiation intercepted by the developing canopy, the greater is the CGR.

The CGR increases with intercepted radiation. When integrating both quantities with respect to time, again a linear relation is achieved (Fig. 10.7). It has been mentioned before (Section 5.2) that on average in the northern hemisphere the insolation is 28.34 MJ m^{-2} day^{-1} reaching the earth's surface after having

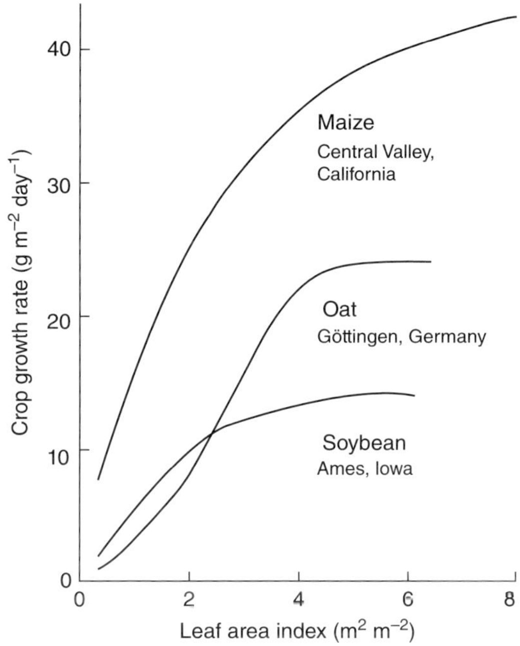

Fig. 10.5. Relationship between leaf area index and crop growth rate for three crops (data from Shibles and Weber, 1965; Williams *et al.*, 1965a; Ehlers, 1991).

passed through the atmosphere. Roughly 50% of global radiation is PAR, about 14 MJ m^{-2} day^{-1}. Therefore during a growing period of 100 days, somewhere in the order of 1400 MJ m^{-2} will be received at the ground in the form of PAR. Part of it will be intercepted by the crop and transformed into growth (i.e. into accumulated dry matter) (Fig. 10.7). The slope of the lines indicates the performance of a crop to build up dry matter per unit of intercepted PAR. This feature is called *radiation use efficiency* (RUE; Sinclair and Muchow, 1999). It seems that the C_4 crops are superior to C_3 crops in RUE and that within the group of C_3 crops, legumes and oil-producing plants have a smaller RUE than small-grain cereals (Fig. 10.7, Table 10.1).

On average RUE is 3 g dry matter MJ^{-1} intercepted PAR (Table 10.1). Based on intercepted global radiation, this figure corresponds to a RUE of about 1.5 g dry matter MJ^{-1}. Plant material contains on average 18 kJ of heat of combustion g^{-1} dry matter (Greef *et al.*, 1993). That is to say that 1 MJ of global radiation produces 1.5 g plant material, containing 27 kJ of chemical energy. These considerations lead to the result that on average *crops use no more than 2.7% of the energy of global radiation.*

When LAI increases, more of the radiation energy is intercepted, causing CGR to increase until radiation interception reaches a maximum. The longer the leaf area stays green, the larger is the *leaf area duration* (LAD; Eqn 9.5), and the greater is the accumulated plant dry matter and the energy equivalent. That is the reason why the above-ground dry matter is related to the *length of the season*

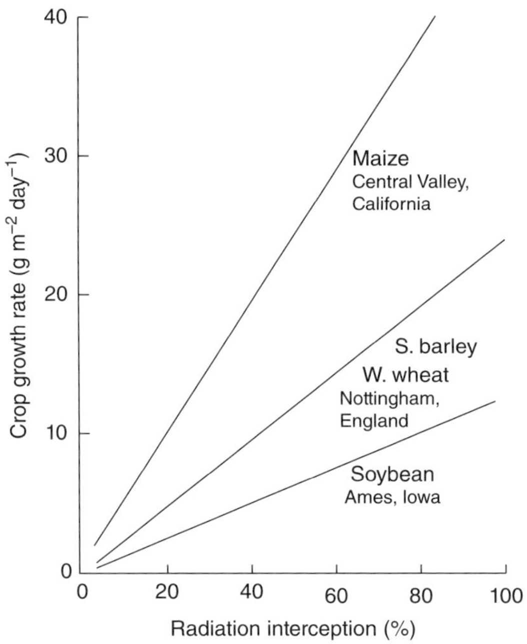

Fig. 10.6. The growth rate of a crop stand depends on the quantity of global radiation intercepted by the canopy. Maize as a C_4 crop, growing in California under conditions of high radiation and daytime temperature, has a faster growth rate than the C_3 crops of winter wheat and spring barley in England. The smallest growth rate is with soybean in Iowa, a nitrogen-fixing C_3 pulse; the pods and, more notably, the seeds being rich in energy (data from Shibles and Weber, 1965; Williams *et al.*, 1965a; Gallagher and Biscoe, 1978).

(Fig. 10.8). Because C_4 plants are more efficient in radiation use, they produce a greater dry matter yield over the season than C_3 plants.

For a long time the RUE (Table 10.1, Fig. 10.7) was considered as a quantity characteristic of a plant species and fairly constant throughout the season (Monteith, 1977). But it has now been found that there are variations among cultivars of a species, for example those of sorghum. The RUE of genotypes varied between 2.1 g dry matter MJ^{-1} PAR and 3.8 g dry matter MJ^{-1} PAR under the growing conditions for a given year in Kansas. Variation in RUE was more closely associated with the amount of biomass produced than with the cumulative total of PAR intercepted (Narayanan *et al.*, 2013). RUE may also vary within the season because of changes in growing conditions. Water shortage can induce a reduction in the rate of net photosynthesis, as stomates will close and a larger part of the assimilate produced may be transported into the root. Rising temperature causes the respiration rate to increase to a larger extent than the rate of gross photosynthesis, after a particular temperature has been exceeded (Fig. 1.4). Thus the rate of net photosynthesis declines. The rate also declines with age, as the physiological activity falls off. All these factors can alter RUE.

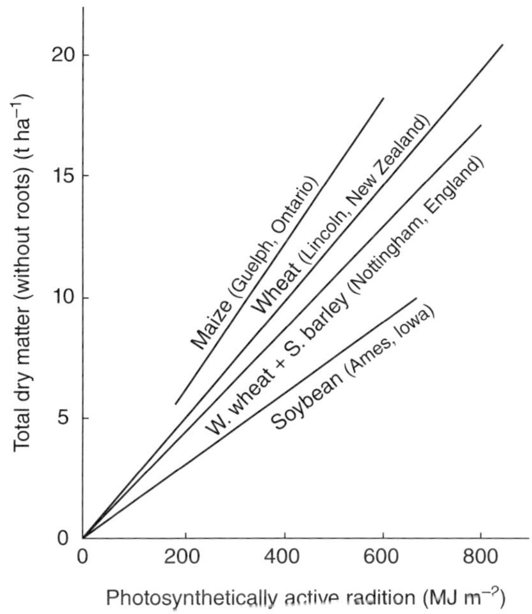

Fig. 10.7. Accumulated above-ground dry matter of maize in Ontario, spring wheat in New Zealand, winter wheat and spring barley in England and soybean in Iowa as a function of intercepted photosynthetically active radiation, accumulated during the season (data from Shibles and Weber, 1965; Gallagher and Biscoe, 1978; Wilson and Jamieson, 1985; Tollenaar and Aguilera, 1992).

Table 10.1. Radiation use efficiency (RUE, g dry matter MJ^{-1} intercepted photosynthetically active radiation) of various crops (Tollenaar and Aguilera, 1992; oilseed rape from Morrison and Stewart, 1995).

Crop	Type of CO_2 fixation	RUE
Maize	C_4	3.3
Sorghum	C_4	3.2
Wheat	C_3	3.0
Barley	C_3	3.0
Rice	C_3	2.9
Peanut	C_3	2.8
Oilseed rape	C_3	2.8

RUE may fall when the saturation deficit of the air increases (Fig. 10.9). One explanation is that some plants react directly to the saturation deficit by stomatal closure at high deficit. On the other hand, plants may react indirectly to high deficit by an imbalance in the plant's water budget, as described earlier. Closing of stomates reduces net assimilation as well as transpiration without necessarily changing their ratio in any fundamental way. Stomatal control does

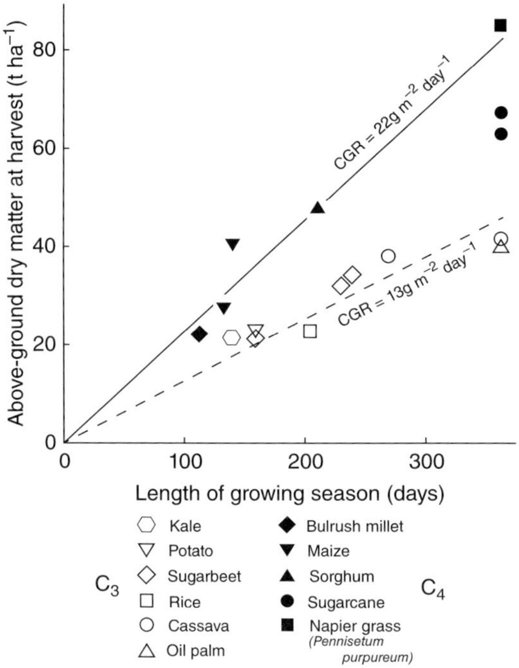

Fig. 10.8. Above-ground dry matter produced at harvest as related to the length of the season for various C_3 and C_4 plants. The CGR is the mean crop growth rate during the vegetation period (after Monteith, 1978).

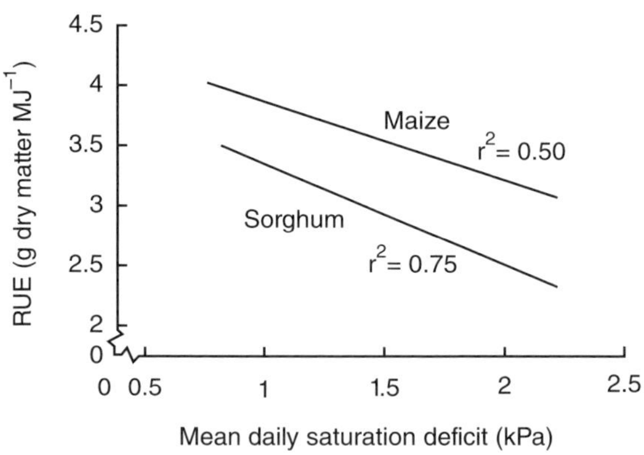

Fig. 10.9. Radiation use efficiency (RUE) of field grown C_4 crops (maize and sorghum) as a function of the mean daily saturation deficit of the air. The deficit was averaged for the time period before anthesis (after Stockle and Kiniry, 1990).

not affect intercepted radiation, leaving aside the possible effect of drought on leaf folding, leaf rolling and leaf abortion. Consequently, when the radiation load hitting the canopy is not reduced under an inadequate water supply, the RUE will be smaller.

11 Water Use and Dry Matter Production

11.1 Relations and their Optimization

As early as 1850, Sir John Lawes of Rothamsted indicated a 'definite relationship' between the water moving through a plant and the amount of substances fixed from the air (Lawes, 1850). There are two fundamental reasons why dry matter production and water use by crop stands are closely related. First, both the processes of CO_2 assimilation and of H_2O transpiration (Eqn 8.6) are strongly dependent on radiant energy. For CO_2 assimilation the relationship holds true until light saturation is approached (Fig. 10.1). But this limitation is more valid for single leaf surfaces than for crop stands with light penetration into deeper leaf storeys. Secondly, during water shortage both processes are reduced by stomatal control.

By the beginning of the 20th century, botanists from various countries, most noticeably agricultural scientists, started to look in more detail at the relationships between water use and dry matter production. Root dry matter was largely ignored at this time. It turned out that experiments with plants growing in deep pots or containers were much easier to control than experiments in the field. In the field some components of the water balance equation (Eqn 3.1) could not be determined at all, or could only be estimated very roughly. Therefore calculation of *evapotranspiration* (ET) proved quite unreliable. The separation of *transpiration* (T) and *evaporation* (E) was simply impossible. In containers, however, variables like seepage and surface runoff (about which there was great uncertainty) either did not occur or they could be measured easily. By properly covering the soil, evaporation could be completely prevented, so transpiration could be measured with greater certainty.

In the USA Briggs and Shantz (1913a,b, 1914) and some other agronomists conducted a series of classical experiments between 1910 and 1920 to explore the dependence of biomass accumulation by food crops and industrial plants on water use. Results are depicted in Fig. 11.1 for durum wheat (*Triticum durum*),

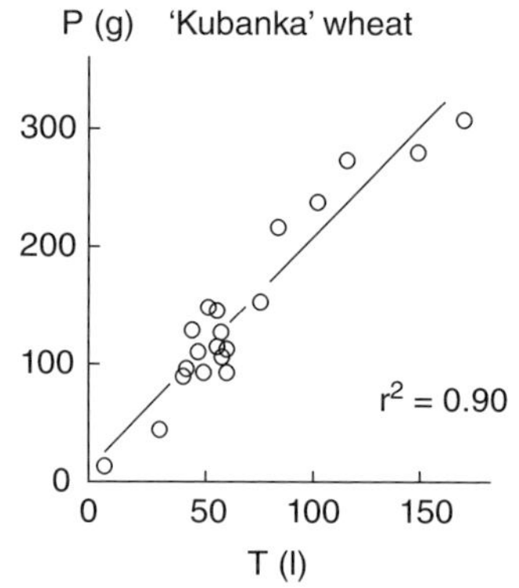

Fig. 11.1. The relationship between cumulative transpiration (T) and accumulated dry matter (P) in above-ground parts of 'Kubanka' wheat (*Triticum durum*), which was grown in galvanized iron containers, 40 cm in diameter over the years 1911–1922. The containers were filled with 115 kg of soil. The sites in the USA were: Akron, Colorado; Mandon, North Dakota; Newell, South Dakota; and Delhart, Texas (data from Briggs and Shantz, cited by de Wit, 1958).

cultivar 'Kubanka'. These results were taken from the comprehensive literature study on 'transpiration and crop yields' by de Wit (1958).

These early findings, shown in Fig. 11.1, demonstrate the close relationship between cumulative transpiration and total above-ground yield. In the experiments with 'Kubanka' wheat, water use and biomass produced were severely altered by the amount of water supplied, which ranged from being optimum over the season to only providing limited support. The resulting regression coefficient was 2.07 g dry matter (DM) l^{-1} of water or 2.07 kg DM m^{-3} H_2O. This is the *transpiration efficiency* (TE) (Tanner and Sinclair, 1983) of 'Kubanka' wheat. The reciprocal is the *transpiration ratio* (TR, 483 l kg^{-1} DM). It seems that TE (and consequently TR) is, to a large extent, independent of the level of water supply and water use. Notice that the start of the regression line coincides almost exactly with the origin of the coordinates, indicating that in these container experiments soil evaporation had been successfully eliminated.

That is not the case with field experiments where, in addition to plant transpiration, soil evaporation is inevitably included in measured water use. Therefore the relationship between water use and dry matter is shifted towards a greater water use, unless the evaporation is determined or estimated separately. If such an estimate of evaporation is available, the evaporation can be subtracted from total water use (ET) (Hanks *et al.*, 1968). Figure 11.2 presents more recent data on water use and biomass from north-east Germany. Forage maize was grown in *lysimeters* with varying supplies of water, allowing determination of the total ET (Mundel, 1992), but not of E. The intercept of the biomass–ET regression line with the abscissa is 194 mm (Fig. 11.2). The intercept may be considered as a first and approximate estimate of cumulative soil evaporation in the presence of plants (Hanks, 1974; Hanks and Rasmussen, 1982; Walker and Richards, 1985). According to Ritchie (1983), the ET-intercept indicates the soil evaporation

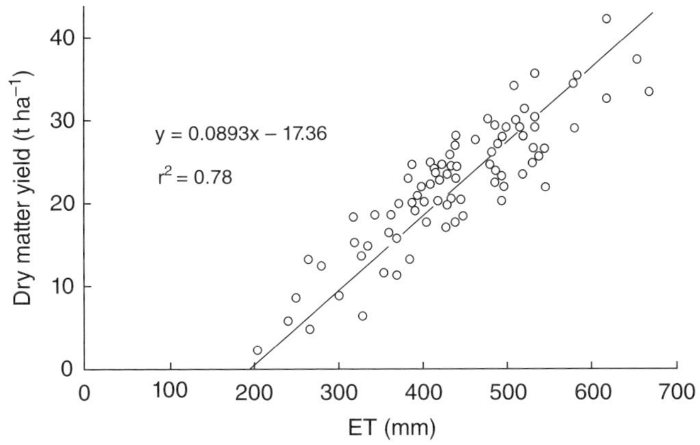

Fig. 11.2. Relationship between cumulative evapotranspiration (ET) and dry matter yield of forage maize. The crop was grown in Paulinenaue, north-east Germany, on lysimeters with a groundwater table 125 cm below the surface (based on Mundel, 1992).

during growth early in the season, when the leaf area index (LAI) is less than about 1. Subtracting this rough estimate of E from ET one obtains an estimate of TE, represented by the slope of the regression line. The estimated TE of forage maize is 0.0893 t ha^{-1} per mm. That corresponds to 0.00893 kg m^{-2} per mm or 0.00893 kg m^{-2} per l m^{-2} or 8.93 kg DM m^{-3} H_2O. This figure for maize TE is four times the value for 'Kubanka' wheat already mentioned. The TR of forage maize is 112 l H_2O kg^{-1}. It is obvious from the regression line in Fig. 11.2 that when evaporation is not taken into account, the efficiency of water for biomass accumulation will be smaller the less dry matter is produced.

This simple example emphasizes that TE or TR can only be determined when the data allow soil evaporation to be estimated. In the case where the value of E as part of ET is unknown, the only option that remains is to relate biomass accumulation to cumulative ET. In doing so we obtain the *evapotranspiration efficiency* (ETE; Tanner and Sinclair, 1983), or *water use efficiency* (WUE), a term widely used. The reciprocal could be called the evapotranspiration ratio (ETR).

From Fig. 11.1 it can be calculated that in 'Kubanka' wheat the individual values of TE vary between 1.5 and 3.0 kg DM m^{-3} H_2O and of TR between 330 and 670 l H_2O kg^{-1} DM, respectively. Briggs and Shantz had already noted in 1917 that the TR of lucerne grown in containers was influenced by the *evaporative demand* of the atmosphere, which they estimated by use of a pan evaporimeter, a device similar to that shown in Fig. 5.6. Starting in North Dakota the TR increased all the way south to Texas (Table 11.1). Similarly, pan evaporation as an estimate of E_p increased, leaving the ratio TR/E_p nearly constant (Table 11.1).

In his extensive review, de Wit (1958) showed that biomass dry matter (P), transpiration (T) and potential evapotranspiration (ET_p) approximated by open pan evaporation, are related by:

$$P = m \cdot T / ET_p^n \tag{11.1}$$

In the equation m is a crop-specific factor and n is a site-specific factor. Factor n is 1 for semi-arid and arid, high radiation regions. For humid, more temperate

Table 11.1. Transpiration ratio (TR) of 'Grimm' lucerne at four experimental stations on the Great Plains, USA, in the summer of 1912. The TR varies with the potential evaporation (E_p) (original data of Briggs and Shantz, 1917; cited by Tanner and Sinclair, 1983).

Location	Period	TR (l kg^{-1})	E_p^a (mm day^{-1})	TR/E_p (m^2 day kg^{-1})	m^b (kg m^{-2} day^{-1})
Williston, North Dakota	29 July–24 September	518	4.04	128	0.0078
Newell, South Dakota	9 August–6 September	630	4.75	133	0.0075
Akron, Colorado	26 July–6 September	853	5.74	149	0.0067
Delhart, Texas	26 July–31 August	1005	7.77	129	0.0077

[a]E_p, Potential evaporation measured with a pan evaporimeter.
[b]m, Crop-specific factor from Eqn 11.1 (de Wit, 1958).

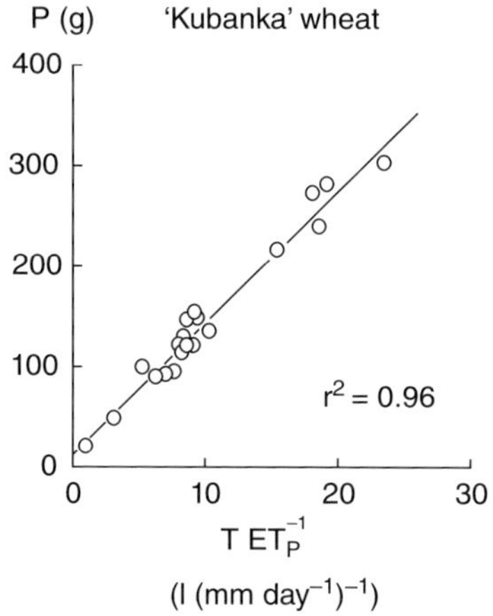

Fig. 11.3. Biomass production (P) of 'Kubanka' wheat, grown in the Great Plains, USA, as a function of cumulative transpiration (T), which is normalized by the potential evapotranspiration (ET_p). The ET_p was determined by evaporation pans. Experimental conditions are given in Fig. 11.1 (after de Wit, 1958).

climates n is zero (de Wit, 1958). For 'Kubanka' wheat, grown in the semi-arid environment of the Great Plains, the relation is depicted in Fig. 11.3. Compared with Fig. 11.1 the coefficient of determination (r^2) has increased by normalizing T using ET_p. The factor m of 'Kubanka' wheat is 13.3 g m^{-2} day^{-1} or 0.0133 kg m^{-2} day^{-1}, which is almost twice that obtained for m of lucerne (Table 11.1). The factor m indicates that under standardized ET_p, the cereal 'Kubanka' wheat has a greater TE than the forage legume lucerne, both crops being grown under the arid conditions of the Great Plains with a high percentage of bright sunshine.

At the leaf surface, stomates control the gas exchange of CO_2 and H_2O. The exchange of the two gases through the stomatal openings is a diffusion process.

Unlike CO_2 exchange, Bierhuizen and Slatyer (1965) showed that H_2O exchange is governed by the *saturation deficit* (SD) of the air ($e_s - e$, in short Δe, see Eqn 5.7). As SD varies during the course of a day as well as from day to day, Bierhuizen and Slatyer (1965) (see Box 11.1) incorporated a mean Δe into their equation, which relates – like de Wit's equation (Eqn 11.1) – transpiration (T) to biomass production (P):

$$P = k \cdot T / \Delta e \qquad (11.2)$$

Box 11.1. The saturation deficit of the air determines transpiration efficiency

Transpiration and CO_2 assimilation are two processes that depend on incoming radiant energy. In 1958, de Wit concluded that the intensity of radiation in a region influences the relation between dry matter (P) and transpiration (T) or the transpiration efficiency (TE, Eqn 11.1). The dependence on climate is made clear in the equation by the changing significance of potential evapotranspiration (ET_p) on TE. Only 7 years later, in 1965, Bierhuizen and Slatyer showed theoretically and practically that TE was linked unambiguously to another meteorological quantity, the saturation deficit (SD) of the air. It was taken for granted that this approach was valid for different climates, making a regional separation according to radiation input and latitude unnecessary.

The arguments are as follows: at the leaf surface the CO_2 and H_2O gas exchange takes place by a diffusion process. Water vapour molecules diffuse from the evaporating surfaces of the mesophyll cells through the intercellular spaces and then through the stomates to the outside atmosphere. At the same time, CO_2 molecules diffuse from the outside air to the intercellular spaces inside of the leaf and then into the mesophyll cells and the chloroplasts. The diffusion of the two gases is driven by the differences in concentration between the inside of the leaf and the outside air.

For the transpiration rate of the leaf (T_L) we can write:

$$T_L = \frac{c_i - c_o}{r_o + r_i} \qquad (1)$$

For the rate of net photosynthesis of the leaf (P_L) the equation reads:

$$P_L = \frac{c'_o - c'_i}{r'_o + r'_i + r'_m} \qquad (2)$$

In the first equation c_i is the concentration of water vapour inside the leaf and c_o the value outside the leaf in the ambient air. The corresponding values of the second equation are the concentrations of CO_2 outside (c'_o) and inside (c'_i) the leaf. It can be noted here that instead of writing the equations in terms of concentration, it is more correct to use the partial pressure, which depends on temperature. When, however, the temperatures of the leaf and the air are the same, it is unimportant which parameter is chosen. The term r stands for the diffusion resistance for water vapour, which is inversely related to the diffusion coefficient (D) and directly related to the diffusion distance (Δz), i.e. $r = \Delta z/D$. The term r′ is the diffusion resistance for CO_2. The resistances at the boundary layer of the leaf surface (r_o, r'_o) and those for stomates and intercellular spaces in the leaf (r_I, r'_I) are connected in series. They are applicable to gas diffusion through a gaseous medium. In the case of CO_2, but not of H_2O, a third resistance in series has to be considered. This is a resistance encountered in the cell wall and the liquid phase of the cytoplasm when CO_2 diffuses to the sites of assimilation in the chloroplasts of the mesophyll (r'_m).

Continued

Box 11.1. Continued

We should add a word on the units in Eqns 1 and 2. When expressing the rate of gas transport in units of g cm^{-2} leaf area s^{-1} and the gas concentration in the unit of g cm^{-3}, then r has the unit of s cm^{-1}. Another consideration leads to the same unit of r: the reciprocal of the diffusion resistance r is r^{-1}, the diffusion conductance, which corresponds to D/Δz. D has the unit of cm^2 s^{-1} and Δz is in cm, ending in cm s^{-1} for r^{-1}.

The following reasoning assumes that to a first approximation the leaf temperature corresponds closely to the temperature of the outside ambient air. It also supposes that the intercellular spaces are completely saturated with vapour. Under these circumstances the difference in vapour concentration $c_i - c_o$ of Eqn 1 can be replaced by the vapour pressure deficit of the air. The SD of the ambient air, $e_s - e$, is taken as an estimate of the driving force of transpiration, which is a diffusion process. Moreover, it has to be taken into account that the molecular weight of CO_2 (44 g mol^{-1}) is larger than the one for water vapour (18 g mol^{-1}). Therefore the D of CO_2 will be smaller and r'_o and r'_1 will be bigger than the corresponding values for H_2O vapour. But the ratio of the resistances of the two gases in diffusion through air inside and outside the leaf will be constant. Also r'_m is accepted as being more or less constant. Under these conditions Eqns 1 and 2 can be combined:

$$T_L / P_L = \frac{e_s - e}{\Delta CO_2} \times k^* \quad (3)$$

In Eqn 3, ΔCO_2 is the concentration difference $c'_o - c'_i$, and k* is a constant factor. The equation indicates that T_L/P_L, which is the transpiration ratio of the leaf, $(TR)_L$, increases directly with increasing SD. When the CO_2 concentration difference increases, however, $(TR)_L$ will decrease.

For individual plant species c'_i is almost constant. It is roughly 210 ppm for C_3 plants, but only about 120 ppm for C_4 plants. This difference in c′ explains the small $(TR)_L$ of C_4 plants compared with C_3 plants. The equation also explains that plants, which close their stomates at noon when $e_s - e$ is high, are able to avoid CO_2 fixation under unfavourable exchange conditions (i.e. when water loss is large per unit of CO_2 intake).

Equation 3 can be adapted to the field situation with crop stands by supposing that ΔCO_2 of a crop species does not vary too much during the course of a day. We can also assume that the LAI influences the crop growth rate (CGR) (Fig. 10.5) and the transpiration rate (Fig. 9.3) in very much the same way. Therefore we can accept what Bierhuizen and Slatyer (1965) proposed, writing Eqn 4:

$$P = k \times \frac{T}{\Delta e} \quad (4)$$

Here P is the biomass dry matter and T the transpiration, produced and consumed within a certain period of time. In this case, neither P nor T is related to leaf area but to the soil surface area. The factor k was assumed to be a crop-specific constant, which may be explored under different climatic situations. The equation indicates that the transpiration efficiency (TE), which is P/T or the reciprocal of TE (the transpiration ratio (TR)) is very dependent on SD.

Expressing the crop stand P in Mg ha^{-1}, T in mm and SD in Pa, the factor k attains the unit of Mg ha^{-1} mm^{-1} Pa. Since 1 mm of evaporated water corresponds to 10^7 g ha^{-1}, we may express k simply in the unit of Pa. Using this simplification, the numerical value of k in Pa is one-tenth of k in Mg ha^{-1} mm^{-1} Pa. Equation 11.2 received considerable recognition in the scientific community. In contrast to de

Wit's equation (Eqn 11.1), which changes from dry to humid climates, it was expected that Eqn 11.2 would be valid for various climates. Equation 11.2 indicates that increasing the SD of the air reduces the transpiration efficiency (TE). A very similar effect of SD on radiation use efficiency (RUE) was shown in Fig. 10.9.

11.2 The Transpiration Ratio and a Related Standard

From the earlier part of this chapter it should have become clear that the estimation of the *transpiration ratio* (TR) or of its reciprocal, the *transpiration efficiency* (TE), of a crop species includes a number of uncertainties. The lack of precision stems from: (i) difficulties in determining the components of the water balance exactly; (ii) the inability to eliminate soil evaporation; and (iii) because weather and climate have a profound effect on the relation between biomass and transpiration.

In the USA, Briggs and Shantz (1913a,b, 1914) published the TR of various field crops. Some of their original TR results, established in the Great Plains, are listed in Table 11.2. These are compared with more recent results from England and Germany, where in most cases it is not TR, but the evapotranspiration ratio (ETR) that has been determined. It can be seen from the table that the US values are

Table 11.2. Transpiration ratio (TR, l H_2O kg^{-1} above-ground biomass) of various C_4 and C_3 crops, estimated in the USA by Briggs and Shantz (1914) between 1911 and 1913 (from Hanks and Ashcroft, 1980). Their data are compared with older and more recent data, collected in the USA, Germany and England. The symbol ø signifies the mean of TR obtained in years as listed.

Crop	TR[a]	Year of comparative investigation	Country	Author	TR
Millet (C_4)	296				
Sorghum (C_4)	308	1966/67	USA	Hanks *et al.* (1968)	333
Maize (C_4)	351	1978–81	Germany	Roth *et al.* (1988)	214[b]
Wheat (C_3)	488	1971	Germany	Ehlers (1976)	388[b]
		1983	Germany	Roth *et al.* (1988)	301[b]
		1975–92 (6 years)	Germany	Gall *et al.* (1994)	272–461 (ø 359)[b]
		1978–82 (5 years)	England	Goss *et al.* (1984b)	155–246 (ø 194)[b]
Barley (C_3)	529	1983	Germany	Roth *et al.* (1988)	224[b]
		1992	Germany	Gall *et al.* (1994)	312[b]
Oat (C_3)	562	1903–04	Germany	von Seelhorst and Fresenius (1904); von Seelhorst and Müther (1905)	266–268
		1976–83 (4 years)	Germany	Ehlers (1989)	192–294 (ø 246)
Sugarbeet (C_3)	394	1971	Germany	Ehlers (1976)	210[b]
Potato (C_3)	624	1974–92 (6 years)	Germany	Gall *et al.* (1994)	155–272 (ø 199)[b]
Lucerne (C_3)	832				

[a]Data of Briggs and Shantz (1914), collected in 1911–1913.
[b]Evapotranspiration ratio (ETR).

two or three times greater than those from Europe. This at first seems surprising, because evaporation is included in most of the data from the two European countries and that should have increased the ratio relative to the TR data from the USA, where only transpiration was considered. Nevertheless, the European results are not larger, but smaller than those from the USA.

The European investigations on evapotranspiration (or transpiration) and biomass relations were undertaken within the last quarter of the 20th century (Table 11.2). What is important is the fact that the impact of *climate* on TR has not always been recognized. For instance, the North American results were accepted as being valid for the humid climate of Germany. The applicability of large TR values was not called into question for many decades (Ehlers, 1997), even though they were collected in the semi-arid steppe climate of the Great Plains. In the steppe climate the *saturation deficit of the air* (Eqn 11.2) is much greater than that in central Europe, giving rise to the observed increase of TR (Table 11.2).

Even within Europe the example with wheat (Table 11.2) raises the suspicion that the TR is smaller in the humid maritime climate of England than in the less humid transitional climate of Germany. The TR data of wheat gave about 190 l kg^{-1} in England, 350 l kg^{-1} in Germany, compared with 490 l kg^{-1} on the Great Plains.

In the semi-arid climate of the Great Plains with warmer temperatures during the cropping season, the TR of C_4 plants is clearly smaller than that of C_3 plants. In Germany, with a more temperate summer season, the difference in TR between plant groups, which follow a different carboxylation pathway, is not quite so obvious (Table 11.2) and may not even exist.

The table also shows that the TR data of crops like wheat, oat and potato are quite variable from year to year. On the other hand, from the values given for sorghum and oat, it is obvious that there is no fundamental and dramatic change in the TR of a crop from the beginning to near the end of the 20th century. Such an improvement of the water use–biomass relation might have been expected from efforts in plant breeding.

For a long time it was generally accepted that in contrast to TR or TE, the factor k of Eqn 11.2 (Eqn 4 in Box 11.1) was basically independent of the climatic conditions during crop growth. The k factor has been determined in field experiments, mainly conducted over the years from 1970 to 1994 (Table 11.3), where assessment of the components of the water balance proved to be more reliable than in the early years of TR measurements. Soil evaporation was either estimated or measured, so that the k values were based on water use as transpiration. Moreover, for the evaluation of k the saturation deficit of the air (SD) has to be measured and assessed properly. As the H_2O–CO_2 gas exchange through the stomates is not taking place during the night (except for plants using crassulacean acid metabolism (CAM)) the *daytime saturation deficit* has to be calculated, but not the deficit of the total day (Tanner, 1981). Generally, the deficit is larger for the daylight period than for the 24-h day (Fig. 9.14). Therefore the k factor during daylight is greater by a factor ranging from 1.43 to 1.59 than that for the whole day (Ehlers, 1989; Haefele *et al.*, 2009).

With the aid of measured k factors, the various crops can be arranged according to their transpiration efficiency (TE), estimated in different climates (Table 11.3). The TE is inversely related to TR. The greater the k value of a crop,

Table 11.3. Factor k of Eqn 11.2, a measure of transpiration efficiency (TE) normalized by the saturation deficit, for various C_4 and C_3 crops (in part from Tanner and Sinclair, 1983). The saturation deficit Δe relates to the daytime period with radiation.

Crop	Type of CO_2 fixation	Year	Location	Δe (Pa)	k (Pa)	
Sorghum	C_4	1970	Kansas	2380	13.8	
Maize	C_4	1974/75	Logan, Utah	2040	8.4	
		1974/75	Ft Collins, Colorado	2090	10.2	
		1974/75	Yuma, Arizona	4420	10.1	
		1974/75	Davis, California	2010	9.9	9.1
		1912	Nebraska*	1900	8.9	
		1912	Nebraska**	3130	9.1	
		1981/82	Elora, Ontario[a]	1000	7.4	
Miscanthus × *giganteus*	C_4	1994	Chelmsford, England[b]	~800	9.5	
Wheat	C_3	1981	Manhattan, Kansas[c]	1806	6.2	
		1981/82/83	Lincoln, New Zealand[d]	690	3.1	
Barley	C_3	Beginning of 1980s	Lincoln, New Zealand[d]	640	4.0	
Oat	C_3	1976	Göttingen[e]	1165	4.2	
		1977	Göttingen[e]	627	3.5	3.5
		1982	Göttingen[e]	920	2.9	
		1983	Göttingen[e]	770	3.5	
Potato	C_3	1972/73/76	Wisconsin[f]	1460	6.5	
		Beginning of 1980s	Lincoln, New Zealand[d]	760	5.9	
Lucerne	C_3	1977	Wisconsin	1040	4.3	
Soybean	C_3	1970	Kansas	2380	4.0	
Pea	C_3	1980–84	Lincoln, New Zealand[d]	640	3.8	
Faba bean	C_3	1983	Göttingen[g]	800	3.1	

[a]Walker (1986); [b]Beale *et al.* (1999); [c]Asrar *et al.* (1984); [d]Wilson (1985); [e]Ehlers (1989); [f]Tanner (1981); [g]Ehlers, unpublished results.
* In humid glasshouse; **in dry glasshouse.

the larger is TE at a given saturation deficit (SD). A large SD causes a small TE, whereas the k factor itself was assumed to be independent of SD. The C_4 plants sorghum (13.8 Pa), *Miscanthus* × *giganteus* (9.5 Pa) and maize (7.4–10.2 Pa) head the group of plants that are very efficient at water use (Table 11.3). Potato follows (5.9–6.5 Pa) as a C_3 crop. The remainder are small-grained cereals (2.9–6.2 Pa), lucerne (4.3 Pa) and grain legumes (3.1–4.0 Pa).

Tanner and Sinclair (1983) re-evaluated the results of an earlier experiment in Nebraska with maize (Table 11.3) carried out by Montgomery and Kiesselbach (1912). In this experiment, maize was grown in two glasshouses. In one glasshouse the air was kept humid and in the other it was kept dry during the main growing period. The TR of maize was smaller (214 l H_2O kg^{-1} biomass) in the humid glasshouse than in the dry one (340 l kg^{-1}), but the two k values, calculated by Tanner and Sinclair (1983), were about 9 Pa (Table 11.3). None the less, one might speculate that k is not completely independent of climatic influence. Maize grown in Ontario, Canada, where the saturation deficit is small (Δe = 1000 Pa) compared with locations in the USA (2000–4000 Pa), gave the smallest k value. Also the wheat in New Zealand (Δe = 690 Pa) has a smaller k than the wheat grown in Kansas (Δe = 1806 Pa). In the Göttingen experiments with oat, the largest k was determined in the dry year of 1976. And the Wisconsin potato (Δe = 1460 Pa) had a larger k than the New Zealand potato (Δe = 760 Pa).

In summary, it seems that the factor k increases a little with increasing Δe, whereas radiation use efficiency (RUE, Fig. 10.9) and transpiration efficiency (TE, Eqn 11.2) become smaller when the air is dry. We do not know the reason for the increase in k. First, we can state that increasing Δe should reduce TE in such a way that k remains unaffected. But instead of being constant, k increases with Δe. We can speculate that it is caused by a slight stomatal closure as a consequence of the greater Δe. Such a response of transpiration reduction to increasing Δe has been verified, for example with a number of sorghum genotypes (Gholipoor *et al.*, 2010). The partial closure of the stomatal openings will affect transpiration more than assimilation. This is because the main resistance in the CO_2 diffusion path, the mesophyll resistance (Box 11.1, Eqn 2), is in the liquid phase, which is hardly affected by stomata closing. There is no such path resistance in the liquid phase for H_2O vapour. Here the total resistance in diffusion is made up solely by resistances in the gas phase (Box 11.1, Eqn 1). Therefore, as a result of stomatal closure, total resistance of water vapour diffusion is increased much more (and hence the diffusion rate is decreased) than is the case for the diffusion of CO_2. While this hypothesis of partial closure of stomates may explain the reduction in transpiration, another argument can explain an increase of CO_2 inflow. It is possible that a slight closure leads to a decrease of the internal CO_2 concentration c'_i and thus to a steeper CO_2 gradient, by which the CO_2 diffusion into the leaf could be promoted (Haefele *et al.*, 2009). Both these explanations, the lower transpiration and the higher assimilation, may explain a gentle rise in k due to stomatal control.

Often a large saturation deficit in a region goes along with a hot air temperature. The elevated temperature can increase the photorespiration of C_3 plants, counteracting the possible effect of stomatal closure on the factor k. As C_4 plants are free of photorespiration, the influence of temperature on k may be less pronounced in C_4 than in C_3 plants. Generally, the temperature range for the maximum assimilation rate is cooler in C_3 than in C_4 plants.

After all the discussion of k we arrive at the question: To what extent will TE be affected by change in Δe?

A more recent and new compilation of data sets (1973–2004) on TE of wheat and maize will answer that question and supports the thesis that the factor k really

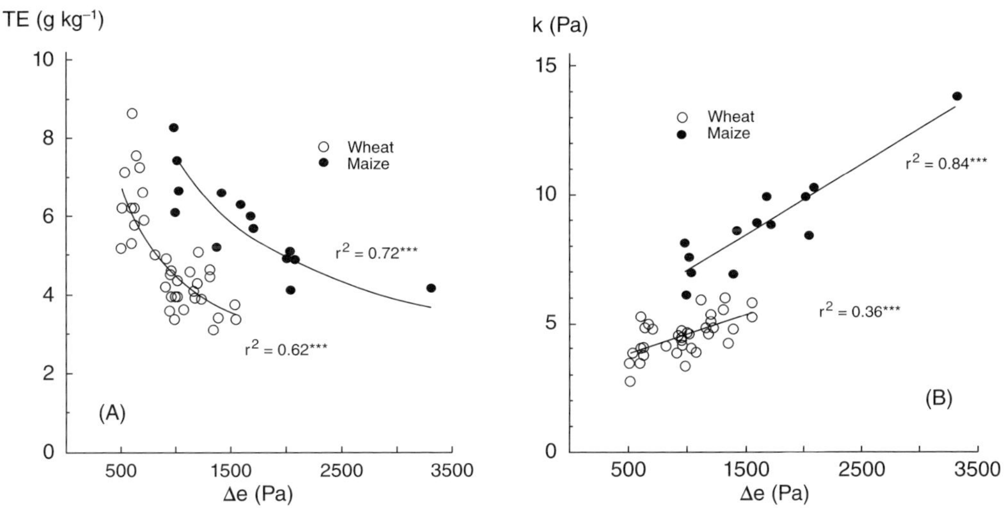

Fig. 11.4. (A) Transpiration efficiency (TE) declines with increasing daytime saturation deficit of the air (Δe) in maize and wheat. (B) On the other hand, the factor k for each of the two crops increases with rising Δe. (Data from Kremer *et al.*, 2008.)

depends on daytime SD or Δe (Kremer *et al.*, 2008). TE declines with increasing Δe in both crops (Fig. 11.4A) as stated earlier with reference to Eqn 11.2. Maize, a C_4 crop, attains the TE values of wheat, a C_3 crop, at a much larger Δe. For a given Δe, the TE of the C_4 crop is about 1.7 times greater than the TE of the C_3 crop. In contrast to TE, the factor k increases with Δe, contrary to what had been expected previously (Fig. 11.4B). Certainly, for both these crops k is not a constant. Nevertheless, despite factor k varying with Δe, distinct differences in the transpiration efficiency between the crops remain.

11.3 Water Use and an Estimate of Dry Matter Production

Using Eqn 11.2 together with the variable crop-specific factor k, taken from Fig. 11.4B, the transpiration efficiency (TE) and the transpiration ratio (TR) can be calculated. Factor k depends on the average daytime Δe of the air for the environment in question. This Δe is considered to be valid throughout the principal growing period. The approximate crop-specific values of k (Fig. 11.4B) are given in Table 11.4 for two locations: (i) Göttingen in the temperate, humid climate of Germany; and (ii) Akron in the arid steppe climate of the Great Plains in Colorado, USA. The influence of the climate is documented quite well by the calculated values of TE and TR. The TE is much larger and the TR much smaller in a temperate climate compared with a continental climate, where the summer months are hot and dry.

Table 11.4 also gives the calculated amount of water used in transpiration that is necessary for achieving a proposed yield goal. The yield is the above-

Table 11.4. Factor k for two crops determined from Fig. 11.4B using the average daytime saturation deficit of the air (Δe) during the main growing period at two locations, Göttingen in Lower Saxony, Germany, and Akron in Colorado, USA. Using Eqn 11.2, the transpiration efficiency (TE) and the transpiration ratio (TR) are first calculated, followed by the transpirational water use for a given yield goal and the total above-ground biomass yield assuming 300 mm water extracted by roots.

Crop	Location	Δe (Pa)	k (Pa)	TE (kg m^{-3})	TR (l kg^{-1})	Yield goal (t ha^{-1})	Transpirational water use (mm)	Predicted yield from 300 mm (t ha^{-1})
Maize	Göttingen	900	6.7	7.4	135	20	269	22.3
	Akron	1900	9.5	5.0	200	20	400	15.0
Wheat	Göttingen	900	4.4	4.9	204	15	307	14.7
	Akron	1900	5.9	3.1	323	15	483	9.3

ground biomass. As the climate becomes more *continental*, the amount of water used in crop transpiration increases. It increases by about 50% between Göttingen and Akron. That is only the water used for transpiration. Additional water will be required for evaporation and possibly for drainage. Conversely, from a fixed quantity of water, stored in the soil profile and replenished by precipitation, only a comparatively small amount of dry matter can be attained in the dry climate with high saturation deficit of the air. Whereas in Göttingen nearly 15 t ha^{-1} of wheat biomass can be produced from 300 mm of water, only 9.3 t ha^{-1} could be obtained in Akron (Table 11.4). With maize the comparable numbers are 22.3 t ha^{-1} and 15 t ha^{-1}, respectively (Table 11.4).

These sample calculations demonstrate emphatically that it is the *climatic condition* that determines the amount of biomass produced for a fixed quantity of water. We have to understand that the supply situation for plants with regard to water as a growth factor is not defined solely by the amount of water available in soil, but also has to be considered from the climatic viewpoint, which is of similar importance. Of the climatic quantities it is the evaporative demand, expressed here as the saturation deficit of the air, that determines the relation between transpirational water use and biomass production.

It seems that the environmental conditions for plant production contain a certain irony. In dry and arid areas with warm temperatures and a high level of radiation for photosynthesis, the amount of soil water stored is normally small; nevertheless a comparatively large quantity of water is needed for the production of biomass. The scarceness of water limits the production level, but the dry atmosphere further reduces the efficiency of transpirational water use. In wet and humid areas, on the other hand, frequently having less incoming radiation, the demand on the water supply is less and the efficiency of water use in crop production is greater. As a result, the actual supply may be so abundant that the surplus can have a harmful effect on plant growth and development.

12 Influence of Nutrient Supply on Water Use and Establishment of Yield

12.1 Yield Dependency on Water and Nutrient Supply

We have already shown that yield is more or less closely correlated with water use. The actual relationship depends on what is considered to be the best indicator of water use: evapotranspiration (Fig. 11.2) or transpiration (Fig. 11.1). The form of the relationship can be modified if climatic variables such as potential evapotranspiration (Fig. 11.3) or saturation deficit of the air (Eqn 11.2) are also considered. It is often claimed that the water use–biomass function is changed by fertility and that the efficiency of water use can be increased by fertilizer application.

We shall discuss this topic of a fertility effect on the water use–yield relationship in some detail later on. The *combined effect* of the growth factors, *water* and *plant nutrients, on yield formation* are discussed first using the results of a *two-factorial field experiment* on irrigation and nitrogen (N) fertilizer, undertaken on the southern Great Plains, in Texas, USA (Eck, 1988). The aim of this experiment was to determine how the independent factors, water and N, affect the dependent variable, here the yield of winter wheat, when applied simultaneously. The first factor tested was the water supply, which was applied at two levels: (i) sufficient water supply maintained by irrigation; and (ii) water shortage throughout the vegetative period (no irrigation). The second factor tested was N fertilizer addition, which was present at four levels. The N rates (kg N ha^{-1}) used were: 0, 70, 140 and 210. They were applied in autumn before sowing the wheat. The results are shown in Fig. 12.1 in the form of a diagram containing three quadrants: A, B and C. Sector D illustrates a special case. The use of a three-quadrant diagram was originally proposed by de Wit (1953) and subsequently highlighted by van Keulen (1986a).

Quadrant A shows the relationship between *N application rate* and *total* above-ground *dry matter yield*; a function that is widely understood in crop production and here presented as the mirror image of the familiar layout. When water supply

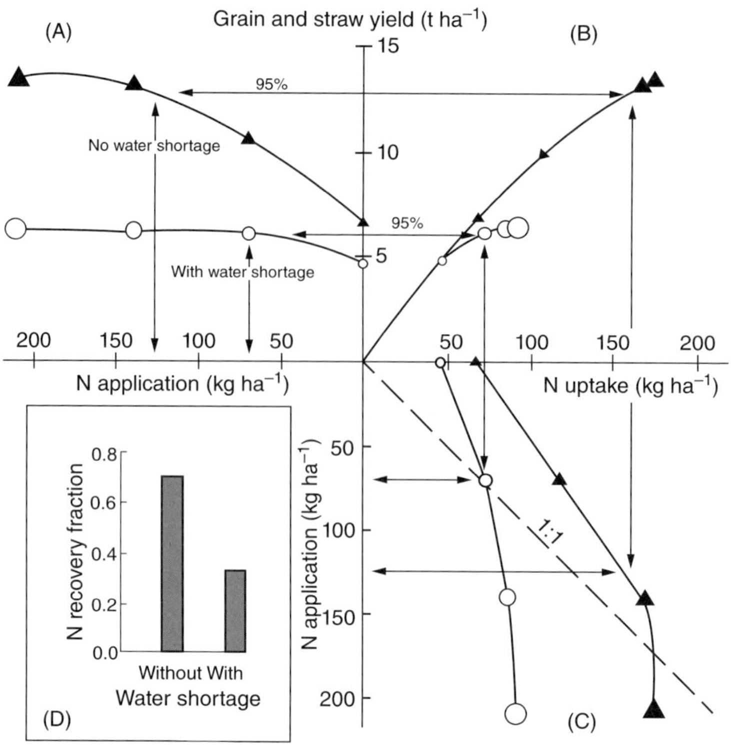

Fig. 12.1. Irrigation and nitrogen fertilizer experiment on the Great Plains, Texas. The results are for winter wheat and are presented in the form of three quadrants. (A) Yield of biomass (grain and straw) as influenced by N fertilizer application and water supply. (B) Relationship between biomass and N uptake at two levels of water supply. (C) Relationship between N fertilizer application and N uptake at two levels of water supply. A separate illustration is presented in (D): maximum N recovery fraction of fertilizer N at two levels of water supply. For more explanations see text (data from Eck, 1988).

is plentiful, yield is increased at all N levels and to a greater extent than when water is short. Even without any N addition (N_0), yield is greater with an ample water supply. At high soil moisture levels, larger amounts of N are available for use by the plants and are utilized in the production of a heavier yield compared with that when soil moisture is limiting. It is likely that *larger quantities of plant-available N* are formed by stimulated *mineralization of soil-borne N*, the large pool of N bound in organic matter. But in addition to mineralization, the *uptake* of mineralized N ($N_{min} = NO_3^- + NH_4^+$) by plants may have been enhanced at greater soil moisture values, supporting *nutrient flux towards roots* by mass flow and diffusion.

It is obvious from quadrant A that at higher levels of soil moisture, one unit of N fertilizer generates a greater yield increase than at limited water supply. Reduced soil moisture will probably not just decrease the nutrient flux from the soil reserves towards the roots (i.e. the *nutrient availability*). This term describes the 'source' of nutrients or – in other words – the capability of the soil to supply a nutrient. In addition, reduced soil moisture will also decrease the 'sink' of nutrients (i.e. the nutrient uptake by a plant – the plant's ability for *nutrient acquisition*). Deficiencies in water and nutrient supply limit the growth of shoot and root in such a way that specific plant properties, important for nutrient acquisition, remain insufficiently developed. For instance, these properties are represented by the density and depth of the root system, the number and length of root hairs and the physiological activity of roots. The activity may be related to the excretion of acids or chelating agents, by which 'insoluble' chemical compounds of the soil nutrient stock are made soluble and consequently available to plants (Section 6.1). But a restricted sink also pertains to poor emergence and shoot development, defective shoot and leaf formation as well as to limitations in the formation of reproductive plant organs.

Quadrant A of Fig. 12.1 represents a good example of the law of the optimum, which is a successor of the *law of the minimum*, the latter being formulated first by the German, C.P. Sprengel (1787–1859) in the year 1828, when he was a professor in Göttingen. That was already 27 years before the famous J. von Liebig (1803–1873) in 1855 reformulated the theory of the minimum growth factor that limits plant yield (Böhm, 1997; van der Ploeg *et al.*, 1999). The *law of the optimum*, however, was published first by E. Wollny (1846–1901) in 1887, when he was a professor of agricultural physics in Munich. G. Liebscher (1853–1896), another professor of agriculture in Göttingen, added to the law of the optimum in 1895 (Böhm, 1997). This theory describes the combining effect of growth factors on yield development (Claupein, 1993). The growth factors, here N and water, act on the plants in combination. Translated from Liebscher's words quoted by Claupein (1993): 'For greater production the plant can use a growth factor, which is available at a minimum level, to a greater extent when the other growth factors are present within the plant's optimum range'.

The mutual dependence of the two growth factors in affecting the yield is depicted in quadrant A by the course of the two diverging curves. The effect of N fertilizer application on yield depends on the water supply and the effect of water depends on the N rate. According to Wollny's view the growth factors are supporting each other, when after starting at a low level that is borderline their supply is increased until the individual optimum range has been achieved. When all the growth factors involved are at their optimum level, maximum yield will be attained (Claupein, 1993). The two N response curves in quadrant A exemplify yet another theory, the *law of diminishing returns*, formulated by E.A. Mitscherlich (1874–1956) from Königsberg. Each additional quantum of N fertilizer causes an additional increment in yield, but with diminishing returns. The returns are generally smaller at limiting than at adequate levels of water supply. This form of response suggests that not all of the growth factors contributing to yield, although not monitored in the experiment (radiation, CO_2, temperature, micro- and macronutrients), are in the optimum range. Furthermore, it is suggested that by optimizing the N supply the relative deficiency of all the other factors becomes more evident.

The shared effect of water and N supply on crop yield can formally be treated as an *interaction*, a concept from statistics: the effect of N on yield depends on water supply, and vice versa; the effect of water (irrigation water, soil moisture) on yield formation is shaped by the N supply (Fig. 12.1).

Quadrant B of Fig. 12.1 demonstrates that the *total yield* changes linearly with the *uptake* of *N* by the crop, when the total acquisition is small. The function starts from the origin of the coordinates and, at lower values of uptake (up to about 50 kg N ha^{-1}), is not influenced by water supply. Such a relationship seems unexpected at first glance. It is only at higher yield levels that the relationship deviates from linearity and shows the influence of water. In this upper range, N uptake or N acquisition increases at a greater rate than does the actual yield. Under these conditions of large N uptake, N is temporarily stored in the form of amino acids, amides and even nitrate, since the rate of protein synthesis does not keep pace with the rate of N uptake. This is the situation where N uptake is no longer paralleled by biomass growth and N becomes less and less the limiting factor for yield. When water supply is poor, *nutrients* become *concentrated* in plant tissue at a comparatively low yield level. When water supply is ample, N also becomes concentrated but this occurs at a greater level of yield. Nevertheless, despite the increase in nutrients, the lack of water produces a lower yield level than when there is an ample water supply.

The initial linearity may be explained by the more or less constant concentration of the nutrients in the plant material at harvest, that concentration being characteristic of a species and which by then is at a minimum level. The concentrations of the macronutrients N, phosphorus (P) and potassium (K) are 0.4%, 0.05% and 0.8% for small cereal straw and 1.0%, 0.11% and 0.3% for grain, the percentage being related to dry matter content (van Keulen, 1986b).

In quadrant C, *N uptake* is related to *N application rate*. In the smaller range of N fertilizer application rates, N uptake increases linearly with N applied. When water is ample, the linear portion ends at larger application rates of N compared with those when the water supply is limited. With sufficient water, N is taken up at greater rates at all levels of N application than with a limiting water supply. Even without any N dressing, the difference in N uptake due to soil moisture is evident. Water supports – if not the N supply from the soil resources – the utilization of N by plants. In the smaller range of N applications, more N is taken from the soil than is applied as fertilizer. The 1:1 line (Fig. 12.1C) depicts the situation when N application rate and N uptake rate are in balance. When water supply is limited, N uptake is less than the amount applied, once the application exceeds 75 kg N ha^{-1}. This critical application rate changes to about 170 kg N ha^{-1} when the water supply is sufficient.

The *recovery fraction* of fertilizer N is the increase in N uptake compared to the increase in fertilizer applied. Hence the recovery fraction is the slope of the function that relates N uptake and N application (quadrant C). The slope is at maximum for small rates of application. The maximum recovery fraction is inserted in section D of Fig. 12.1. It is obvious that the maximum fraction is significantly greater when water is available than when water is in short supply. Water supply not only enhances uptake from the N pool of the soil, but also uptake of fertilizer N.

Now let us suppose that a farmer wishes to contribute to the *efficiency of resource use* in agriculture (de Wit, 1991) by deliberately restricting the yield level slightly. With this kind of restraint he intends to reduce any possible *environmental pollution*. His contribution is the restriction of crop yields to 95% of the maximum obtained either with low or ample water supply (Fig. 12.1). If the water supply is sufficient, he can reduce the N rate to 125 kg ha^{-1}, but if water is in short supply, the N rate can be lowered still further to only 70 kg ha^{-1} (quadrant A). In both cases luxury consumption of N in terms of N accumulation within the plant biomass is largely avoided (quadrant B). In no case will N be applied at a greater rate than will be taken up by the crop (quadrant C). And finally the recovery fraction of applied N will be relatively large (section D).

12.2 Influence of Nutrient Supply on the Relationship between Water Use and Yield

When the growth factors are supplied to plants in non-limiting quantities, one can expect that both growth rates and final yield will be at a high level. How then does the optimum combination of yield-forming factors influence *the relationship between biomass production and water use*? It is often suggested that a non-limiting nutrient supply will reduce the water requirement and use necessary to obtain a given biomass yield. Conversely, it is expected that nutrient deficiency will increase the amount of water required to obtain a given yield level.

To check these preconceptions, we need to look at empirical results on *water use*. In the experiment with irrigation and N fertilization in USA (Fig. 12.1), the water use of the wheat crop was evaluated for the N_0 and N_{210} nitrogen application rates. The assessment was based on the soil water content down to 180 cm depth at the beginning and the end of the investigation period and on the amount of precipitation and irrigation water supplied during the season (Eck, 1988). Consequently, in addition to transpiration the calculated water use included two other possible losses – soil evaporation and deep seepage – and one gain – the capillary rise into the root zone. Runoff was not considered so presumably it did not occur. The results are depicted in Fig. 12.2A for total (above-ground) biomass and in Fig. 12.2B for grain yield.

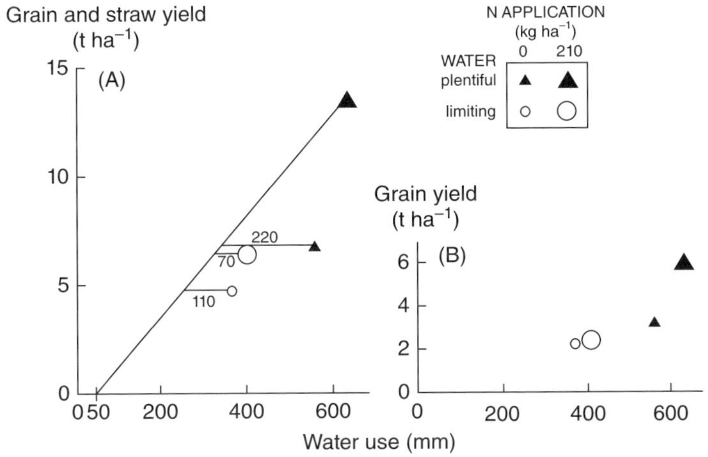

Fig. 12.2. The relationship between water use and the above-ground biomass yield (A) and grain yield (B) in winter wheat as influenced by water supply and N fertilizer application. The 'auxiliary lines' are explained in the text (data from Eck, 1988).

It can be seen that the *water use* has increased with the increase in total yield and in grain *yield*. In the irrigated treatments, whether or not they receive N, the average water use is about 600 mm. That is roughly 50% more than in the non-irrigated treatments, which on average used about 380 mm (Table 12.1). It is important to note that water use is high in the 'wet' treatments and smaller in the 'dry' treatments, but that within these water application treatments *the effect of adding N fertilizer on water use is minor*. In contrast, the *effect of N on* grain *yield* and total yield is *much more evident* than it is on water use, especially when water supply is plentiful (Fig. 12.2).

In this experiment the *harvest index* (HI) can be calculated. It is defined as the ratio of economic yield (here grain yield) to total above-ground yield of dry matter (Donald and Hamblin, 1976). The harvest index may be considered as a measure of *assimilate distribution* within the plant. When averaged over the irrigation treatments the addition of N fertilizer reduces the harvest index from 0.47 to 0.42 (Table 12.1). This is the *first main effect* of N fertilizer application on harvest index in this experiment. Averaged over the two N doses (N_0 and N_{210}), the addition of irrigation water increases the harvest index from 0.42 to 0.45. This is the *second main effect* in this experiment where the water was supplied by irrigation. In the particular combination of N_{210} and limited water supply, the decline in harvest index is rather marked (Table 12.1). There is a strong *interaction* between N fertilizer application and water supply on harvest index. Particularly under conditions of water shortage, increased N fertilizer application favours the deposition of assimilates in shoot and leaves, but adversely affects their transfer into the grains of the ear. This interpretation is appropriate when harvest index is the basis of the measure of *relative* assimilate distribution. In this experiment it is true in *absolute* numbers, as the straw yield (straw yield = total yield minus grain yield, see Table 12.1) increases by 1.45 t ha^{-1} with additional N input when water is in short supply and grain yield rises by only 0.19 t ha^{-1}. However, with a sufficient water supply the N input stimulates the production of an additional 3.88 t ha^{-1} straw, and an increase of 2.82 t ha^{-1} grain.

It is quite surprising that in the irrigated treatment the *total yield* and the *grain yield* were nearly doubled by *N fertilization* (Fig. 12.2) but that *water use* was increased by no more than 69 mm (see Table 12.1). Also, in the non-irrigated treatment, addition of N increased water use only slightly (36 mm), but in this case the yield increase was smaller.

Table 12.1. Water use, total yield, grain yield and harvest index of winter wheat grown in Texas, as influenced by water supply and nitrogen fertilization (Eck, 1988). The w^+ stands for irrigated, the w^- for non-irrigated. The $\bar{x}$ symbolizes the average.

N application (kg ha^{-1})	Water use (mm)			Total yield (t ha^{-1})			Grain yield (t ha^{-1})			Harvest index (grain/total yield)		
	w^+	w^-	$\bar{x}$	w^+	w^-	$\bar{x}$	w^+	w^-	$\bar{x}$	w^+	w^-	$\bar{x}$
0	562	364	463	6.68	4.75	5.72	3.11	2.20	2.66	0.47	0.46	0.47
210	631	400	516	13.38	6.39	9.89	5.93	2.39	4.16	0.44	0.37	0.42
$\bar{x}$	597	382	490	10.03	5.57	7.81	4.52	2.30	3.41	0.45	0.42	0.43

In 1962 Klapp reported that crops provided with 'complete fertilizer' (N, P, K, calcium (Ca) and farmyard manure) showed heavy yield increases relative to that of an unfertilized control, but with no great increase in water use (Fig. 12.3). In this study there was no irrigation treatment, and crops relied completely upon precipitation and the water stored in the 1-m deep loess soil profile. Of the various field crops, sugarbeet and lucerne with three cuts used the greatest amount of water, with sugarbeet being the highest yielding. Figures 12.2 and 12.3 demonstrate that optimizing nutrient supply with fertilizer can increase the *water use efficiency* (WUE) noticeably. This is true for WUE based on total shoot mass (Figs 12.2A and 12.3) and for the agronomic WUE, which in the case of winter wheat is simply based on the mass of the grain (Fig. 12.2B). The effect is more distinct in the case of an ample supply of water than when water is limiting production (Fig. 12.2). A relatively small value of WUE is obtained when a plentiful water supply is combined with the omission of fertilizer application (Table 12.2).

Similar to the wheat experiment in Texas that we discussed earlier, work at the *Dikopshof*, the University of Bonn's farm near Bonn, Germany, found that fertilizer addition increased WUE of winter wheat and winter rye (*Secale cereale*). The crops were not irrigated and relied on the natural water supply from the soil, a loamy silt derived from loess (Table 12.3). The 1-m profile, overlying gravel, stored 330 mm of water at field capacity (FC). The available water at FC was large (cf. Table 9.2). The plant-available water totalled 230 mm (Schulze and Schulze-Gemen, 1957). The WUE, related either to shoot mass or to grain mass, was increased by manure application, and still more by application of mineral fertilizer over that of the unfertilized control. The greatest increase in WUE, however, was achieved by addition of 'complete fertilizer', which consisted of farmyard manure and mineral fertilizer (Table 12.3). In contrast to the Texas experiment (Table 12.1) the harvest index was increased slightly by fertilizer in this German study (Table 12.3), even though the results were not clear cut.

In Section 11.1 it was emphasized that the relationship between dry matter production and water use is generally linear (Fig. 11.2). Now it is surprising that this plausible and simple relationship seems not to hold under conditions where growth is limited by other factors (Figs 12.2 and 12.3). The question arises as to how the tremendous *growth promotion by fertilizer application* can be explained when the water use does not increase accordingly.

Table 12.2. Water use efficiency (WUE) of winter wheat in Texas as influenced by water and nitrogen supply. WUE is based on total and grain yield. The data (including the averages) are calculated from published data (Eck, 1988) presented in Table 12.1. The w^+ stands for irrigated, the w^- for non-irrigated. The $\bar{x}$ symbolizes the average.

N application	WUE (total yield)(kg m^{-3})			WUE (grain yield) (kg m^{-3})		
(kg ha^{-1})	w^+	w^-	$\bar{x}$	w^+	w^-	$\bar{x}$
0	1.19	1.30	1.24	0.55	0.60	0.57
210	2.12	1.60	1.92	0.94	0.60	0.81
$\bar{x}$	1.68	1.46	1.59	0.76	0.60	0.70

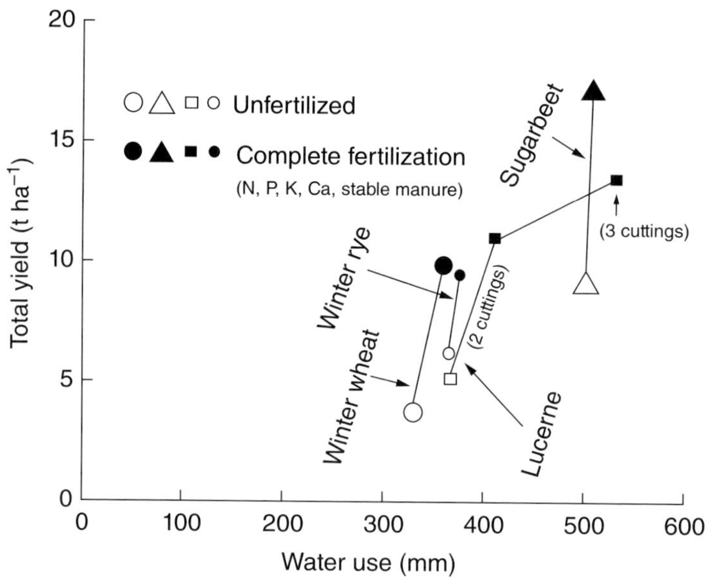

Fig. 12.3. Water use and total above-ground yield of various crops as influenced by fertilizer application. The experiment was conducted at Dikopshof, the University of Bonn's farm, Germany. The soil is a loess-derived haplic Luvisol. Data are averages over several years. Lucerne, cropped as a fertilized or unfertilized ley with two cuts, was also grown with fertilizer to give three cuts (data from Klapp, 1962).

One part of the explanation stems from the assumption that the leaf area index (LAI) and the leaf area duration (LAD) of the crop stand are increased by the enhanced nutrient supply when water is abundant relative to the potential atmospheric demand. In semi-arid Texas, abundant water supply means that irrigation is provided (Fig. 12.2), whereas in the humid climate of Bonn, Germany, it is associated with good water storage in the loess soil (Fig. 12.3). Increased LAI and LAD will increase the daily amount of intercepted radiation, depending on leaf inclination (Fig. 10.2) and the duration of radiation interception. Fertilizer addition will also alter the leaf colour from pale green (nutrient deficient) to dark green, reducing reflection and increasing absorption of radiation. These effects on the canopy will improve the crop growth rate (Fig. 10.6) and the accumulation of biomass (Fig. 10.7).

So far, we have given reasons for *growth promotion* through an improved nutrient supply. Now we will consider the other part of the explanation, of why the *water use* is not substantially altered by fertilizer application. When water supply is not limiting, the total evapotranspiration (ET) will be derived from the separate contributions of *soil evaporation* (E_a, Fig. 9.13) and *plant transpiration* (T, Fig. 9.3), depending on the size of the LAI, which is changed by fertilizer addition. Hence it is important to note that the ratio of E to T may change *without affecting total ET* to any extent, though the ET will vary with potential evapotranspiration. Plants suffering from nutrient deficiency with a small LAI, a thin crop density and short in height (Table 12.3) will allow a greater E and a smaller T (Table 12.4), because a greater percentage of radiation reaches the soil surface (Fig. 9.18). When the soil surface dries as a consequence of evaporation (eventually due also to root water

Table 12.3. Water use, total yield, grain and straw yield, harvest index and water use efficiency (WUE) of winter wheat and winter rye, grown near Bonn in Germany, as influenced by fertilizer application. The maximum height and the final density of the stand are also indicated. The experimental year was 1954 (after Schulze and Schulze-Gemen, 1957).

	Yield (t ha^{-1})					WUE (kg m^{-3})		Stand	
Fertilization	Total	Grain	Straw	Harvest index	Water use (mm)	Total	Grain	Height (cm)	Density (culm m^{-2})
Wheat									
Nil	4.60	1.73	2.87	0.38	360	1.28	0.48	90	205
Manure	5.99	2.40	3.59	0.40	366	1.64	0.66	95	235
N, P, K, Ca	9.20	3.72	5.48	0.40	413	2.23	0.90	120	313
N, P, K, Ca + manure	10.31	4.10	6.21	0.40	420	2.45	0.98	120	353
Rye									
Nil	7.06	2.55	4.51	0.36	365	1.93	0.70	150	334
Manure	9.17	3.61	5.56	0.39	386	2.38	0.94	170	354
N, P, K, Ca	9.75	3.54	6.21	0.36	385	2.53	0.92	180	411
N, P, K, Ca + manure	11.03	4.32	6.71	0.39	390	2.83	1.11	185	450

Table 12.4 Influence of mineral fertilizer on soil evaporation within a crop stand of winter rye. Soil evaporation was measured over 8 h on 6 May 1941 after some rainy days. 6 May was a sunny day with an air temperature of 19 °C and a relative humidity of 38% at noon (Gliemeroth, 1951).

	Mineral fertilizer	
	NPK	None
Stand density (number of shoots m^{-2})	706	586
Plant height (cm)	69	51
Evaporation (mm)	0.55	1.38

extraction from the soil below the surface), *sensible heat* is formed between the sparsely leafed plants. The heated air rises and is carried to the leaves in the form of *advective energy*, as a result of which plant transpiration is enhanced (Ritchie, 1983; Ritchie and Johnson, 1990). This process of increasing T without raising the net assimilation (NAR) but leaving NAR at the level set by radiation is called the '*clothesline effect*' (Tanner, 1957). As a result of increasing temperature, NAR may even fall. The clothesline effect will increase the '*unproductive transpiration*' of plants to the same extent that the soil surface becomes desiccated and soil evaporation declines.

The greater the crop density and the LAI developed as a consequence of a sufficient nutrient supply, the smaller will be the amount of radiation reaching the soil surface, the less the contribution of E to ET and the smaller is the clothesline effect responsible for unproductive transpiration.

In summary we can say that the level of LAI can influence the crop growth rate and the final yield much more specifically than it can affect the actual evapotranspiration rate. The explanations given so far will now be supported and expanded by referring to the inserted *auxiliary lines* in Fig. 12.2A. We will assume that the greatest LAI and LAD were attained in the treatment receiving irrigation and N fertilizer. The result of these treatments was a fairly small value for soil evaporation (E), just 50 mm. This intercept with the abscissa (Fig. 12.2) represents, according to Ritchie (1983), the part of soil evaporation that occurs principally during juvenile growth, when LAI is small (LAI < 1). We shall hypothesize that during the total vegetative period, E totalled no more than 50 mm. Based on this supposition the diagonal auxiliary line shown in the figure determines the *transpiration efficiency* (TE, cf. Fig. 11.2). In comparison with this treatment, where plants are well provided with water and N, the *unproductive water use* in the irrigated N_0 treatment is greater by 220 mm (Fig. 12.2). This unproductive water consumption is attributable on the one hand to more soil evaporation, and on the other to increased unproductive transpiration. But it might also be that in this irrigated but N-deficient treatment, there is more water loss by seepage. If it was just this redistribution in the components of the water balance that needed to be considered within a fixed quantity of ET and modified by different LAI, the water use of the irrigated N_0 and N_{210} treatments could be the same. But, in fact, with N_{210} it is greater by 69 mm (Table 12.1).

This enhanced water use will now be dealt with. Even when the meteorological conditions, such as radiation, saturation deficit, temperature and wind, recorded at the site are much the same across the experimental field, the evaporative demand may be greater for the crop stand well supplied with N compared with the poorly supplied one. As well as a larger LAI, the height of the crop stand (Tables 12.3 and 12.4) will cause a greater retardation of the wind speed (Eqn 8.4). The taller the crop, the greater is the so-called 'roughness length' z_0 (Eqn 8.5). And in addition to the change in z_0, the leaf area (and possibly the number of leaf storeys above z_0 through which the wind blows) increases and as a result the vapour exchange of the canopy will be accelerated (Fig. 7.7). Also the interception of advective energy, originating from neighbouring fields, will be intensified. The '*oasis effect*' will be more effective, the smaller the oasis area (here the plots receiving N fertilizer and irrigation, Fig. 12.2, or the plots with complete fertilizer addition, Fig. 12.3) compared with that in the surrounding area, and the more the oasis and surrounding stands differ in species composition, crop development and plant height.

In the treatment deficient in water and nitrogen (Fig. 12.2), the additional unproductive water use is 110 mm compared with the watered N_{210} plot. This is only half of the additional unproductive water use, estimated for the watered N_0 plot. Presumably, by far the majority of the 110 mm is unproductive transpiration as the soil is left unirrigated. In the treatment with N fertilizer but without irrigation, however, the unproductive water loss is seemingly countered by a somewhat lusher plant development (70 mm). The extra water use of 36 mm by N fertilization in the non-irrigated plots (Table 12.1) has not necessarily to be traced back to a wind and oasis effect, because N will cause fewer differences in LAI and crop height when water is scarce than when it is supplied abundantly. In a situation of water scarcity, however, N fertilizer may contribute significantly to the stimulation of *deeper rooting*. Deep rooting means that a larger reserve of the available field capacity (FC_{av}; Section 9.4) stored in the soil can be extracted, compared with that in the non-irrigated, N-deficient soil.

12.3 Transpiration Efficiency and Fertilizer Application

In discussing the relationship between nutrient supply and WUE we argued that in addition to the effect on biomass production, the proportion of soil evaporation (E) to transpiration (T) was shifted by fertilizer application in favour of T. When E declined under the plant canopy due to drying of the soil near the surface, T was enhanced by the clothesline effect. This additional T was an unproductive part of the total transpiration, as it did not contribute to the synthesis of assimilates.

If it were possible to set up an experiment that eliminated E and stopped advection as far as possible, the relationship between biomass yield and water transpiration as a function of nutrient supply should become evident. As early as 1958, de Wit had argued that the *production–transpiration function is unaffected by the nutrient status*, provided the nutrient level is not 'too low'.

We will clarify the relationship between transpiration and yield under different N applications with reference to a *pot experiment with sugarbeet* (Bürcky, 1991, 1993). Several pairs of plants were grown, each pair in a 30-l container. The wall of the pots included an isolating air cushion. The rooting medium consisted of sand, used for construction work, enriched by micro- and macronutrients. The N application per pot varied between 2 g (control) and 16 g N. The pots were placed in a mobile greenhouse, where the glass cover could be rolled away completely during the day. The pots were irrigated according to the measured water use. The seedlings emerged on 22 March. Starting on 8 June the plants were harvested fortnightly until 17 August; altogether plants were sampled six times (Bürcky, 1991). The results are shown in Fig. 12.4.

Figure 12.4A demonstrates two features of special note. Apparently the use of the sand culture succeeded in forcing the *evaporation* component towards zero, because all the lines start more or less exactly at the *origin of the coordinate axis*. Also it is significant that *yields*, obtained *on successive dates*, follow the same trend. This trend in the yield–water use relationship can also be identified just by taking the final yield data as modified by the N supply (see number 6 in Fig. 12.4A, representing four N levels at final harvest). For each N fertilizer application rate the transpiration efficiency (TE) can be determined from the slope of the yield–water use line. TE is smallest at the control rate of 2 g N per pot. It amounts to 5.26 kg total biomass m^{-3} water. In addition, the dry matter produced is quite small. Evidently N deficiency limited growth. Applying 4 g N per pot, on the other hand, gives the greatest TE – 5.88 kg m^{-3}. The level of production is twice as big as in the control. At still greater N fertilization rates (8 g and 16 g N per container) admittedly the largest total biomass yield is obtained, but the values of TE are only in the middle range. All in all, the *variation of TE* with N fertilizer is shown to be quite *small* and inconsistent.

The picture is quite different when the TE, modified by N application, is related solely to beet production (Fig. 12.4B). The relationship between beet dry mass and transpiration over time is linear only at the lowest N rate, the control. But this regression line does not meet the origin of the coordinates. At higher N rates, however, the function proves to be non-linear. At the higher N rates of 8 g and 16 g N per pot, the TE is distinctly smaller than at 4 g N per pot, with the TE for 16 g N per pot being the least. The explanation is that per unit of transpired water, large N rates favour the growth of the *top* more than that of the *beet*. Nevertheless, heaviest yields of beet are obtained by applying the higher N rates, although the yield is somewhat greater with 8 g than with 16 g N per pot.

Also the greatest *yield of white sugar* is achieved with high N application rates (Fig. 12.4C). It has also to be noted that high N rates can increase the percentage of amino-nitrogen (indicative of poorer quality) and can lower that of purified sugar (Bürcky, 1991). Nevertheless, this disadvantage is more than offset by the heavier beet mass. The largest yield of white sugar is obtained when the N rate is 8 g per pot, though the biggest yield of white sugar per unit of transpired water is obtained at the lower N rate of 4 g N per pot.

Thus there is some supporting evidence that N fertilizer does not necessarily 'improve' TE when the N status of the control is not 'too low'. What has been shown for N has also been demonstrated with potassium (K) (Walker and Richards, 1985). Lucerne was grown on a range of soils, varying 'widely in K-supplying power' as determined by extraction with 1 M ammonium acetate. Deficiency of plant-available K in soils did not reduce the TE of the test plant

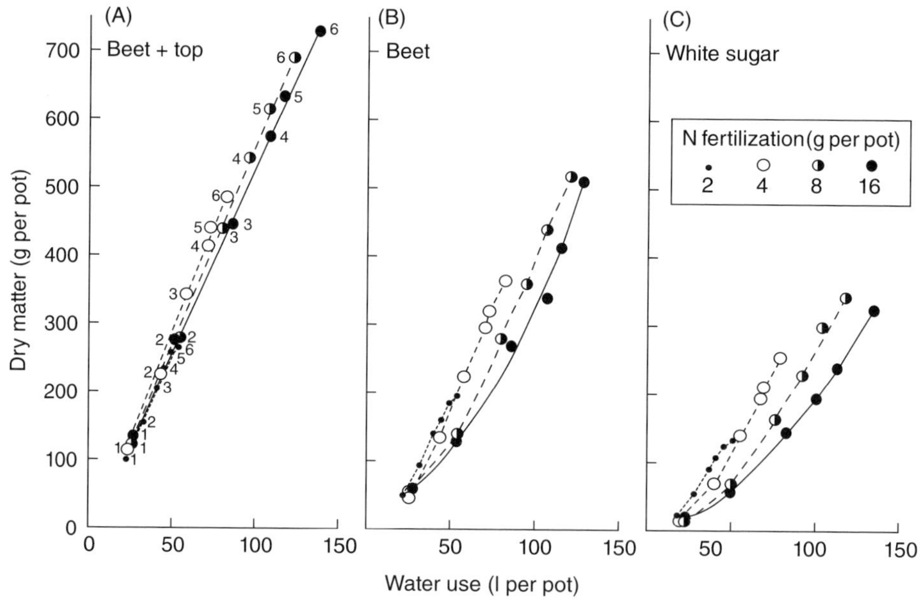

Fig. 12.4. Effect of nitrogen fertilizer on the relationship between dry matter and water use in sugarbeet. The dry mass is related to: beet and top (A); beet alone (B); and white sugar produced (C). The numbers 1–6 signify the harvest dates between 8 June and 17 August 1989 (data from Bürcky, 1993).

Lucerne, even when the plant suffered severe deficiency – confirmed by tissue analysis. Altogether, however, there is a shortage of experiments dealing systematically with transpiration efficiency and its dependence on soil fertility, or with the availability of specific nutrients. In many arid regions there is not only a shortage in water, but the soils also lack *chemical fertility*. All too often major problems with the supply of fertilizer impact producers, especially small farmers away from major centres of population. For instance in the African Sahel, crop production is unpredictable or completely in jeopardy owing to unreliable, erratic precipitation. Moreover, production is often curtailed by poor phosphorus (P) availability (Payne *et al.*, 1991). Therefore, it is argued that in many semi-arid regions not only is the level of crop production low, but what is more, even at this low level, plants will use the water from the transpiration stream for biomass accumulation less effectively than plants growing on fertile soils. That is to say that severe P deficiency worsens the transpiration efficiency of crops. With this idea in mind, Payne *et al.* (1991, 1992) investigated the growth of pearl millet (*Pennisetum glaucum*) grown on a P-deficient, sandy soil (3 mg 'Bray 1-P' kg^{-1}) in Texas, very 'similar to the sandy millet fields of Niger, Senegal and Mali' (Payne *et al.*, 1992). The soils were contained in 75-l pots, which were placed outdoors and were – in addition to N and K – fertilized with 0.00 g, 0.16 g, 0.47 g and 1.08 g P (rounded up in Fig. 12.5), corresponding roughly to 0 kg, 10 kg, 30 kg and 70 kg P ha^{-1}, respectively. The plants were kept under water-stressed or non-water-stressed conditions. In both cases, soil evaporation was prevented. When related to above-ground shoot biomass, the TE of pearl millet increased with P addition from 3.3 kg to 4.4 kg biomass m^{-3} of water, without providing evidence of any distinct effect of water shortage on TE (Fig. 12.5A). Taking root mass into account as well as shoot mass, TE was of course greater (Fig. 12.5B). Moreover, P fertilizer improved not only shoot-related TE, but also whole-plant-related TE. Finally, TE related to shoot and root was larger when water was short (Fig. 12.5B).

The physiological background for *TE decreasing when P supply is limited* is not fully understood. With pearl millet it was found that the net CO_2 assimilation rate of leaves was reduced when the plants received no P fertilizer (Payne *et al.*, 1992). P deficiency also lowered the level of light saturation (Fig. 10.1), which may appear to explain the observed inhibition of the photosynthetic rate. Apparently the inhibition of CO_2 assimilation is caused by a malfunction somewhere in the metabolic pathway, triggered by P deficiency. On the other hand, measured leaf transpiration of millet was increased by P deficiency when the net photosynthetic rate was small (Payne *et al.*, 1992). From this latter observation it seems reasonable to accept that pearl millet does not close the stomates due to P deficiency but keeps them open, even if they do not open any wider.

In recent years, plant physiologists have provided support for the idea that when nutrients are short in supply, plants can enhance leaf transpiration. Intensified transpiration increases water flow in the soil and thus nutrient transport by mass flow towards individual roots. By this mechanism plants may increase the accumulation of a variety of nutrients in the tissue (Barber, 1995). Mass flow depends on the nutrient concentration in the soil solution and in many soils that will be greater for N, Ca and magnesium (Mg) than for P (Barber, 1995). On the other hand, growing pearl millet on P-deficient sandy soil may indicate

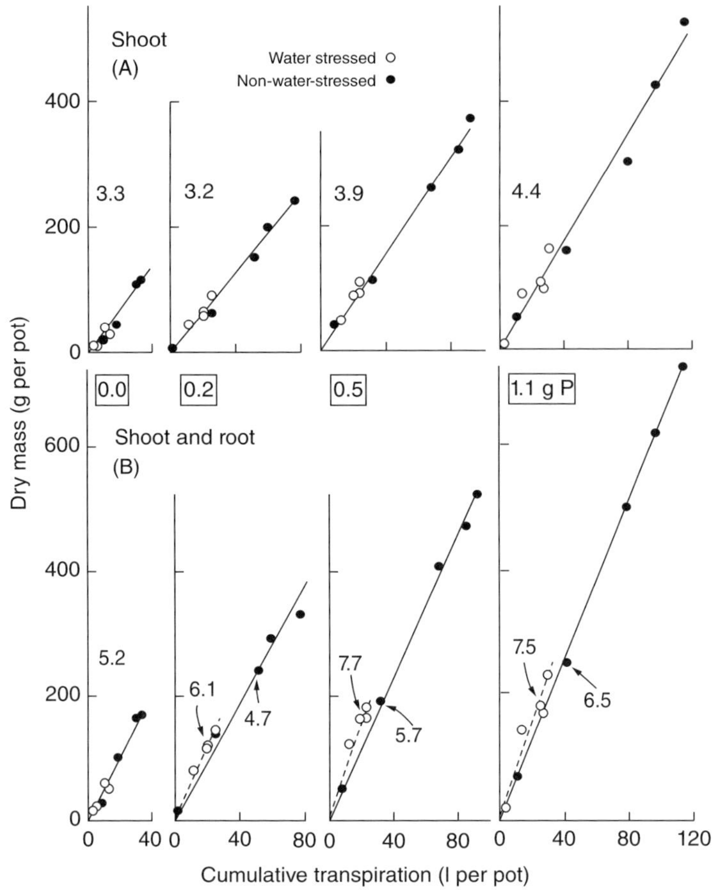

Fig. 12.5. Shoot dry matter (A) and whole plant dry matter (B) as related to transpirational water use of pearl millet, measured at several times during the season. The sandy soil, deficient in phosphorus, was fertilized with increasing P rates (g P per pot). Watering was modified to induce water-stressed or non-water-stressed conditions. For calculation of root dry mass, root decay as observed during the second half of the season is excluded here. The numbers indicate the slope of the regression line, which was forced through the origin. The slope is the transpiration efficiency in g l^{-1} or kg m^{-3}. Calculations based on data from Payne *et al.* (1991, 1992).

that at such a poor location the role of diffusion and interception for P acquisition is of minor importance compared with mass flow (Cramer *et al.*, 2009). Without P fertilizer pearl millet has a low TE but it increases with P supply (Fig. 12.5). Inefficient transpiration based on biomass production can be offset by P addition. Such an effect has been reported for several species in the plant physiology literature (Raven *et al.*, 2004).

Pearl millet *stressed by water shortage* has a greater, not a smaller, TE than the same crop grown without water stress, at least when the P deficiency is not too extreme (Fig. 12.5B). Water shortage influences the distribution of assimilates within the plant, favouring a redistribution to the roots (Hamblin *et al.*, 1990; see Fig. 13.3). But this fact is not sufficient to explain why TE related to whole-plant biomass increases with water stress. It is conceivable that a partial but not complete closure of the stomates causes the increase of TE. A partial closure impedes the water vapour flux out of the leaf (Fig. 12.5A and B) to a greater extent than the CO_2 flux into the leaf (Nobel, 1983; Larcher, 1994). Admittedly, partial closure will increase the diffusion resistance for both gases. But in contrast to H_2O, for incoming CO_2 the highest diffusion resistance is located within the liquid part of the diffusion pathway (i.e. in the cytoplasm of the mesophyll cells; Section 11.2). And the cytoplasm resistance is left unchanged by stomatal aperture.

We can conclude from this chapter that *TE is largely independent of the fertility status* of the soil. Adding fertilizer to the soil can greatly increase the *yield level*, but TE is hardly affected. Only when the fertility status is extremely low will fertilizer increase TE, as shown for phosphorus. Apart from essentially poor soils, the application of fertilizer will have no beneficial effect on transpiration efficiency. It is much more plausible to accept that in soils of reasonable fertility status fertilizer application will enhance *crop growth* and thus increase *water use efficiency* (Section 12.2).

13 Development of Economic Yield under Inadequate Water Supply

13.1 Physiological Reactions and Assimilate Partitioning

Terrestrial plants, provided with a root system for water uptake and a leaf canopy for water emission, regularly experience slight *water shortage* in their tissues. The shortage is caused by the *time lag* between leaf transpiration and root water uptake, a prerequisite for water flow in the soil–plant–atmosphere continuum (SPAC). Plants experience a more distinct temporary water shortage during the course of a day, often in the early afternoon, when at a larger scale the rate of water uptake does not compensate for the accelerated rate of water loss. If rain does not replenish the soil water extracted by roots, the plant water shortage becomes more severe. Such a water shortage of limited duration can be observed easily, for example in sugarbeet after canopy closure during the summer months, when around midday the leaves become flaccid (cf. Fig. 9.14). When the water shortage is not just temporary but continues over a fairly long period because the soil water supply becomes depleted (Fig. 8.4), plants react with clearly defined *physiological and anatomical changes*. In the end these changes become evident because of reduced yields that can be identified if a yardstick for comparison is available, such as the irrigated treatment in Table 12.1.

During water shortage, plants experience a state of tension with associated physiological consequences. These responses have enabled plant species to adapt to dryness by developing various strategies over the course of evolution (Section 1.2). The state of tension is called *stress*. A factor that causes stress is called a *stressor*. When acid rain or ozone (O_3) put pressure on the plants, then sulfur dioxide (SO_2) and O_3 are examples of stressors. Sometimes, too, the lack of an essential growth factor is considered to act as a stressor. When water is in short supply, however, the technical term is not water shortage stress, but simply *water stress*. Some researchers prefer not to talk about the deficient factor, preferring to use a word that has the opposite meaning, namely drought (Section 1.2). Thus, instead of water stress they prefer to talk about *drought stress*. After noting these issues of terminology, we will stay with the term *water stress*.

So when water supply falls short of requirement, plants will experience a water stress that will become more prominent the longer the shortage continues. The water stress can be quantified by measurements of the total water potential, or its component potentials, and the water content of the tissue (Fig. 7.1). Water stress in plants causes disorder among physiological processes (Fig. 13.1), but at what level of stress the malfunction starts depends on the nature of the specific process. It seems that plants have developed a differential *susceptibility to stress* by maintaining specific *physiological processes*. For example, cell enlargement in plants is quite susceptible to water stress. Even at a relatively high water potential, *cell elongation* slows when the potential drops (Fig. 13.1). The process of *cell division*, on the other hand, seems to be less susceptible (Kramer, 1983). Cell elongation and tissue growth require positive turgor (Eqn 7.1, Fig. 8.5), and the turgor changes in very much the same way as total water potential (Fig. 7.1). In contrast, the process of *assimilate translocation* within the plant is only susceptible to very large stresses (Fig. 13.1). An example of this process occurs in cereals when a large part of the assimilates produced in leaves (the *source*) is transported into the head for grain filling (the *sink*).

The processes of *photosynthesis* and *transpiration*, moderated by variation in stomatal opening, are less susceptible to water stress than the process of cell enlargement (Fig. 13.1). It is the H_2O control circuit (Section 7.4) that protects plants from desiccation and restricts CO_2 assimilation. When transpiration is limited by stomatal activity, the cooling effect is reduced. The plant gets warmer, which temporarily stimulates *respiration*. If water stress continues to increase, the respiration rate drops noticeably (Fig. 13.1). The increased respiration stimulated by the temperature rise leads to losses

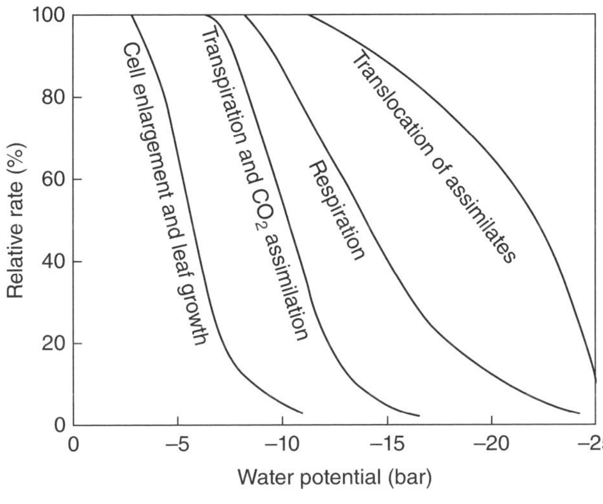

Fig. 13.1. Examples of plant physiological processes that depend on total water potential. The susceptibility of the plant to water stress with regard to specific processes declines from left to right. Semi-schematic diagram (based on Kramer, 1983 and Gardner *et al.*, 1985).

of accumulated dry matter (Fig. 1.4). This is an undesirable result of water stress for plant production, but it helps to explain the comparatively small accumulation of dry matter during periods with warm nights over the growing season. When there is an extended period of stress, premature *senescence* of the vegetation is initiated. This early ageing may be explained by the severe reduction in all the physiological processes essential for life. Continued water stress limits dry matter production and yield and causes untimely death.

When crops are supplied with water steadily throughout the season, but at an inadequate rate for luxuriant growth, water stress develops right from the beginning. Under such environmental conditions, plant growth and organ development will become retarded early in the season. A good example of this can be seen with the chilli pepper (*Capsicum annuum*) cultivar 'New Mexico No 6-4', which is also known as Chile pepper. It was field-grown in New Mexico, USA, and differentially watered by *drip irrigation* (Fig. 13.2). The water-stressed treatment received 610 mm in the form of irrigation water and precipitation, and the non-stressed treatment had ample water – 846 mm in total. In the stressed plants, the development of *leaf area index* (LAI) was slower and the final value was smaller than in non-stressed plants (Fig. 13.2A). The capsicum plants receiving the meagre supply had fewer leaves, which were smaller but thicker and had a greater stomatal density (Horton *et al.*, 1982). By limiting the *source of assimilates*, the *crop growth rate* (CGR) was restricted virtually from the beginning (Fig. 13.2B). Correspondingly, the final dry matter accumulation in the shoot was only 74% that of the treatment with an ample water supply (Fig. 13.2C). Apparently the

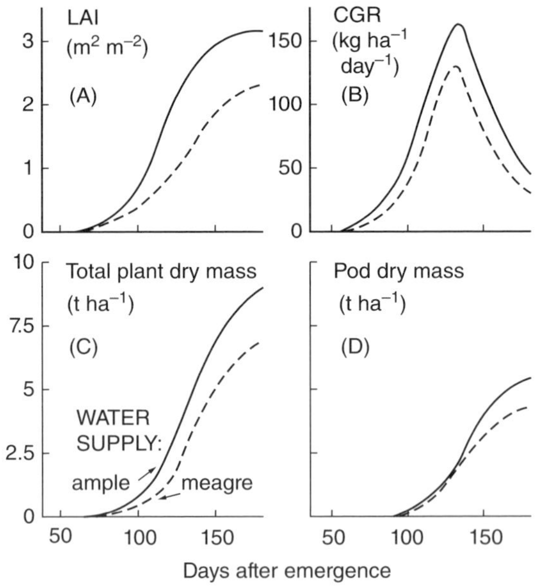

Fig. 13.2. Influence of ample or meagre water supply by drip irrigation during the season on leaf area index, LAI (A), crop growth rate, CGR (B), total plant dry mass (C) and fruit (pod) dry mass (D) in chilli pepper, grown in New Mexico, USA (after Beese *et al.*, 1982).

translocation of assimilates from leaves to the fruits (the sinks) was less hindered under the condition of a steady but meagre water supply until harvest than was the production of the above-ground total dry mass: in the stressed treatment fruit dry mass was 79% of that attained in the non-stressed treatment (Fig. 13.2D). Accordingly, the *harvest index* (HI) was slightly greater under the restricted water supply (0.62) than with an ample supply (0.59). Limited water supply by meagre drip irrigation caused chilli plants to adapt their phenotype (Fig. 13.2) to that of a water saver (Fig. 1.1). Under these conditions the leaf water potential during the day (up to −2 bar) and night (about −0.5 to −1 bar) was slightly more negative in the treatment with a meagre water supply compared with that, which was amply supplied (Horton *et al.*, 1982).

Water stress supports the translocation of assimilates from the shoot to the root. When water is short, first the new growth of plumules is reduced. Moreover, the growth of those leaves that have already formed is slowed down. Therefore, the surplus of assimilates not used in the shoot is transported into the root, a process controlled by plant hormones. The *shoot–root ratio* becomes narrowed and the root mass per unit area may increase slightly. When, however, a significant water shortage prevails during the whole season, root mass may increase considerably, as shown by investigations in the wheat belt of Western Australia, a region with a Mediterranean climate (Fig. 13.3). In an average year with 335 mm annual precipitation, the shoot–root ratio of spring wheat increased from close to 1 (20 days after sowing) to almost 11 at harvest (cf. Table 6.1). In a dry year and with a soil profile lacking available water right from the beginning of the season, however, the ratio was smaller throughout the vegetation period. The maximum value did not exceed 1.9 (Fig. 13.3). The numbers given in the figure for the date of harvest indicate that the wheat plants invested quite a large quantity of *assimilates into the root* during the extremely dry cropping season. In absolute figures,

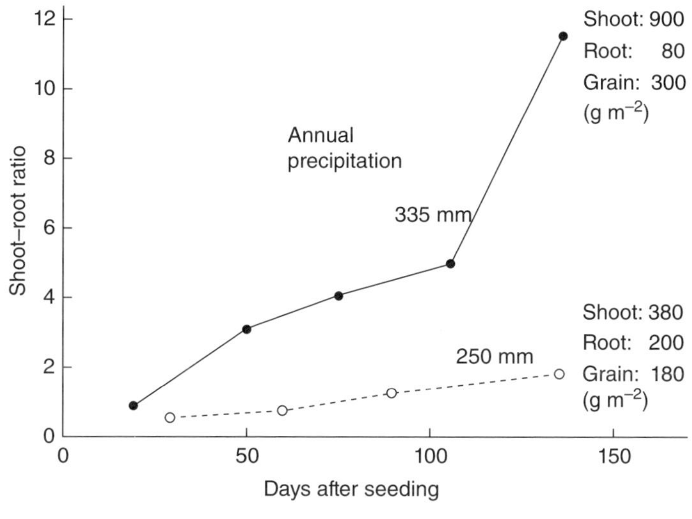

Fig. 13.3. Influence of annual precipitation on shoot–root ratio during a season for a spring wheat variety growing in Western Australia. The figure also indicates shoot, root and grain dry matter yield at harvest (data derived from Hamblin *et al.*, 1990).

two-and-a-half times more dry matter was invested in roots under dry conditions than in an average year. Instead of the normal 80 g, the dry matter translocated to roots was 200 g m^{-2} of ground surface, or instead of only 8% of the total plant mass (including roots) being in the roots, it was 34%. There is *a price to be paid for water stress*, in the form of an investment of assimilates into that part of the plant that is often not perceptible as it is hidden in the soil. The effect of preferential translocation of assimilates into the root can still be strengthened by other stressors. These stress factors are nitrogen (N) deficiency and mechanical impedance – a consequence of soil compaction within a traffic pan (Hamblin *et al.*, 1990).

Water stress may favour *root dry matter* (Fig. 13.3) and *total root length*, but not *rooting depth*. In the Western Australian investigation, water-stressed wheat restricted rooting depth to less than 1 m. Within this 1-m profile, root length density (L_v) was comparatively large, enabling the limited soil water supply to be fully exploited. However, in the year with more precipitation roots extended down to 1.5 m, though with a smaller L_v at comparable depths (Hamblin *et al.*, 1990).

In addition to limited soil water, a restricted N supply may also direct assimilate partitioning in favour of roots. On the other hand, as early as the beginning of the 1950s Gliemeroth (1951) reported that it was *NPK fertilizer application* and not nutrient deficiency that improved rooting depth and water extraction from subsoil layers. Therefore it seems likely that the depth of root growth is another story and is not solely a function of dry matter accumulation in roots. It seems that stress factors are responsible for roots gaining a larger share of assimilates. But it also seems that in the absence of stressors, root growth to depth is consequent on enhanced shoot growth.

Water stress throughout the season causes a larger reduction in shoot than in grain dry matter (Fig. 13.3). As a consequence, the HI is greater in a dry (0.47) than in an average year (0.33). However, when calculating HI on the basis of total of above- and below-ground dry matter rather than on total above-ground dry matter alone, the indices are basically the same: in the dry year the HI is 0.310, and in the average year it is 0.306. Therefore, because in the dry year a larger fraction of assimilates is directed towards the roots, it follows that the investment of assimilates into reproductive organs – into grain – is greater only on a relative basis (i.e. when the basis of HI is shoot dry matter, but not shoot and root dry matter).

13.2 Economic Yield

It has been shown that a water deficit generally has an adverse effect on net primary production of terrestrial ecosystems (Table 1.1, Fig. 1.2A) and crop stands (Figs 11.1, 11.2 and 12.2A). Water deficit may or may not cause an even greater reduction (on a percentage basis) in *economic yield* (i.e. in the yield of the marketable produce) as compared with the above-ground *biomass yield* (total above-ground production). A decline in harvest index (HI) due to water shortage has been observed with winter wheat in Texas (Table 12.1), but not with chilli pepper in New Mexico or spring wheat in Western Australia (Section 13.1). In principle HI must decline when water supply falls short. To get an idea of this concept, we can imagine an irrigation experiment with the water supply varying from ample to very meagre. When there is sufficient water during the season, annual crops will generally grow vegetative biomass as well as the marketable part. But when affected by drought, the plants may not undergo the transition from the vegetative to the reproductive phase, and then *fertilization of the inflorescence* does not occur. That may happen, when either the procarp has not completely developed and reached the flowering stage (small-grained cereals get 'stuck' during stem elongation or jointing), or when the process of fertilization itself is hampered (for instance in maize, see Section 13.3). In addition, during a long-lasting drought, flowers or even fruits may form but do not reach maturity, and are shed prematurely (*abortion*, for instance in grain legumes). Therefore, we can surmise that as well as the formation of a very small biomass there will be no harvestable product at all in situations of extreme dryness. In such a case HI would be zero.

In 1923 Cole and Mathews reported on yield relations in *Triticum durum*, cultivar 'Kubanka' (see Fig. 11.1). This variety of durum wheat had been grown for several years at various experimental stations of the USDA (United States Department of Agriculture) on the Great Plains under conditions of *dryland farming*. Under extreme dryness the above-ground biomass yield was less than 1.5 t ha^{-1}, and in some cases of small biomass yield the crop failed to produce any grain yield (Fig. 13.4). In some other cases – according to the comment of the agronomists:

> when affected by drought the wheat crop seems to spend its last energy in producing grain, and that if there is any chance at all, it will produce some yield of grain. This study indicates that a high yield of straw means a high yield of grain. There have been a few cases when exceptionally favorable weather enables wheat to fill so well (with assimilates) that the yield of grain was out of proportion to the yield of straw. These years are very infrequent, and as a whole the yield of grain and straw are nearly proportional.
>
> (Cole and Mathews, 1923, cited in Howell, 1990).

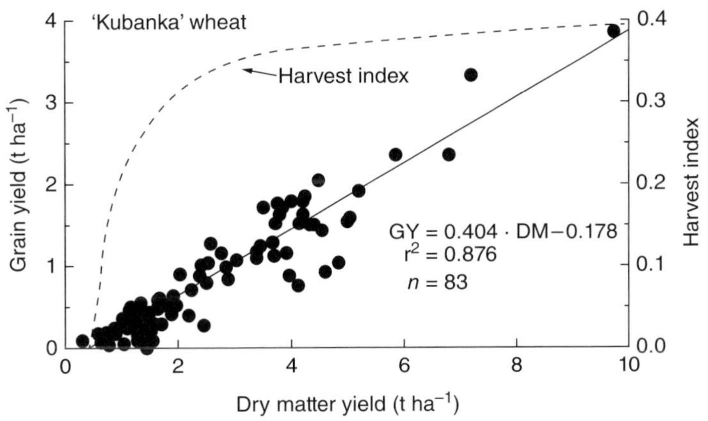

Fig. 13.4. Relationship between grain yield (GY) and above-ground dry matter yield (DM) of 'Kubanka' wheat. This durum variety was cultivated in the Great Plains, USA under conditions of dryland farming (Howell, 1990). Data are from Cole and Mathews (1923). The figure also contains a graph showing harvest index as a function of above-ground yield. The curve is based on the regression line presented.

When grain and straw yield are 'nearly' proportional, there is also a linear relationship between grain yield and total dry matter yield (Fig. 13.4). From the equation of the regression line, given in the figure, the average curve relating HI to total yield can be derived. This relation is shown in Fig. 13.4 and indicates how dramatically HI changes, as biomass yield increases from very low levels. In contrast, when starting at a higher yield level, HI remains almost untouched by improvements in biomass yield.

At the same time, the individual values plotted on the graph indicate that for a given total yield, grain yield may vary over a wide range (Fig. 13.4). For instance, when the total yield is 4 t ha^{-1}, grain yield varies between 0.7 t ha^{-1} and 1.8 t ha^{-1}. Correspondingly, HI changes between 0.18 and 0.45. This variation is not inconsiderable and demonstrates that HI is not just dependent on the average quantity of water supplied over the season. Rather, it can be influenced, to a greater extent, by how much water is provided during individual *phenological stages* (Section 13.3).

As the economic yield and biomass yield are more or less closely related (Figs 13.2 and 13.4) and as biomass yield depends strongly on water use, all else being equal (Figs 11.2 and 11.3), then we should expect that *economic yield and water use* are related too (Fig. 1.3). These interconnections are presented in Fig. 13.5, using data from an irrigation experiment in Utah, USA. In this field experiment with spring wheat the level of water supply was gradually changed. This was accomplished

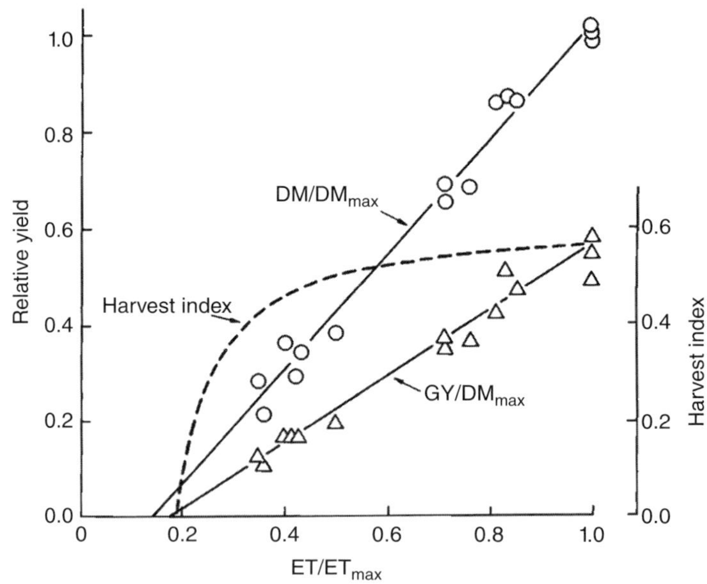

Fig. 13.5 Yield of above-ground dry matter (DM) and of grain (GY) in spring wheat, grown in Utah, USA. The yield is expressed relative to total dry matter of 'non-stressed' wheat (DM_{max}) and plotted as a function of relative evapotranspiration (ET/ET_{max}). The maximum ET was ascertained directly at the line source of the irrigation water (Sharratt *et al.*, 1980; from Musick and Porter, 1990). The figure also contains a plot of the harvest index.

by a simple, but brilliant, method. The actual evapotranspiration (ET) of the crop, grown on either side of a fixed sprinkler line source, was estimated from measured irrigation rate and was found to decrease with distance from the source. Directly at the emitter, water was applied at a rate for maintaining optimum growth conditions, taken as the maximum rate of actual evapotranspiration (ET_{max}). For each location the relative evapotranspiration (ET/ET_{max}) was determined. The relative yields of grain and of biomass were calculated in the same way. The relative yield was plotted as a function of relative evapotranspiration as a measure of water deficit (Fig. 13.5). Both grain and biomass yield were linearly related to water use.

Generally with an increase in total yield, grain yield also increases, depending on water use. When the water use increases from a low production level, the HI at first shoots up (Fig. 13.5). When evapotranspiration is in the relatively high range, however, HI hardly shows any change. At the beginning of the 20th century, 'Kubanka' wheat (*T. durum*) had attained a maximum HI of about 0.39 (Fig. 13.4). Later, in the 1970s, the HI of spring wheat (*Triticum aestivum*) was much higher, about 0.55 (Fig. 13.5). Although different species of wheat are being compared in this case, real progress in plant breeding is indicated, and we will deal with that in Section 16.1.

13.3 Water Shortage at Different Phenological Stages

So far we have focused on how water shortage reduces dry matter production of plant stands. Under extended dry conditions the mass of roots may increase, not just in relation to shoot mass, but even in absolute terms (Fig. 13.3). But inevitably the *total biomass will decrease due to water stress*. No matter what, when a crop stand is hit by periods of water shortage during the season, the *maximum biomass yield* is no longer attainable. The maximum yield can be attained only when, under otherwise identical conditions, any water stress is avoided. This statement is important and valid for the total biomass as well as for the above-ground biomass. But the question remains as to how the *economic yield* is affected by situations of water shortage. Like biomass yield, normally the yield of marketable produce also declines with water shortage. But as will be shown later, there are exceptions to this rule. The extent to which economic yield in particular suffers from water shortage will depend on several circumstances: (i) the intensity of stress; (ii) the duration of the dry period; (iii) the moment that the water shortage occurs during the different developmental stages of the crop; or (iv) how the capacities of sinks and sources for assimilates are affected by water stress. Finally, the extent to which plants can recover from the water stress experienced is very important. When periods of favourable water supply follow dryness, this may enable partial compensation for a reduced assimilate production. More specifically this means, for example, whether plants are able to produce new leaves by budding in order to produce *new sources of assimilates*, and whether plants are enabled in the course of an extended flowering period to form new fruits as *additional sinks* for assimilates, enhancing the capacity that was left after the period of water shortage.

Such a flexible pattern of growth is the particular attribute of grain legumes (cf. faba bean, Fig. 9.1) and chilli pepper (Fig. 13.2) for instance, allowing the plants to adapt best to periods of water shortage whenever they occur. These plants show an *indeterminate growth* habit. The growth of the shoot is not terminated at the stage of flowering as occurs in cereals (Fig. 9.1). In fact, shoot growth in these indeterminate plants generally continues even to the end of the season, as long as growth conditions are favourable. New nodes with developing leaves and internodes are formed (Fig. 13.2A). As a rule, the growth of stem and leaves goes on in parallel with the growth of fertilized reproductive organs over a longer part of the season (Fig. 13.2A and D). The period of flowering is usually not restricted to a period of only a few days, but will be drawn out over weeks as is the case with faba bean (Fig. 9.1, see timescale of phenological stages).

The development of plants with *determinate growth* follows a very different path, where the flowers are built from a terminal bud or apical cone. With these plants the stages of growth of vegetative shoot and of reproductive organs, of fruits with seeds, are separated from each other much more strictly in terms of time. With small-grained cereals it is not until the flag leaf has appeared that the ear emerges, followed within a very few days by flowering, pollination and fertilization. From thereon, grains act as a sink and start to fill, while the leaf canopy no longer expands, but goes into decline. Finally the flag leaf will act as a primary source of assimilates, but there are also glumes and awns that can act as additional sources, using solar radiation for photosynthesis during grain filling.

Like small-grained cereals, *maize* also belongs to the group of plants with determinate growth. Whereas small-grained cereals have bisexual flowers, the two inflorescences of the maize plant are unisexual. After the formation of the flag leaf the male inflorescence, the *tassel*, appears first. Shortly afterwards the silk will become visible at the female inflorescence, the *cob* or ear.

The effect of water stress on maize was tested in plants grown in buried 76-l containers. Stress due to water shortage was established once during the season, by interrupting irrigation at one of nine different growth stages in such a way that the plants showed clear signs of water stress for a period of 4 days. The stress became apparent by wilting of the uppermost, fully expanded leaf. When water supply was restricted at tasselling and up to 3 weeks after silking (i.e. after pollination), *grain yield* was drastically reduced compared with the control with sufficient water supply (Fig. 13.6A). During this period the number of *grains set per cob* was reduced (Fig. 13.6B) either by an insufficient development of the female flower primordia, by disturbing pollination or by stunting of the young pollinated ovules. During this stage of transition from vegetative growth to grain formation, maize has been found to be very *prone to water stress*, definitely more susceptible than the small-grained cereals with bisexual flowers. In the example given for maize, the yield loss due to water shortage at silking was slightly more than 50%!

Poor grain setting in the ear may possibly be compensated by a heavier kernel weight. However, this compensatory effect did not appear when water stress occurred during silking (Fig. 13.6B). During this stage and for some time after, the *stalk* proved to be a strong competitor for assimilates. As the capacity of the kernels to compete for assimilates was reduced (i.e. the number of sinks in the cob had been reduced) an additional sink was opened up – the stalk (Fig.13.6C).

Water shortage during the vegetative stages of development and during the late stages of ripening affected grain yield only to a relatively small extent (Fig. 13.6A).

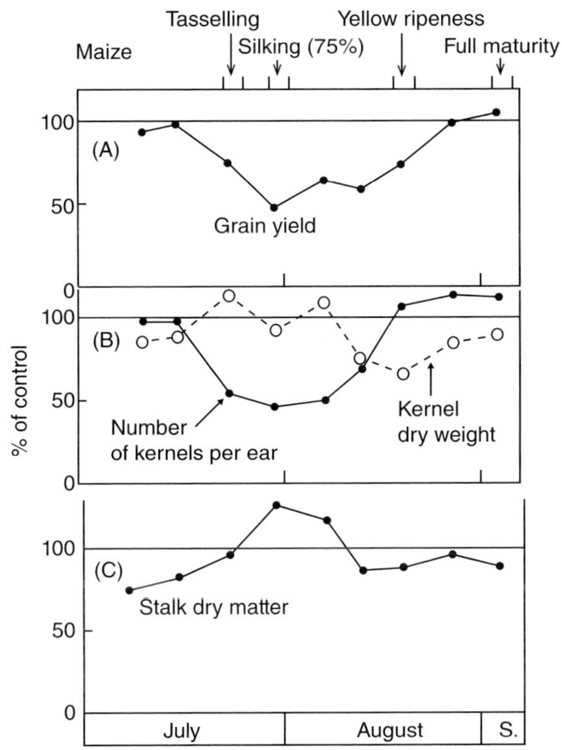

Fig. 13.6. Maize plants are exposed to water shortage at different developmental stages: influence of single water stress on final grain yield (A); yield components (B); and stalk dry matter (C) (from Claassen and Shaw, 1970a,b).

During grain filling, grain setting was no longer negatively influenced by water stress, but *kernel weight* was (Fig. 13.6B). It follows, therefore, that during this late stage after flowering, water stress did not impair sink capacity but had more effect on the strength of the sources. Probably fewer assimilates were produced by the remaining leaves. To compensate for the reduced assimilatory performance of the leaves, reserve material stored in the stalk was then mobilized and translocated into the grains. At this late phase of water stress the stalk took on the role of an assimilate source. In complete contrast, water stress at an earlier phase, represented by the first two dates in July, not only reduced the growth of leaves but also that of the stalk (Fig. 13.6C). The smaller dry mass accumulation in the stalk was possibly the reason for the reduced translocation of assimilates from the stalk to the kernels (Fig. 13.6B). Before anthesis, photosynthetic products are generally stored not just in the stalk but throughout the plant. In non-stressed maize almost 10% of this stored material contributes to the formation of the kernel mass (Simmons and Jones, 1985).

Maize hardly ever undergoes significant tillering. Therefore, under practical field conditions, sowing density determines the stand density. This is different for the other cereals, where crop density, expressed as the number of eared culms per unit area, is quite independent of the seeding rate. Here the ability to form tillers, which is largely influenced by environmental factors, can allow compensation in the number of shoots per unit area. In *spring barley*, for instance, dry periods, especially during tillering and jointing, reduced *total biomass* (above ground) as well as *grain and straw yield* compared with the non-stressed control (Fig. 13.7A). *Harvest index* was little affected (Fig. 13.7B). The yield reduction at this stage was primarily a result of the smaller *plant density* (Fig. 13.7C). During stem elongation, the density of culms generally becomes reduced, but conversely some other components of yield are developed during that stage. These components, the number of *spikelets* per spike (or panicle) and the number of *florets* per spikelet, may react negatively to water stress. In barley there is one floret per spikelet, which does not change with water stress. A possible yield reduction by water shortage therefore takes place in the barley spike due to the number of fertile spikelets per spike, which is – in addition to crop density – accordingly smaller (Fig. 13.7D). The smaller *grain set* in the head can be compensated for by a

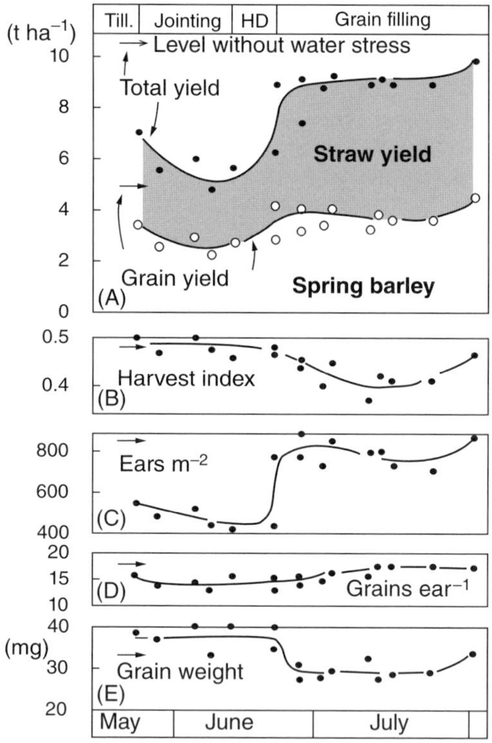

Fig. 13.7. Effect of temporal water shortage, induced at different stages of development in distichous spring barley. The effect is indicated for total yield and grain yield, and for straw yield by difference (A), for the harvest index (B) and for various yield components (C, D, E). The periods of dryness lasted between 5 days and 15 days. The dots stand for the end of a dry period. The level of yield and yield components of the control plants without water stress are indicated by the insertion height of arrows. Till., Tillering; HD, heading (after Mogensen, 1980 and Mogensen and Jensen, 1989).

larger *grain weight* (Fig. 13.7E) when, later in the season, the stage of grain filling takes place in the absence of water stress. Under such conditions the *yield of the single head* (number of kernels per ear × kernel weight) may be greater (in the current example 550 mg) than in a situation when water shortage develops later after grain filling has started (490 mg). This kind of enhanced compensation for early water stress is probably a consequence of a more favourable exploitation of the spacing factors like radiation, nutrients and water by the single culm, when the stand density is low (Fig. 13.7C). And this effect of a greater yield per head, caused by early water stress, may be more pronounced in a distichous barley (as in this experiment) than in a hexastichous barley.

One possibility for compensatory growth important for the survival of a cereal has not yet been mentioned. When, in the current example, water stress ended before the start of heading, barley was able belatedly to form *lateral shoots* or suckers by *second growth*. But these shoots ripened 2 weeks later than the normal shoots. Therefore, they were not considered in Fig. 13.7. A farmer too, would not wait for the maturity of the late shoots, as there would be a fear that grain would be shed from the normal shoots if harvest is delayed for too long. That is likely to happen, but there can also be an issue caused by the unripe grain on the suckers if harvest is not delayed.

If water shortage does not begin during vegetative development but only later during grain filling after heading and anthesis, yield may be reduced to a much smaller extent compared to the control. While the straw yield remained almost unaffected, the grain yield dropped from 5 t ha^{-1} to 4 t ha^{-1} or slightly less (Fig. 13.7A). Consequently HI was reduced (Fig. 13.7B). At this late stage of development, water stress hardly influenced those yield components that were set early in the season (Fig. 13.7C and D). But as in the case of maize, the filling of the grain was negatively affected (Fig. 13.7E). Therefore, it is correct to assume that the source capacity, the photosynthetical performance, of the flag leaf was strongly reduced by late water shortage. In contrast to early water stress, stress during grain filling caused a smaller yield reduction (Fig. 13.7A). It is possible that just as in maize (Fig. 13.6), a larger translocation of assimilates from culm and leaves to the kernels has contributed to this effect. The yield drop was smallest when water stress was only induced late in the period of grain filling. The late stress was no longer effective and all the yield components remained near the control level (Fig. 13.7).

In summary we can conclude that for crops with determinate growth, water stress during the vegetative part of growth before anthesis can reduce the strength of sources and the capacity of sinks for assimilates in individual plants. The smaller source strength can be partially compensated for by better utilization of the growth factors, which because of the sparser stand density become optimized for a single plant within an enlarged growing space. From this it follows that reduced yield per unit area is primarily a consequence of the smaller stand density. Water stress in the phase of reproduction, however, does not have much effect on the capacity of the sinks, but has considerable impact on the strength of the sources. The consequence of a reduced rate of leaf photosynthesis for the yield is moderated by intensified translocation processes. Maize is strongly susceptible to water stress, as pollination is considerably hampered during tasselling and silking.

Soybean was originally a plant of indeterminate growth, but there are now many determinate cultivars that have been introduced by plant breeders. In indeterminate soybean, vegetative growth takes place simultaneously with the growth of reproductive organs for quite a long time. Within the phase of reproductive growth the stages of flowering, pod development and bean filling overlap, each lasting for several weeks (Fig. 13.8). In this example the *seed yield* of the field crop was less affected by water stress during the main period of flowering than by that induced at later stages from the middle of pod development to the middle of bean filling (Fig. 13.8A). This was true for the seed yield per plant as well as for the seed yield per unit area, as the planting density was kept constant and did not change with early water shortage (Shaw and Laing, 1966). When soil drying occurred early in flowering, the number of *pods per plant* was scarcely reduced (Fig. 13.8B). Admittedly a larger number of flowers were aborted, but later flowering compensated for the loss, so that the number of pods stayed almost the same. However, water stress in the middle of flowering near the end of July reduced pod formation significantly by the *abortion of flowers and pods*, but a slight compensation in yield was attained by two counteracting yield components, the *number of seeds per pod* and the *seed weight* (Fig. 13.8C, D). Water stress at a relatively late stage, in the

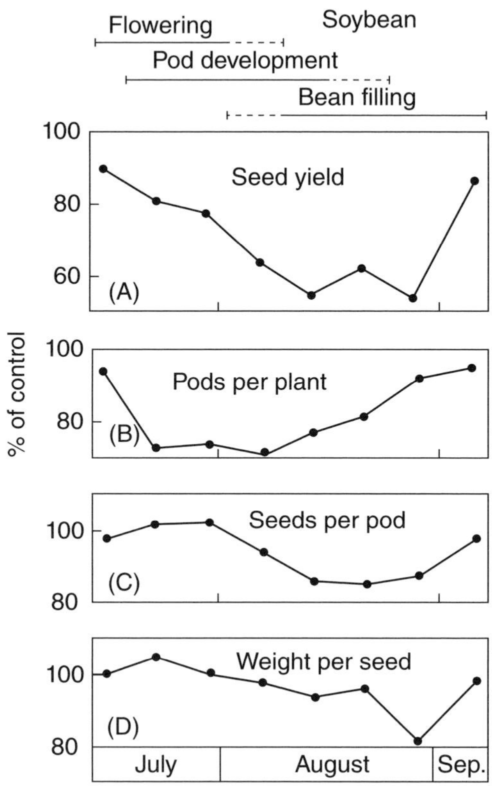

Fig. 13.8. Effect of a single, temporal water stress at various growth stages on seed yield in soybean (A) and on yield components (B, C, D) (after Shaw and Laing, 1966).

middle of seed filling, greatly reduced the possibility for compensatory growth. Now all the yield components were negatively influenced by water shortage. Pods were shed, fewer seeds were formed per pod and the seed weight also declined markedly below that of the control because the assimilate inflow was small. Sinks for assimilates were undoubtedly lost, but also the sources became less active in producing assimilates. Just as for maize and spring barley, stress at the end of seed filling in soybean had the least effect on yield.

While a short period of water stress may affect the economic yield to a greater or lesser extent, a longer lasting period of dryness will reduce the yield more drastically (Figs 13.2, 13.3 and 13.5). In the case of prolonged dryness the strength and capacities of sources and sinks become smaller depending on the time of onset and the duration of the dry period. Moreover, the possibility of compensatory growth through the formation of new reproductive organs and leaves and by assimilate translocation is less and less likely.

Up to this point we have been working on the idea that a crop stand, adequately supplied with moisture for potential transpiration, will produce the largest yield in terms of both total dry matter and grain. In the next example we will consider a variation on this theme. The subject deals with *faba bean*. In this example, the soil water content was varied by irrigation or by applying a protective cover to the soil. For the control of soil water, the agronomists selected two periods: (i) *during flowering*; and (ii) *after flowering*. When grain legumes were irrigated during and after flowering (Fig. 13.9, i–i: irrigated), and when the soil water content was correspondingly high, then the greatest above-ground *biomass yield* was attained (control =100%). However, when the plants remained unirrigated during the two stages (u–u: unirrigated), or the soil water content dropped to an even lower percentage (10–30% of available field capacity, FC_{av}), the total yield was smallest. If we accept that soil moisture content can be a rough indicator of the quality of water supply and therefore of actual transpirational water use, then we might consider that Fig. 13.9 provides confirmation of the relationship between cumulative dry matter and cumulative transpiration.

The effect on the *seed yield* of the faba bean is a different matter. The economic yield increases with the soil's water content during flowering until a *transition* water content is reached. An additional increase in soil water then makes the seed yield drop again! The water content after flowering, however, reflects our experience that a ready water supply supports seed yield. The greatest seed yield is therefore produced by the treatment that combines a medium moisture level during the flowering period with a higher moisture level (40–50% FC_{av}) after the end of that period, treatment u–i. What is the underlying explanation?

Like soybean, faba bean also has an indeterminate pattern of growth, provided that this has not been modified by plant breeding activities. A luxuriant water supply during flowering stimulates shoot growth and the formation of leaves. The growth of shoot and leaves consumes a great deal of the newly formed assimilates. The shoot apex with the freshly growing leaves represents a highly *attractive sink*. The flowers and young pods, inserted at the lower nodes (node numbers 1–11), have to compete with the apex for assimilates, and do so without too much success. That is the reason they are *aborted* (Grashoff, 1990b). When irrigation is continued after flowering through the period of pod development, those

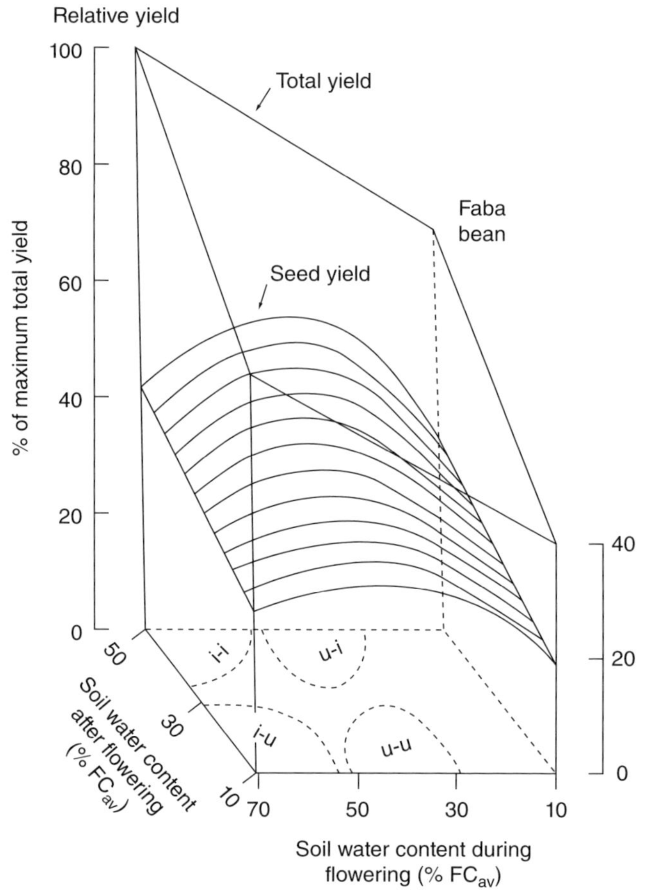

Fig. 13.9. Total yield and seed yield of faba bean as influenced by the soil water content, which was varied during and after flowering. The sequence of letters indicates the sequence of irrigation during (first letter) and after (second letter) flowering. The two letters stand for: i, irrigated; u, unirrigated. The treatment i–u, for instance, means that during flowering (i) the soil water content was maintained at about 55–70% of the available field capacity (FC_{av}). After flowering, irrigation was stopped (u) and the soil water content dropped to about 10–30% of FC_{av} (after Grashoff, 1990a).

flowers that have not been shed and which are located primarily in the upper leaf storeys (node number > 11) will continue to develop into pods with seeds. The yield obtained under i–i condition is achieved with an unfavourable harvest index.

Even a 'mild' water stress during flowering retards the elongation of upper internodes and leaves, while CO_2 assimilation and the distribution of assimilates are reduced to a lesser extent (Grashoff and Verkerke, 1991; cf. Fig. 13.1). The mild water shortage effectively slows down new growth of leaves and therefore eliminates highly competitive sinks for assimilates at the tip of the shoot. Hence a greater proportion of assimilates is ready for translocation into those young pods located at the lower nodes. When the establishment of these pods (taking place during a period of mild water stress at flowering) is followed by sufficient water supply thereafter (see treatment u–i in Fig. 13.9), assimilates then will be used for the formation of pods and seeds in the upper nodes. In the cultivar used for this experiment, the growth of the shoot tip came largely to a standstill, despite irrigation being restarted during pod development.

This faba bean cultivar required a water supply that was variable with time if the maximum seed yield with the greatest harvest index was to be attained. Specifically, this cultivar required a slightly reduced water supply during flowering and a more plentiful supply thereafter.

There are many reasons why an over-ample water supply to soils and plants can be harmful to plant growth. Wet conditions may support the spread of harmful organisms and troublesome weeds. Soil warming can be slowed down. Hypoxia may hamper root growth and N mineralization. But a luxurious water supply may also induce an unwanted stimulation of shoot growth, as a result of which the seed yield may be limited, as shown in the faba bean experiment.

13.4 Relation between Water Use and Economic Yield in Principal Crops

From a number of examples we have learned that plants produce an economic yield whenever the basis is strong and healthy, allowing for flourishing and grain filling. But it is also true that after attaining a certain biomass level the realization of the economic yield can turn out to be quite variable (Fig. 13.4). Yield development depends on climatic conditions, water supply and fertilizer application, on water shortage at different phonological stages, influencing capacities of sources and sinks. There is a connection between biomass yield and economic yield, but both are not closely and unambiguously related. That means that the harvest index (HI) changes accordingly to the conditions that dominate crop growth during the season. Examples were shown for fertilizer application (Table 12.1, 12.3) and irrigation (Table 12.1). In accordance water use efficiency (WUE) related to economic yield, here grain yield, varied with cultivation methods (Table 12.2, 12.3).

Recently the Food and Agriculture Organization of the United Nations (FAO) published a compilation of data on economic (grain) water use efficiency (WUE_G) and transpiration efficiency (TE_G) of world leading crops (Sadras *et al.*, 2011), shown in Table 13.1.

The data show a broad variation in WUE between 0.1 kg m^{-3} and 2.3 kg grain m^{-3} of water. Because of unproductive water losses the WUE numbers are generally smaller than those for TE, which vary between 0.7 kg m^{-3} and 3.7 kg m^{-3}. It seems that in general both measures of efficiency are greater with C_4 than C_3 crops, and winter crops are more efficient than summer-grown crops. Moreover, the efficiencies seem to reflect differences in the energy content of the grain. Carbohydrates are the product of photosynthesis (Section 1.1), and occur initially in the form of the energy resource glucose. The sugar can be converted into starch, which is the main product of cereals. For 1 g starch a quantity of

Table 13.1. Economic (grain) water use efficiency (WUE_G) and economic (grain) transpiration efficiency (TE_G) of major annual crops under rain-fed conditions, based on seasonal water use and seasonal transpiration (data from Sadras *et al.*, 2011). Authors presented data in the unit kg grain ha^{-1} mm^{-1}. For the conversion in kg grain m^{-3} see Section 11.1.

Crop	Type of CO_2 fixation	WUE_G (kg m^{-3})	TE_G (kg m^{-3})
Millet	C_4	0.1–1.2	1.7
Sorghum	C_4	0.5–2.1	2.0–3.0
Maize	C_4	0.6–2.3	3.0–3.7
Wheat	C_3	0.5–1.0	2.0–2.2
		0.5–1.0[a]	
Rye	C_3	0.7–1.1[b]	
Rice (rainfed lowland)	C_3	0.2–0.8	1.5–2.2
Soybean	C_3	0.6–1.0	0.8–0.9
Winter pulses (faba bean, chickpea, lentil, lupin)	C_3	0.2–1.6	0.9–2.0
Cotton	C_3	nd[c]	0.9 (seed)
Sunflower	C_3	0.3–0.5	0.7–0.9
Winter oilseeds (*Brassica* spp.)	C_3	0.1–0.8	1.2–1.5

[a]Tables 12.2 and 12.3.
[b]Table 12.3.
[c]nd, not determined.

1.2 g glucose is needed. But 1 g protein, as formed by pulses, requires 1.6 g glucose, and finally 1 g fat as in sunflower or winter oilseeds demands not less than 2.7 g sugar (Sadras *et al.*, 2011). The energy content of the grain reserve material follows the sequence fat > protein > starch and this order seems to be inversely reflected in the order of WUE and TE, with cereals > pulses > oilseeds.

14 Water Stress in Plants

14.1 Measuring Water Stress in Plants

Plants lose water from their tissues during periods of water shortage. With the loss of water they experience a state of stress or tension that can be measured with suitable equipment. Sometimes, however, it may be sufficient to evaluate the level of tension by scoring the degree of leaf wilting as a first approach. For instance, with sugarbeet the visible *degree of leaf wilting* can be assessed easily. With some species of the grass family the estimated extent of *leaf rolling and folding* can serve the same purpose. On the other hand, powerful and meaningful measuring methods and devices have been developed since the 1960s. We will mention the most important ones, focusing on those that work reliably under field conditions. With this aim in mind, we will differentiate between methods that characterize the water balance of the plant itself, and those that indicate the effects of the stress on physiological qualities and processes.

The soil moisture characteristic curve (Fig. 5.1) relates soil water content (a property of *capacity*) to soil water potential (a property of *intensity*) (Section 4.2). To measure *plant–water relations*, we can take a very similar approach. First, there is the cell or tissue water content (capacity) that has to be measured, and then there is the total plant water potential (intensity), which can be split up into the partial potentials, including pressure potential, osmotic potential, matric potential and gravitational potential (Eqn 4.5). Differences in total water potential (ϕ) determine the direction of water movement within the soil–plant–atmosphere continuum (SPAC). Therefore ϕ is of primary importance for the assessment of water flow within SPAC (Table 7.4). This quantity also allows the assessment of water stress and an estimate of the impact of stress on physiological processes (Fig. 13.1).

For the determination of the *total plant water potential*, one method has proved to be very successful as it produces reliable data under field conditions (Steiner and Hatfield, 2008). The technique relies on the *pressure apparatus*, developed by Scholander *et al.* (1965). The apparatus is also called a 'pressure chamber' or, in honour of the inventor, the 'Scholander bomb'. The technique together with some of the underlying basic assumptions will be explained in some detail. A leaf, composed of tissue and cells (cytoplasm, vacuole, cell wall, intercellular spaces, xylem vessels and phloem sieve tubes), is hydraulically connected with the rest of the plant. Due to differences in ϕ, water flows through the vessels into the leaf (Table 7.4). At any time of the day selected for measurement, the 'pull' or suction, causing the water to flow, is interrupted by cutting the leaf with the aid of a sharp razor blade. When investigating crop plants, we often take the youngest, uppermost, fully expanded leaf that is exposed to the sun's radiation. However, to deal with high trees, a twig from the upper part of the canopy may have to be shot down! The suction prevailing at the time the leaf is removed for analysis pulls the 'xylem sap' into the vessels (Fig. 14.1C). The leaf (or leaf blade with cereals)

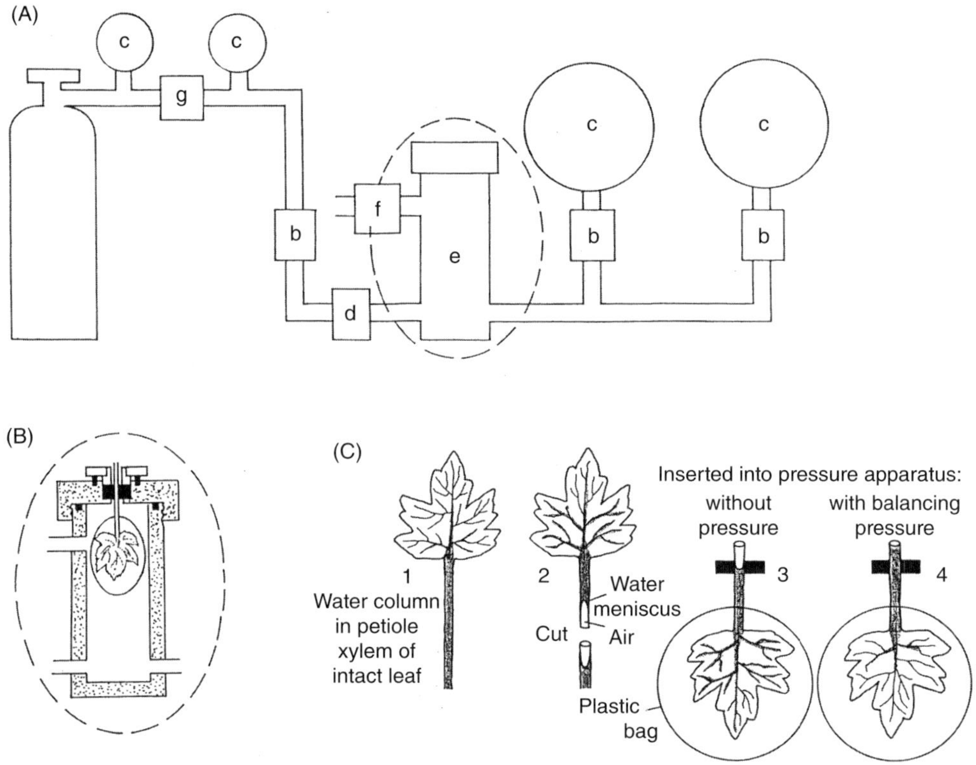

Fig. 14.1. Diagram of pressure apparatus. (A) The total assembly: a, gas cylinder containing pressurized air or nitrogen; b, switch-off valve; c, pressure gauge; d, flow valve; e, pressure chamber; f, outlet valve; g, pressure-reducing valve. The two pressure gauges on the right-hand side serve different areas of pressure, from 0 bar to 20 bar and from 20 bar to 70 bar. In (B) a cut leaf is shown, fixed into the pressure chamber. The leaf is wrapped in a polyethylene bag or self-adhesive foil (from Turner, 1981). Today the devices, commercially available, are small, handy and suitable for field work. In (C) the continuous thread of sap is shown in the stalk of the uncut leaf (1). After cutting, the thread of sap has retreated into the vessel (2). The leaf is inserted into the pressure chamber (3), which is pressurized up to the moment when the xylem sap appears at the cut surface (4).

is wrapped in self-adhesive, airtight foil immediately after it is excised to prevent continuation of water vapour loss through the stomates and cuticle. Sometimes the leaf is wrapped even before cutting. The wrapped leaf is promptly fixed into an opening in the lid of the pressure apparatus. For that purpose the uncovered leaf stalk is put into a sealing ring, often a half-cut silicone stopper. The stopper should hold the stalk tightly, but not squash it. With the stopper in place, just a short end of the stalk protrudes from the upper side of the lid (Fig. 14.1B). Immediately after the leaf is fixed in place the lid is tightly clamped to the apparatus. Thereafter the chamber is slowly pressurized by use of compressed air or nitrogen gas (Fig. 14.1A). At the beginning the pressure has to be increased at a rate not greater than 1 bar s^{-1}. Approaching the end of measurement, the rate should be less than 0.2 bar s^{-1}. The end of measurement will be attained precisely when the xylem sap just begins to appear at the cut surface, visible as a glistening, flattened surface (Fig. 14.1C). The appearance of the sap can be observed precisely with the aid of a magnifying glass. At the moment the sap becomes visible, the gas supply is turned off and the final pressure is read and recorded. The entire procedure from leaf cutting to pressure reading should not take more than 2–3 min.

The question that has to be answered is what *kind of potential* do we actually measure with the *pressure chamber*? Before cutting, the leaf is transpiring during daylight hours, and hence the total potential in the mesophyll is lower than the potential in the xylem of the stalk. After cutting and wrapping the leaf, transpiration and consequently sap flow within the xylem is interrupted. By restricting further losses as far as possible, water simply gets redistributed within the cut leaf. After redistribution has taken place, the total water potential will be the same at all locations within the leaf. The potential is equilibrated throughout the leaf. Then, by applying an excess pressure to the chamber, the xylem sap is forced back to the cut surface. The *positive pressure*, read from the manometer, indicates the *pressure potential of the xylem sap* that existed before cutting. The pressure potential in the xylem bears a negative sign (Table 7.4). The pressure potential in combination with the negative *osmotic potential of the xylem sap* represents a close approximation of the *total potential* that existed in the xylem sap before the leaf was excised. At the same time, the total potential of the xylem sap is also an approximate value of the total potential within the mesophyll cells at the time when the leaf was still hydraulically connected to the rest of the plant. In fact, the total potential value determined is an average value of the *leaf water potential*. This value (after pressure chamber equilibration) is somewhere between the total water potential of the xylem sap, which is slightly higher (less negative), and the total potential in the mesophyll, which is slightly lower, when water or 'sap' was still flowing within SPAC (Hsiao, 1990). Quite often, determination of the osmotic potential of the xylem sap is neglected as the content of osmotically active substances is generally small. The osmotic potential corresponds usually to about –0.5 to –2 bar (–1 bar in Table 7.4), but not to lower values. It may be much more negative for instance in the xylem of halophytes and some xerophytes.

The total potential can also be measured in the laboratory by use of more delicate equipment, the *thermocouple psychrometer*. A thermocouple is formed from two wires, made from different metals or alloys, which are joined in a loop by two soldered junctions. The thermocouple psychrometer measures the equilibrium value of the relative humidity within a closed chamber that contains a cut-out disc sample of the leaf being investigated. Before the humidity measurements can be made, one of the junctions of the thermocouple has to be wetted with a droplet of water. The droplet can be introduced from outside or produced internally by employing the 'Peltier effect'. An electric current is passed through the thermocouple loop, which results in the cooling of one of the junctions. Water vapour condenses at this junction and forms a droplet of water. The liquid water on the 'wet' junction evaporates at a slower or faster rate depending on the relative humidity within the chamber, which is dominated by the total water potential of the sample of leaf tissue. Depending on the rate of evaporation, the wet junction cools down. The temperature difference between wet and dry junction causes a voltage in the loop, which is measured. Voltage, temperature reduction, evaporation rate and relative humidity are all interrelated. The relative humidity is a measure of the total potential ϕ of the leaf tissue sample (Eqn 14.1):

$$\phi = \frac{RT^*}{\bar{V}} \ln \frac{e}{e_s} \tag{14.1}$$

In this equation R is the universal gas constant, T* is the absolute temperature and $\bar{V}$ is the partial molal volume of water. The water vapour pressure in the closed chamber (e) is in equilibrium with the water of the plant tissue. The term e_s is the saturated vapour pressure in air above pure water at T* (compare equation in Table 7.5). The numbers in Table 7.5 and Fig. 5.9 make clear that the method operates in the range of high relative humidity. Therefore it is absolutely necessary to keep the temperature around the thermocouple constant, a requirement that makes field application nearly impossible. Various other laboratory methods for measuring plant water potential are described by Turner (1981) and Hsiao (1990).

The total water potential changes during the course of a day (Fig. 9.14). The question arises at which *time of the day* the measurements should be made to characterize the state of tension, the water stress in plants. Plants generally experience the *greatest stress* in the early afternoon (Fig. 9.14). This low potential should be recorded daily, if possible, and these measurements should be continued throughout the vegetative period (Fig. 9.15). On the other hand, the potential is greatest in the early morning hours, right before sunrise (Fig. 9.14). As long as the total potentials in soil and plant equilibrate overnight when soil moisture is sufficient (Fig. 8.4), the plant water potential in the early morning hours can serve as an *indicator of soil water availability*. The potential characterizes the remaining ability of the soil to provide water in extractable form to the root system. The plant water potential measured before dawn can therefore be regarded as the mean value of the water potential prevailing in those rooted soil layers that still have residual moisture available for uptake (Meyer and Green, 1980). When the rooted soil profile dries and potentials in plant and soil are no longer equilibrating overnight, the pre-dawn plant water potential loses its significance as an indicator of accessible soil water left in the rooted profile (Fig. 8.4).

As well as total plant water potential, the *water content of the tissue*, too, can serve as an indicator of plant water stress. The water content has to be determined in the laboratory. Therefore the method will be mentioned only briefly. Often the relative water content (RWC) is determined, applying the following equation:

$$RWC = \frac{FW - DW}{TW - DW} \tag{14.2}$$

FW stands for the present fresh weight of the plant or part of the plant, harvested for investigation, DW is the dry weight, and TW is the weight after the sample has reached turgidity following immersion in water. Plants, well supplied with water, will attain an RWC value at noon that is 0.88 or higher (Hsiao, 1990). When the plant water deficit is sufficient that the turgor approaches zero, the RWC is usually between 0.72 and 0.88. At this range of RWC, the leaves are visibly wilted or rolled, and photosynthesis has been reduced to a considerable extent. When RWC is further reduced to 0.5–0.6, the plant will die, even if the desiccation only lasts for a few hours (Hsiao, 1990).

The *component water potentials* are much easier to measure in the laboratory than in the field. The *osmotic potential* can be ascertained after destruction of the plant tissue. The destruction of the cell walls exposes the cell content of osmotically active compounds. For that purpose the tissue is either frozen and thawed, or boiled. Afterwards the osmotic potential is measured with the aid of a thermocouple psychrometer or some other technique. The *pressure potential* is obtained by difference from the total potential ($P = \phi - O$). The pressure potential of individual plant cells can also be measured directly by a sophisticated technique, where a microcapillary syringe is inserted into the plant cell (Hüsken *et al.*, 1978).

The osmotic potential can be ascertained, too, by use of the Scholander pressure chamber, preferentially in the laboratory. The following procedure does not provide an instantaneous measurement for a fixed time of the day, but will generate a *water potential–water content* (or cell volume) *function*, as shown in Fig. 7.1. Such a moisture release curve is called a *pressure–volume curve*. A common procedure is to harvest leaves in the evening hours. The leaves are taken to the laboratory, immersed in water and recut below the water surface. They are left in water overnight to become water saturated. The next morning, a leaf in 'full' turgor is wiped dry and the fresh weight is taken immediately (TW in Eqn 14.2). Then the leaf is placed into the pressure chamber and the xylem sap, which exudes at successive pressure stages, is collected. The sap is stored in a small glass tube, containing cellulose as an absorbing material. The tube is always weighed after the leaf has attained equilibrium at a given pressure, and the weight of sap collected over a particular pressure step can then be determined. Sap collection ends when the highest pressure has been applied. The fresh weight of the leaf, attained after the final application of pressure, is also recorded. The difference between TW and the final leaf fresh weight is the total weight of the sap that has been squeezed out. However, the total weight of the sap actually collected will be slightly less because of losses, for instance by evaporation. These losses are shared out among the pressure stages by calculation. Thus a corrected fresh weight is obtained for each pressure step (FW in Eqn 14.2). Finally the dry weight (DW in Eqn 14.2) is determined.

When applying high pressure (15–20 bar and higher) to the chamber, plant cells will lose water to such an extent that the turgor will be zero. Over this pressure range the (positive) pressure in the apparatus corresponds to the (negative) total potential, which equates to the (negative) osmotic potential. The small contribution from the matric potential (Fig. 7.1) is ignored. Under such conditions the following equation holds (Turner, 1981; Hsiao, 1990):

$$\frac{1}{P_c} = \frac{V_t - V_a}{RT^*n_s} \tag{14.3}$$

In the equation P_c is the pressure applied to the chamber, V_t is the total volume of water in the leaf at P_c, V_a is the volume of 'apoplastic water', R is the universal gas constant, T* is the absolute temperature (K) and n_s is the number of moles of solutes. The apoplastic water is the water in the xylem and cell walls. From the apoplastic water the symplastic water is differentiated, the volume of which is $V_t - V_a$. This liquid is contained inside the cells, where the concentration of solutes is greater than in walls and xylem. The liquid is enclosed by the semi-permeable plasmalemma and contains the vacuole in addition to the cytoplasm with the nucleus and plastids.

According to Eqn 14.3 the inverse value of the chamber pressure ($1/P_c$) can be plotted versus V_t. This plot results in a linear relation for the range of high pressures, provided that V_a and n_s do not change. Such a relationship is shown in Fig. 14.2 for oat, where V_t is replaced by a normalized water volume, the RWC (Eqn 14.2). Within the linear part of the relationship, P_c is greater than 17 bar (1.7 MPa) and RWC is smaller than 0.78. Deviations from the straight-line relationship arise, when RWC (or the cell volume) is greater than 0.78 and the exerted pressure is less than 17 bar or so (Fig. 14.2). At this transition point the total water potential (ϕ) is −1.7 MPa. The deviation of the curved part from the linear part at RWC greater than 0.78 is caused by the positive turgor or pressure potential. Such a relation with a straight and a curved part presents the pressure–volume curve in total. The relationship resembles the moisture characteristic curve of a soil (Fig. 5.1). It allows the evaluation of ϕ and of the component potentials O and P in relation to RWC (see caption of Fig. 14.2 for details).

The plant water status can be indicated directly by measuring total water potential (ϕ) and the component potentials O and P as well as the RWC. There are also indirect ways to register the water status of plants. Increasingly powerful equipment has been developed that allows plant water assessments to be made in the field (Steiner and Hatfield, 2008). These techniques indicate the effect of the water status under normal or stressed conditions on physiological properties and metabolic turnover. The *porometer* measures the rate of leaf *transpiration* momentarily and calculates *leaf stomatal conductance* to water vapour under standardized conditions (Fig. 9.14). The conductance is an indirect measure of stomatal opening. An advanced type of porometer incorporates gas exchange equipment that allows short-term measurements of the rate of water vapour release together with the leaf's *net CO_2 assimilation rate* (Schulze *et al.*, 1982; Blanke and Bower, 1991). Hence a modern porometer not only accounts for stomatal responses to water stress, but it may also serve to characterize the instantaneous or intrinsic (Condon *et al.*, 2002) *transpiration efficiency* based on the exchange of two gases at the leaf surface (cf. Fig. 15.7). At the time of measurement a cuvette is clamped to the leaf, covering a fixed area of the blade. Environmental factors, such as CO_2 concentration and relative humidity of the air within the

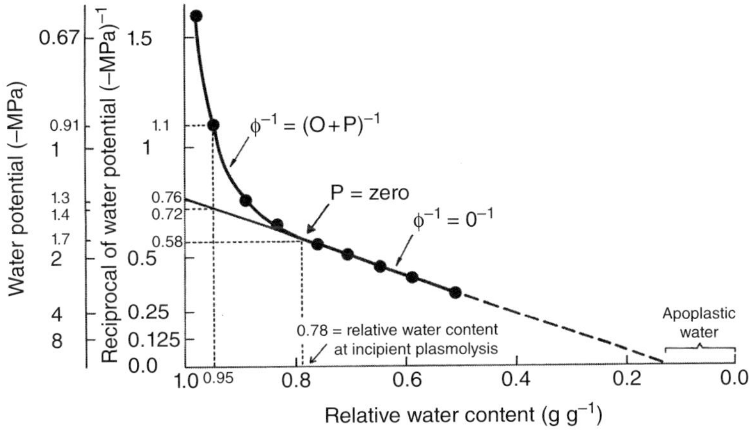

Fig. 14.2. Pressure–volume curve of an oat leaf, measured with the pressure chamber. The water volume in the leaf tissue is presented by the relative water content (RWC). The leaf water potential is the total water potential (ϕ). The potential is negative and its value is identical to the positive equilibrium pressure, exerted stepwise on the leaf tissue in the pressure chamber. The reciprocal of ϕ is plotted as a function of RWC. The relationship is linear at small values of RWC (< 0.78). In this range ϕ is made up solely by the osmotic potential (O) as the pressure potential of the leaf tissue (P) is zero. At RWC > 0.78 the relationship is non-linear. In this range ϕ is composed of O and P. The deviation of measured values from the extrapolated linear regression is caused by P. At RWC = 0.78 incipient plasmolysis is attained, where P approaches zero and ϕ becomes equal to O. Here the reciprocal of ϕ, ϕ^{-1}, is –0.58 MPa^{-1}, and thus ϕ is –1.7 MPa (see auxiliary ordinate on the far left side for non-reciprocal ϕ). The intersection of the extrapolated linear part of the relation with the ordinate (–0.76 MPa^{-1}) indicates the reciprocal of O at maximum P (at 'full' turgor). O here is –1.3 MPa. When extrapolating the linear part to the abscissa, an estimate of the part of apoplastic water in the tissue is obtained. At RWC > 0.78 total potential and component potentials are determined as follows. For instance at RWC = 0.95 the value of ϕ^{-1} = –1.1 MPa^{-1}. It follows that ϕ is –0.91 MPa. For O^{-1} we read the value –0.72 MPa^{-1}. It follows that O is –1.4 MPa. According to the equation P = ϕ – O, we calculate P = –0.91 MPa – (–1.4 MPa) = +0.49 MPa. In this way we obtain a potential diagram as shown in Fig. 7.1 (after K. Grimme, Göttingen, personal communication).

crop stand are recorded and mimicked inside the cuvette. Other environmental variables such as temperature and radiation are measured but not controlled. It is the principle of the 'steady state porometer' to measure gas exchange at the leaf surface under natural conditions, which are not changed by the measuring device. This idea is also facilitated by restricting the period of cuvette attachment to about half a minute, which allows just a few seconds for the measurement. The results of porometer measurements cannot reasonably be extrapolated to obtain integrated values appropriate to a single plant or plant stand, nor to a longer time span of hours or a day.

Water stress can be measured indirectly by recording the *leaf temperature* (Tanner, 1963). When plants are well supplied with water and transpire at a high rate, the leaf temperature (T_L) will be cooler than ambient air temperature (T_A). When the water supply falls short, the stomates will tend to close more. The cooling effect is reduced and the difference $T_L - T_A$ becomes less negative. The difference is determined shortly after noon, when T_A is highest. T_L can be recorded from far distant satellites (remote sensing) or directly in the field. The direct measurement of canopy temperature is greatly facilitated by the use of portable, hand-held *infrared thermometers* that operate without direct contact. Summing up the daily temperature differences (evaluated shortly after noon) over distinct growing periods leads to a measure of plant water stress (Idso *et al.*, 1977). This stress indicator is related to the yield capacity as affected by the level of water shortage. The more the water supply is limited, the less negative is the daily temperature difference, the higher is the stress–temperature parameter and the smaller is the yield.

The maximum transpiration of crop canopies, well supplied with water, depends on evaporative demand (Fig. 8.6) or more specifically on the water vapour saturation deficit of the air (Eqns 8.3 and 8.4), a major driving force in setting the potential evapotranspiration. Therefore the temperature difference $T_L - T_A$ not only depends on the actual transpiration controlled by soil water supply, but will also be influenced by the vapour pressure deficit of the air (Idso *et al.*, 1981; Idso, 1982). Figure 14.3 contains data on $T_L - T_A$, measured for lucerne and plotted as a function of the saturation deficit of the air, $e_s - e$. The crop was well supplied

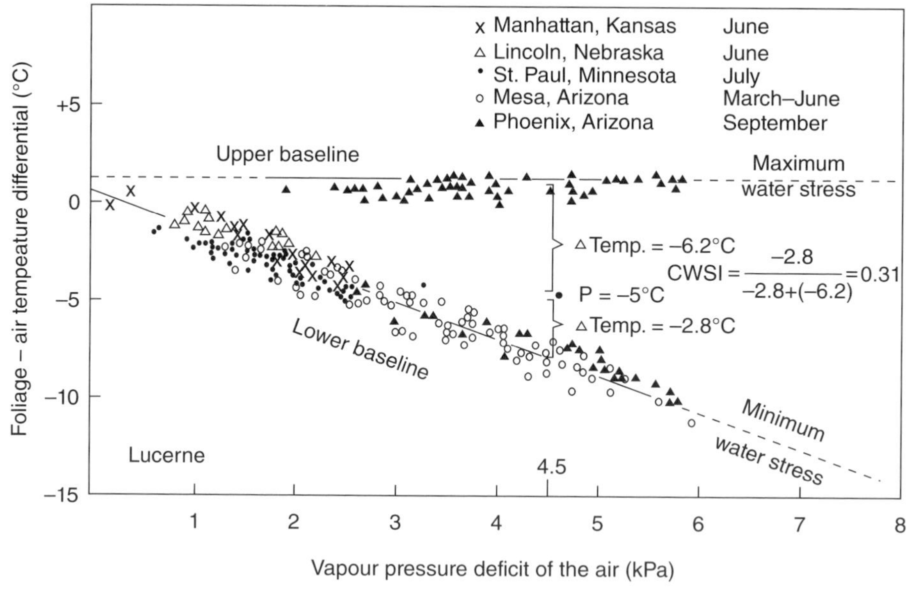

Fig. 14.3. Derivation of the crop water stress index (CWSI). At increasing vapour pressure deficit of the air, the foliage temperature of fully transpiring lucerne becomes more and more reduced compared with the ambient air temperature (lower baseline). The upper baseline temperature results from foliage in dry periods without transpiration. This temperature indicates maximum water stress. It is slightly higher than the temperature of the surrounding air. Data were obtained at five locations in the USA. For the example calculation of the daily CWSI see text for explanation (expanded from Idso, 1982).

with water and therefore at maximum transpiration. With increasing saturation deficit (SD) the foliage temperature decreases due to increased transpiration. The differential $T_L - T_A$ forms a '*lower baseline*', indicating that the lucerne has no, or is at 'minimum', water stress. The line is also called the 'non-water-stressed baseline'. Under conditions of maximum water stress, when stomates are completely closed and transpiration is zero, a second baseline is observed, the '*upper baseline*' (Fig. 14.3). This line does not change with SD and is positioned a little above air temperature. After experimental establishment of upper and lower baselines, a *crop water stress index* (CWSI) can be derived. For the explanation of the index we will assume that the vapour pressure deficit, measured in early afternoon, was 4.5 kPa (Fig. 14.3). Such a high deficit will rarely be attained in the mid-west of the USA (Minnesota, Nebraska, Kansas, see Fig. 14.3) or in Germany (Fig. 9.14), but occurs much more frequently in regions such as Arizona in the south-west of the USA (Fig. 14.3). We will also assume that at the SD of 4.5 kPa the leaf temperature measured was –5°C compared with the air temperature and –6.2°C compared with the upper baseline (P in Fig. 14.3). It is obvious that under the conditions assumed, the lucerne crop must have partially closed its stomates. If not, P should have been positioned at the lower baseline, and the temperature difference to the upper baseline should have been –9°C. Now the difference is taken between the potential reduction in foliage temperature (–9°C) and the actual reduction (–6.2°C), which is –2.8°C. This difference is then related to the potential reduction, which results in 0.31, the value of CWSI for that day. The daily CWSI values are either summed for a period or the mean of that period is taken, arriving at a period-related CWSI.

When CWSI is averaged over defined growth periods, it is negatively related to crop yield. Negative relationships have been established for total above-ground yield as with forage plants and for the grain yield of cereals (Fig. 14.4). Originally, CWSI was derived for dry, arid areas. But the index has also been successfully used in more humid climates for forecasting the yield of various crops under water stress (Keener and Kirchner, 1983; Hattendorf *et al.*, 1988). By now the CWSI is widely accepted as one of the principal methods for identifying water stress in crop plants (Steiner and Hatfield, 2008).

14.2 How Plants Perceive Water Stress

The previous discussions started with the assumption that water shortage in plants is triggered by an imbalance in water uptake compared with water loss. We supposed that water shortage in the tissue causes water stress, accompanied by physiological responses, which ultimately result in a smaller net assimilation rate and reduced dry matter production. There are a few questions left unanswered. Where in the plant is the *sensor* that 'feels' or registers a signal, indicating that water stress is likely in the near future? What kind of a *stress signal* causes the sensor to react? And is there another kind of *message signal* by which the sensor communicates with the stomatal guard cells to induce stomatal closure during periods of stress?

According to what has been said before, it is the *shoot* with the transpiring leaves exposed to the atmosphere that senses a reduction in water potential. This

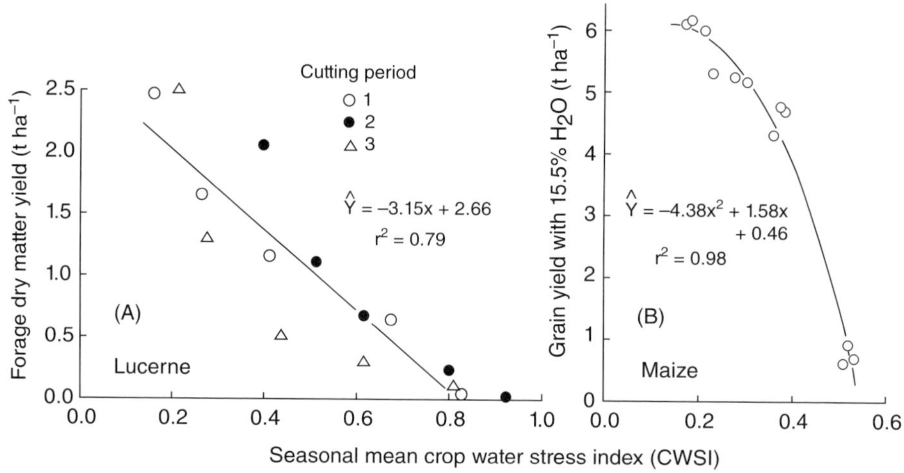

Fig. 14.4. Forage yield of lucerne (A) and grain yield of maize (B) as a function of the crop water stress index (CWSI). The mean CWSI is the average of daily values, measured in early afternoon, for a 2-week period before cutting lucerne grown in New Mexico, USA (A). For maize (B) grown in Antalya, Turkey, daily CWSI values were averaged from early pollination to mid-September (based on Abdul-Jabbar *et al.*, 1985; Irmak *et al.*, 2000).

reduction indicates that a situation of water shortage does develop. Guard cells enclosing the stomates (Fig. 7.6) will lose turgor pressure whenever water potential and relative water content are getting reduced. Turgor loss induces the closing of the stomates. But situations are also conceivable in which plants, well supplied with water, transpire water (and assimilate CO_2) at a high rate without any restrictions by the H_2O control circuit (Section 7.4) as a result of high evaporative demand. These high rates are made possible, because of (or in spite of) a greatly lowered leaf water potential (Fig. 8.2). In contrast, if water supply is limited, the leaf water potential may drop to exactly the same value but then the stomates will be closed. These conflicting responses seem to suggest that the nature of the *message signal* for triggering guard cell movement does not obey *hydraulic–physical* laws exclusively.

It has already been mentioned (Section 7.4) that changes in turgor pressure, which result in the opening and closing of stomates, are accompanied by an active transport of solutes between epidermal or subsidiary cells and the guard cells. An essential solute for stomates control is K^+. Modern views now consider that the signal for an emerging water stress to the guard cells is not solely of a physical nature but also includes the transfer of a *biochemical* component. If so, the question arises again about where the sensor is localized within the plant. There is good evidence that the role of the *root as an early sensor* of the onset of water stress was overlooked for a long time (Davies and Zhang, 1991). For more insight into this field of plant physiology an experiment will be described that had an unexpected result.

Apple trees (*Malus domestica*) were grown in soil-filled containers using the '*split-root*' technique. From the beginning of growth the root system of every tree was split into two parts, with each part growing in a separate container. The technique allowed a different supply of water to the two parts of the root system.

Where the water supply to one of the containers and its associated roots was interrupted, but the other container and its roots continued to be well supplied, the growth of leaves was markedly reduced. The reduction was roughly 40% relative to the leaf growth in the treatment where both containers were watered throughout (from 160% to 100% in Fig. 14.5). Leaf growth slowed down when half of the roots were in drying soil, although a significant water deficit could not be detected in the shoot. Furthermore, it was not just leaf growth that was hampered; the formation of leaf buds was also impaired.

After the 'pretreatment' period of 24 days the actual experiment started (day 0 in Fig. 14.5). The soil and the roots that had been subject to drying were rewatered. Immediately the plants started to accelerate leaf growth again, reaching the growth rates of the trees with regular supply after 10–12 days (Fig. 14.5). This kind of plant reaction implies that the roots in the dry soil must have sensed the dryness in some way, and that they transmitted a chemical signal to the well-hydrated shoot. The result so far appears to be more or less evident, but what was unexpected was that when the roots in the dry soil were not rewatered but were simply cut off instead, those trees also recovered from 'water stress'. After pruning off half the root system the leaves started to grow again at a higher rate, almost reaching the daily increment of the well-watered trees (Fig. 14.5). It is likely that with the excision of the dry roots both sensor and signal transmitter have been removed, which before had retarded leaf formation and growth in a fine-tuned interaction (Gowing *et al.*, 1990).

At first it does seem to be implausible that plant roots thriving in moist soil will sense arising water shortage sooner and more sensitively than the shoot, growing

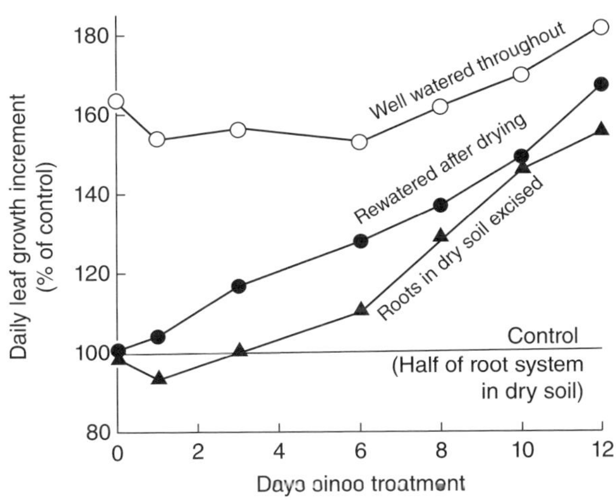

Fig. 14.5. Investigations in apple trees cultivated in containers by use of the split-root technique. In the 'control' half of the root system, growing in one of two containers, the soil dried. The daily leaf growth increment of the control plants was compared with the performance of trees well watered throughout. After 24 days (day 0 on the abscissa) some of the apple trees, with half of the root system dried, were fully rewatered. With some other trees the roots in the dried container were excised at day 0 (after Gowing *et al.*, 1990).

in a dry atmosphere (Kramer, 1988). But, by taking a look at Fig. 9.12 it becomes apparent that some part of the root system in the upper profile is exposed to a soil environment with highly negative water potentials beyond the permanent wilting point. That is true at least for the later part of the season and when the soil has suffered high evaporative water losses. It may well be that in a soil profile, drying from the surface, the *root tips* sense the dryness.

The sensing tips may react differently in two ways. First, they can adapt to soil dryness by *osmotic adjustment*. The adjustment means that an accumulation of osmotically active compounds is taking place within the cells of a root tip. These substances are solutes, mainly organic but also in inorganic form. The solute concentration causes a decrease in osmotic potential and an increase in pressure potential or turgor at a given cell water content. The increased turgor supports the growth of root tips to depth and thus the exploration of additional water reserves (Turner, 1986). Secondly, drying tips can send a *chemical message signal* to the shoot through the xylem, enabling the shoot to undergo an early and '*flexible*' *adaptation* according to the water status of the soil. In consequence, the stomatal cells can respond actively and restrict the transpirational water losses (Box 14.1). The fine-tuning of guard cells in faba bean (Fig. 9.13) indicates such an active response. In this way, by osmotic adjustment and by signalling, turgor and growth of root and of shoot can be supported and maintained when a water deficit starts to develop. Up to now it has been definitely demonstrated that *abscisic acid* is produced in the root, serving as a signal to the shoot for controlling water loss and plant growth (Davies and Zhang, 1991).

But what kind of a *stress signal* is *sensed by roots* when the soil dries and water stress develops? There are three parts to the answer. First, the root tissue will undergo a reduction in water potential and water content. This makes the root shrink in a drying soil, as a result of which the *root hairs* and the *rhizoderm* may *lose contact* with the surrounding soil particles. Secondly, it has to be recognized that elongation of root tips will respond very sensitively to changes in the *mechanical impedance* of the soil, which heavily increases with diminishing soil water content (see Fig. 17.8). And finally, it has to be considered that soil dryness will cause a change in *nutrient flux* from the soil towards the root surface, where it is still in contact with the soil. The rate of nutrient supply will be reduced and the nutrient composition of the influx will be modified. These chemical changes may serve as a signal that water stress is beginning to build up (Davies and Zhang, 1991).

Box 14.1. Signalling between roots and shoots

The common observation that there is a distinct relationship between the growth of roots and shoots led to the concept of *sources and sinks* within the plant, with photosynthesizing leaves acting as sources and sinks being the new leaves, stems, floral parts, fruits, seeds and roots. The studies on plant growth controlling substances – hormones – showed that root and shoot apices were common sites for the synthesis of these compounds or their precursors, and that these compounds relayed messages between the above- and below-ground organs to ensure coordination of growth.

Continued

Box 14.1. Continued

A number of studies have shown that growth of the shoot is adversely affected as the soil dries because of changes in the water relations of leaves, particularly the reduction in leaf water potential and in turgor. However, plants that are growing in drying soil may show no change in leaf water potential but still exhibit stomatal closure and slower growth rates (Passioura, 1988). One explanation is that the plant perceives the extent of the drying of the soil, and reduces the aperture of its stomates accordingly. The suggestion was made in the early 1980s that *chemical signals* were transmitted between the root and shoot to effect a broad range of responses in the shoot, including regulation of stomates, and leaf initiation and expansion (Jones, 1980; Cowan, 1982). At the same time, Goss and Russell (1980) argued that the response of plants to mechanical constraint, such as experienced by roots growing in compacted soil, was mediated by hormonal signals originating in the root apex. Masle and Passioura (1987) showed that increased mechanical constraint induced by soil drying or compaction resulted in similar effects on leaf development and expansion. Tardieu and co-workers systematically investigated the growth of maize in compacted and non-compacted soil, with and without irrigation (see Tardieu, 1994). Plants grown in compacted soil closed their stomates once the top 10 cm of soil was depleted of water, despite the fact that the remaining rooted soil layers contained extensive reserves (Tardieu, 1994).

The experiment by Gowing *et al.* (1990), which we described in Fig. 14.5 and associated text, was the first to demonstrate the transmission of a positive signal from roots to shoots when one part of the root system was experiencing drying soil. Tardieu *et al.* (1992, 1993) showed a close relationship between stomatal conductance and the concentration of abscisic acid (ABA) in the xylem sap of field-grown maize, although the sensitivity of stomates to ABA varied according to leaf water potential (Tardieu *et al.*, 1993).

Tardieu (1994) concluded that the effect of soil compaction on stomatal regulation was also mediated by the synthesis of ABA in the roots. *Phytohormones* other than ABA could be involved in signalling between roots and shoots. For example, Vaadia and Itai (1969) reported a decrease in the production and transport of cytokinin by sunflower roots subjected to drought or salinity stress. If this group of phytohormones was involved in the response to developing water stress, it would be an example of a *negative signal*, since the absence of the compound would result in changes in shoot activity. However, Davies and Zhang (1991) have pointed out that cytokinins can counter the closing of stomates induced by ABA, so that a decline in their production could enhance the impact on shoots of increased ABA release by the roots. Mahouachi *et al.* (2007) reported evidence that accumulation of the phytohormone jasmonic acid could be a trigger for the formation of ABA in papaya (*Carica papaya* L.) seedlings. However, there was no evidence that indole-3-acetic acid (IAA) played any part in the response to drought. Borel *et al.* (2001) have obtained evidence of a chemical signal that acts together with ABA when *Nicotiana plumbaginifolia* was subjected to drying soil. In investigations with leaves of wheat, Munns and King (1988) concluded that ABA was not the compound involved in reducing the rate of transpiration from detached leaves, since its removal from xylem sap did not overcome the anti-transpirant activity. The formation of a large protein molecule was found to be induced by desiccation of barley embryos (Walbot and Bruening, 1988). This and several molecules of similar characteristics (called *dehydrins*) appear to allow plants to withstand a degree of dehydration, possibly by maintaining the tertiary structure of macromolecules or supporting the recovery of damaged ones (Close, 1996). A large molecule found in the xylem sap of droughted plants resulted in dehydrins being produced by unstressed plants that had been treated with the sap (Chandler *et al.*, 1993).

Continued

Box 14.1. Continued

Although there are clearly some important differences between species, in maize and a number of other plants ABA seems to be the most important signal compound. The question is whether the chemical signal is sufficient to explain the level of stomatal control observed in field-grown plants. Tardieu and Davies (1993) argued that both chemical and hydraulic signals combine to modify stomatal conductance. They developed a model that takes account of the transport of water to the root surface, through the plant and into the atmosphere. The concentration of ABA in the xylem sap is considered to be a function of the water potential of the root and the water flux through the xylem. Stomatal conductance (g_s) is given by sum of the minimum value of the conductance (g_{min}) and the exponential function of the product of the ABA concentration in the xylem (X_{ABA}) and an exponential function of leaf water potential (ϕ_L):

$$g_s = g_{min} + \alpha \exp\left\{X_{ABA} \cdot \beta \exp\left(\delta\phi_L\right)\right\}$$

where α, β and δ are fitted parameters. The value of $g_{min} + \alpha$ is equal to the maximum stomatal conductance.

They found that the model simulated field measurements of the typical daily pattern of stomatal conductance (Fig. 9.14), and also showed the lack of any unique relationship between conductance and leaf water potential, suggesting the contribution of the chemical signal ABA. The model also simulated the response of stomatal conductance to the changing resistance to water flow as the soil dried. The simulated results were again similar to those observed in the field experiment.

Tardieu and Davies (1993) suggested that the short-term linkage between stomatal response and changes in the soil environment would be provided by the variation in sensitivity of guard cells to ABA concentrations, which are relatively well buffered against transient changes in hydraulic conditions such as those due to the passage of individual clouds. Longer-term effects on the control of developmental processes within the plant would result from the steady increase in ABA concentration of the xylem due to the drying of the soil.

The degree of integration of hydraulic and chemical signals may differ between plants. For example, there is evidence that in some trees hydraulic signals are the most important for the control of stomates. Equally, the control of signal production by roots may also differ. This has been explored using transgenic lines of *N. plumbaginifolia* that vary in their ability to synthesize ABA, together with wild-type plants and a mutant which lacks the gene to produce zeaxanthin epoxydase (the enzyme that is important for the final stages of ABA synthesis). Although the transgenic lines vary in their ability to accumulate ABA in response to drought, the sensitivity of their stomatal guard cells to exogenous ABA introduced into the xylem remained the same. None the less, there were differences between transgenic lines in the effect of soil drying on stomatal conductance (Borel *et al*., 2001).

To gain insights into shoot to root signalling, Vandeleur *et al*. (2014) investigated the influence of leaf transpiration on root hydraulic resistance (L_0), a flow resistance mainly located in the root cortex (Section 8.1, Fig. 8.1). They focused on root conductance, which is the reciprocal of root resistance and commonly expressed per unit surface area of root. However, Vandeleur *et al*. (2014) used root dry mass to normalize L_0, resulting in a unit of measurement of kg H_2O s^{-1} MPa^{-1} g^{-1} root. When light intensity varied, L_0 was linearly related to leaf transpiration in soybean and grapevine. Subjecting soybean and maize to leaf pruning or partial excision, sufficient to reduce leaf area by about 30%, caused L_0 to decline sharply within an hour by about 50% depending upon the initial magnitude of L_0.

Continued

Box 14.1. Continued

In soybean, turgor pressure in the root cortex also declined following excision of the shoot tip, including the first fully expanded leaf. Over the same time interval, there was a reduction in the expression of several genes responsible for the formation of aquaporins (Section 7.1) in the plasma membrane of both grapevine and soybean. In soybean, the expression of one aquaporin protein was reduced five- to tenfold and the effect was maintained over the time that L_0 was affected. This strongly suggests that signal from the shoot to the root involved aquaporins, which are greatly influenced by ABA (Mahdieh and Mostajeran, 2009).

The concept that *hydraulic and chemical signals* are integrated in the plant to control the movement of water and water balance within the plant has important implications. First, the plant response is governed by the source, the variation in supply from the soil, and not simply reacting to changes in demand. Secondly, the response is genetically controlled and therefore there are likely to be differences between plants in the relative importance of the two types of signal. Finally, the dual control suggests that the stomates should have variable sensitivity to the signal or signals produced by the roots.

15 Climatic Factors Influencing Yield

15.1 Growth-limiting Climatic Factors

Once a day the earth rotates on its axis, causing *day and night*. The axis of spin is not positioned perpendicular to the earth's orbit around the sun or to the incoming rays from the sun. Rather the axis is tilted by an angle of about 23°. On 21 June the upper part of the axis, the part through the northern hemisphere, is at an angle of 23° towards the sun's rays, exposing the North Pole to uninterrupted radiation and causing the spectacle of midnight sun. More generally, at this time the days are long, with more than 12 h of radiation north of the equator. At the same time there are shorter days with less than 12 h of radiation being received in the southern hemisphere. By 21 December, however, the southern part of the axis is at an angle of 23° towards the incoming rays, now causing 'long' days in the southern hemisphere and 'short' days in the northern hemisphere. The reason for this *shift in day length* during the year is that the 23° inclination of the axis is kept fixed within the space coordinates during the orbit of the earth around the sun, lasting for 1 year. In short, we can say that the daily rotation of the earth causes the regular succession of day and night, while the annual passage around the sun and the tilting of the axis result in changes in the length of day and night from one day to the next throughout the year. The amplitude of this annual variation in day and night length is small at the equator, but it increases at higher northern and southern latitudes. The same is true with daily *solar radiation* (Fig. 15.1). There is a slight variation at the equator with two inconspicuous maxima and minima. At higher latitudes radiation peaks in summer and is least in winter. It may also be noticed that the global radiation flux received during one day at higher latitudes surpasses the daily flux at the equator during the summer months. Of course this is much more a matter of day length than of instantaneous radiation flux density. A distinct annual variation in sunshine duration and intensity is responsible for the phenomenon of *seasons*.

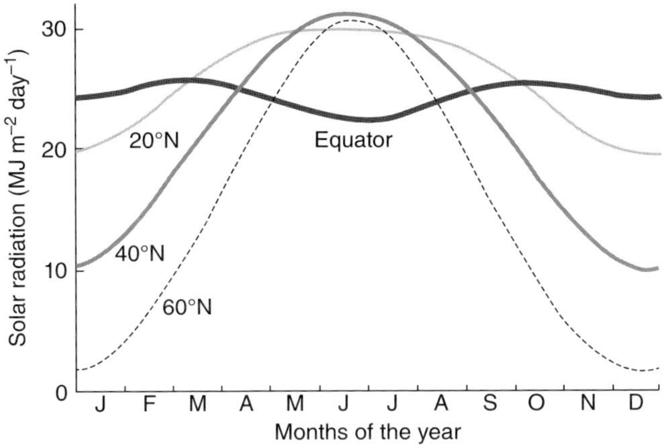

Fig. 15.1. Annual variation in global radiation at different northern latitudes and at the equator. The global radiation reaching the soil surface was calculated for the maximum possible duration of sunshine, which varies with latitude and time of the year. The attenuation of radiation by a cloud cover was excluded (calculation based on data tabulated in Withers and Vipond, 1974).

Radiation is an important factor that limits plant growth during the winter season in Central Europe and other areas at higher latitudes of the northern or southern hemisphere. In the 'dark' season, radiation flux may be so small that gross photosynthesis is drastically reduced as compared with respiration (Fig. 10.1). Because of limited radiation the months during the winter represent the time of colder *temperatures*. Cool temperatures will reduce the rates of both assimilation and respiration (Fig. 1.4A). A large number of plant species from these areas have adapted to the unpleasant conditions of life. Some perennials, for instance, *shed* their *leaves*, and winter annuals have developed the need for *vernalization*. A stimulus of low temperatures for some period of time is actually required to induce the formation of reproductive organs. That stimulus is experienced during the winter months. Such adaptations help avoid luxuriant shoot growth and the initiation of flowering ahead of the cold season. The likelihood of surviving and overcoming the ravages of winter is enhanced, as is the possibility of successfully completing reproduction during the subsequent summer season. Until they are vernalized those plants remain in a rosette state during winter, and the apical meristem remains near the soil surface. This way the meristem is comparatively well protected from damaging frosts. The cultivation of *winter forms* of cereals like barley, rye and wheat and of oilseed rape is widely spread in regions with a temperate climate. However, in regions with a more continental climate the winter forms of these small-grain crops are not grown as much. During long persisting periods of very low temperatures and in the absence of deep snow cover, crops have little chance of survival. In these areas *summer forms* are preferred, being free of the need for vernalization. These forms are also cultivated in countries at lower latitudes, where the winter is usually mild, such as in Mediterranean climates, or where the winter is absent, as in the subtropics and tropics. In these regions it is not so much the radiation and the temperature that determine the start of the growing season, it is much more the availability of *soil water* for germination and growth, and that is dictated by the onset of the *rainy season*.

Generally speaking, the cropping season starts when the climatic factors become favourable. In regions of higher latitude the limitation of vegetative growth by low temperature and radiation gradually disappears during spring. The increase in intensity and in daily length of radiation makes the temperature of air and soil rise. Crops develop leaf area, and an ever-increasing percentage of radiation is intercepted by the canopy (Fig. 10.4). Accordingly crop growth rates increase (Figs 10.4, 10.5, 10.6).

In spring as the *day length* increases in areas of higher latitude, *long-day plants*, such as the small-grain cereals of temperate zones, undergo a *developmental acceleration*. The plants pass through developmental processes much more rapidly under the conditions of long-day radiation. This effect of day length is called *photoperiodic induction*. The developmental stages affected include: (i) the formation of leaf primordia; (ii) the unfolding of leaves; (iii) flower formation; (iv) extension growth; (v) anthesis; and (vi) fruit formation. Rising *temperatures* in spring stimulate the acceleration in development (Baeumer, 1992). The temperature effect can be explained by the concept of *thermal time*. The warmer the temperature (above a base temperature) that plants experience during a period of growth, the shorter is the period needed to attain a particular growth stage. Thermal time is expressed in temperature-time units, and is the integral of temperature over time. The unit is for example °C × day. Sometimes thermal time is referred to as 'heat units' or 'growing degree days'.

It is therefore a consequence of photoperiodism and temperature that late sowing of summer annual crops in spring may induce a considerably accelerated passage through vegetative development, and a very short time to build up shoot mass. Plants with a small shoot mass normally set few fruit. The fruit that is set is not well supplied with assimilates as the number and area of sustaining leaves remain small. Shortage of assimilates required for fruit growth is accentuated due to the fact that developmental acceleration also causes early ripening. The German saying that 'May-oat is chaff-oat' is indicative of late seeding causing a shorter growing season for various reasons, with the consequence that the capacity of the crop to yield well will remain poor.

Maize originates from the low latitude region of Central America. The original forms and cultivars are *short-day plants*. Exposed to long-day radiation when spring seeded in temperate climates, plant development becomes retarded. The modern cultivars now used for commercial production have been transformed into *day-neutral plants*, which are more or less unaffected by day length. In maize, rising temperatures in spring increase growth rate more than the developmental rate. As it is a C_4 plant, maize assimilates CO_2 very efficiently and unlike C_3 plants it does not exhibit photorespiration. Therefore maize, like many other C_4 plants, has superior dry matter accumulation at higher temperatures compared with C_3 plants. But this superiority can be achieved only (Fig. 10.1) at high radiation flux density (Baeumer, 1992).

Carbon dioxide is an essential unit in the production of plant biomass (Fig. B1.1). Although its concentration in the atmosphere is steadily increasing due to human activities, it still remains a growth-limiting factor. Therefore one may expect that an ongoing increase in CO_2 concentration will improve the CO_2 exchange rate at the leaf surface and the growth rate of crops. These relations will be discussed in more detail in Section 15.3.

From the foregoing it is clear that, in addition to temperature, it is radiation that limits plant growth and production at higher latitudes during winter. But we are particularly concerned about if and when *water supply* becomes a growth limiting factor, and whether growth limitation because of water shortage occurs, even in the temperate climates of regions within Central Europe. During the transition from spring to summer, *air temperature* and *saturation deficit* of the air (Fig. 15.2, p. 228, Göttingen, Germany) increase just like potential evapotranspiration. Therefore one might assume that the water supply will become more critical during the summer months because evaporative demand is large, even though this period also coincides with a peak in monthly precipitation (Fig. 15.2). So we end up with the question as to whether these climatic conditions are sufficient to generate periods of water shortage and possibly water stress in plants during the course of a year. The question is posed without taking into account the water storage capacity of soils, a quality that is certainly of prime importance for balancing periods with little rainfall. In other words, are there any months during the year when plant production is limited not so much by *radiation* but by periods with insufficient *rainfall*? These considerations will not be restricted to just one climatic zone, but will be extended by comparison to *five types of climate* (Fig. 15.2).

In the upper row of Fig. 15.2 five *climatic diagrams* are depicted according to the approach of Walter (1970). The abscissa contains the months of the year. The left ordinate gives the long-term (often 30 years) mean monthly temperature, and the right ordinate the mean monthly rainfall. Both the ordinates are arranged in a form that the scale for temperature corresponds to twice that for precipitation, i.e. 20 °C corresponds to 40 mm. Whenever the rainfall curve is positioned above the temperature curve, a period of *humidity* is indicated, emphasized by the hatched area. This kind of diagram with the period of humidity extending throughout the year is typical for a temperate, sub-oceanic climate. There are climatic zones, however, where periods of humidity alternate with periods of *aridity* (Fig. 15.2). During these dry periods the rainfall curve drops below the temperature curve. The corresponding area is stippled. Dry periods exist in regions with Mediterranean, monsoon and savannah climates. The steppe climate is arid in autumn, and the degree of humidity is less developed throughout the year than in temperate regions; the hatched area being much smaller. Regions with monsoon and savannah type climates may experience periods of both aridity and excessive rainfall (the latter being indicated by the black area). Notice the discontinuity in the rainfall scale.

In changing from temperate to a tropical climate the mean annual temperature increases (Fig. 15.2, upper row). At the same time the annual variation in temperature between winter and summer declines. In Cotonou, Benin, air temperature cycles annually like radiation at the equator (Fig. 15.1) with two slight maxima and minima (Fig. 15.2). Here the relative humidity (RH) is high throughout the year, ranging between 80% and 88%. That is the reason why the *daytime saturation deficit* (SD) is comparably small in spite of high temperatures (Fig. 15.2, middle row). On the other hand, in the Indian city of Nagpur, the RH

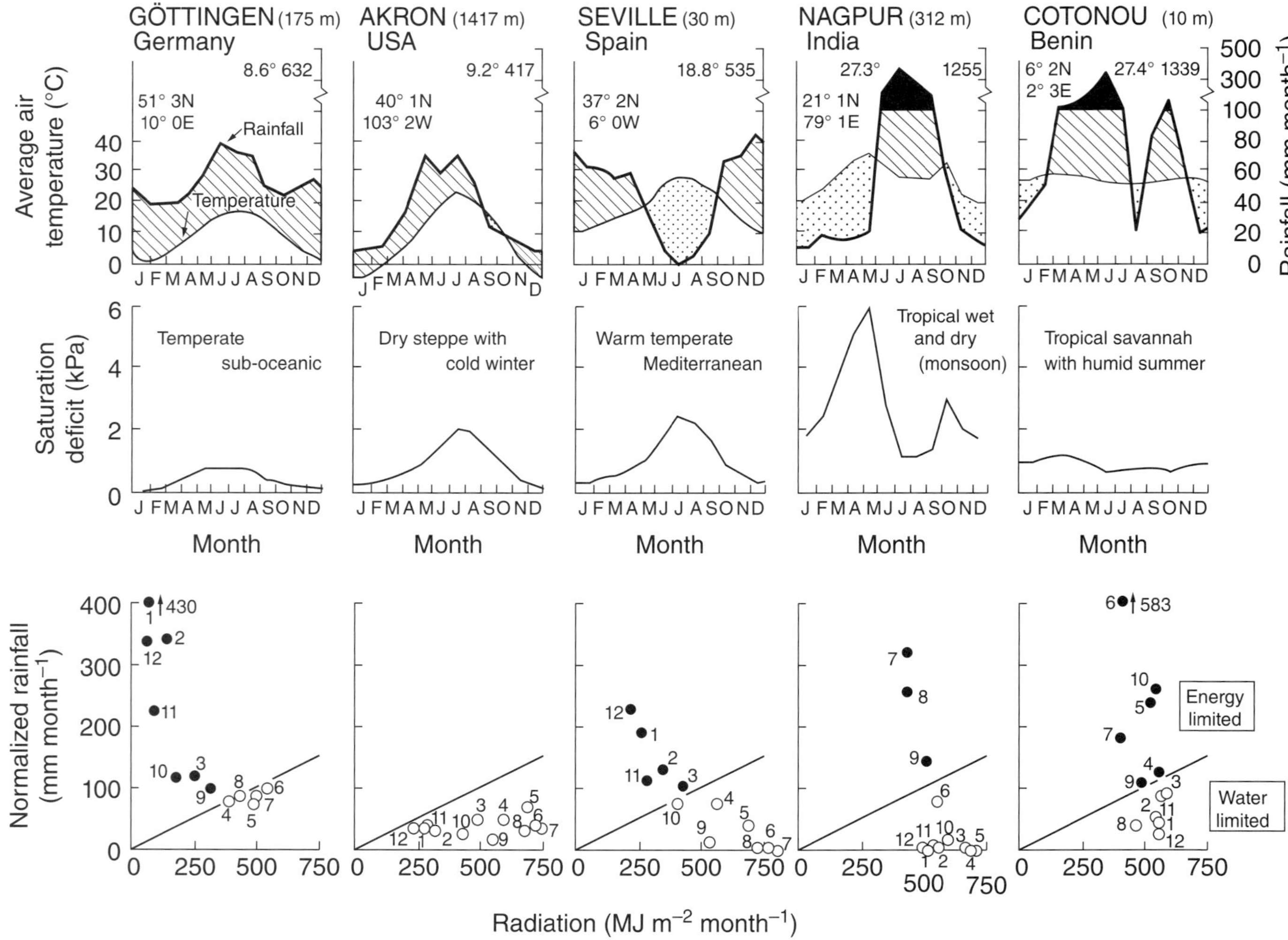

Fig. 15.2. Upper row: Climatic diagrams for five locations, based on the concepts of Walter (1970). The graphs depict the annual variation of monthly average temperature and rainfall (including snow). A temperature of 10°C corresponds to a mean monthly rainfall of 20 mm. The

varies dramatically from 27% in May to 78% in July. Temperature drops from 35 °C to 27 °C. Accordingly, the daytime SD (Section 11.2) drops considerably from the extremely high value of 6 kPa at the end of the dry season to only 1.1 kPa during the monsoon months. In general for tropical climates the SD changes between dry and rainy seasons. In the temperate, steppe and Mediterranean climates the SD varies from winter to summer. In this sequence the mean SD and its annual variation increase (Fig. 15.2, middle row). The large summer SD observed in the Mediterranean climate, as illustrated by Seville in Spain, develops during the dry months, whereas in the temperate and the steppe climates it happens despite the fact that summer rainfall is greater than that in winter. During the summer months the daytime SD in Akron, Colorado, is about twice that in Göttingen, Germany (Sections 11.2 and 11.3).

The next concept we introduce, leading to the bottom row of graphs in Fig. 15.2, is based on the idea of the Scottish agricultural meteorologist, Monteith (1990). From his experiences in the Indian monsoon climate he concluded that *growth promotion during the rainy season* can only be partially attributed to increased *soil moisture*. 'About one third of the benefit of rain in terms of crop production can be ascribed to a decrease of *vapour pressure deficit*' (Monteith, 1986b, cited in Monteith, 1990). An explanation for this phenomenon seems reasonable, based on Eqn 11.2. Such a drop in SD is clearly evident at the start of the wet season (Fig. 15.2, Nagpur). The ordinate of the graphs in the bottom row of Fig. 15.2 depicts the '*normalized*' *monthly precipitation* (P_n), which makes allowances for the influence of SD. P_n is defined as:

$$P_n = \frac{P}{\Delta e} \cdot (\Delta e)_o \tag{15.1}$$

In the equation P is the long-term average monthly rainfall and Δe is the corresponding SD, measured during the daylight hours. The term $(\Delta e)_o$ is the 'standard SD'. (When Δe is measured in kPa like here, then $(\Delta e)_o$ is 1 kPa. When Δe is given in mbar, then $(\Delta e)_o$ attains the value of 10 mbar.) The term $(\Delta e)_o$ is introduced

Fig. 15.2. Continued.
altitude (in brackets), the mean annual temperature (°C) and mean annual precipitation (mm) are indicated. In addition, the latitude and longitude of the site are listed. Hatched areas represent humid seasons, dotted areas represent arid seasons. Black areas stand for seasons of monthly rainfall >100 mm.
Middle row: Annual variation in mean monthly 'daytime' saturation deficit of the air. The type of climate is indicated.
Bottom row: The monthly rainfall is 'normalized' by the monthly daytime saturation deficit and related to monthly global radiation received at the location in question (Monteith, 1990). The numbers 1–12, added to filled and open circles, stand for the corresponding months. The straight line is based on two plant inherent features: its slope is defined by the ratio of radiation use efficiency and a measure of transpiration efficiency (factor k of Bierhuizen and Slatyer, 1965). Crop production is energy limited during those months when the normalized rainfall exceeds that indicated by the line, marked by filled circles. Production is water limited during months located below the line (open circles). The climatic data originate from Müller (1982) and the German Weather Service (1982–1998). For more explanations see text.

simply to keep the unit of normalized precipitation P_n in the unit of P (i.e. mm). In the graph P_n is plotted as a function of the long-term average global radiation received per month. The radiant energy was calculated from measured values of sunshine duration and meteorological factors, as explained by Withers and Vipond (1974). Data on sunshine duration were taken from Müller (1982) and supplied by the German Weather Service (1982–1998).

The graphs in the bottom row of Fig. 15.2 contain a straight line. The slope of the line (j) is the result of two plant parameters that are already well known to us:

$$j = \frac{RUE}{TE \cdot \Delta e / (\Delta e)_o} \tag{15.2}$$

In Eqn 15.2 RUE is the *radiation use efficiency* (see Table 10.1) and TE is the *transpiration efficiency*. The quantities Δe and $(\Delta e)_o$ are the same as in Eqn 15.1. The product TE × Δe is equivalent to factor k in the Bierhuizen–Slatyer equation (Eqn 11.2). Data on k for various crops were given in Table 11.3 (see also Fig. 11.4B). As both RUE and k can be considered to be *crop-specific properties* that are constant for a given Δe, then the ratio used in calculating j should behave similarly.

To obtain a value of j, we will use a value for RUE of 3 g plant dry matter (DM) MJ^{-1} intercepted photosynthetically active radiation (PAR) (Table 10.1), which corresponds approximately to 1.5 g DM MJ^{-1} intercepted global radiation. We will use the maize value of 8 Pa for k of a C_4 crop, and for C_3 crops like small-grain cereals we will use 4 Pa (Table 11.3, Fig. 11.4B). Then $j(C_4)$ is: 1.5 g DM/MJ divided by 8 Pa/1 kPa. The latter fraction is 0.008. We can convert each of these components to a more convenient number by multiplying by the ratio kg [DM] per (kg [H_2O]/m^2). The first fraction is then 1.5 g DM/(MJ m^{-2}), and the second fraction is changed to 0.008 kg DM/mm H_2O or 8 g DM/mm H_2O. The slope $j(C_4)$ is then: 1.5 g DM/(MJ m^{-2}) divided by 8 g DM/mm H_2O, resulting in a value of approximately 0.2 mm H_2O per MJ m^{-2}. The slope of the straight line in the bottom row of graphs of Fig. 15.2 has been calculated in this way. The value of j for C_3 plants is about 0.4 mm H_2O per MJ m^{-2}, but will not be considered in the following section.

Having followed the explanations given so far, the point of the approach will become obvious. With the aid of the straight line we can recognize months of the year when crop production is characterized by an abundant rainfall relative to the evaporative demand of the atmosphere. The transpiration of the crop is equal to the potential when the canopy is fully developed, covering the ground. Dry matter accumulation and transpiration are set by the energy supply in the form of radiation, intercepted by the plant stand. The storage of excess water in the soil is not considered. The crops receive more rain than needed. Crop production is *energy limited* (EL environment). Months with EL production are characterized by filled circles, situated above the line with a slope equal to j.

Depending on the type of climate and the season, crop production may also be influenced by a shortage of water, supplied to and extracted by the growing root system. Water uptake may be inadequate, at least during part of the day, as compared with normalized rainfall. Plants then suffer from water stress and close their stomates, restricting CO_2 uptake. Thus water shortage reduces the utilization of radiation for biomass production. Plants may handle extended periods of drought by leaf rolling, folding or shedding, reducing transpiration as well as the energy load hitting the canopy. Both transpiration and production are *limited by water* (WL environment) supplied in rainfall. Months with WL production are depicted as open circles. They are located below the line in Fig. 15.2 (Monteith, 1990).

In a *temperate climate* such as in Göttingen (Fig. 15.2), there is an apparent shift from distinctive EL conditions in winter months to slight WL conditions in summer months (May–July). Radiation is very much reduced in December and January (< 100 MJ m^{-2} $month^{-1}$). It is much greater in summer, but does not exceed 530 MJ m^{-2} $month^{-1}$ (June). The yearly duration of sunshine, added for comparison, is 1473 h.

In the *steppe climate* of Akron, Colorado, USA, the duration of sunshine is very much greater, about 3000 h $year^{-1}$. Rainfall is two-thirds that of Göttingen. But whereas in Göttingen the monthly mean RH is always higher than 65%, it varies in Akron between 50% and 57% throughout the year. The air in Akron is quite dry. The high SD is the main reason why the normalized rainfall is so small in the steppe climate as compared with the temperate climate. The radiation reaches 750 MJ m^{-2} $month^{-1}$. The high level of radiation and low level of normalized rainfall are the cause of WL conditions during the entire year, which are accentuated during the summer months.

Total sunshine per year in the *Mediterranean climate* of Seville, Spain, is 2878 h. Radiation changes during the year are similar to those of the steppe climate, but peaks at greater values in June and July, when rainfall is almost nil. There is a seasonal shift from EL conditions in winter (being less extreme than in a temperate climate) to WL conditions in summer (being more extreme than in the steppe climate).

In the *monsoon climate* of Nagpur, India, the sun shines for 2611 h $year^{-1}$. The monthly radiation varies much less during the year than in the three types of climate mentioned so far. Rainfall is considerable (>200 mm $month^{-1}$) during the monsoon period from June to September. Precipitation during these 4 months accounts for 87% of the yearly total. June is actually the transition month from the dry period to the rainy season. The early days of the month are usually free of rain, and the SD remains high. Thus the normalized rainfall of the transition month turns out to be small relative to the real monsoon months. But it is higher than that from October to May. During the monsoon, rainfall and normalized rainfall are abundant. It is more than needed to satisfy the potential demand set by radiation. It is an EL condition. During the rest of the year, each month is characterized by severe WL conditions.

The *tropical savannah climate* in Cotonou, Benin, experiences 2302 h of sunshine per year. Maximum monthly radiation is less than that received during high radiation months of the monsoon, the Mediterranean and the steppe climates. But as in a tropical monsoon climate, annual variation in monthly radiation is also small in this type of tropical climate. It is characterized by a bimodal distribution pattern of rainfall. The main rainy season lasts from March to July with 67% of annual precipitation. From May to July EL conditions exist. After a dry spell in August there is a shorter rainy season from September to November with 23% of total rainfall per year. In October, crop production is EL. But half of the year, from November to March and the month of August, is characterized by greater or lesser WL conditions.

In summarizing this section on climatic diversity we may conclude, based on the evidence of Fig. 15.2, that in *temperate climates* water supply from rainfall is quite evenly distributed throughout the year. Peak rainfall in the summer months makes sure that radiation can be used by crops without too much concern over water limiting conditions. This is especially true on soils that store a large volume of *available water* in the *effective rooting depth* (Section 9.4). Water supply may become short during June and July or perhaps even in August, especially in sandy soils that only contain a small amount of extractable water (Table 9.2). In the *steppe climate* water supply through rainfall is much more restricted. Because of very low winter temperatures arable cropping is not possible during the cold season and the fields are left bare. This is the time to collect water in the soil in preparation for the coming vegetative season, in part by capturing snow. Nevertheless, water supply through rainfall is so severely restricted that crops cannot take full advantage of the high level of radiation during summer. To make use of the radiation, additional water through irrigation will be necessary. In a *Mediterranean climate* rainfall in winter can be used directly for crop production, as the temperatures do not fall excessively. Nevertheless temperature and radiation are suboptimal for crop production. Only with the return of spring can high crop growth rates be realized, but water will soon become the factor that limits production. Irrigation becomes necessary if the radiation is to be exploited fully. Under a *monsoon climate*, temperature and radiation are favourable to plant production throughout the year, which is also true for the *savannah climate*. But in the monsoon climate there is a long uninterrupted period of dryness, which makes irrigation necessary. It is the most prominent climate for irrigated, mostly flooded lowland or wetland rice, with two or three crops per year. Because of the bimodal rainfall in the savannah climate, the need for irrigation will be less. Here lowland rice may be grown without additional irrigation water, often alternating with a short-season cash or food crop within the same year.

So far we have looked at some types of climates, characterized by the long-term average of various meteorological variables. But in particular years, the climatic conditions may depart quite markedly from normal. Less than normal precipitation accompanied by high SD may lead to disastrous situations in crop production in those years. That is especially true, and ironically so, for areas where even the mean annual rainfall is small. In many of these regions plant growth cannot be aided by irrigation, as the natural and technical prerequisites are lacking. Hence, in vast areas of the world cropping depends solely on natural rainfall and its annual variation. Field production without the support of irrigation, relying completely on natural precipitation, is called *'rainfed' agriculture*.

The departure in climatic variables from those of normal years may be considerable, even in regions of temperate climate, causing definite WL conditions. In Göttingen, the year 1989 (Fig. 15.3, upper row) was warmer by 1.5 °C and drier by 145 mm than the average year (Fig. 15.2). From May to July the SD was higher by about 0.2 kPa (Fig. 15.3, middle row). According to the Walter diagram (upper row) there was aridity in May and July. May was a month with unusually high radiation, and during the May–August period crop production was apparently WL (bottom row). Did the shortage in normalized rainfall affect yield?

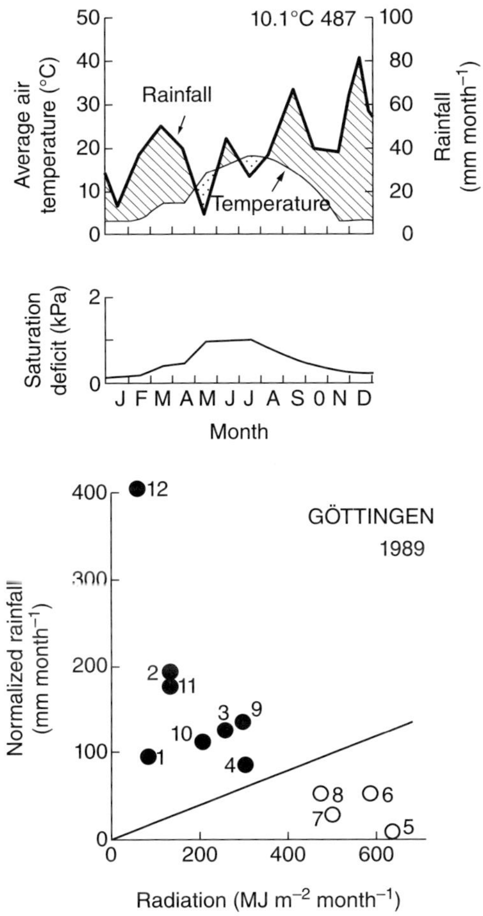

Fig. 15.3. Climatic diagram for Göttingen, Germany, after Walter (1970) in the dry year 1989 (upper row). The saturation deficit (middle row) and the normalized rainfall as a function of radiation (bottom row) are shown as well. More explanations are given in Fig. 15.2 and in text.

The answer is evident in Table 15.1. The mean yield of three cereal species, grown in the district of Göttingen, was definitely smaller in 1989 compared with the 10-year average from 1983 to 1992. The months from May to July are important for setting the yield. In 1989, all were warmer and drier than the 10-year average (Table 15.1), as a result of which the decline in yield was apparently induced. In contrast with 1989, conditions in 1986 resulted in comparatively higher yields. In 1986, annual rainfall and rainfall in May and July were greater than average. Also temperatures in May and June were warmer, but were not harmful to yield due to balancing rainfall. In 1983, however, rainfall did not compensate for the high temperatures in June and July, which probably caused the yield depression in that year.

In the district of Göttingen soils are predominately loamy or clayey in texture with a medium to large amount of available water (Fig. 3.2). In hilly areas they

Table 15.1. Weather (meteorological station Göttingen) and grain yield (district of Göttingen) in three individual years compared with the 10-year average. Temperature and precipitation are given as annual means and for three specific months of the vegetation period. This period is important for grain formation in cereals (data from German Weather Service, 1982–1998; Federal Office of Statistics, 1983–1992).

Period	1983	1986	1989	1983–1992
	Temperature (°C)			
May	11.6	14.6	14.4	13.1
June	16.8	16.3	16.1	15.4
July	20.4	17.6	18.3	17.9
Year	9.6	8.0	10.1	9.1
	Precipitation (mm)			
May	79	83	8	54
June	35	51	45	66
July	48	66	27	57
Year	588	687	487	624
Crop	Grain yield (t ha^{-1} and relative)			
Winter wheat	5.6 (88)	6.9 (108)	5.4 (84)	6.4 (100)
Spring barley	3.2 (80)	4.8 (120)	3.3 (83)	4.0 (100)
Spring oat	3.7 (84)	6.2 (118)	3.0 (68)	4.4 (100)

are represented by Luvisols derived from loess, Vertic Cambisols from argillaceous rocks and Rendzinas from shell-limestone. Gleysols and Fluvisols have developed in the flood plains. Therefore one should expect a reasonable capability of the soils to compensate for limited rainfall and large evaporative demand by the supply of stored water. In areas with more sandy soils of lower available field capacity (FC_{av}, Table 9.2) the supplying capacity should be lower and the yield variability between years much higher. When sugarbeet, a crop with a long vegetative period and greater water use (Table 9.3), is grown in Germany on these lighter soils, additional water supply by irrigation is common.

What is striking about the yield data in Table 15.1 is that, when comparing single-year data with the 10-year average, the yield difference for spring cereals (i.e. barley and oat) is greater in both absolute and relative terms than for winter wheat. Such a differential response of spring and winter grown crops to changes in annual weather patterns seems to be a common phenomenon (Hlavinka *et al.*, 2009). The better *yield stability* as well as the higher yield level of winter wheat indicate that winter crops can make use of spring soil moisture during April and May (Fig. 9.5), much better than spring-sown crops (Figs 9.5 and 9.6). Those are the months with transition from EL conditions to WL conditions, when crop growth is limited by a climatic water deficit (Figs 15.2 and 15.3). But the SD and the deficit in normalized rainfall build up gradually and do not affect yield formation immediately. The longer the time of transition is extended, the deeper the root system grows into the subsoil and the greater is the FC_{av}. In contrast to winter wheat, the growth and water use of spring-sown crops are retarded at this time. Therefore roots are growing at shallower depths and the main period of water use may be delayed until June and July (Figs 9.4, 9.5 and 9.6), when the SD is generally high and the expenditure of water for biomass production has increased unfavourably. Under these circumstances the threat from a climatically induced water deficit is greater and even more likely, as the date of harvest is normally later than in winter cereals.

15.2 Climate Change

'The greatest experiment on earth has begun. We are changing the parameters that drive the planet's living systems' (Rogers *et al.*, 1994). More than 20 years have passed since the release of this statement. To the large majority of scientists it has become obvious that the global climate is greatly changing and that the change is taking place rapidly, caused by the influence of increasing human activities. The number of researchers who reject anthropogenically induced climate change as a fact has steadily decreased. Global climate change has emerged as a serious challenge and threat to mankind. There is both the hope that and a fear as to whether mankind is intelligent enough to develop effective countermeasures in time to overcome the accelerated change. In the worst-case scenario climate change develops a momentum of its own, where all efforts to stabilize and control the system prove to be in vain. For any successful strategy the basic prerequisite is that mankind opts for more self-restraint in attitude and expenditures. As in Europe, nations worldwide have to come together, have to cooperate, have to develop common policies and give up striving for national dominance. The fact is that up to the present, global decision makers have not taken serious account of, nor acknowledged, responsibly for the magnitude and consequences of the threat to the climate. And this is in spite of the increase in extreme weather events hitting both poor and rich nations alike. Fundamental countermeasures have not yet been employed on a global scale.

Global climate change is being caused by alterations to the gas composition of the atmosphere. For example, the concentration of *carbon dioxide* (CO_2) near ground level has increased from 275 ppm in the pre-industrial era to about 400 ppm, a value recorded for the first time in May 2013. That is an increase (Fig. 15.4) of not less than 45.5%. In addition to CO_2 there are *trace gases* affecting the global climate. These are *methane* (CH_4), *nitrous oxide* (N_2O, laughing gas), *water vapour* (H_2O) and *ozone* of the troposphere (O_3). The troposphere is the layer extending from ground level to about 11 km in height, and conditions there determine the weather we experience (Fig. 15.5). Although all of these trace gases in the atmosphere are produced by *natural* ecosystems, each of them have experienced enrichment within the past 100 years because of human activity. In addition there are some *anthropogenic* gases, belonging to the group of *chlorofluorocarbons* (CFC, Fig. 15.5), which have been produced and released to the atmosphere only since the mid-20th century.

The *ozone of the troposphere* is formed by photochemical oxidation of *carbon monoxide* (CO), CH_4 and some *volatile organic compounds* (VOC). The oxidation is catalysed by *nitrogen oxides* (NOx). Since pre-industrial time the tropospheric

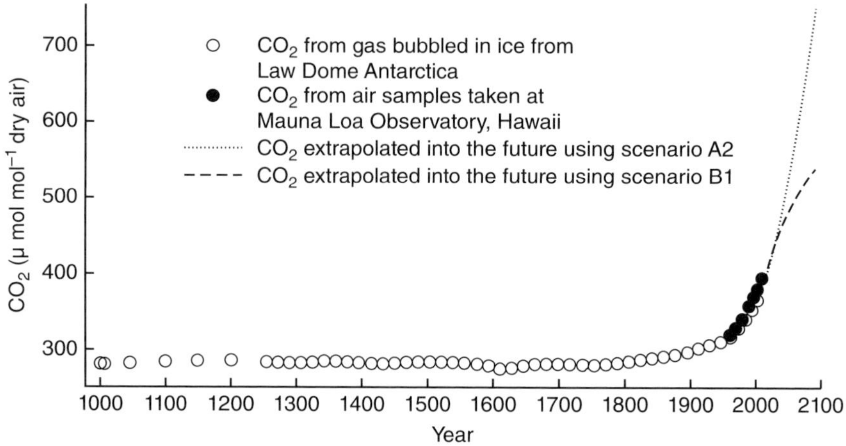

Fig. 15.4. Temporal development of the CO_2 concentration in the atmosphere over the last 1000 years with two projections for further changes over this century. The early CO_2 concentrations were measured in gas bubbles sampled from Antarctic ice (MacFarling Meure *et al*., 2006). From 1959 onwards the atmospheric CO_2 was directly measured in Hawaii, 3800 m above sea level (Keeling *et al*., 2009). From 2020 onwards the CO_2 rise was projected assuming two different socio-economic conditions: B1, cooperation of nations, environmental sustainability, population growth from 7.1 billion in 2013 to 9 billion in 2050 and then declining; A2, independent, self-reliant nations, regional economic development, continuously increasing population (IPCC, 2000, 2013a).

O_3 concentration has more than doubled, but still does not make up more than roughly 10% of the total amount of ozone. The tropospheric O_3 not only contributes to climate change, but at the same time it is a serious *air pollutant* at ground level, affecting flora, fauna as well as human beings. About 90% of total ozone, however, is contained in the stratosphere, being positioned above the troposphere. The stratosphere extends from roughly 11–50 km above the ground. *Stratospheric ozone* is concentrated in the *ozone layer* (Fig. 15.5), forming a natural protective filter against ultraviolet (UV) rays. This *UV filter* gets broken down by halogen radicals like *chloromonoxide* (ClO) and *bromomonoxide* (BrO) and by NO_x, under the influence of UV rays (Fig. 15.5). The action of these compounds causes an enlarging *hole in the ozone layer*, as a result of which the amount of UV radiation reaching the ground is increased, particularly in the southern hemisphere. The destructive agents originate in the troposphere from *CFCs*, which are unnatural gases released in industrial processes, or from N_2O, enriched by human activity (Enquete-Kommission, 1995). However, the rate of destruction of the ozone layer has now slowed because of successful restrictions on the release of CFCs, so that the layer may recover gradually (UNEP Ozone Secretariat, 2010–2011; Schönwiese, 2013).

The rise in concentration of trace gases, not just CO_2 in the troposphere has a serious consequence. The gases amplify the so-called *greenhouse effect*. Anyone using a greenhouse will recognize this effect. The short-wave solar radiation passing through the glass is converted inside the greenhouse into long-wave

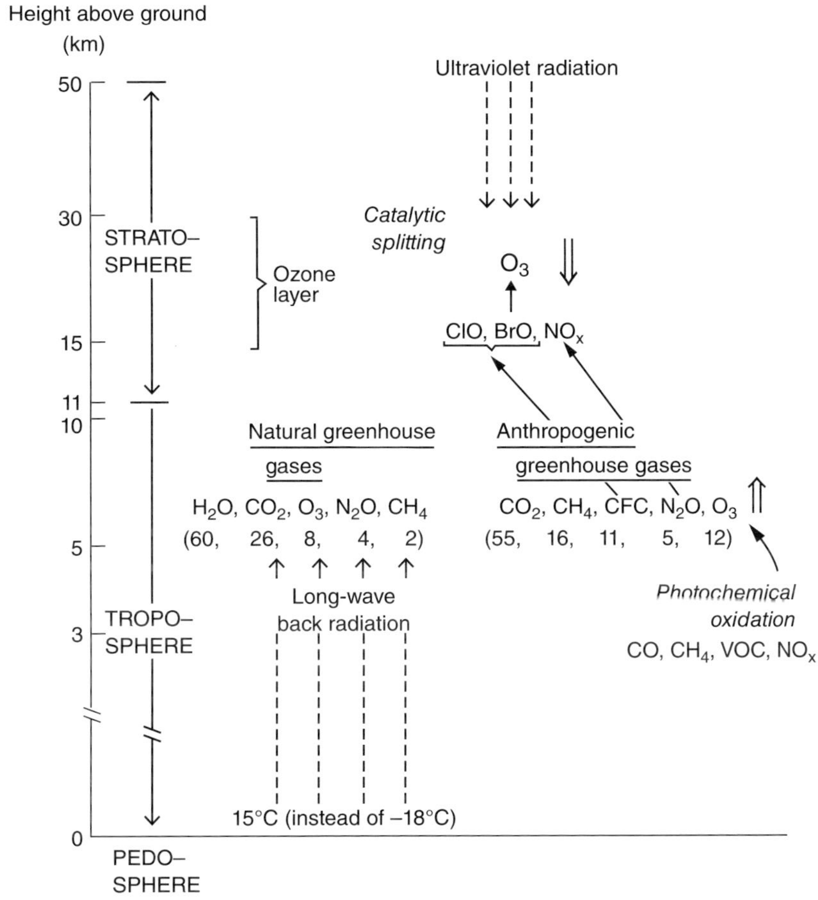

Fig. 15.5. Extension of troposphere and stratosphere from the ground (logarithmic scale). The troposphere contains natural greenhouse gases. Their relative contribution to the natural greenhouse effect is given as a percentage of the total (numbers in brackets). The natural warming of the globe is from –18 to +15°C. Moreover, the gases of the troposphere contributing to the anthropogenic greenhouse effect (contribution expressed as percentage of the total given in brackets) are listed. The tropospheric ozone concentration is increasing (upward arrow). The increase is caused by photochemical oxidation of trace gases, originating from human activity. In contrast, the stratospheric ozone concentration is decreasing (downward arrow) as it gets destroyed by catalytic splitting. The destructive agents originate from gases in the troposphere. CFC, Chlorofluorocarbons; VOC, volatile organic compounds. Chemical symbols are explained in the text (derived from Enquete-Kommission, 1994, and Schönwiese, 2013).

thermal radiation. The converted radiation does not pass back through the glass easily, but is retained within the glasshouse. Earth is not surrounded by a glass roof, but by the troposphere. The natural gases (CO_2, CH_4, N_2O, H_2O and O_3) of the troposphere reduce long-wave back-radiation, the radiation that normally returns to space (Fig. 5.5). The amount of radiation that is returned to space determines the earth's heat balance. Under 'natural' conditions an average global temperature of about 15°C is maintained near the ground. Without the retaining shield the average global temperature would drop to −18°C (Fig. 15.5). At present the *natural greenhouse effect* is reinforced by an *anthropogenic greenhouse effect*, resulting in *global warming*. The additional input of trace gases is the result of various human activities. These include the burning of fossil fuels, clearing of forests and setting fire to the trees, or burning of other natural vegetation. It is estimated that energy consumption (industry, housing, traffic) – mainly by industrial nations – contributes roughly 50% to the anthropogenic greenhouse effect. CFCs contribute another 20%, and forest clearing a further 15%. The latter proportion (15%) is also the estimated *contribution of agriculture* to global warming, although estimated values for the contribution of agriculture in individual countries range up to 60% (Table 15.2) (Enquete-Kommission, 1995; UNFCCC, 2013; FAOSTAT, 2014; WRI, 2014). Agricultural enterprises contribute the gases CH_4 (lowland paddy fields, cattle farming), N_2O (organic and inorganic fertilizer application), and make a minor contribution of CO_2 (direct energy consumption and indirect energy expenses in branches of industry serving agriculture). It is evident that industrial nations are responsible for the main generation of anthropogenic greenhouse gases. Most of these nations are located in medium to high latitudes. Ironically they will be less affected by climate catastrophes than most of the developing countries situated more in the lower latitudes. These nations will suffer greater impacts and have fewer financial resources available to rectify damage and losses (Lotze-Campen and Schellnhuber, 2009).

The *global warming* that took place over the 20th century (1901–2000) averaged 0.06°C per decade, with the rate from 1956 to 2005 averaging 0.13°C per decade (IPCC, 2007). The projection of the rate of rise in the earth's surface temperature over the rest of the 21st century ranges between 0.1°C and 0.5°C per decade for CO_2 emission scenarios B1 and A2, respectively, (see Fig. 15.4). Such a forecast is based on the results from complex circulation models, which aim to

Table 15.2. Estimated total greenhouse gas (GHG) production by agriculture and as a proportion of the national annual production for various countries (data from UNFCCC, 2013;[a] FAOSTAT, 2014 and WRI, 2014[b]).

Country	Total GHG emissions for 2010 (Gg CO_2 equivalent)	GHG from agriculture (Gg CO_2 equivalent)	Agricultural GHG as percentage of total (%)
Australia[a]	511,939	84,143	16
Belgium[a]	118,903	4,497	4
Brazil[b]	2,136,210	392,641	18
Canada[a]	789,058	53,925	7
China[b]	10,081,530	664,303	7
France[a]	446,870	92,589	21
Germany[a]	925,830	70,360	8
India[b]	2,304,390	609,102	26
New Zealand[a]	59,295	34,412	58
UK[a]	553,148	51,882	9
USA[a]	5,797,284	461,497	8

simulate regional weather patterns under changed environmental conditions. According to model predictions, warming will be more severe over land than over oceans (IPCC, 2013b; NASA Goddard Institute for Space Studies, 2014). As a result of the warmer average temperature, fewer frosts and more nights that are warmer than was normal are forecast. Little impact is expected on the overall frequency of very warm days. It is suggested that the increases in the number of warmer nights will be located predominantly in the south-eastern states of the USA, the land surrounding the Mediterranean Sea and in central Asia. However, it will be in the subtropics and mid-latitudes that the frequency of warmer days will increase. Summer temperature extremes are likely to be warmer and mid-latitude regions will also become drier. In the northern hemisphere, this will be a particular feature of central and southern Europe. It appears that there is a climate feedback between soil moisture and regional surface temperature that will drive the changes in the number of warmer days in the northern hemisphere. A different feedback, this time being generated by the difference in temperature between the land and the ocean, will drive hotter and drier conditions in Australia. The rise in temperature will intensify the global *water cycle*: larger quantities of water will be evaporated, and rain will fall in increased amounts and likely with greater intensity. Also the spatial *distribution of precipitation* will change, with increasing precipitation at higher latitudes, particularly in the northern hemisphere above 50°. There is also a projection of more rainfall over equatorial Pacific and Indian Oceans but a decrease in the subtropics. However, the changes in the subtropical regions will probably not be outside the existing variation. Importantly, the decreases in precipitation anticipated for the Mediterranean, Caribbean, Central America, south-western USA and South Africa are predicted to differ significantly from the current variability over the next 200 years. In the mid-latitudes of the northern hemisphere there is also likely to be more winter rainfall. To some extent it can be argued that changes in rainfall will follow a pattern of the wet areas getting wetter and dry areas getting drier. Australia, a country that regularly faces drought, will become drier, except for the northern most areas (IPCC, 2013b). Southern Europe and the south-western states of the USA are also expected to become drier and warmer, which will further threaten yield stability. In North America, the increase in frost-free days may provide opportunities for increased use of winter crops and longer growing seasons in the northern parts of the Great Plains.

Can we already recognize local climate change in *actual weather data* recorded over the past decades? The trends of the mean annual temperature and precipitation, recorded in Göttingen for six decades, is presented as an example (Fig. 15.6). To reduce scatter, the data have been smoothed by averaging over 3 years in sequence. It can be seen that the smoothed weather variables change cyclically as a function of time. Minimum as well as maximum of annual temperature are staggered with a time interval of 7–10 years. For precipitation the temporal cycle is less distinct. The smoothed averages of annual temperature can vary within a cycle by more than 2°C and those of precipitation by almost 300 mm. During a period of 60 years the annual temperature has increased by about 1.2°C. In contrast to the data on temperature, there is no significant time trend for precipitation. The Göttingen weather data coincides with other reports from Germany (Schönwiese *et al.*, 2005). On a global scale the annual temperature rose by 0.85°C between

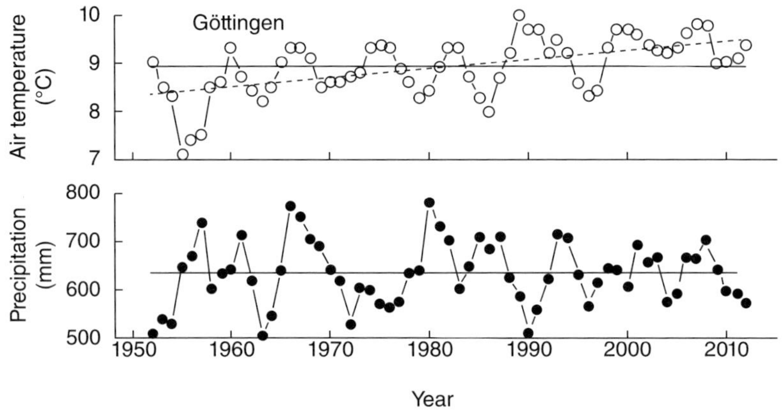

Fig. 15.6. Course of annual temperature and precipitation since the beginning of the 1950s in Göttingen, Germany, presented as average values, smoothed over 3 years in sequence (weather records of the Institute of Agronomy and Plant Breeding, data from the local weather station and from the German Weather Service, 1982–1998; 1999–2000; from 2001 to 2013 all the data was supplied by the German Weather Service, personal communication). The horizontal lines are averages for the recorded period. For temperature the significant time trend (dashed line) is also depicted ($r^2 = 0.32$).

1880 and 2012 (IPCC, 2013b). Depending on the production rate of released GHG and the kind of models used, a rise in temperature between 1.1°C and 5.4°C is predicted for 2100. With scenario B1 (see Fig. 15.4) the range should vary between 1.1°C and 2.9°C, with A2 between 2.0°C and 5.4°C at the end of the 21st century. Limiting the rise in temperature to no more than 2°C above the pre-industrial level (Meinshausen *et al.*, 2009) would appear to be acceptable and essential if we are to 'prevent dangerous anthropogenic interference with the climatic system' (UN, 1992). Any further increase is rated as 'non-acceptable' because of life-threatening environmental catastrophes. But it is essential to recognize that about half of the acceptable temperature increase has already taken place.

15.3 Plants, Soils and Cropping Pattern in a Changing Environment

In which ways will plant growth and development respond to a change in climatic factors? If the trend really continues at the present scale, a climate change is about to happen with a rate never experienced before on earth. *Climatic zones* would probably be shifted to higher latitudes in only a few decades. *Natural vegetation*, with its associated flora and fauna, would find it difficult to follow these changes. In any case, a restructuring and displacement, and eventually even the extinction of species have to be considered. These processes may happen all the more decisively as growth conditions and competitive interactions are being altered. For *plants under cultivation*, farmers can control some of the growth conditions more or less successfully. But even under crop production, changing climatic factors will affect plant responses. Furthermore, it will be impossible to predict quantitatively the precise effects on plants and plant communities. The uncertainty has to do with the fact that the exact trends in the climate are still uncertain, and that we do not fully understand the interaction of climatic factors and the way growth and yield potential are affected (Rötter and van de Geijn, 1999).

In C_3 plants the anticipated *rise in temperature* will speed up the rate of development more than that of biomass production (Section 15.1). Therefore growth stages will be passed through more quickly, leaving only a shorter period for dry matter accumulation and the establishment of yield. On the other hand, the yield of C_4 plants generally is promoted (Section 10.1). However, rising temperatures increase the saturated vapour pressure and the saturation vapour pressure deficit (SD, Fig. 8.3) and therefore the potential evapotranspiration (ET_p, Eqn 8.6). Larger ET_p causes greater rates of *transpiration*. Therefore a given soil water supply will be depleted earlier in the season than would be the case at cooler temperatures. This effect will be accentuated because the rate of soil evaporation will also be increased. Thus *water stress* conditions and their negative impact on yield formation are more likely. The increasing SD suggests that the transpiration efficiency will decrease (Fig. 11.4). Increase in temperature is thus a cause of rapid water use, lower yield levels and less economic water use in crop production.

Since 1900, annual *precipitation* has increased over a large part of northern Europe with its temperate climate; changes have mainly occurred in winter and spring but not in summer and autumn. However, in those countries of southern Europe with a Mediterranean climate, annual precipitation has declined (Schönwiese *et al.*, 1993; IPCC, 2007). If this trend continues as forecast, it suggests that the modest increase in precipitation in temperate Europe will not be sufficient to balance the effect of temperature on increasing the *deficit in extractable soil water*. Of course in the Mediterranean part of Europe the deficit in soil water will be even more pronounced (Enquete-Kommission, 1994). In the USA early climate models projected an increase in annual precipitation. But more recent models maintained this trend just for the north-east, whereas in the south-west annual precipitation was projected to decrease. For the major field-cropping regions changes are thought to be small. But looking at summer rainfall, a decrease is predicted for the major cropping areas. Only a strip near the Atlantic coast shows no significant change (Sinclair, 2011).

Climate and soils interact to cause a deficit in soil water, sufficient to affect plant performance. Coarse-textured sandy soils have a comparatively small water storage capacity, so these soils are easily replenished in temperate or Mediterranean climates during winter and after the start of the rainy season in wet tropical regions. The remainder of water that has infiltrated the soil will, however, be drained away. Because of their small *field capacity* (FC) these lighter soils cannot store a surplus of precipitation for transpirational water use during the main growing season, when ET_p is usually greatest. In this respect loamy and silty soils have an advantage for plant production. The higher FC enables them to provide a larger quantity of rainfall for plant water use during the next season, especially when the *available field capacity* (FC_{av}) and the *effective rooting depth* ($z_{r\ eff}$) are both large (Table 9.2). On the other hand, the favourable water storage characteristics of loamy soils can become effective only if the seasonal precipitation is sufficient to fill the whole soil profile with water up to FC. In many drier parts of the world, soils will fail to reach FC to any appreciable depth because rainfall is insufficient. Even in a temperate climate this happens occasionally, as occurred during the late 1980s when there were four consecutive years of reduced precipitation in Eastern Germany and in the region around Göttingen (Fig. 15.6).

In summary we can state that the capability of a soil for storing extractable water depends both on its texture and, more importantly, on the attendant conditions (i.e. the amount of rainfall).

Early warming in spring is an important precondition for *timely sowing* of summer crops. But at the same time, timely seeding demands early secondary tillage to prepare the seedbed, which may be difficult when the soils do not dry because of excessive rain. Seedbed preparation can become a difficult task on heavy clay soils, such as vertic Cambisols, if *temperatures rise in winter* and there is little or no frost. Frost causes the building of ice lenses in the topsoil. These ice lenses can cause clods to shatter into smaller aggregates, an ideal state for germinating seeds and referred to as *natural tilth*. When frost fails, this sort of tilth does not develop. Then a tilth has to be created solely by mechanical means, for instance by use of power-take-off-driven rotary harrows, which is a costly procedure and sometimes has a poor rate of success. Early seeding will increase the potential yield of crops by *extension of the vegetation period* and provision of the opportunity for early leaf development and light interception (Chapter 10). The early seeding of long-day plants slows the rate of development because of photoperiodism, which again may have a positive effect on the attainable yield.

Most meteorologists are convinced (and insurance companies are sure) that global warming will cause the atmospheric currents (the air streams and the wind speed) to increase. In general, more *unstable weather conditions* are anticipated for the future than are now the norm. At higher latitudes the question arises: Will weather instability lead to sudden temperature drops and to spring frosts, which may be extremely harmful to early sown summer crops like sugarbeet and maize? At lower latitudes heat waves can diminish potential yield by water shortage and heat stress. The important question for all agricultural areas is: 'What will be the amount and, in this context, the intensity of rain in future?' Will the common rainfall events turn into short-duration rains with larger drop size and greater intensity? Will the *erosivity* of rainfall increase and will *soil erosion* be accelerated? And will global warming reduce the *organic matter content* of soils? Does this affect good tilth and decrease aggregate stability, thereby increasing *erodibility* and the potential erosion rate? If so, farmers have to adapt to rough weather conditions by adopting or developing appropriate methods of soil tillage, seeding and cropping to strengthen soil conservation measures. Inversion tillage has to be replaced by reduced tillage, mulch tillage or zero-tillage: practices that are covered by the term *conservation tillage*.

Due to rising temperatures, winter-sown crops like oilseed rape or small-grain cereals will have accelerated growth and development. What are the consequences for weed infestation, parasite infection and fungal attack? Will the need for vernalization be satisfied under the higher temperatures? Will farmers be encouraged to grow the winter form of crops that are now at high risk in temperate climates because of winter temperatures being too cold? The winter forms, for instance of oat or faba bean, could escape summer drought because of their early maturity. Are rising temperatures the key to growing soybean at higher latitudes? Will that strategy also work for those later and larger types of maize, which have a higher potential yield than the early but smaller types that are now commonly cultivated in such regions? All these questions reflect some of the ideas on how the *cropping scenarios* in a region may be modified in the near future by climate change. Climate change could affect the choice of tillage systems, require measures for erosion control and the selection of adapted species and cultivars. Rising temperatures and reduced precipitation will necessitate modifications in *crop rotations*, calling for crops that require less water during the season. At the same time the integration of *cover crops* into the crop rotation has to be reconsidered from the standpoint of water saving and of soil conservation. In areas where summer rainfall is reduced, the risk that major crops fail to achieve their potential yield is increased by growing stubble crops and winter catch crops. Tillage and seedbed preparation become more difficult and the risk to proper seeding, successful germination and emergence is enhanced. The same may be true for the establishment of early winter crops like oilseed rape and barley, which already have to be seeded in late summer or early autumn.

Under climate change summer-sown crops could be endangered by drought or by *torrential rains*. Such heavy rains can displace seeds and erode soil from upslope positions, burying seeds and small plants preferentially in more downslope positions. Generally the seedbed quality of the topsoil may deteriorate to such an extent that the seeds fail to emerge quickly, uniformly and at a high rate. When a slaked or puddled topsoil layer turns into a hard crust as a result of drying, then field emergence may be drastically diminished, especially when the seeds are small and of low vigour.

While it is true that higher temperatures together with reduced precipitation indicate declining yields in the future, this does not necessarily apply if the *rise in atmospheric CO_2 concentration* is also considered (Supit *et al.*, 2012). If the trend caused by human activity is not stopped in the near future, then the concentration of CO_2 is going to double sometime in the second half of the 21st century (Rogers *et al.*, 1994; see Fig. 15.4). The gas is an essential constituent for synthesizing carbohydrates in plants but in spite of atmospheric enrichment it is still available only at a 'minimum' level and is therefore yield limiting. CO_2 is a growth factor, to which the law of the optimum applies (Section 12.1). With CO_2 enrichment of the atmosphere all the other growth factors can be better utilized by plants. This stated view applies primarily to C_3 plants. For these plants, the double function of ribulose diphosphate carboxylase/oxygenase (Section 10.1) shifts towards *CO_2 assimilation* by an increase in the CO_2 concentration, and photorespiration is reduced. On the other hand, for C_4 plants, phosphoenolpyruvate carboxylase has a strong affinity for CO_2. Therefore the enzyme is saturated to a large extent with CO_2 even at present atmospheric concentration. Hence it is expected that CO_2 assimilation by C_3 plants will increase by up to 60% within the coming 50–70 years, but with C_4 plants the increase will turn out to be comparatively modest (Taiz and Zeiger, 1991).

Even in past years, the increasing CO_2 concentration (Fig. 15.4) will have made some contribution to the *yield increase* of crops. In contrast to the proud boast of how successful humans have been in developing crops with a high yield performance because of progress in plant breeding, fertilizer utilization, protection and agronomy, the unintentional contribution of civilization has, in the main, been concealed. For quite a long time it has been known that the gas limits plant growth. This knowledge has been put into action in the case of 'CO_2 fertilization' in glasshouses. For a quantitative assessment of CO_2 concentration effects on crop yield technical research facilities have been developed keeping the gas concentration at a constant level independent of the ambient air. These systems consist of closed growth chambers or more recently of larger open facilities outside in the field. These latter facilities are named FACE, free-air carbon dioxide enrichment. In FACE facilities CO_2-enriched air is blown continuously from bars into a circular field area that otherwise experiences the same weather conditions as the surrounding field (Ainsworth and Long, 2005). Earlier and more recent experiments indicated that a doubling of the CO_2 concentration would increase the yield level of C_3 plants by an average of 33%, but only by about 8% in C_4 plants (Kimball, 1983, Hatfield *et al.*, 2011). Because of the greater assimilation rate the mass of shoot and fruit increases. But it also appears that the partitioning of assimilates to roots is increased in many crops (Madhu and Hatfield, 2013). From these observations one may conclude that plants with increased photosynthetic rate have the potential for a greater nutrient and water uptake. It is another question whether or not the greater capability of plants for nutrient and water acquisition is paralleled by an increase in the capacity of the soil to store these essentials in plant-extractable form within rooting depth.

Just as in the case of temperature, an increase in CO_2 concentration tends to speed up the *developmental rate* (Cleland *et al.*, 2007) although changes in wheat tend to be small (Kimball *et al.*, 2002). The duration of the season is shortened and the timing of anthesis and ripening is brought forward.

Increase in CO_2 concentration causes a partial *narrowing* of the *stomatal pores*. Stomatal closure reduces the rate of H_2O loss per unit leaf area more strongly than the rate of CO_2 intake. That is because – in contrast to H_2O – the greatest diffusion resistance for CO_2 is located in the liquid phase in the mesophyll (Box 11.1). It is in the mesophyll that the CO_2 is used for the synthesis of carbohydrates, either within the chloroplasts of the mesophyll cells (C_3 plants) or within the bundle sheath cells (C_4 plants). And this major resistance is not at all affected by stomatal closing. As closure hinders the transport of water vapour more than that of CO_2, the *transpiration efficiency* (TE, Section 11.1) is greatly improved (Fig. 15.7).

The decrease in the H_2O exchange rate relative to that for CO_2 is indeed only one reason for the rise in TE. Another reason is the increase in the CO_2 concentration itself, which accelerates the CO_2 transport rate at the leaf surface. With C_3 plants, the increase in TE is more a consequence of increasing the CO_2 exchange rate and not so much the suppression of the H_2O exchange rate (net photosynthesis increase to transpiration reduction is about 90:10). The relationship in C_4 plants is quite different. The ratio is 27:73. Here the suppression of H_2O release is of much higher relevance than the direct CO_2 effect.

The partial closure of the stomates causes an opposite reaction with respect to the suppression of water vapour release by CO_2 enrichment. Reduced canopy transpiration induces an increase in canopy temperature, which in turn modifies the vapour pressure inside the leaves and consequently the transpiration rate. Therefore any application of measurements based on single leaves (Fig. 15.7) to field situations needs careful consideration.

When the transpiration rate per unit leaf area is lowered by partial closure of stomates, this does not mean that the rate is also reduced if it is based on unit soil surface area (Fig. 15.8B), because the leaf area index (LAI) may increase with increasing CO_2 concentration at an accelerating rate. And so in this field experiment with winter wheat, the cumulative transpiration was not seriously changed within an irrigation treatment when the CO_2 concentration was raised. (Soil evaporation was largely but not completely reduced by covering the soil surface with black plastic film.) Nevertheless, grain yield increased dramatically. At any CO_2 concentration, the economic yield level was greater in the irrigation treatment with ample water supply (Fig. 15.8A). With limited water supply and high CO_2 concentration almost the same yield level was attained as with sufficient water and 'normal' CO_2 conditions. Hence it seems that water shortage can be compensated for by CO_2 enrichment within certain limits. To put it another way, one can argue that if *the deficiency factor is water* it can apparently be utilized by plants more effectively when CO_2 becomes *enriched* (law of the optimum). It is noteworthy that in both irrigation treatments the transpiration ratio was lowered by increasing CO_2 concentrations (Fig. 15.8C). Accordingly, the economic TE (Section 13.4) was raised. Quite similar results were attained later in FACE studies (Hunsaker *et al.*, 2000).

It remains uncertain whether the yield increase due to CO_2 enrichment, as documented by pot and field experiments, will indeed materialize as climate changes. That depends on possible interactions between CO_2 and some other climatic factors in affecting plant growth, factors which are expected to change in future such as temperature and precipitation. But the uncertainty has also to do with the impossibility of predicting the change in the concentration of harmful

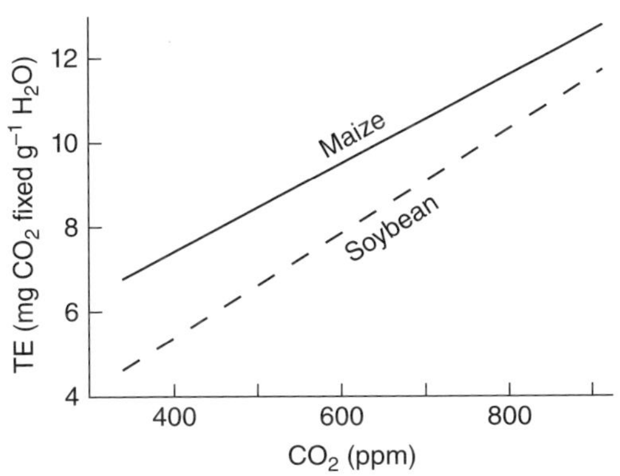

Fig. 15.7. Transpiration efficiency (TE) of maize (C_4 plant) and of soybean (C_3 plant) as affected by atmospheric CO_2 concentration. In this experiment TE was determined by measuring the gas exchange of an intact leaf (after Rogers *et al.*, 1994). It is the instantaneous or intrinsic TE (see Section 14.1).

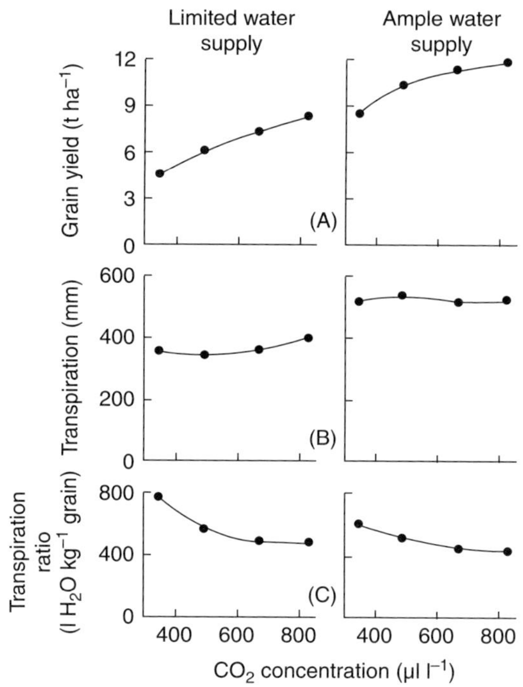

Fig. 15.8. Influence of increasing CO_2 concentrations in the atmosphere on grain yield (A), transpiration (B) and transpiration ratio (C) of winter wheat grown under two irrigation treatments by using plexiglass chambers in Kansas, USA. The results are the mean of data from 2 years (based on Chaudhuri *et al*., 1990).

gases like ozone or to foresee the severity and frequency of catastrophic weather conditions. These weather conditions include sudden falls in temperature, late frost, intensive storms, hail, drought and heat waves. In addition, some scientists are of the opinion that the stimulus of CO_2 enrichment will not last for long. They consider that in the long run plants might adapt to higher CO_2 concentrations, for example by reducing stomatal frequency or the activity of the enzyme ribulose-1,5-bisphosphate carboxylase/oxygenase (RuBisCO). In view of all these uncertainties, the question of plant growth under changed climates will probably only be answered during the course of real events. Only then will we become certain whether a unit of soil water will translate into economic yield more efficiently than it is today. But even so, given all the uncertainties identified, specialists in modelling try to characterize general trends in crop yields under the expectation of a growing world population that needs to be fed. What does seem to be agreed is that in northern Europe winter crops, such as winter wheat, could benefit more from climate change than spring crops, such as potato and sugarbeet. Crops, such as maize, planted in late spring may even be subject to high temperatures and drought (Supit *et al*., 2012).

16 Breeding for Yield and Water Use

16.1 Comparing Old and New Cultivars

Commonly, newly released cultivars of crop species have a greater *yielding ability* than older varieties. In those cases where the yield increase is achieved by lengthening the *growing season*, we would expect that there would be a larger interception of global radiation and a greater cumulative transpiration from the canopy. This in return suggests that the transpiration efficiency (TE) would stay more or less constant. In making this assumption we neglect, of course, the influence of the vapour pressure deficit (SD) on TE, which could easily change when the growth period is extended. Amazingly enough, this method for increasing yield has seldom been followed with success by plant breeders (Gill, 1994). Rare examples may well be oilseed rape, late potato, sunflower and some grain legumes like faba bean and field pea. In any case the water supply from rain and soil must be sufficient to ensure yield in full measure. On the other hand it is quite common to conserve and exploit the length of the growing period of those cultivars with a given genetic make-up. For that purpose various agronomic measures are employed aimed at preventing premature senescence of the green, photosynthetically active tissue of leaves and other organs like glumes and awns. Some of the common measures adopted are: (i) late top dressing with nitrogen; (ii) fungicide application to leaves, fruits and heads; and (iii) pest control. Where water is available, irrigation may be advised under dry conditions. All of these measures help to maintain green leaf area during grain or fruit formation. More recently, cultivars of maize and sorghum have been released with the special trait to 'stay green', when suffering from water shortage at late development stages. These cultivars prolong the post-anthesis grain-fill period by postponing early senescence induced by terminal drought.

Thus, by making full use of the growing season, TE will not change too much. This statement is for TE related to the biomass. However, when TE is related to the economic yield, that is the yield of marketable products such as grain (TE_G), rather than to biomass or biological yield then this statement may no longer be correct. This is because the harvest index (HI) usually increases with greater exploitation of the growing season by lengthening the period of grain or fruit formation. Under such conditions, water may be used more efficiently per unit of *marketable* yield.

As a rule, plant breeders have not changed decisively the length of the growing season of small-grain cereals. Therefore the above-ground biomass has remained at a more or less constant level. This conclusion can be evaluated by determining the yield, under present growing conditions, of old and modern cultivars that have been released over the past 150 years (Fig. 16.1A). Yield comparisons made in this way allow for differences in cropping intensity and soil fertility between earlier experiments and today's conditions. The effect of increased carbon dioxide (CO_2) concentration (Section 15.3) is also taken into account. Of course, this kind of comparative experiment requires that sufficient seed of the old cultivars is still available, and that a seed stock has been built up by replanting ahead of the experiment. The outcome of such an experiment with winter wheat and spring barley is depicted in Fig. 16.1A. Clearly, *grain yield* has definitely increased during 150 years of genetic improvement, although *total yield* remained at more or less the same level. The HI was improved considerably by breeding for high economic yield (Fig. 16.1B), whereas *culm height* was shortened (Fig. 16.1C).

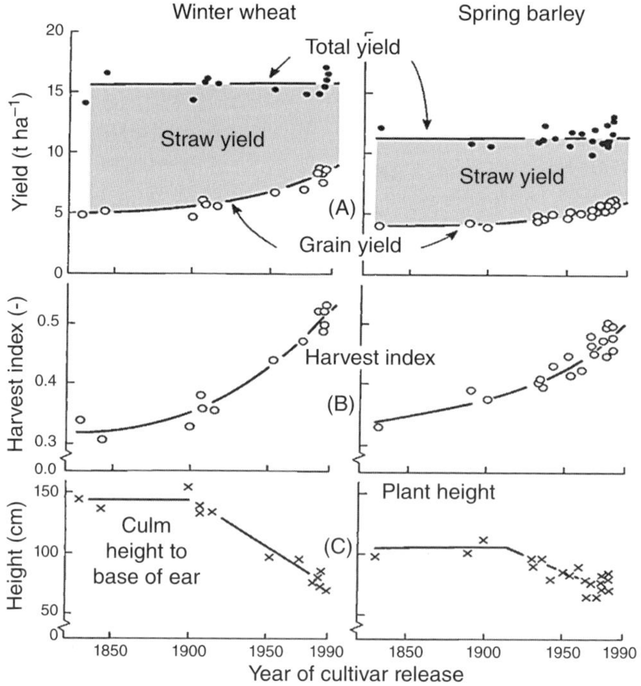

Fig. 16.1. Dry matter of total and grain yield and of straw yield by difference (A), harvest index (B) and culm height (C) of winter wheat and spring barley as affected by the year of cultivar release in the UK (data from Riggs *et al.*, 1981, and Austin *et al.*, 1989).

The apical meristem of shorter plants produces a smaller *number of leaves*. Those leaves that open later rather than early in the season (i.e. during stem elongation) are positioned at the upper nodes, where they take on a comparatively *vertical inclination*. Short-strawed varieties of cereals have a greater *resistance to lodging*. They form *longer ears* with a larger number of fertile florets and spikelets. The double-ridge stage, indicating the transition from leaf to ear formation, occurs several days earlier. Similarly, the period to the end of ear formation (i.e. for wheat this is when the terminal spikelet has developed) is shorter, but by fewer days, so that the total *period of ear formation* is slightly extended. Anthesis too, occurs earlier in the UK by about a week (Austin, 1999), which may result in the *period of grain fill* being extended by several days. Despite reduced tillering some modern cultivars may produce slightly more *ear-bearing culms per plant* compared with older varieties. That applies also to the *number of grains per ear*, whereas the *kernel weight* may remain unchanged. In summary, we can conclude that modern cultivars can attain a larger seed yield with a smaller vegetative mass. The secret behind these cultivars is the greater leaf area index (LAI) and the longer *leaf area duration* (LAD) *after anthesis*. The LAD is the integration of LAI with time. A larger value for LAD signifies a comparatively greater source strength for producing assimilates. This LAD is characterized by the particular features of the *flag leaf*. But the source can be efficient in producing assimilates only when there is simultaneously a sufficiently large sink capacity in the ear for the absorption of photosynthates. Otherwise there will be an accumulation of photosynthetic products, and a feedback mechanism then reduces their production. Capacities of sources and sinks have to be consistent with each other, and in modern cultivars this prerequisite is largely met after flowering.

In the UK, HI for wheat and barley greatly increased over one and a half centuries (Fig. 16.1). This development now seems to have come to an end; the HI of modern wheat cultivars is not much in excess of 0.5 (Austin, 1999), with an upper limit of 0.6 (Fischer, 2007). There is speculation (Blum, 2011), that even in the past, plant breeding was not directly aimed at increasing HI, but at selecting for marketable yield. With this goal in view there was 'a selective pressure on morphology and assimilate partitioning process but not on basic plant production (biomass)'. Therefore, future yield increase will be more dependent on selecting for biomass and grain yield while maintaining HI (Blum, 2011).

After this digression on the breeding success with modern 'semi-dwarf' cultivars we shall return to the consequences for water use. As neither the growing period nor biomass production have changed decisively in modern cultivars, the cumulative transpiration as well as the TE are left more or less untouched. On the other hand, as HI has been increased dramatically (Fig. 16.1), the TE related to grain yield (TE_G) generally has improved.

Despite being shorter and having fewer leaves, modern cultivars may *shade the soil surface early*, as shown by an example from the wheat belt in Western Australia, where there is a Mediterranean climate. Typical of the climate is a concentration of rainfall in autumn and winter (Fig. 15.2). After the autumn sowing (end of May) of a spring wheat variety, transpiration (T) increases rapidly after canopy development during June–July (Fig. 16.2). This happens earlier with the new variety than the old one. At the same time, the soil surface is shielded from radiation, thus reducing soil evaporation (E). Again, this occurs earlier with the new cultivar (Fig. 16.2). When cumulated over the total season, results show that E was 12 mm less for the new cultivar than from the old one (Table 16.1). The rapid development, an earlier start to transpiration and the rapid suppression of soil evaporation is therefore one 'recipe for success' of the modern cultivars in this type of environment (compare also E/ET in Table 16.1).

Compared with the old variety, the modern cultivar also matured earlier (Table 16.1), which may be advantageous in Mediterranean climates where water supply becomes restricted in spring (Figs 15.2 and 16.2). In Western Australia 4 mm of water was saved by reduced T when the semi-dwarf variety was grown (Table 16.1). In a humid climate, however, earlier maturation may not necessarily be the trait needed for a modern cultivar. Nonetheless, it is quite surprising that the modern cultivar produced a larger above-ground biomass despite the lower T (Table 16.1). One explanation is the fact that the crop invested less dry matter in the root system than the old cultivar (Siddique *et al.*, 1990a). Another explanation is based on more fundamental principles. Compared to the 'costs' of water the new, faster-growing variety was able to 'buy' the CO_2 at a more favourable 'exchange rate' than the old cultivar could. In the Mediterranean climate, temperature and vapour pressure deficit both increase at the end of winter (Fig. 15.2), and the monthly rainfall total declines (Fig. 16.2). Under such conditions the modern cultivar, which moves more quickly through its development stages and hence reaches anthesis and maturity earlier (Table 16.1), was able to produce a bigger biomass per unit of water transpired. The *CO_2:H_2O gas exchange rate* at the leaf surface was apparently more favourable, which is clearly reflected in the larger TE (Table 16.1).

Due to the faster development, the modern wheat cultivar used less water before anthesis than did the old. Hence a *larger proportion of soil water* and rainfall was left

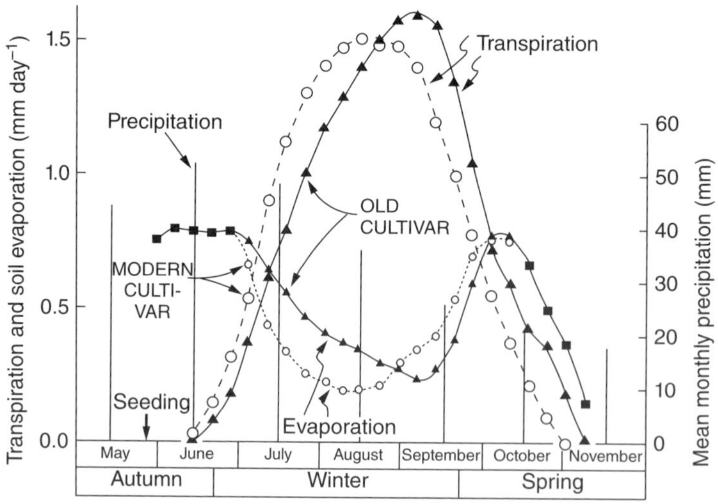

Fig. 16.2. Seasonal variation of transpiration and soil evaporation associated with an old ('Purple Straw') and a modern ('Kulin') cultivar of spring wheat, grown in the wheat belt of Western Australia with a Mediterranean-type climate (after Siddique *et al.*, 1990b). The mean monthly precipitation is also shown (Turner, 1993).

Table 16.1. Phenology, yield and water use of an old and a modern spring wheat cultivar and of a modern spring barley cultivar. Crops were grown in the wheat belt of Western Australia with Mediterranean climate. Explanation of abbreviations: ET, evapotranspiration; pra, pre-anthesis; poa, post-anthesis; ETE, evapotranspiration efficiency; G, grain; E, evaporation; T, transpiration; TE, transpiration efficiency; DAS, days after sowing; DM, dry matter (based on Siddique *et al*., 1990b).

Crop	Wheat		Barley
Cultivar	Old	Modern	Modern
Year of release	1860s	1986	1984
Date of anthesis (DAS)	125	106	102
Date of maturity (DAS)	167	148	145
Biomass yield (t DM ha^{-1})	4.8	5.1	6.2
Grain yield (t DM ha^{-1})	1.2	1.9	2.1
Harvest index (–)	0.25	0.37	0.34
ET (mm)	209	193	187
ET_{pra} (%ET)	84	78	78
ET_{poa} (%ET)	16	22	22
ETE (kg DM m^{-3} H_2O)	2.3	2.6	3.3
ETE_G (kg DM m^{-3} H_2O)	0.6	1.0	1.1
E (mm)	87	75	64
T (mm)	122	118	123
E/ET (%)	42	39	34
TE (kg DM m^{-3} H_2O)	3.9	4.3	5.0
TE_G (kg DM m^{-3} H_2O)	1.0	1.6	1.7

for the *period of grain filling* after anthesis (Table 16.1). With the old cultivar, on the other hand, this part of the water use was small, both in total and as a proportion of the whole. Less water was available to the crop after anthesis. Furthermore, this water was transpired under less favourable conditions of exchange with CO_2, since the season was more advanced due to slower crop development.

As the old wheat cultivar developed comparatively slowly, and the date of anthesis lagged behind the new variety by about 3 weeks (Table 16.1), the crop suffered more from water stress during grain fill than did the new one. For that reason CO_2 assimilation was reduced, and less photosynthates were available for grain production. This disadvantage was partly compensated for by the fact that – in proportion to the final grain mass – relatively more reserve material was remobilized after anthesis. This material was moved from stem and leaves into the florets to support grain fill (Siddique *et al*., 1989).

In Table 16.1 data on the two wheat varieties are given along with those of a modern barley cultivar. Compared with the new wheat variety, barley still had a faster development. Earlier leaf opening and a greater *leaf area ratio* (LAR, assimilatory leaf area per unit of plant material present) and greater *specific leaf area* (SLA, leaf area per unit leaf weight) were responsible for a distinct reduction in unproductive water loss by soil evaporation. Barley protected the soil surface earlier from radiation by 'thinner' leaves and developed a higher competitive capacity against *weeds* that also consume water. In barley all the *indices of efficient water use* (ETE, ETE_G, TE, TE_G) were most favourable. Barley most effectively combined the strategy of evaporation control and transpiration improvement with that of drought escape (Section 16.3) due to early maturity.

Many of the efforts in breeding cultivars for water-limited environments have been summarized in an equation, targeting economic or grain yield (Y_G). The Eqn 16.1 (Condon *et al.*, 2004, based on Passioura, 1977) consists of several components that make up a yield framework aimed at intelligent use of water as a growth-limiting factor. The equation also provides a tool to establish objectives for genetic improvement. The equation reads:

$$Y_G = ET \cdot (T/ET) \cdot TE \cdot HI \qquad (16.1)$$

All the components of the equation, not truly independent of each other, have been quantified in the example given in Table 16.1 and can be used to recalculate Y_G for the two wheat cultivars and the barley cultivar. Grain yield, Y_G, depends on the cumulative evapotranspiration, ET, specifically the proportion that is transpiration, T. The more that T dominates ET, and that depends on the particular circumstances prevailing, the larger the potential Y_G. In addition, Y_G is governed by biomass transpiration efficiency, TE, and finally 'how effectively the achieved biomass is partitioned into the harvested product', the HI (Condon *et al.*, 2004).

16.2 Future Strategies in Plant Breeding

We showed in Section 16.1 that in a warmer Mediterranean climate with limited water, faster crop development may lead to an improvement in both evapotranspiration efficiency (by reducing evaporation) and transpiration efficiency (by a more profitable gas exchange at the leaf surface). In addition to the length of the growing season and the longer period of radiation use, these are the reasons why it is usual for *autumn-sown crops* not only to exceed corresponding *spring-sown crops* in terms of yield, but also to use water more efficiently. This is true for all climates with a winter season (Fig. 15.2) and has been demonstrated, for instance with sunflower in Spain (Gimeno *et al.*, 1989) and chickpea (*Cicer arietinum*) in Syria (Keatinge and Cooper, 1983). However, for continental climates, such as the steppe, *frost hardiness* is essential for the cultivation of winter-annual crops. Crops without sufficient winter hardiness can be severely damaged or even killed during winter. In North America for instance, the northern boundary of the winter wheat belt coincides well with the −12°C isotherm for daily mean temperatures in January and February. North of this isotherm, spring wheat is more important (Martin *et al.*, 1976). Apart from physiological injury, *damage by winterkill* can be caused by frost heave of the soil, by frost desiccation, by hypoxic conditions either below a covering of snow that has formed an impermeable crust or due to stagnant moisture, and finally by fungal attack or pest infestation (snow mould, mildew, take-all, eyespot, fruit-fly, wheat-bulb fly, crane-fly). Breeding for resistance and tolerance, and the application of pesticides to deal with biological attack, are the prerequisites for mitigating the negative effects of injuries that threaten the winter survival of crops.

The rate of all vital processes that contribute to growth, such as photosynthesis, increases with temperature until an optimum temperature is reached (Fig. 1.4). Such activation, however, requires that a particular minimum or *base temperature* is exceeded. Hence, if fostering growth over winter is the goal, this might be achieved successfully by breeding for a lower base temperature. Of course, a conventional breeding programme aimed at that goal requires *genetic variability* of that trait within the species concerned (Gill, 1994). Possibly the *vigour* of crops during winter could be strengthened by lowering the base temperature for physiological processes. For instance, winter-grown faba beans stand out from summer-grown forms, not so much for their frost hardiness, but probably for their greater viability. These cultivars are capable of forming a number of lateral shoots, once the apical meristem of the main axis has been killed by frost (Lawes *et al.*, 1983). Increased vigour may be related to the size of the embryo in seeds – one way, in addition to LAR, that barley differs from wheat (Richards *et al.*, 1993).

Just as for winter-sown crops, early development of summer-grown crops can be important for achieving higher yields, greater evapotranspiration efficiency and a better yield stability from year to year by *escaping from summer drought*. From the breeder's perspective, the tolerance for low temperatures could be taken up as a goal for improvement. By *advancing anthesis*, the period of seed formation can be extended, and should be associated with a lower mean saturation deficit of the air. In grain legumes the period of reproduction could be curtailed by breeding for *determinate shoot growth*. Determinate pulses eliminate the additional sink for photosynthates, formed by those leaves that develop late in the season.

Earliness in a crop provides an opportunity to remodel the water balance at a site so as to reduce *seepage losses*. This aim can be achieved by breeding for *rapid root growth to depth* without necessarily increasing the mass of roots (and thereby the requirement for additional assimilates below ground). In part this could be achieved by having roots with sparser branching in the topsoil but more in the deeper rooted layers (Wasson *et al.*, 2012). Enhanced branching in deeper layers could not only reduce drainage below the rooting zone but also increase water uptake from these layers, which typically do not contribute to water uptake to any great extent (Figs 9.12 and 9.13). Because of the difficulty in directly assessing roots hidden in the soil, such breeding objectives can be achieved only with considerable effort. In any case, examination of the root system would be necessary, and that can degenerate into a never-ending task, given the large number of individual plants that have to be checked in the field. One aspect of changed root morphology that would be amenable for initial investigation under controlled environments would be a modification of the radial and longitudinal water transport pathways. This would include increased length and frequency of root hairs and selecting for xylem vessels of larger diameter (Wasson *et al.*, 2012). However, it seems to be even more difficult to measure the water use efficiency (or *transpiration efficiency*) of plants as *a trait* directly in the field *for breeding* purposes. And the problem remains whether water use is related to total biomass yield or to economic yield, because it is difficult to assess the amount of water extracted by roots. In contrast to the determination of *nutrient uptake* by direct analysis of plant material, the measurement of *soil water uptake* is far more cumbersome. It is even more so for field investigations than for container experiments. And container trials are of limited value, if soil structure is modified and soil depth, layering, volume and temperature are different from the situation in the field (Box 9.1).

Over the past three decades a new and promising technique has been developed that allows screening for transpiration efficiency (TE). The method is based on the identification of the two stable *carbon isotopes* ^{12}C *and* ^{13}C in plant material by mass spectroscopy. Both isotopes are present in plant tissue as they are taken up from the atmosphere in the form of CO_2. In the air, the bulk of CO_2 is $^{12}CO_2$ (98.9%), and about 1.1% is $^{13}CO_2$. Plants distinguish between the two isotopes, C_3 plants being more discriminating than the C_4 plants. The 'lighter' $^{12}CO_2$ is preferentially absorbed than the 'heavier' $^{13}CO_2$. In other words, C_3 plants discriminate against $^{13}CO_2$ in favour of $^{12}CO_2$ (Park and Epstein, 1960). The *discrimination of* ^{13}C ($\Delta^{13}C$) 'is a measure of the $^{13}C/^{12}C$ ratio in plant material relative to the value of the same ratio in the air on which plants feed' (Condon *et al.*, 2002). The discrimination is preferentially measured in leaf tissue. It 'provides a time-integrated measurement of the plant's transpiration efficiency over the period during which dry matter is assimilated' (Wang *et al.*, 2013). The equation defining $\Delta^{13}C$ is:

$$\Delta^{13}C = \frac{(^{13}C/^{12}C)_{\text{atmosphere}}}{(^{13}C/^{12}C)_{\text{plant material}}} - 1 \qquad (16.2)$$

The greater the discrimination against ^{13}C, the more positive is $\Delta^{13}C$. The discrimination of the C isotope with higher mass in favour of the 'common' ^{12}C is linked to the plant's capability to reduce the CO_2 partial pressure inside the intercellular spaces (p_i) in comparison to the partial pressure (p_o) outside in the atmosphere (Box 11.1). The smaller the ratio of p_i/p_o in plants due to the amount and activity of the carboxylase, the less the plants discriminate between the two isotopes. Therefore in C_4 plants (with p_i/p_o about 0.3) the discrimination is far less than in C_3 plants, where p_i/p_o is around 0.7. Therefore, C_3 and C_4 plants may be identified and separated from each other by examination of $\Delta^{13}C$.

Within the C_3 plants, however, it was found that the 0.7 value for p_i/p_o is indeed only an approximation. Within a plant species there may be a variation in this ratio, which causes a change in $\Delta^{13}C$, as shown for the relation in wheat (Farquhar and Richards, 1984):

$$\Delta^{13}C = (4.4 + 22.6 p_i / p_o) \cdot 10^{-3} \qquad (16.3)$$

According to Eqn 16.3, when p_i/p_o is 0.7, then $\Delta^{13}C$ (in short Δ) is about 20.2×10^{-3} or 20.2‰ (per thousand). In wheat genotypes Δ varies roughly between 19‰ and 23‰ (and p_i/p_o between 0.65 and 0.82), the range being a little bit smaller with barley (Fig. 16.3). Under water-limited conditions it was found that *the smaller* p_i and hence *the smaller* Δ, *the greater* was the *TE* of a genotype (Fig. 16.3). When water was non-limiting, both biological and grain yield were positively correlated with Δ (Condon *et al.*, 1987). From this it seems that Δ is not only a *selection criterion* (a trait) for TE, but also for the yielding ability of crop cultivars.

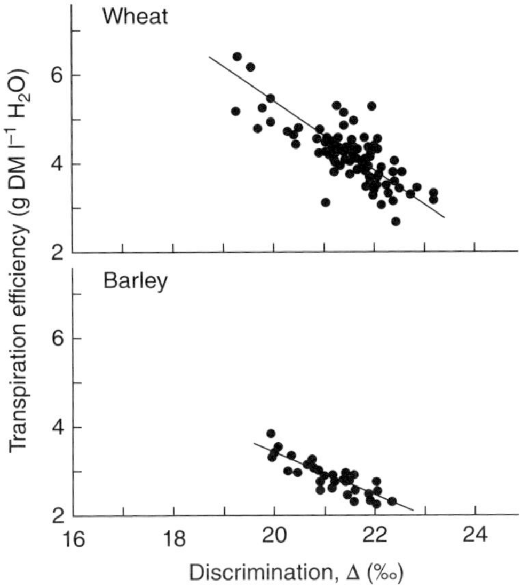

Fig. 16.3. Relationship between transpiration efficiency (related to biomass) and discrimination (Δ) of the carbon isotope ^{13}C during CO_2 assimilation in wheat (Farquhar and Richards, 1984) and in barley (Hubick and Farquhar, 1989). Each relationship is based on traits of various genotypes within a species.

Over the years of intensive research predominantly in Australia it turned out that a relation between Δ or TE (Fig. 16.3) and grain yield of wheat genotypes could not definitely be 'scaled up' (Condon *et al.*, 2004). That is because the variation in p_i/p_o or Δ (Eqn 16.3) and hence in TE can generally be achieved by two different physiological mechanisms: p_i can be lowered by partial stomata closure reducing stomatal conductance or by higher photosynthetic capacity (Condon *et al.*, 2002). Both mechanisms can also act together. Genotypes tending to limited conductance (small Δ) may produce a smaller biomass and grain yield than those with high conductance (large Δ) in the absence of soil water deficit due to sufficient rainfall or irrigation. That is because in this case small Δ and high TE are associated with less photosynthetic activity, reducing net assimilation rate (NAR) and crop growth rate (CGR) (Condon *et al.*, 2004). On the other hand, in environments with limited water resources, these genotypes may be superior for yield. In wet environments, however, genotypes with a large Δ and low TE proved to grow faster and that translated into larger grain yields. In such cases Δ was 'positively correlated with grain yield and dry matter production' (Condon *et al.*, 1987).

The inconsistencies related to the interactions of physiology and environment represent a great challenge for incorporating Δ into breeding efforts (Condon *et al.*, 2004). However, despite all the difficulties and drawbacks it steadily became apparent that Δ was not only a *selection criterion*, a trait, for TE, but also provided a basis for predicting the yielding potential of crop cultivars in environments with varying climate and water supply. In Australia two new wheat cultivars were bred combining special traits and Δ relations derived from appropriate genotypes. These cultivars had yield benefits compared to current cultivars, and were released for use in New South Wales during 2002 and for the northern cropping region in 2003 (Richards, 2006). The Δ technique has now been applied in a variety of C_3 crop plants in addition to wheat and barley, including rice, *Phaseolus* bean, cotton, tomato and sunflower (Xin *et al.*, 2009).

The evaluation of carbon isotope discrimination requires special techniques that may not be easily available, and are costly when used on a large scale. Therefore plant breeders look for simple traits that can be assessed more rapidly and inexpensively. For peanut, for instance, it was shown that *specific leaf area* (SLA) (see Section 16.1) was positively correlated with Δ. As Δ is negatively correlated with TE, as shown for cereals in Fig. 16.3, SLA also proved to be negatively related to TE (Wright *et al.*, 1994). That again indicates that genotypes with 'thicker' leaves have a greater TE, probably as a consequence of higher photosynthetic capacity. SLA seems to be predictive of TE and may serve as an indirect selection criterion for TE. This conclusion was drawn from experiments with peanut (Wright *et al.*, 1994), pearl millet (*Pennisetum glaucum*; Brown and Byrd, 1996), a tropical forage legume (*Stylosanthes scraba*; Thumma *et al.*, 1998), and a C_4 grass, switchgrass (*Panicum virgatum*; Byrd and May, 2000). However, that linkage may not apply to other crops like cowpea (Ismail and Hall, 1993) or to specific environmental conditions. The trait in barley that contributed to the comparatively large TE of this crop in addition to its faster phenology (Table 16.1), was a high SLA, not the low SLA suggested by the above results. Therefore, it might well be that a large SLA is good for saving water by reducing evaporation early in the season, and also good for early light interception. A large SLA may also be important for early transpiration, when the atmospheric vapour pressure deficit is low, and hence TE is favourably high. But later in the season, at a higher radiation level, when soil water supply may fall short and when vapour pressure deficit may also be high, a lower SLA (thicker leaves) might be advantageous for CO_2 fixation at relatively low transpirational costs.

16.3 Application of Molecular Biology to Improve Crop Performance Under Drought

The first crop plants to be developed by modifying their genetic make-up using the newly emerging techniques of molecular biology were released for commercial use in the mid-1990s. The first to be released successfully carried *genetic information* enabling the maize plant to make a toxin from the bacterium *Bacillus thuringiensis* that provided protection against insect pests. Since the beginning of this century, there has been increasing interest in using molecular biology to protect plants from *abiotic stresses*, particularly cold and drought. In 2012 and 2013 a number of genetically modified 'drought tolerant' maize cultivars became available. Early reports suggest that the yields of these new cultivars were somewhat variable in the response to the prevailing drought conditions across the USA.

One commercially available cultivar carries the *CspB* gene from the bacterium *Bacillus subtilis*, which codes for a *cold shock protein* that maintains the structure of *RNA* in a form that it can continue to function within cells that are under stress, including that due to drought (Castiglioni *et al.*, 2008). The *transgene* apparently operates throughout the life of a maize plant and influences crop yield through kernel number. Although all the physiological changes in a plant resulting from the genetic modification are uncertain, it is suggested that the somewhat slower growth of plants helps conserve water in the soil so that it is available for grain fill (Edmeades, 2012).

In an analysis of changes induced by genetic modification of plants in relation to drought tolerance, Lawlor (2013) concluded that slower growth was commonly reported. In consequence such plants can have smaller leaves and leaf areas than unmodified plants at a set time after seedling emergence. Stomatal conductance also is smaller in many genetically modified plants. This can be the result of a reduction in the size of stomatal pores or the number per unit area of leaf, which may be associated with increased abscisic acid (Lawlor, 2013).

There are very many traits associated with plant performance under adverse conditions of water supply that are being or have been investigated using the tools of molecular biology (Lawlor, 2013). However, Passioura (2007) and Lawlor (2013) have questioned whether all these traits are commonly found under field conditions or are simply symptoms associated with short-term withdrawal of the water supply in the laboratory. In addition, Lawlor (2013) has criticized many of the experimental protocols used to evaluate the impact of genetic manipulation on drought tolerance because insufficient attention is paid to ensuring similar leaf areas and architecture when comparing water use by wild-type and transgenic plants. In addition he pointed out that the use of small volumes of coarse-textured rooting medium can lead to an underestimation of the possible benefits from the ability of plants to function at lower leaf and root water potentials. This is because once soil begins to dry the soil water potential declines rapidly but the volume of water released for a given change in potential is much less than occurs when it is moist (Fig. 5.1). The large volume of soil that roots can exploit in the field means that the drying is more gradual and transport through the soil to the root surface becomes more important in regulating supply (Section 7.2).

For real progress in developing crops that show true resilience when water is in short supply, the selection of traits and the experimental protocols to test them have to be enhanced. To achieve this requires increased collaboration between molecular biologists, plant physiologists, plant breeders, agronomists and soil scientists (Passioura, 2007).

17 Controlling the Soil's Water Balance by Soil Management

17.1 Which of the Balance Components can be Changed?

In water-limited areas, a major objective for soil management is to modify the field water balance for the benefit of plant production. Plants, and for us it is especially crop plants, rely on the steady supply of soil water, flowing to the roots in response to demand over the entire season. Therefore arable soils have to be conditioned to take in *rainwater*, the sole source of soil water, at a rate equal to the rainfall rate. This prevents water from being lost as runoff. But there are also instances in a dry climate, where the goal is to concentrate rainwater in discrete patches by inducing runoff from surrounding areas, so-called 'water harvesting'. In such cases the intake rate at the watered spot must be considerably higher than the rainfall rate. These cases are very special and are described in Box 17.1 and Section 19.2. Water supply from *groundwater* to crops will be important only in locations where the soils are characterized by a shallow water table, such as Fluvisols, Gleysols and some Histosols (lowland moor). Here the water moves up into the rooting zone by *capillary rise*. The rate of supply from that source can be controlled within limits by regulating the groundwater level. Raising the level can improve the water supply of crop stands during the growing season. On the other hand, it is necessary to lower the level of the water table where the goal is to improve trafficability or workability, as well as aeration and the warming of the soil. These are essential facets of arable fields with respect to the management of soils for plant development. In the past it was not infrequent for some 'amelioration' to have been excessive. In such cases drainage of water was so intense and the level of groundwater so deepened that the soil actually became droughty.

We will restrict this discussion to soils in which the groundwater is so deep that capillary rise does not contribute to root water uptake. Here it is exclusively rainwater that replenishes the water-filled pore spaces in the rooted profile after it has become depleted by plant water use during a dry period between rainstorm events or seasons. Maximum exploitation of stored soil water for use in transpiration requires unproductive soil evaporation to be reduced to a minimum. For water *the soil needs to be managed so that it acts like a trap.* The soil has to allow water intake at a sufficient rate that it absorbs all the water provided in rain. Following the capture of rainwater the soil should retain it and prevent its escape back to the atmosphere. Measures to *encourage infiltration* have to be accompanied by measures to *control evaporation.*

Hence the soil water budget has to be managed in such a way that the water supply is kept at a high level while the water loss is kept as small as possible. In other words, any build-up of surface water or runoff must be avoided and evaporation must be minimized. Where these goals are successfully achieved, the efficiency of evapotranspiration (ETE) will be improved.

We shall contemplate this fact by considering the following. The term ETE was introduced in Section 11.1. It is defined as:

$$\mathrm{ETE} = \frac{\mathrm{P}}{\mathrm{ET}} \tag{17.1}$$

where P is the above ground biomass dry matter production of the crop per unit area and ET is the actual cumulative evapotranspiration. Transpiration efficiency (TE) is defined as P/T, where T is the transpiration of the crop plant (see Eqn 11.2), so that P = TE · T. We can therefore use this to substitute for P in Eqn 17.1:

$$\mathrm{ETE} = \frac{\mathrm{TE} \cdot \mathrm{T}}{\mathrm{ET}} = \frac{\mathrm{TE} \cdot \mathrm{T}}{\mathrm{T} + \mathrm{E}} = \frac{\mathrm{TE}}{1 + \mathrm{E}/\mathrm{T}} \tag{17.2}$$

Equation 17.2 was introduced by Cooper (1983) and shows that ETE increases to the extent that E is constrained relative to T. In the case that E approaches zero, ETE will simply have the same value as TE.

In earlier chapters we mentioned that in addition to E, other quantities that contribute to unproductive water loss, such as *subsoil drainage* (D) and *runoff* (R), are seldom measured separately in field studies on crop water use (Eqn 3.1). However, just like evaporation, these quantities normally contribute to total water use but the relative amounts are site specific and generally unknown. For this reason, Gregory (1989a) amended Eqn 17.2 by adding these loss components. The modified equation reads:

$$\mathrm{ETE} = \frac{\mathrm{TE}}{1 + (\mathrm{E} + \mathrm{R} + \mathrm{D})/\mathrm{T}} \tag{17.3}$$

The more effectively soil management succeeds in reducing the three *loss-making components* E, R and D in relation to T, the greater will be the value of ETE (Eqn 17.3). Strictly speaking, however, extending the equation by R and D takes us away from the original concept of efficiency, that is from ETE, how much water is used by evapotranspiration in relation to biomass production. Rather, the efficiency is now related to the total use of water at a site associated with plant dry matter production. Therefore, all of the unproductive water losses, namely E, R and D, are assessed in proportion to the only form of productive water loss, which is transpirational water use of the crop (T). Based on these concepts we finally arrive at what is called the *water use efficiency* (WUE) of plant production (cf. Section 11.1). Hence we can rewrite Eqn 17.3:

$$\mathrm{WUE} = \frac{\mathrm{TE}}{1 + (\mathrm{E} + \mathrm{R} + \mathrm{D}) / \mathrm{T}} \qquad (17.4)$$

In Eqn 17.4 (as in Eqns 17.2 and 17.3), TE represents a quantity that is more or less constant for a specific plant species at a given vapour pressure deficit (see Section 16.2). Taking TE as a 'constant' numerator, we can easily recognize that the greatest WUE will be obtained when T is maximized in relation to all the other loss-making quantities. On the other hand, it is not only WUE that will increase. When T increases in absolute terms during the season, because E, R and D decrease, then P also will improve.

Improving the water supply to roots, consequently increasing transpiration, hopefully means that occurrence of water shortage and stress can be avoided, providing an opportunity to produce larger yields of biomass and marketable products. However, an increase in T through enhancement of the water supply, for instance by irrigation, will increase WUE only if the rise in T is proportionately more than that of E, R and D together (Gregory *et al.*, 2000).

So far we have considered the loss-making components of the soil water balance equation (Eqn 3.1) that contribute to unproductive water use. But there is also a 'productive' water use that can minimize the crop WUE. This is T_W, the transpirational water use by weed populations (Turner, 2004; Hillel, 2011), producing undesirable biomass in competition to the crop. To be thorough T_W has to be incorporated in Eqn 17.4, arriving at Eqn 17.5:

$$\mathrm{WUE} = \frac{\mathrm{TE}}{1 + (\mathrm{E} + \mathrm{R} + \mathrm{D} + \mathrm{T_W}) / \mathrm{T}} \qquad (17.5)$$

In the following sections, quite different strategies will be presented for optimizing the water supply to crop plants. First, measures of soil management will be discussed. These are followed in Chapter 18 by appropriate strategies for plant cultivation. Whenever possible, we will refer back to Eqn 17.5 to explain the objective of individual practices.

17.2 Controlling Infiltration

As rainwater is the sole positive quantity of the natural soil water balance equation, when groundwater is very deep below the root zone, all techniques of field husbandry have to be directed towards the maintenance of a high level of *infiltrability* in the soil (Section 5.3). Maintaining or increasing the soil's infiltrability is associated with less puddle formation and runoff of surface water (R in Eqn 17.5) or, when possible, to prevent this form of water loss entirely. We have to work harder to reach this objective the more that infiltrability is restricted by properties of the soil, such as texture (medium-sized silt), weak aggregation, lack of lime or gypsum and small organic matter content or the greater is the intensity of rainfall, the longer that rainstorms last, and the more sloping is the land form on the farm. Generally, it is the runoff of water that does not infiltrate that causes overland flow and *soil erosion* (i.e. the detachment of soil particles and their removal from a site by water). Water erosion may often pass unnoticed because the loss occurs more or less uniformly across the land area, just as occurs with erosion by wind. This form of surface erosion by water is called *sheet erosion*. Soil erosion is the consequence of heavy rainfall in conjunction with decreasing infiltrability during soil wetting (Fig. 5.11), or inherently low infiltrability. The kinetic energy of falling raindrops makes aggregates at the soil surface disintegrate or slake into individual particles. The impact of the drops distributes these particles over the soil surface. The surface becomes levelled and smoothed, depressions that allowed puddles to form and retain water are filled up by detached soil material of fine grain size like silt, often interlayered by clay particles and organic matter. Large pores, initially open at the soil surface, become clogged by solids. A skin of fine particles covers the surface. The rate of water infiltration decreases drastically, as does the capacity to store surface water in depressions. Surface water starts to flow downhill, and its drag force causes the topsoil to be swept away. Soil erosion does not just cause an irretrievable loss of arable land and a field's cropping capacity; a loss summed up by the phrase *'on-site' damage*. Importantly, soil organic matter, plant nutrients, pesticides and other contaminants can be removed, both in the runoff water and attached to the flushed soil material. These substances are deposited somewhere else causing *'off-site' damage*. Watercourses, creeks and rivers become shallow by sedimentation and siltation, and rivers as well as sea bays and the open sea get polluted and become eutrophic. Therefore, controlling infiltration has a *double significance*. It optimizes the plant-available water content of the soil and reduces soil erosion caused by unrestrained water.

The secret of high infiltrability is the *roughness* and *openness* at the soil surface (Dixon and Peterson, 1971). Preserving roughness and openness is particularly important, when the upper horizon of the arable soil, the A_p horizon, is intensively tilled. Roughness is created by soil aggregates, formed and exposed at the soil surface. At best these aggregates are not just mechanically formed by breaking larger-sized massive clods into granular peds with tillage implements, but are created by the joint action of soil microbes and plant roots that forms the mineral and organic particles into porous, stable peds like *crumbs* (Goss and Kay, 2005). Crumbs show a certain degree of resilience and of *'intra-aggregate'* porosity for water storage. Openness means that *large soil pores* start at the soil surface between individual peds. Their orifices are 'open', unblocked and hydraulically connect the soil surface with the underlying topsoil, and even more importantly with the subsoil. These large *'inter-aggregate'* pores ensure the soil's *capability to accept rainfall* at a high rate and to distribute and drain the water within the entire profile. The infiltrability is retained as long as the hydraulic connection is not disrupted. The maintenance of pore openness requires a sufficient *stability of the aggregated soil* to withstand shear failure or plastic deformation by *mechanical forces*. These forces are exerted by raindrops or, for instance, by *agricultural machines* when passing over the land. The forces penetrate and are distributed within the soil, and may change aggregate porosity, soil porosity and soil structure when the external forces exceed the soil's internal stability. The requirement for soil stability presents a dilemma for any *tillage system* aimed at loosening the soil. That is especially true for deep inversion tillage with the *mouldboard plough*. Any loosening action reduces the number of intergranular points of contact between soil particles and thus lowers the soil's internal resistance to external forces. Some negative consequences of soil loosening are poor trafficability, soil re-compaction, destruction of large pores and decreased infiltrability the next time the soil is subjected to mechanical stress.

In a Luvisol derived from loess, mouldboard ploughing increased the porosity within the A_p horizon to 50% by volume (Fig. 17.1A). The saturated hydraulic conductivity (K_s) used as a measure of infiltrability was almost 1000 cm day^{-1} (Fig. 17.1B). When compaction decreased the porosity to 40% by volume, K_s was reduced to about 10 cm day^{-1}. From the figure it is evident that soil loosening increased considerably the volume percentage or porosity of large pores > 60 μm (i.e. pores that drain at pF < 1.7). Furthermore, the effect on the porosity of smaller pores < 60 μm was only marginal (Fig. 17.1A). The large pores represent essentially the inter-aggregate pores, and the finer pores are those within the aggregates – the intra-aggregate pores. Whereas the *loosened soil* was *subangular-blocky* to *granular*, the *compacted soil* was *massive*, containing no discernible individual peds. In the compacted soil the inter-aggregate porosity was small (Fig. 17.1A).

Hence it seems that the porosity associated with pores > 60 μm is largely determined by *soil structure*, whereas the porosity of finer pores < 60 μm is much more influenced by *soil texture* (Fig. 17.1A), that is the mechanical composition, how much sand, silt and clay it contains (cf. Fig. 5.1). This is very similar to the circumstances surrounding *saturated and unsaturated hydraulic conductivity* (cf. Fig. 5.3). Creating large pores by tillage that loosens the soil, shifts the conductivity function to higher values at pF < 1.7 (Fig. 17.1B). Therefore,

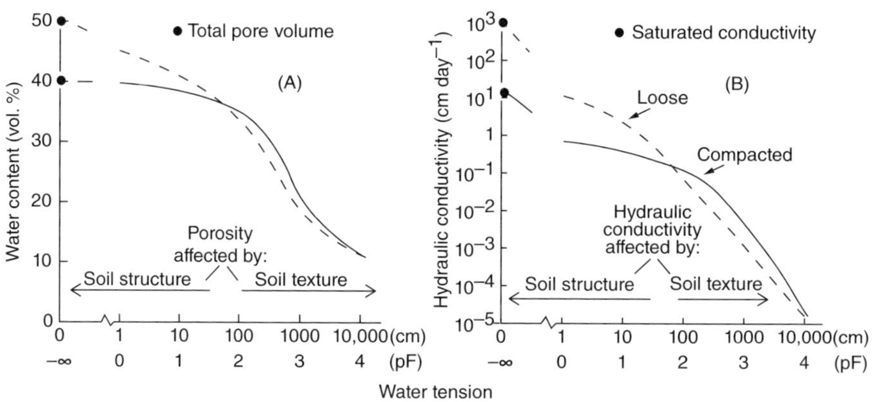

Fig. 17.1. Soil moisture characteristic curve (A) and hydraulic conductivity function (B) of a Luvisol derived from loess. On the one hand, the soil was loosened by tillage and contained subangular-blocky to granular peds. On the other hand, the soil was compacted by mechanical loading, resulting in massive soil structure. For more explanations see text.

soil loosening is a way to increase infiltrability. But this condition of rainfall acceptance will only persist as long as the state of looseness and porosity does not change. For instance, whenever the soil gets re-compacted in the course of *machine traffic*, porosity and infiltrability can be dramatically reduced. Wheel tracks directed downhill are often the starting point of the so-called *rill or gully erosion* by water.

From these considerations conclusions can be drawn about infiltration control: (i) unavoidable machine traffic should not be routed downhill; (ii) whenever possible operations necessary for crop production should be combined; and (iii) the proportion of the land that is wheeled should be reduced by choosing implements that cover a wide area and are not too heavy. To keep the re-compaction effect at minimum, wheel tracks should be confined as far as possible to repeatedly used tramlines.

Over an extended period of time the soil's infiltrability will remain sufficiently high, when soil aggregates, exposed at the soil surface, are stable enough to withstand the mechanical stress exerted by machines and the impact of raindrops. *Stable aggregates* resist disruption and surface levelling during heavy rainfall. Maintaining roughness and openness at the soil surface is an essential feature of productive soils for infiltration control. Soils with these properties can maintain large infiltration rates over an extended period of time. And indeed, considerable intake rates are required when heavy rain is pelting down during a thunderstorm. In many semi-arid and tropical regions the number of high intensity rain events is greater than experienced in temperate regions. But even in Central Europe, it is not unusual for a 30 mm precipitation to fall within a time as short as 30 min. For the absorption of rain, the infiltrability or the value of K_s must not be lower than 144 cm day^{-1} (an intensity of 30 mm 0.5 h^{-1} is the same as 144 cm day^{-1}). This value corresponds to a 'very high' conductivity, characteristic of soils with a 'very well' pronounced soil structural development (Arbeitsgruppe Bodenkunde, German study group of Soil Science, 1982). According to the Soil Science Society of America, the conductivity value of 144 cm day^{-1} is rated as medium class, typical of sandy materials (Klute and Dirksen, 1986).

In general, *sandy soils* have an infiltrability large enough to receive even intense rain. With *loamy and silty soils*, however, the infiltrability often declines during the vegetative period due to the weakness of aggregates. Then the infiltrability has to be re-established for the next crop by tillage operations that increase soil porosity. But heavy rain may cause *aggregate breakdown* and loss of porosity, causing infiltrability to drop drastically again. Additional measures have to be taken to increase aggregation and reduce the susceptibility to surface puddling or silting-up. Soil structure can be *stabilized physico-chemically* by liming or adding gypsum, and by applying stable humus in the form of *organic amendments*, like farmyard manure or compost. These amendments are usually processed from organic residues by rotting during storage in pits. The processing involves the respiration of lightly decomposable substances and at the same time the formation of new organic macromolecules. This fraction of stable organic matter is relatively resistant to microbial attack and therefore persists for some years when added to the soil. On the other hand, *plant residues* left in the field after harvest can be degraded rapidly by soil microbes, especially when the residues are rich in N, such as those from a green manure crop with a C:N ratio of around 20. These residues make an insignificant contribution to stable soil humus. The contribution is greater with cereal straw with a C:N ratio of around 100. Generally, it is the quick turnover of organic residues that induces a temporary *microbial stabilization* of soil structure, be it for just a few weeks or months (Monnier, 1965).

The canopy of living plants protects the soil from the kinetic energy of falling raindrops and thereby against aggregate slaking. A similar *physical protection* is afforded by a cover of dead plant material, called *mulch*. Like living plants, mulch provides this positive umbrella effect to maintain soil infiltrability at a high level (Fig. 17.2). The presence of mulch will reduce the potential of the water entering the soil surface, which favours adsorption into the finer pores of the soil matrix and so decreases the amount moving downwards by preferential flow (Box 5.2). For high intensity rainfall of short duration this can result in more water being stored in the surface layers than if the soil is bare. But there is another important aspect of the soil protection provided under a mulch cover. Mulch evens out fluctuations in soil temperature and moisture, and stimulates microbial growth. Microbial activity *stabilizes soil structure* against slaking. So there are two positive effects of mulch on the maintenance of infiltrability (Triplett *et al.*, 1968): (i) physical protection; and (ii) aggregate stabilization.

There are particular 'heavy' *clay soils* that are by nature tightly packed. Their mineral particles stick closely together. The particles form a cohesive mass, lacking planes of weakness. Prominent individual peds are not visible. These clay soils are massive. The constituent *clay minerals*, such as kaolinite, illite and hydrous mica, are *not* (or only partially) *expandable*. They lack a distinct shrink–swell characteristic

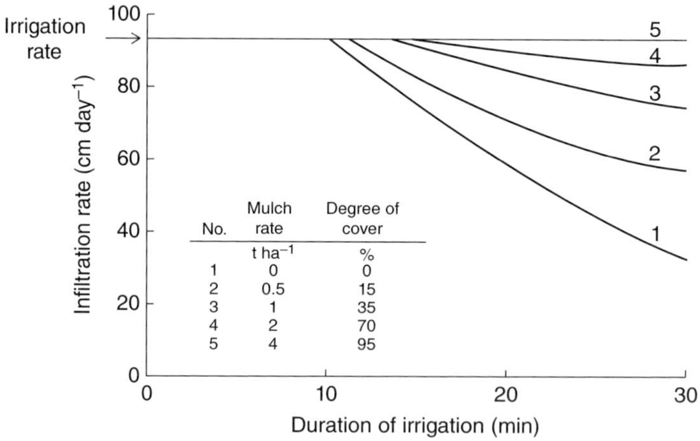

Fig. 17.2. Influence of chopped straw as a mulch cover on water infiltration into a Luvisol derived from loess with 5% slope. The irrigation rate was 92 cm day^{-1}. At a mulch rate of 4 t ha^{-1}, corresponding to a degree of cover of 95%, the infiltration rate was equal to the irrigation rate throughout the experimental period of 30 min. Without mulch cover the soil surface slaked easily and the infiltration rate dropped early. Even small mulch rates improved the soil infiltrability and the formation of surface water was delayed (after Roth, 1992).

with the build-up of fissures, cracks and peds. These inactive, 'structureless' soils may benefit from organic manure as an additional means to improve infiltrability. The desired effect here is not so much caused by physico-chemical action or by enhancement of microbial activity. It is a physical *dilatation effect* of the organic matter, separating the mineral particles to some extent. The bulk density is reduced and spaces and pores are left free between solids for rapid water intake and aeration. The *friability* (ease of crumbling) of these soils is improved. The lower plastic limit is also called the friability limit, and at water contents greater than this value the clay soil is *plastic* and sticky. Organic matter shifts the friability limit to greater water content (Kohnke, 1968). What we expect from organic manure seems to be perfectly achieved under a perennial crop stand of lucerne or a grass–clover mixture. Plant litter contributes to the friability and porosity of the surface soil without any need for a mechanical incorporation of residues by tillage. The taproots of the legumes penetrate the top- and subsoil. After the roots decay, open pores or small channels some millimetres in diameter are left that allow drainage of tension-free surface water.

Besides those 'additional' measures of soil management, the infiltrability can be regulated directly by mechanical tillage operations when the soil is dry enough and therefore of *friable* and not of *plastic consistence*. Any *loosening operation* aims to create pores and cavities between soil aggregates. The aggregates themselves are supposed to stay porous and should not be compacted, in spite of the shearing, pressing and slicing action of tools during tillage. By increasing the volume percentage of *inter-aggregate pores* the saturated hydraulic conductivity and the soil infiltrability are increased (Figs 17.1 and 17.3). In this way the uncompacted aggregates, containing intra-aggregate pores, can be supplied with percolating water at a high rate. Dry aggregates will absorb the water like a sponge and will store it temporarily. The water will be retained in available and unavailable form, the proportions varying according to the soil's textural composition (Fig. 3.2).

Clayey soils have only a narrow range of water content where their *consistency* is *friable*. Within this optimum range, clayey soils can be tilled relatively easily with only light draught. The *tillage effect* associated with breaking up a coherent, non-structured soil mass into smaller aggregates is then greatly promoted. These clay soils are called 'hours-and-minutes-soils' in Germany, and are often too dry or too wet for successful tillage. Clay soils, tilled at the right time and therefore well aggregated, may have a larger saturated hydraulic conductivity (Fig. 5.3) at a given macroporosity than weakly aggregated clay soils (Fig. 17.3). The reason for this discrepancy is the way that the large pores are interconnected, either with strong constrictions or with straight links. It is the *continuity of pores* that creates differences in infiltrability at a given macroporosity. *Plant roots* and *earthworms* create large pores of an extremely high continuity when pushing aside soil particles or when hollowing the soil matrix. This system of *biopores* is destroyed again and again by tillage, especially by deep inversion tillage.

Soil loosening increases the *infiltrability*. But this does not mean that the infiltrating water is transmitted readily into the non-loosened soil horizons below. Instead, it may well be that the water is at first *temporarily stored* within the cavities of the upper layer formed by tillage. From these interstices the water will flow into the unloosened subsoil at a rate that is much less than the infiltrability.

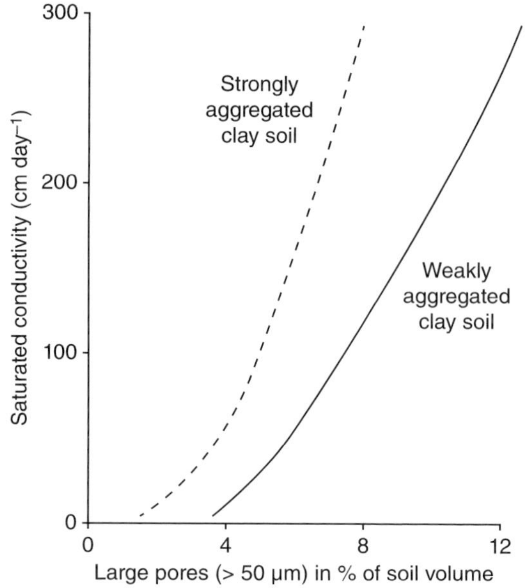

Fig. 17.3. The volume percentage of large pores, for the most part positioned between aggregates, influences the saturated hydraulic conductivity. In the weakly aggregated clay soil the inter-aggregate pores are poorly connected hydraulically. In the strongly aggregated clay soil, on the other hand, the connection is much more effective in allowing water flow. Here the continuity of the large pores is very much improved compared with the weakly aggregated soil (after Warkentin, 1971).

The slowing down may cause stagnant moisture. *Waterlogging* will be the more severe, the greater the precipitation rate and the smaller the saturated conductivity of the unloosened soil (van Duin, 1955). In sloping areas water can be diverted laterally, flowing downhill over the top of the impeding layer; a process called *interflow*. At some point in the lower slope position the infiltrated water may re-enter the soil surface. *Ploughpans or tillage pans*, created by soil deformation and compaction, have only a *low percentage* of large pores (Fig. 17.1A), which are highly *discontinuous*. The pans can delay water seepage for several days. These are the periods of *oxygen deficiency* at the bottom of the A_p horizon and in the subsoil.

In sloping areas it is essential to form additional *storage space* for the *intake of excess water* to prevent *runoff*. Such techniques include contour ploughing, bench terracing, furrow dyking and creation of microbasins, just to mention a few, with quite a variation according to local conditions and available technology.

Loamy and silty soils slake very easily, which may happen readily during winter or during the rainy season when left uncropped. To counteract rapid aggregate breakdown, farmers tend to till the soil so that a *coarse furrow* slice is formed. Coarse clods will disintegrate more slowly than finer granules.

Slaked surfaces reduce soil infiltrability dramatically because they form the boundary layer between the atmosphere and the soil, and because the 'skins' lack large pores for rapid absorption and infiltration of rainwater. The effect of such a 'seal' on water infiltration during heavy rain was studied with the aid of a computer simulation, based on the equations of Darcy and Richards (Section 5.3). At a rainfall intensity of 48 cm day^{-1} the pre-ponding period (Fig. 5.11) was restricted to only 20 min, before surface water was built on top of the slaked loess-derived Luvisol (Fig. 17.4). During this short period, when only 7 mm of rain had fallen, the infiltrability decreased below the precipitation rate. Weather statistics show that in the Göttingen district, such a rainfall event can be expected to occur at least twice a year. In the case of a non-puddled, open surface but in the presence of a plough pan, the pre-ponding period lasted for almost 2 h (114 min). Up to that moment 38 mm of rain had infiltrated the soil (Fig. 17.4). From then on, water was backing up from the pan at 22 cm depth throughout the plough layer to the surface. What is not indicated in Fig. 17.4 is the fact that the holding of water by the pan will promote aggregate breakdown and slaking at the soil surface.

For the cultivation of a crop a farmer will usually disturb the soil twice mechanically. In addition to the various other aims of tillage (the incorporation of residues, shed seeds, fertilizer and lime; killing weeds (Eqn 17.5); elimination of soil-borne pathogens) one purpose is to re-loosen the soil after a period of settling and consolidation. What is called *primary tillage* for the next crop often starts directly after harvest of the previous crop by chopping the stubble. *Stubble tillage* is said to make the soil crumbly. The soil surface becomes rough and open again, good for water infiltration and soil water storage. Stubble tillage prepares the soil for the next and most noticeable form of primary tillage, which again is thought by farmers in many regions to be beneficial and even absolutely necessary for growing

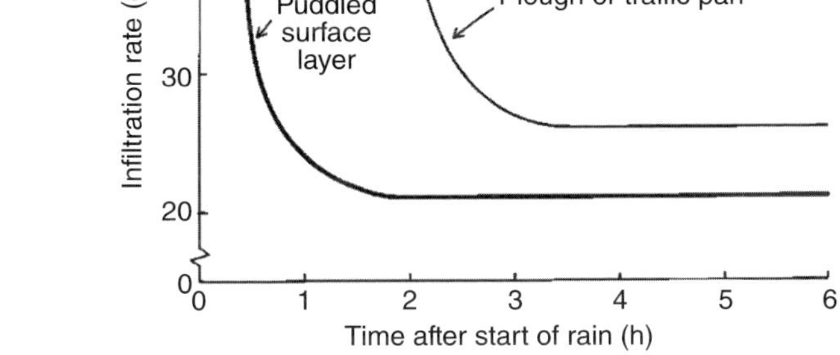

Fig. 17.4. Computer simulation of water infiltration into a Luvisol derived from loess. The soil had either a slaked surface layer or a plough or traffic pan that restricted water infiltration after some time. The precipitation rate was 48 cm day^{-1}. Without a restricting layer the water was infiltrating the soil completely for at least 6 h (after Ehlers *et al.*, 1980b).

the succeeding crop. This tillage operation is a more or less *intense* and *deep soil manipulation* that makes use of various tillage tools like the mouldboard plough, disc plough, cultivator, rotavator or others. Either this major part of primary tillage is performed immediately ahead of seeding the next crop, or the cultivated soil is left bare (and unprotected) for some time, even for months over the winter period. *Secondary tillage* is soil manipulation mostly for seedbed preparation. Often the objective is to prepare a seedbed in one pass to avoid excessive re-compaction. Depending on local tradition, mechanization level, soil, climate and crop selection there is an immense range of *tillage methods* applied during the seasons of a year. In addition to tillage method, the *tillage systems* too will vary from region to region according to the dominant tillage tool used for primary tillage. Those tillage systems that aim to leave a protective mulch cover at the soil surface are collectively covered by the term '*conservation tillage*'. Usually it is expected that, when it is time for the seeding of one crop the soil will still be covered by residues from the previous crop, the degree of cover being at least 30%. This objective calls for non-inversion, less intensive and shallower soil manipulation. The extreme form of conservation tillage is where soil tillage is completely omitted, confined just to the action of the seed-drill. This extreme system of firm-soil husbandry (van Ouwerkerk, 1974) is called '*zero-tillage*' or no-till (Section 17.5).

17.3 Controlling Evaporation

As long as the water flux to the soil surface is not limited by the hydraulic properties of the soil, the rate of evaporation is largely governed by external meteorological conditions. These are net radiation, vapour pressure deficit and wind speed. During this first stage of evaporation the rate of water loss can hardly be manipulated by altering soil properties (Section 5.2). However, the length of this stage, where water loss is at maximum, can be shortened by a *loosened topsoil* (Fig. 5.8). This way the reduction of the unproductive water loss (see Eqn 17.5) can be induced earlier, giving rise to improved WUE, and by an increase in T to increased P. But one has to bear in mind that the loosening action should never be started before the topsoil layer has *dried sufficiently*. Farmers have to delay tillage until the lower plastic limit is attained and even exceeded. Only after some drying is the soil of *friable consistency* and will crumble when cultivated. When, however, the soil is wet and therefore of plastic consistency, any mechanical operation will cause soil compression and distortion. Larger pores are destroyed by kneading and smearing. Therefore we conclude that when a soil has been slaked by rain, tillage might be necessary. But the operation must be *postponed* until the soil has dried off. Cultivation will be a helpful evaporation control, not the moment that the soil is tilled, but shortly thereafter. It is directed towards shortening the period of first-stage drying after the next rainfall event.

A loosened soil, for instance one with fine subangular-blocky structure, shows a distinct hydraulic conductivity function, quite different from the function of a compacted soil without visible peds and considered to be massive (Fig. 17.1B). On drying, the decline in conductivity is more pronounced in the aggregated soil, similar to a sandy soil. This response is in contrast to that in the massive, compacted soil, which behaves more like a clay soil (Fig. 5.3) with a much smaller fall in conductivity. The saving in soil water achieved by loosening the soil and reducing evaporation is dependent on time (Fig. 17.5). Suppose the soil stays dry for a longer period of time because of the lack of rainfall. Then the water savings due to early evaporation reduction decrease during later intervals because after some time the evaporation rate of the loosened soil remains slightly greater compared with that of a dense soil without early evaporation control.

Roughness and openness of the topsoil, a principle of infiltration control, is also valid for evaporation control. But the reason is a different one. For infiltration control, loosening is essential to increase the saturated hydraulic conductivity but for evaporation control it is important to reduce the unsaturated conductivity in the higher tension range (Fig. 17.1B). Farmers would argue that loosening results in *cutting off the capillarity* of finer pores.

Stubble cultivation was mentioned in Section 17.2 as a means of infiltration control. Aiming at the increase in soil water content, stubble cultivation supports two mechanisms: one for controlling *water gain* and one for controlling *water loss*. When the soil has dried under the previous crop, loosening is important to increase infiltrability and thus the actual gain of infiltration water, even under heavy rain. Moreover, loosening induces an early reduction in the evaporative water loss. The upper loosened soil layer becomes dry, as it is cut off from the capillary pores of the consolidated soil layer beneath. The dried top layer separates the water-conducting bottom layer from the vapour-absorbing atmosphere. Water then has to pass through the desiccated top layer, mainly by the process of vapour diffusion. The effect of surface drying and evaporation reduction is sometimes referred to as the '*self-mulching effect*'. Both modes of action are strengthened by *mulching of organic residues*, which is a characteristic feature of conservation tillage systems (Section 17.5).

We will give a short description of the effect of a residue cover on evaporation. A *mulch cover* on the soil surface changes the physical conditions for water evaporation from the soil surface. The cover reduces the energy load

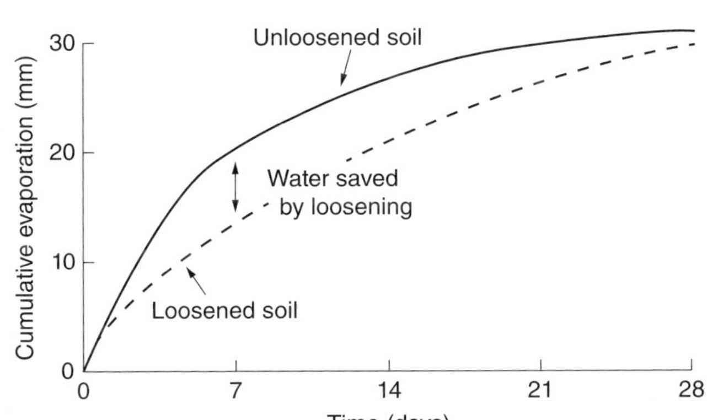

Fig. 17.5. Effect of soil loosening on the cumulative evaporation of a moist soil (diagrammatic presentation) (from Ehlers *et al.*, 1986a, 1987).

available for vaporization by reflecting and absorbing incoming net radiation. At the same time the cover lowers the temperature, the vapour pressure deficit of the air and the turbulent motion of the air near the soil surface. These modifications are the principal reasons why, in contrast to soil loosening, a mulch cover suppresses the evaporation rate right from the beginning (Figs 17.6 and 17.7). But even so, the actual water saving depends on the interval between rainfall events. The shorter the interval, the more water is saved (Fig. 17.6).

Stubble tillage in connection with mulch residues left at the soil surface is a tillage system that is well known in many arid or semi-arid regions of the world. It is commonly called *stubble mulch tillage* and forms an essential part of *dryland farming*. Stubble mulch tillage is a form of conservation tillage, provided that a considerable quantity of residues is maintained at the soil surface despite repeated soil manipulation. Tillage is repeated to keep the soil surface rough and open within the period between harvest of the fore-crop and seeding of the after-crop. The operations aim at both evaporation and infiltration control. Moreover, weeds that use water and volunteer crops are killed, often by application of a power-take-off (p.t.o.)-driven tillage tool, a '*rod-weeder*'. Quite often though, in dryland farming the sequence of crops is interrupted by a *fallow year*. The aim of fallowing is to concentrate the rainwater of 2 years for the support of cropping in only 1 year. In Section 17.5 it will be shown that the actual water savings of fallowing are quite small because of severe water losses, mainly by evaporation.

Although the principles of evaporation and factors that offer some measure of its control seem to be straightforward, there are numerous interactions that can occur that result in apparent contradictions. The interactions can result from changes in soil properties and in boundary layer conditions

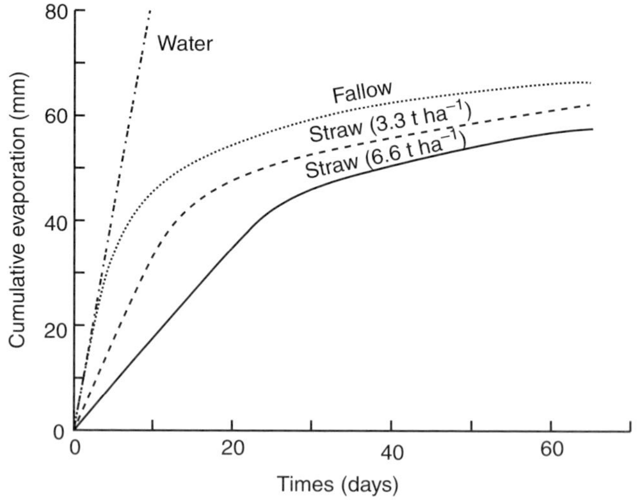

Fig. 17.6. Effect of straw mulch on the cumulative evaporation of a moist soil. For the purpose of comparison, a curve for the evaporation from an open water surface has been plotted (after Bond and Willis, 1969; published in Ehlers *et al.*, 1986a, 1987).

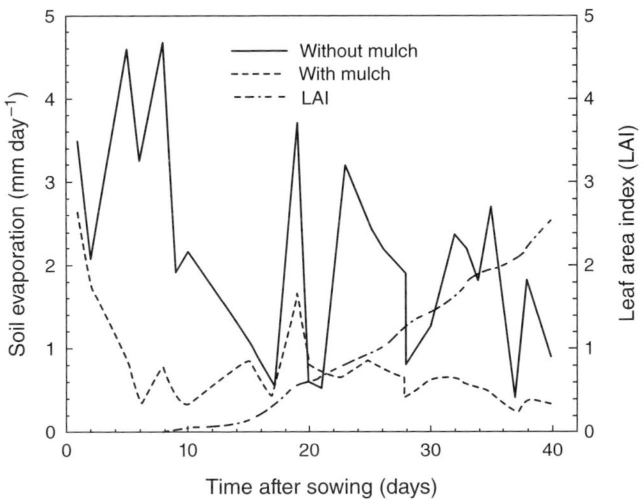

Fig. 17.7. Daily soil evaporation rate with maize under wheat straw mulch and without mulch. The maize was double cropped with wheat as preceding crop. The maize growing season started at the beginning of June. Mulching markedly reduced soil evaporation from the onset and this continued for the early period of crop development. During the time course of 40 days after sowing the leaf area index of the young maize increased from 0 to 2.5. As leaf area increased the daily evaporation rate declined, especially on the unmulched treatment. Results from 1999 in the North China Plain (Zhang *et al.*, 2005).

that vary with management practices, including tillage and residue handling, as well as soil type and climate (Basch *et al.*, 2012).

17.4 Increasing the Quantity of Extractable Soil Water

The water in the soil profile, extractable by crop roots, is a function of the available water at field capacity, in short the 'available field capacity' (FC_{av}), and the effective rooting depth (Section 9.4). If farmers could increase FC_{av} by *enrichment of the soil organic matter content*, seepage losses could be reduced under otherwise unchanged conditions. The water use efficiency and the yield potential of the site could be improved (Eqn 17.5). *Anthrosols* are considered to be rare examples of soils where humans have so heavily concentrated organic materials in the soil that the FC_{av}, the nutrient supply, and hence the yield capacity markedly increased. The soils are found in Ireland, the Netherlands and in north-west Germany. The uppermost topsoil, covered with grass or heather, was removed from distant grassland or forests. This sod, which was rich in organic matter, was used as bedding in the animal barns and then applied to the fields in the immediate vicinity of the villages as a manure–compost mixture for the amelioration of the sandy or loamy soils. This was not a short-term process and quite a few generations were involved in reshaping the soil, beginning in the late Middle Ages and only ending during the 19th century. The high level of amelioration attained could never have been achieved within a single generation.

Under the conditions of modern field husbandry *with inversion tillage*, the possibility of increasing the organic matter content of arable soils to any significant depth by manure or fertilizer application seems to be rather modest (Lehne, 1968; Rauhe *et al.*, 1968). Even with conservation tillage systems (Section 17.5) the concentration of soil organic matter is more or less confined to the 0–10 cm layer (Benjamin *et al.*, 2007; Angers and Eriksen-Hamel, 2008), contributing almost nothing to the FC_{av} of the soil profile. On the other hand, it is true that the cultivation of virgin soils leads to a rapid and substantial *decrease in organic matter* (Jenny, 1933). Yet it is only where the adoption of conservation tillage has resulted in continuous cropping due to better water capture that significant increases in soil organic matter within the profile have been reported (Janzen *et al.*, 1998; West and Post, 2002). From that we may conclude that in arable soils the FC_{av} will not be greatly improved by soil management directed towards an increase of organic matter.

Nowadays, many arable soils are trafficked by heavy machines like tractors and harvesters. When the soils are wet and therefore within the plastic range, the bearing capacity and trafficability are limited and the soils become compacted (Hamza and Anderson, 2005). In consequence, the depth of the rooted soil profile may be reduced. With mouldboard ploughing usually one front and one rear wheel of the tractor run in the furrow, and in conjunction with a blunt ploughshare this results in the formation of a *plough sole* or a *traffic pan* just below the ploughing depth. *Subsoil compaction* to 50–60 cm depth and deeper is caused by field machines that have steadily increased in weight with time, even up to the present day (Horn *et al.*, 2000; Berisso *et al.*, 2012). For instance, the *gross weight* of a self-propelled six-row sugarbeet harvester is now up to 50 t and even more when the bunker is filled. This corresponds to the weight of a modern battle tank! The *wheel load is 8–10 t*, which is twice the value of 5 t recommended by an international working group as an upper limit that should never be exceeded. In sensitive soils or under wet conditions the tolerable wheel load should definitely be less than 5 t (Håkansson *et al.*, 1987). Agricultural engineers try to combat the high wheel-load effect by mounting extra broad tyres, but the stress transmission into the *subsoil* is relieved only to a smaller extent (Söhne, 1958). The three-dimensional stresses that originate from *powered, slipping and heavy-loaded wheels* induce soil deformation whenever they exceed the internal soil strength. When this happens to a soil for the first time, the stresses cause harmful subsoil compaction by compression and shearing. Such a state of soil compaction and pore destruction may survive for many decades (Berisso *et al.*, 2012) and may even get worse over time. This is especially true when the stresses from agricultural machinery are applied year after year. These stresses counteract any tendency of physical or biological re-loosening. They act cumulatively in compacting the soil over time until the final state of compression is obtained under given conditions.

Soil compaction reduces the *rootability* of the soil (see Fig. 17.9), the effective rooting depth and the amount of extractable water (Table 9.2). Compaction affects rooting essentially in two ways. Increasing bulk density (BD) raises the mechanical strength of the soil, in this case the strength to resist deformation by the growing root tip (Eqn 6.1), sometimes called *penetration resistance*. Compaction lowers the percentage of large pores (Fig. 17.1A). Constriction of pores and loss of pore continuity inevitably reduce gas diffusion. Insufficient aeration or limited *oxygen supply* is the second reason for slowing down root growth to depth, as the metabolic processes become impaired (Boone, 1988). Aerobic respiration is constrained, possibly resulting in anaerobic conditions, such that microbial activity promotes production of greenhouse gases such as N_2O (Berisso *et al.*, 2012).

These relationships are illustrated in Fig. 17.8. First, it appears that the strength of the soil to resist root penetration increases as the soil water content decreases. The assessment of the penetration resistance is made by imitating root action using a metal cone that is fixed to a rod and pushed through the soil; a device known as a penetrometer. Water acts like a lubricating film between soil particles. Hence, when the water content decreases, the ease of particle displacement is impaired and the soil resistance increases. Secondly, the soil resistance increases with increasing BD. The closer the particles have been pushed together, the bigger is the number of intergranular contact points and the more difficult it is to displace them. In the following analysis we suppose that root growth is limited by three physiological factors: (i) a soil *water content* close to the permanent wilting point (PWP, 7% by weight; too dry); (ii) a high *penetration resistance* near 3.6 MPa (too firm); and (iii) a soil *air content* of less than 10 vol.% (too wet). This latter boundary for 'too wet' was calculated from the total porosity minus 10 vol.% of aerated pores, taken to be necessary for adequate aeration. The volumetric water content obtained was then converted into the unit weight%. All three limiting sectors are depicted in the figure. One can easily see that the '*least limiting water range*' for optimum root growth (da Silva *et al.*, 1994) becomes constrained the more the soil is subject to compaction, indicated by increasing values of BD. At a BD of 1.3 g cm^{-3} the range is about 23 weight%. At a BD of 1.5 g cm^{-3} the range has narrowed to about 6 weight% and at a BD of 1.6 g cm^{-3} the 'window' of optimum conditions has been completely closed. Arriving at such a high BD means that the

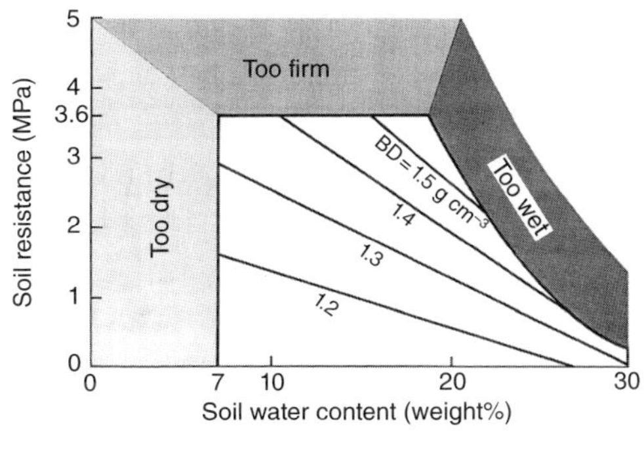

Fig. 17.8. The change of soil resistance, measured with a penetrometer, as a function of soil water content and bulk density (BD). Three sectors are depicted that limit root growth: too dry, too firm and too wet. The border lines are low water content (< 7 weight% = permanent wilting point or PWP), high mechanical resistance (> 3.6 MPa) and low air content (< 10 vol.%) at high water content. The increase in BD narrows down the least limiting water range for optimum root growth. The data are valid for a loess-derived Luvisol (after Ehlers, 1993).

least limiting water range with respect to root growth is no longer available. In the compacted soil, root growth is restricted at all water contents because of either insufficient aeration or too great a mechanical impedance, or both. On the other hand, at a smaller BD of 1.2 g cm^{-3}, root growth never becomes limited by 'too wet' nor by 'too firm' soil conditions. In uncompacted soil, root growth will be impaired only when the soil becomes 'too dry' (Fig. 17.8).

For a sustainable soil husbandry it is important to keep the water that is stored in deeper soil layers accessible for extraction by roots. In the long term such a strategy of *soil care* and protection will support crop growth to a greater or lesser extent depending on additional factors such as soil and climate. By adopting techniques from the range of options indicated earlier, seepage loss and possibly runoff (Fig. 17.4) can be reduced (Eqn 17.5). For pursuing the goal of soil conservation it is imperative to pay much more attention to the origin and distribution of harmful subsoil compaction. The *prevention of subsoil compaction* is more cost-effective in any case than a costly *subsoil reclamation* to promote deep rooting and water extraction (Fig. 17.9). This is particularly true given that the effect of subsoil loosening is often short lived. Re-compaction takes place quite quickly, unless the organization of operations for soil management is changed drastically.

What we outlined with respect to access of soil water in the deeper layers of the soil profile is also true for the acquisition of plant nutrients available in the subsoil. These include P, K, Ca, Mg and nitrate-N. Nitrate can be leached easily into the subsoil and groundwater (cf. Fig. 2.2). These supplies of nutrients remain unexploited by crops and are lost to the environment if the penetration of the root system to depth is impeded.

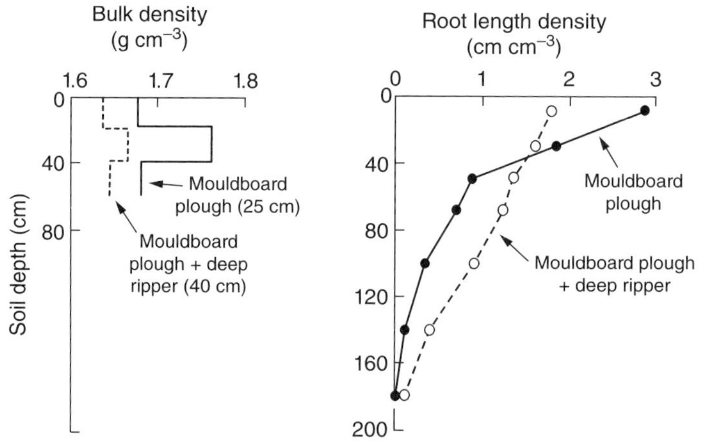

Fig. 17.9. Influence of subsoiling on bulk density of a loamy sand and on root length distribution of spring wheat 99 days after sowing. A tillage-traffic pan hindered root growth into the subsoil in the mouldboard-ploughed soil. The cumulative evapotranspiration (ET) was 570 mm in the presence of the pan. Subsoiling increased ET to 640 mm and improved grain yield and water use efficiency (after Bennie and Botha, 1986).

Box 17.1. Collection of water south of Sahel: a man, a hoe and barren land

The Sahel is a 6000 km long belt of semi-arid land stretching across Africa from Senegal in the west to Eritrea in the east. In the western part it is up to 800 km wide and separates the Sahara desert in the north from the thorn savannah in the south (Fig. 1.5). The very northern part of Burkina Faso, an interior state of West Africa, forms part of the Sahel. The climate of the Sahel is harsh with daily high temperatures exceeding 30°C over the whole year and a small annual rainfall, on average not more than 400–500 mm, concentrated on two summer months. In some years the wet season fails completely. In combination with the high temperatures the summer rainfall is insufficient to permit a closed plant cover. This is the area of *banded vegetation*, the 'tiger bush' (Section 2.2). Longish ranges covered by vegetation, alternate with elongated strips with no, or very meagre, plant growth. Green and bare bands develop almost parallel to contour lines in staggered formation. Nature provides a good example of how to deal with limited resources to support life, even under these poor conditions. The bare soil diverts the rainwater to the vegetation bands, where it optimizes conditions for plant growth. The soil surface of the bare band is crusted and hard. The crusted soil with limited wettability is also hot and hostile to life. The saturated conductivity and the final infiltrability (Section 5.3) are small, and as macropores are absent in the bare surface soil, infiltration cannot be enhanced through preferential flow (Box 5.2). Rain usually falls in the form of heavy (thunder) showers. As the water cannot enter the soil at a high rate, surface water builds up and flows downhill along gentle slopes and passes into the vegetation belt. Here the soil surface is rough and covered by litter. The soil is rich in humus and builds a habitat for flora and fauna. Among the soil fauna, animals, such as termite, ant and earthworm, actively loosen and move organic and mineral particle associations. The soil becomes structured and particles aggregated. Covered and shaded by plants, the soil allows rapid infiltration of the floodwater from the bare area. Heavy erosion is avoided, and the organic and mineral sediments shifted by surface water from the bare strip, add to the chemical and physical quality of the covered soil. The water enables faunal and microbial activity, turnover of organic matter, release of nutrients and a lush growth of plants with their root system. One may conclude that the banded vegetation represents an autonomous, self-sufficient system that is oriented towards the support of life.

The Sahel is bordered to the south by a zone with a Sudan-Sahel climate. In Burkina Faso, this zone covers the middle of the country, the Central Plateau. The rainy season lasts for 4–5 months, usually from June to October, with an annual rainfall varying between 500 mm and 800 mm. Before the rains, a dust-laden desert wind, the *Harmattan*, stops at the end of the dry season and temperatures increase up to 45°C from March to May. Thereafter the warm and humid monsoon winds start and temperatures fall to between 29°C and 32°C in August, the coldest month. At a first glance, the climatic conditions seem to be superior for agricultural production compared with those of the Sahel. In Burkina Faso about 80% of the population live outside the cities and depend on subsistence farming. But in the 1970s and 1980s peasant farmers and their families suffered heavily from less than average rainfall and drought. Groundwater level in wells fell by 50–100 cm year^{-1}. Famine spread, and a lot of people died of starvation. Many of the undernourished survivors moved to the cities, and villages emptied. Cultivated land became wasted and eroded by wind and water. The process of *desertification* was supported by human misuse as a consequence of increased population: (i) bushfire; (ii) clearing from trees; (iii) overgrazing of rangeland mainly by goat; and (iv) overuse of groundwater. All these facts contributed to the infertility of large areas and the disintegration of the rural society.

Around 1980 in Yatenga, a province in the north-west region of the state with a Sudan-Sahel climate, a peasant farmer started to cultivate the land with his hoe. He persevered

Continued

Box 17.1. Continued

and demonstrated complete conviction in his beliefs, despite the terrible conditions of heat, dryness, wind, dust, thirst and hunger. It may well be that he was inspired in his work after seeing and appreciating the 'tiger bush' and its conformity to natural laws that are directed to the creation of a natural fertility that supports plant growth. We do not know the real facts but his idea must have sprung from his imagination rather than from an intellectual evaluation of known facts as he is illiterate. This great man, who started 're-greening' the land, is named Yacouba Sawadogo. A documentary film on his achievements bears the title 'The Man Who Stopped the Desert' (www.1080films.co.uk/yacoubamovie). When he was a boy he attended a Qur'an school in Mali. But he was not successful and, as he was not able to learn elementary knowledge, he was sent home. Nevertheless, the head teacher, a sheikh, prophesied that he would become a wise man and people would adopt his insights, with intelligent people coming from faraway countries to seek his advice. It took a long time for this prophecy to come true and people called him a madman at the beginning, when he started to follow his conviction (Zeit, 2012). Every single morning he went to the fields to cultivate the grim-looking land. He used the well-known method of hoeing a pit into the hard soil, a 'Zaï' in the native language. A period of 60 days and 5 h of hard work per day (300 person-hours) are needed to prepare 1 ha of barren land with 12,000–25,000 pits. He performed this painstaking work daily during the dry season, not during the wet season, as it was the tradition. In doing so he opposed the existing law and he got into trouble with the authorities. But he still continued.

In the traditional method, which had failed too often during what had been a series of dry years, once a pit is prepared seeds are placed at the bottom and lightly covered with loosened soil. Over time, Yacouba Sawadogo's innovations included altering the size of the pits. Compared with the traditional practice, his pits became larger, up to 60 cm in diameter and about 30 cm deep. He varied the size according to soil quality and the plants to be seeded. He placed the excavated soil downhill bordering the pit (Fig. B17.1) to enlarge the water intake area and the storage capacity of the pit. A mixture of compost, manure, crop residues, organic debris and ashes were placed at the bottom of the hole and then covered by a thin layer of soil (Fig. B17.1). Then the seeds were sown into the soil. The addition of organic matter attracted termites, which started to mix thoroughly the materials that had collected in the pits. These animals also create channels into the subsoil (Fig. B17.1) that support root development and the infiltration of water by preferential flow. Windblown soil, litter and leaves captured by the pit add to the soil fertility. Importantly, the preparation of the pits has to be completed in good time before the beginning of the rainy season to ensure the germination of the crop plants.

The water supply per pit depends on the extent of seasonal rainfall and how effectively water can be collected in addition from the surrounding crusted area (Fig. B17.1) that is not hoed. The water delivery to the Zaï can be deliberately increased by discharge of water 'harvested' on a larger area, possibly outside the Zaï field. *Water harvesting* is a technique already known from ancient Palestine (Section 19.2) and is a key of the success for the formation of banded vegetation (Section 2.2, and cf. above). For this kind of harvest fist-sized stones are piled along contour lines. The *stone bunds* slow water runoff downhill and facilitate water infiltration into the soil. With time these bunds become solidified and sealed by silt material and debris. Collecting water in a catchment area by water harvesting is one important step to support vegetation and crops. A step further on is the combination of the Zaï system with the stone lines. Yields are increased and complete crop failure is minimized. Less runoff and increased infiltration are supposed to have benefitted groundwater recharge in some areas.

Continued

Box 17.1. Continued

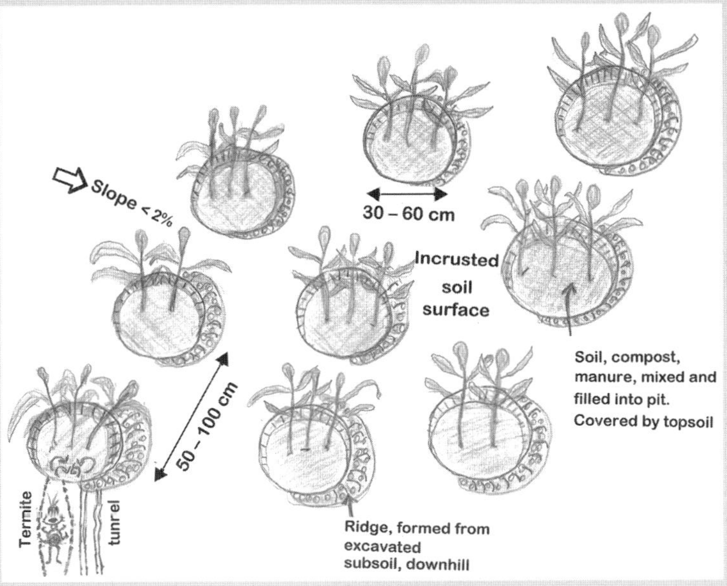

Fig. B17.1. The Zaï technique developed in Burkina Faso. The planting pits form a micro-catchment, concentrating water supply and enhancing soil fertility on the spot. The pits are dug in staggered rows to increase the area of water runoff on incrusted soil for 'feeding' the Zaïs.

Yacouba Sawadogo became more and more successful over the years, whereas his fellow citizens were in danger of malnutrition. When he started, the people were distrustful and even hostile. But when they recognized that he was not just a hard-working farmer but an inventive and successful genius, he gained their proper respect and recognition. They tried to emulate his techniques and he was helpful with advice. His fame and that of some progressive colleagues of Yatenga province spread over Burkina Faso, Mali, Niger and other states touched by the Sahel. As their reputation increased, they received more and more inquiries. In response they began the use of market days to explain methods and to exchange not only ideas but also commodities, for instance seeds of species and cultivars. They built a 'Zaï school' close to the roadside and tried to transplant their knowledge and experience into the heads and hearts of those eager to adopt and try out their techniques.

Yacouba Sawadogo improved the Zaï technique and grew sorghum and pearl millet. Then he started to raise several species of local trees. Eventually he became the first guardian of a property covered with 30-year-old forest that was surrounded by unproductive and unadopted land. And again the authorities from the capital of Yatanga took a look at that airy island of peace and birds, so uncommon in the region. Now, after years of bitterness, the conflict with the city administration seems to be settled.

Mr Sawadogo provides an excellent but rare example of the importance of local initiative in battling against difficulties and problems on the spot. He demonstrates that it is possible to modify traditional practices that have lost their relevance and as a result

Continued

Box 17.1. Continued

achieve huge success without help from the outside; not from government, from international institutions, known for their assistance through financial support, or from scientific advisors. Yacouba Sawadogo has been an inspirational leader in effecting the rehabilitation of 200,000 or 300,000 ha of barren land on the Central Plateau of Burkina Faso. Based on his achievements crops are being grown again, reasonable yields achieved and agroforestry developed in an area subject to overuse by people and hit by climatic change. Efforts to rehabilitate the soil still continue.

We end this tribute by adding new data on sorghum yields attained just south of the home province of Yacouba Sawadogo. The results show how effective independent efforts can be even though they are just carried out by hand without the use of any machinery (Table B17.1).

Table B17.1. Sorghum yield (grain kg ha^{-1}) on farms in Provence Zondoma, Burkina Faso, as influenced by two soil rehabilitation measures, stone bunds and Zaï, in 3 years with variable rainfall during the growing season (Sawadogo, 2011).

		Measure of rehabilitation						
	Annual rainfall	None	Stone bunds			Zaï		
Year	(mm)	Control	Farm 1	Farm 2	Farm 3	Farm 1	Farm 2	Farm3
2006	359	436	892	772	932	1200	975	1312
2007	336	319	723	819	856	1100	1230	922
2008	449	642	1065	1012	1039	1600	1508	1456
Average for 2006–2008	381	466	893	867	942	1300	1238	1230

17.5 Conservation Tillage

The term *conservation tillage* is a term that was coined in the USA after the 'dust bowl' in the 1930s. The USA experienced severe problems with both *wind erosion* and *water erosion* on prairie land that had been mouldboard ploughed regularly. So soil conservation became an important issue, and later on the same problems developed in various other countries. Conservation tillage means a radical turning away from the *'clean' inversion tillage* associated with the *mouldboard plough*. Instead of the mouldboard plough, tillage implements are favoured that do not invert the soil but instead keep organic residues on the surface. The *disc plough* has to be mentioned as a first step towards crop residue mulching, but in many instances the tillage effect is still regarded as too intense. Loosening and mulching without inversion is best achieved for instance with a *chisel plough*, a *winged blade cultivator*, or similar tools. Conservation tillage is now applied to any tillage systems that leave *mulch covering the soil surface by 30%* or more at the time the subsequent crop is seeded under conditions where the concern is water erosion. But, if wind erosion is the major risk, systems must maintain the *equivalent* of at least 1120 kg ha^{-1} of flat small-grained residues (Uri, 1999). With this definition in mind, it is clear that a tillage system where soil 'tillage' is reduced just to the inescapable soil disturbance required for effective placement of seed by a seed drill has to be included within the concept of conservation tillage. This is indeed a very extreme form of a tillage system without any purposeful primary or secondary tillage. The system is known as *no-till* or *zero-tillage*. Zero-tillage became feasible at the beginning of the 1960s after the invention of non-selective *herbicides* with only a short-term residual effect. These chemicals allow the immediate control of weeds but do not affect the emergence of crop plants only a short time later. In addition to economic reasons, such as the savings in time and energy, the principal idea of zero-tillage was to develop a soil structure similar to that under grassland, with stable aggregates that are resistant to erosion. The structure is generated by natural physical and biological processes, essentially without any influence from mechanical operations. Eventually, apart from the very top few centimetres of the soil, the bulk density is usually greater than that of ploughed soil. Therefore no-till is a land husbandry practised on *firm soil*. In addition to relying on the effectiveness of herbicides, another prerequisite is the availability of *special seed drills*, adapted to insert the seed into the firm soil covered with residues. For that purpose the coulters are often fluted or are arranged in groups – triple-disc coulters.

Depending on climate (wet), soil (poorly drained), crop rotation, crop and available technology, zero-tillage might not be the best choice to achieve sufficient crop establishment for adequate yield. Often it will be absolutely necessary to create a more favourable seedbed that encourages rapid and uniform field emergence of a high percentage of seeds. Weed control and large amounts of residues (a difficult condition for precision seeding) can prove to be an additional impetus to perform at least a shallow tillage operation, restricted to 5 cm or 8 cm depth or so. As the soil is 'scratched' just superficially by use of *sweep cultivators* or p.t.o.-driven *rotavators* or *rototillers*, this system of minimum or *reduced tillage* may still be regarded as a firm-soil husbandry.

Mulching and marginal soil disturbance are the keys to control erosion by wind and water, and to save precipitation for crop production, by favouring infiltration, minimizing evaporation and trapping snow by the stubble. Snow trapping is at its best in standing residue left by zero-tillage (Hansen *et al.*, 2012). The effect of soil management on soil water storage is illustrated in Table 17.1, presenting data from Akron, Colorado in the Central Great Plains with steppe climate (Fig. 15.2). The data describe the historic development of tillage systems in dryland agriculture with fallowing, and their effect on water storage, fallow efficiency and winter wheat yield.

Over the course of time, tillage operations became less intense with respect to frequency and depth (Table 17.1). It is clear that within the wheat–fallow sequence only a marginal part of the precipitation during the 14 months of fallowing was stored in the soil. Nevertheless, reducing the tillage intensity increased the *fallow efficiency*. In fact it was nearly doubled. Improved water storage increased grain yield of winter wheat to roughly 250%.

The precipitation retained by the soil during the fallow period is largely lost (Table 17.1). Presumably the greater part is not lost as runoff because of inadequate infiltrability, but mainly as evaporation. Also significant seepage loss can be excluded in these silty prairie soils, under a semi-arid climate (Fig. 15.2).

Table 17.1. The development of tillage systems for wheat–fallow cropping over the 20th century, together with their effects on water storage, fallow efficiency and winter wheat yield at Akron, Colorado on the central Great Plains (data from Greb, 1970; Wittmuss *et al*., 1973; Greb *et al*., 1979).

Tillage system	Years in use	Number of tillage operations	Average annual precipitation (mm)	Precipitation during fallow (mm)	Soil water stored (mm)	Fallow efficiency[a] (%)	Wheat yield (kg ha^{-1})
Maximum tillage: plough and harrow (dust-mulch)	1915–1930	7–10	439	531	101	19	1070
Conventional tillage: shallow disc, bare soil, rod-weeder	1931–1945	5–7	401	467	112	24	1105
Modified conventional: disc, chisel, rod-weeder	1946–1960	4–6	416	507	137	27	1725
Stubble mulch: sweep, rod-weeder	1957–1970	4–6	–	–	–	27–33	–
Minimum tillage: herbicides replace one or two tillages	1968–1977	2–3	389	476	157	33	2165
Zero-tillage: herbicides only	Since 1975	0	–	–	183	40	2690

[a] Defined as percentage of soil water storage of precipitation received during the fallow season, lasting from wheat harvest in mid-July to wheat seeding in mid-September of the next year (14-month fallow period).

Evaporative water losses will increase whenever precipitation falls in the form of light showers. Then a greater percentage of total rainfall is retained by the surface mulch, and the soil surface is wetted to only a small extent. The water returns directly to the atmosphere from the retaining mulch. Heavy and persistent rain showers are required before the water infiltrates to some depth in the soil. The deeper the profile is wetted, the smaller is the chance that the water will be lost again by evaporation from the soil surface. The thicker the mulch cover the greater the reduction in evaporation (Fig. 17.6), because it insulates the soil from radiation and from the evaporative demand of the atmosphere.

Changing the tillage system from inversion ploughing to reduced tillage or zero-tillage (i.e. giving up the tradition of turning the soil 'upside down') induces a change not only in bulk density, but also in *soil structure, aggregation* and *tilth*, which is a process of development that takes some time, probably some years. *Organic matter* accumulates in the soil near the surface, and it includes carbon (Fig. 17.10) and nitrogen fractions. Soil *microorganisms* are supported in their activity; enzyme activity also increases as well as aggregate stability near the soil surface (Fig. 17.10). Roughness and openness at the surface are created in well-drained middle-textured, loamy-silty soils by the processes linked to the stratification and concentration of organic matter (Fig. 17.10). At best, the soil starts to crumble. After omitting inversion tillage some clayey soils develop an open and rough tilth, when they contain active clay minerals of the expandable type. These clays create aggregation by their swelling and shrinking activity, and the type of structure is angular to subangular blocky.

The mulch cover protects the soil from desiccation and from large fluctuations in temperature, which creates more favourable environmental conditions for *earthworms*. Moreover, some species of earthworms (the deep-burrowers) feed on the residues, whereas others (the shallow-dwelling worms) prefer the enriched

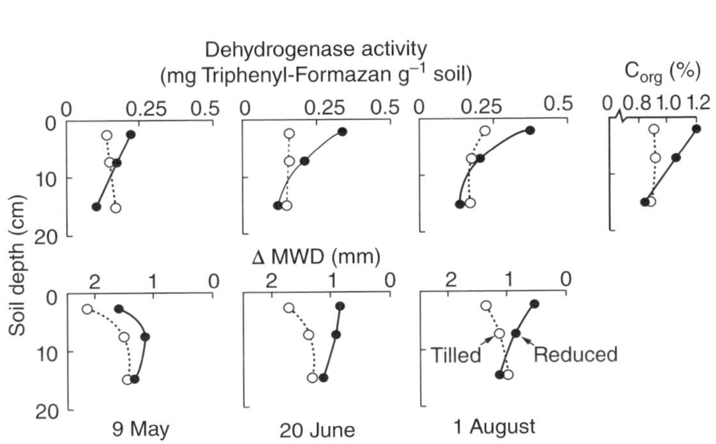

Fig. 17.10. Profiles of organic carbon (C_{org}), dehydrogenase activity and aggregate stability in a loess-derived Luvisol during 1985, either with mouldboard ploughing or with reduced tillage. The reduced-tillage treatment was last mouldboard ploughed 12 years before. The term Δ MWD means change in mean weight diameter of aggregates after wet sieving as compared to dry sieving (from Ehlers and Claupein, 1994).

soil organic matter. Earthworms promote the decomposition of organic material, support aggregate formation by their casts, and reduce crusting at the soil surface. And they make another conspicuous contribution to the physical performance of the soils: worms puncture the soils. Deep-burrowers create permanent channels mainly in the vertical direction, connecting the soil surface with the subsoil. In German Luvisols the burrows of *Lumbricus terrestris* can reach a depth of 2 m. The channels serve as a preferred growth medium for roots. They present essentially no physical resistance, and their walls are enriched with plant nutrients and organic matter.

Inversion *ploughing* reduces the *abundance of earthworms*. In plots with mouldboard ploughing the population density is usually significantly smaller than in plots with reduced till (Fig. 17.11). The same is true for the channel density per square metre (Fig. 17.11). There is evidence that deep loosening of the soil to 50–55 cm depth, even using a non-inverting 'paraplough', can be harmful to the animals (Fig. 17.11). A density of up to 200 channels m^{-2} seems to be quite a large area occupied by the animal openings, but this is a wrong impression. In most instances the area (or the volume) is less than 1% (Fig. 17.12) occupied by burrows.

Mouldboard ploughing can be harmful to earthworms and will destroy *worm channels* in the inverted topsoil. But over one season the channels are usually re-built if the worms survive. Nevertheless, due to regular mechanical destruction the density of channels in total is smaller than in zero-tilled land, where the channels stay preserved to a much greater extent (Fig. 17.12). Even

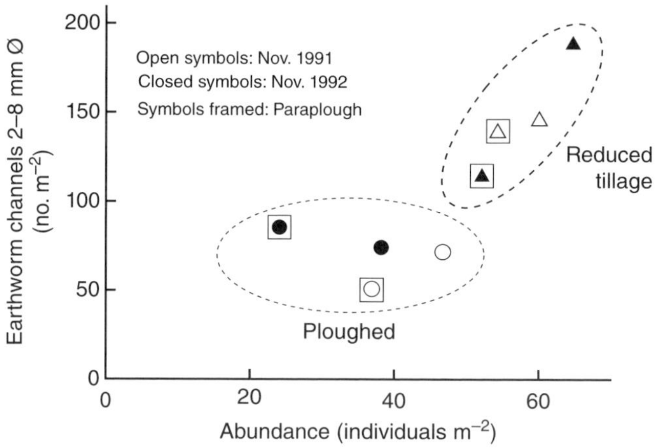

Fig. 17.11. The abundance of earthworms and the burrow density per unit area are influenced by the tillage system (mouldboard ploughing and reduced tillage) and by deep loosening with the paraplough. The soil was deep loosened in 1990, the year when reduced tillage was introduced on formerly mouldboard-ploughed land, using a rotary spade harrow (stubble mulching) and a power-take-off (p.t.o.)-driven rotary harrow. The burrows were counted on soil samples taken from near the surface. The soil is a loess-derived Luvisol from the edge of the Solling mountains in Germany (based on Fenner, 1995).

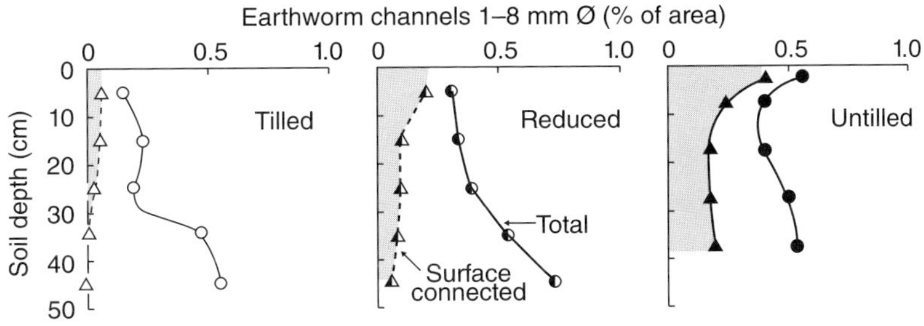

Fig. 17.12. The effect of conservation tillage (zero-till, reduced till) and of mouldboard ploughing on the percentage area of earthworm channels (1–8 mm diameter) in total and on surface-connected channels. Surface connection was ascertained by infiltration of a slurry made of plaster of Paris. Last ploughing in no-till – 11 years; on reduced till – 12 years before. The soil is a silt-loam Luvisol near Göttingen, Germany.

in the reduced-till topsoil the density is greater than in ploughed soil. The substantial increase in the total number of channels per unit area in the subsoil is due to the fact that here a proportion of burrows are old and no longer inhabited by the present earthworm population. In contrast to the tilled topsoil these channels remain preserved for many decades.

Not all the channels are *hydraulically connected* with the *soil surface* (Fig. 17.12). The number of connected channels is extremely small in the ploughed soil. And the connection to the subsoil is almost completely disrupted. This is not the case in the reduced-till and the zero-tilled soils. Here quite a number of channels are open to the soil surface, connecting the soil surface continuously with the subsoil (see also Goss *et al.*, 1984a). During heavy rain these channels serve as a *natural drainage system*, conducting tension-free surface water to the subsoil, thus preventing runoff and erosion and possibly increasing transpiration (Eqn 17.5). It was estimated that the open channels in zero-tilled land are capable of receiving infiltrating surface water during a thunder shower at a rate of more than 1 mm min^{-1} (Ehlers, 1975b).

Switching over to conservation tillage means that the mechanical stress exerted on the upper layer of the subsoil is brought to an end. With mouldboard ploughing this stress is exerted regularly, often once a year, deforming and compacting the soil layer right below the ploughed horizon. The stress originates from the ploughshare and the tractor wheels running in the furrow. Omitting inversion ploughing, the steady development towards the formation of a *tillage-traffic pan* ceases. When there is no pan present in zero- or reduced-tilled soil, *root growth to depth* is not restricted any more. Rather it is promoted by *earthworm channels*, into which the roots grow and then follow from the topsoil to the subsoil (Fig. 17.13). In soil that is subject to mouldboard ploughing and a pan is present, root growth to depth can be restricted (Fig. 17.13). In the example, fewer roots entered the earthworm channels as passages with little mechanical resistance leading to the subsoil. Apparently, the growth into the channels was restricted to some extent by the pan.

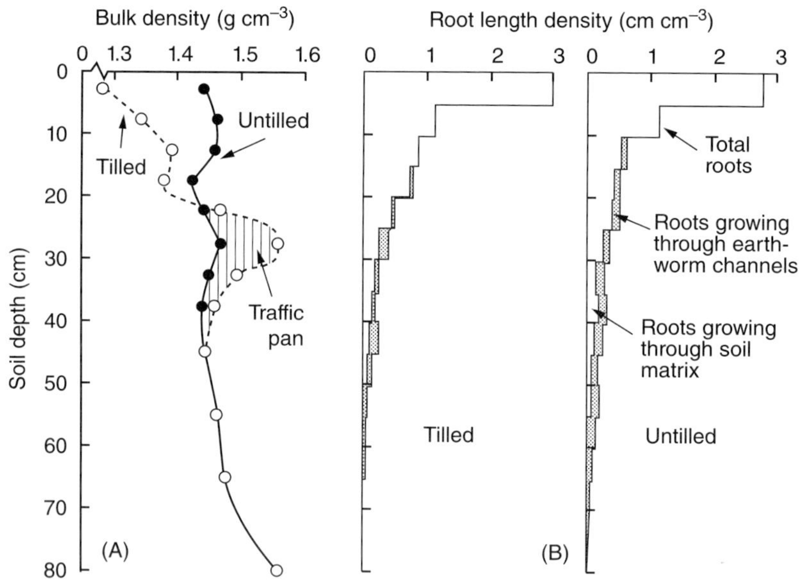

Fig. 17.13. (A) Effect of mouldboard ploughing and zero-tillage on bulk density profile and pan formation in a silty, loess-derived Luvisol. Zero-tillage started from mouldboard ploughing 6 years before. (B) Profiles of root length density of oat in the middle of stem elongation for the two tillage systems. Roots in total are separated into those growing through the pores of the soil matrix and those penetrating through earthworm channels into the subsoil (based on Ehlers *et al.*, 1983).

Tillage-traffic pans may be ameliorated under conservation tillage, punctured by soil animals or disrupted by mechanical means. A sandy loam in Alabama, either mouldboard ploughed or zero-tilled, contained an abrupt pan between15 cm (depth of disking) and 32 cm depth with a bulk density of 1.57 g cm^{-3}. Breaking the pan mechanically by *subsoiling* stimulated *deep rooting* of maize in ploughed and zero-tilled land and improved *water uptake* from deeper soil layers as well as total water use (Fig. 17.14). But even in *zero-tilled* land without subsoiling, maize took up more water and had better water use than in ploughed soil, although in both cases the pan was present. This unexpected result gives rise to the supposition that in zero-tilled soil the root-impeding effect of the pan had disappeared in some way. There must have been planes of weakness, fissures, cracks or burrows that allowed the roots to enter the subsoil, although the uptake directly below the pan in 30–50 cm remained small compared with the other treatments (Fig. 17.14).

Subsoil exploration for water by roots can be an important issue in dry climates where water supply is limited (Kirkegaard *et al.*, 2007). When water limits yield, *deep rooting* will support the development of *yield* components by crops as occurred in the maize experiment in Alabama. The smallest yield was attained in the ploughed, non-subsoiled treatment with lowest water use. The highest yield was from the no-till subsoiled land with greatest water use (Weatherley and Dane, 1979).

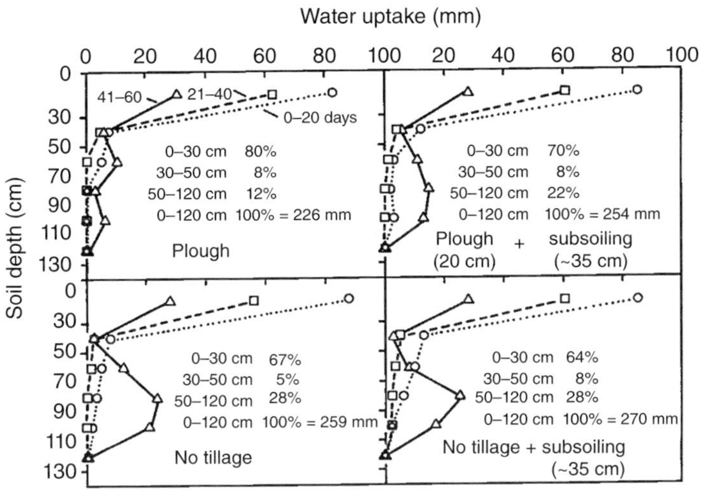

Fig. 17.14. Water uptake by maize roots as a function of depth and time (days after start of measurement on 15 May) and total water extraction including evaporation, as influenced by the tillage system (mouldboard plough and zero-till) and by subsoiling. The soil is a sandy loam, an Orthic Acrisol from Alabama (after Weatherley and Dane, 1979).

Conservation tillage may help farmers to *conserve soil moisture* for crop production. It increases infiltrability, lowers evaporation, stimulates activity of microbes and soil fauna, may assist subsoil rooting and increase water extraction. But *not all soils react positively* to conservation tillage in yield formation, especially when water is abundant and does not limit yield. Managing wet *clayey soils* by conservation tillage might turn out to be a difficult job, as seedbed quality remains poor even after conservation tillage has been followed for some years. Aggregation at the surface may be induced in clayey soils with active, expandable clay minerals. But clayey soils with non-expandable minerals may stay dense and sticky for long periods of time. They do not have the capability of an 'internal aggregate formation'. For these inactive, heavy soils mechanical loosening to depth might be essential to increase aeration porosity. Some *mixed grained sandy soils*, on the other hand, may heavily compact in mechanized agriculture when not deeply tilled. In these soils conservation tillage might necessitate a deep loosening with a heavy cultivator or paraplough once in a while.

Generally we can say that the better the soils are drained and the lower precipitation is, the better will be the outcome of conservation tillage with respect to crop production – and vice versa. And we have to add, the drier the climate and the higher the rainfall intensity, the more essential is conservation tillage for water storage and soil protection.

18 Controlling Water Use by Crop Management

18.1 Crop Rotation

Crop rotation is a planned temporal cyclical sequence of crops in an arable field, often with each one being grown as a *pure stand*. The crops may be annual, biennial or perennial. Perennial crops that have been integrated into a rotation include forage grasses, leguminous forage crops, or grass–clover *mixtures*. Generally a farmer will decide on the cropping sequence in such a way that, under the local conditions and opportunities, *environmental factors* like radiation, water and nutrients are used to the full. This goal will be pursued while taking care not to exhaust the fundamental components that contribute to *soil fertility* and *soil quality*: two of the preconditions for *sustainable land use*. By the skilful combination of *preceding crops* with *successor* or *following crops* within the entire sequence, a farmer will try to make full use of the beneficial effects of a previous crop on the performance of the following crop. *Favourable effects* come for instance from *legumes* fixing nitrogen, from *leafy crops* promoting biological activity and tilth by shading, or from *forage crops* suppressing weeds because of a high stand density. On the other hand, the farmer also tries to exclude *unfavourable effects* such as pests and diseases on susceptible following crops. Appropriate rotations can prevent diseases like basal rot (a fungal disease of cereals) or root disorders caused by specific nematodes that belong to the group of animal pests. However, the well-tested principles of organizing crop rotations have been largely abandoned over the last few decades, and increasingly have fallen into disuse. There are certain 'economic constraints', pushing farmers towards unbalanced rotations. And there is a wide selection of agrochemicals available to facilitate the transition from longer rotations with a diversity of crops to shorter ones with few crops being included.

What is the association of ideas between the establishment of crop rotation and water as a growth factor? Crops with a long *growing season* can consume the *supply of soil water* to a considerable depth, like lucerne and sugarbeet (Fig. 12.3) or maize (Fig. 11.2). A large water use by the previous crop may negatively affect the water supply for the following crop. The *previous crop–following crop effect* will be more critical, the less *precipitation* there is in a dry climate, the greater the *available field capacity* (FC_{av}) in the rooted profile, and the closer the harvest date of the preceding crop is to the sowing date of the next crop. The less the precipitation and – relative to the saturation deficit – the drier the climate, the more the preceding crop depletes the available soil water supply. The more the depletion, the less likely it is that the *soil water supply* will be *replenished* by precipitation, either within a short period or over the winter or during the next rainy season. Even in the temperate climate of Germany, it occasionally happens that the productive loess-derived silty-clayey Phaeozems of the eastern states do not return to field capacity (FC) during the winter. However, when FC and hence FC_{av} are small, as is the case on lighter, more sandy soils (Fig. 3.2), it is more or less the rule that the pore volume is fully replenished with water to FC by the next spring or after the rainy season. To be precise, the total soil profile will be easily replenished down to the maximum rooting depth. On the coarser-textured soils, therefore, the *preceding crop effect on water withdrawal* will be less dramatic compared with that on the finer-textured soils with a larger FC_{av}, as the maximum deficit resulting from root water uptake will only be comparatively small in absolute terms. Such a small deficit can be replenished easily, more or less unaffected by any slight differences in deficit that might be caused by the actual root water uptake of the specific preceding crop. Because there is little difference in water extraction between crops on coarser-textured soils, it follows that water will be less critical as a growth factor in any combination of preceding and following crops. However, as an additional factor, a low FC_{av} may nevertheless exclude crops that require lots of water from cultivation on sandy soils if irrigation is not available.

When the *establishment of a crop* is put at risk by a large soil water deficit, *seeding time* may or may not have to be delayed, depending on circumstances. In temperate or continental climates, where there is some freezing of the soil over winter, dryness in the autumn may inhibit farmers from the timely sowing of *winter crops*. Here late germination of autumn-sown crops may put them at risk of frost-kill, so the alternative is to plant *spring crops* instead. In contrast, in warmer climates sowing of the next crop often takes place at the end of the dry season. The seedbed can possibly be prepared more easily with the onset of the rainy season, and certainly in a timelier manner, and seeds placed in the soil at this time are ready for immediate germination.

Water is a prime prerequisite for *germination* and *emergence in the field*. Before the embryo can make use of the reserve material stored in the endosperm or cotyledons, the seed has to imbibe water, which causes the *seed* to *swell*. The water enables the action of *enzymes* to break down the reserve material into mobile compounds, such as sugars, fatty acids and glycerol, and amino acids. These compounds are then transported by diffusion to the embryo for respiration and the new growth of cells and tissue. Even slight rain showers may cause a re-wetting of the topsoil, stimulating germination. But if the subsoil is not replenished by water, the young seedling may become *stunted* or even die. The chance of sufficient recharge will be less the closer the sowing date follows the harvest of the previous crops. This happens in some areas where oilseed rape or barley is grown. These crops can be seeded in late summer after wheat has been grown as the preceding crop. The same can be true with *intermediate (cover or catch) crops*, grown between full-season crops for green manure purposes or for animal fodder.

In dry regions such as the semi-arid steppe climate, with little precipitation and a large saturation deficit (Fig. 15.2), water shortage can be severe, even on prairie soils formed from loess. It may be so severe that the *annual precipitation* is not sufficient to provide enough soil water to meet the requirements for transpiration and the development of any kind of grain yield in every season (Fig. 13.4). It is estimated that 180–250 mm of soil water is required on the Great Plains, USA, as a basal supply for the cultivation of winter wheat before any grain yield is produced (Willis, 1983). This quantity of soil water serves the fundamental need of the crop to build biomass and the unavoidable requirement for some evaporation from the soil. In South Australia with a Mediterranean climate (Fig. 15.2), about 130–160 mm of rainwater during the growing season (April–October, autumn–spring, cf. Fig. 16.2) can be regarded as the essential water use by crop and soil that just produces wheat biomass but no grain yield. This is an estimate based on the published data of French and Schultz (1984a,b). In the Great Plains, each additional 100 mm of soil water will increase grain yield by roughly 2 t grain ha^{-1}. A similar figure can be estimated for South Australia, provided that production is not hampered by nutrient deficiency, diseases or other shortcomings in soil and crop management (French and Schultz, 1984b). So, in a climate with *small and erratic precipitation*, the annual rainfall may be too little in some years for any substantial grain production. To reduce the *risk of a crop failure* in arid regions, a *fallow* is often incorporated into the crop rotation. Often the fallow is followed by either one or two crops, largely depending on climatic conditions. The goal of fallowing is to *accumulate the precipitation* that falls during an extended crop-free period, and therefore avoid extraction of this water by plants. The total crop-free period commonly lasts for more than a year. The precipitation is targeted to the production of the crop or crops that follow. In the wheat–fallow rotation at Akron, the period of fallowing lasts for 14 months while that of wheat growth covers 10 months (Table 17.1). But as the table shows, most of the precipitation received during the fallow period disappears without contributing to wheat production. This was true especially in the first half of the 20th century, when fallow efficiency was only about 20%, and 80% of the water was lost, a loss stimulated by intensive tillage.

Fallowing can be efficient. For example a 2-year wheat–fallow rotation can give a yield that is more than double that from annual wheat production, where in some years there may be no grain produced at all. To make this a realistic comparison, production costs have to be taken into account. And these are certainly greater for annual cropping. Hence fallowing will be advantageous in dry areas where there is a high probability that annual cropping will fail during the year. The push to introduce fallowing in a region will increase depending on how small is the annual rainfall received and how great is the fallow efficiency that can be achieved (Table 17.1).

However, it is the poor efficiency of fallowing that causes problems in dry areas like those with a steppe climate, which includes a cold winter but substantial *rainfall* during the *warm summer months* (Fig. 15.2). On the other hand, annual precipitation is not sufficient in the Great Plains to allow annual production of the principal crop, which is hard red wheat (*Triticum aestivum*), winter- or spring-planted. It is surprising, but most of the rainfall accumulated in the Great Plains is lost again over the summer months during the fallow (Table 18.1). 'Paradoxically, fallow is not only inefficient but most inefficient during the periods when precipitation is most substantial (i.e. summer)' (Farahani *et al.*, 1998). This is quite different from the situation in a Mediterranean climate, where *rainfall* is concentrated during the *cool winter months* (Fig. 15.2). The winter months are warm enough to allow a retarded crop development. On the other hand, they are cool enough and the saturation deficit is sufficiently small to allow both limited transpiration and some water to be stored in the soil. After the winter rains it is the stored water that acts as the primary supply for the remainder of the cropping season, during which there is less precipitation but warmer temperatures and greater saturation deficits.

The *paradox of poor fallow efficiency* during the summer rain period provides important lessons for crop management. Table 17.1 shows an example from the early period of dryland cropping at Akron between 1915 and 1930. Over the 14 months of fallowing when the precipitation totalled 531 mm, there was a very large loss (531–101 = 430 mm) of water by evaporation. This loss of 430 mm during fallowing was almost equivalent to the annual rainfall of 439 mm! In the northern Great Plains the fallow lasts for about 21 months with spring wheat production, and in the central and southern Great Plains with winter wheat the fallow period is 7 months less (Farahani *et al.*, 1998). In both these fallow systems the *greatest storage efficiency* was attained over the *period from wheat harvest to next spring* (Fig. 18.1). In the spring wheat area, soil water storage during that autumn and winter period amounted to roughly 60–70% of the total plant-extractable water. The total is the amount of stored water ultimately attained after 21 months fallowing when the spring wheat was seeded. In the winter wheat area, however, the water stored between harvest and the next spring was up to and even exceeded 100% of that finally attained (Fig. 18.1).

For winter wheat production in Colorado, 59% of precipitation was stored during the '*overwinter*' fallow (Table 18.1). In the initial period after harvest and in the final period before sowing the *precipitation storage efficiency* was much less, because of a large *soil evaporation*. By 11 May, after the end of the overwinter period and the early spring fallow, a total of 133 mm water had been accumulated in the soil, which is actually 8 mm more than measured at the end of the whole fallow season (cf. Fig. 18.1). Hence water storage decreased during that part of the fallow period that received largest rainfall (Tables 18.1 and 18.2), that is during the late fallow period from spring until the seeding of wheat in autumn – an ironic situation indeed.

Modern tillage systems with residue retention, including reduced tillage and zero-tillage, have improved the fallow efficiency (Tables 17.1, 18.2). The efficiency increases with the amount of residue cover (Unger, 1984). However, there has been no fundamental change to having large *unproductive losses of water from the summer precipitation* late in the fallow. One reason for this is the limited production of residues, given the available water (Farahani *et al.*, 1998).

A fundamental solution to this challenging problem comes from '*cropping intensification*', which is quite contrary to the original concept on which fallowing was introduced to the American Great Plains by early settlers. The first step in intensification is an expansion of the 2-year wheat–fallow rotation by inclusion of a

Table 18.1. Precipitation (P), change in soil water storage (ΔS), precipitation storage efficiency (PSE) and daily actual evaporation rate (E) for three specific periods of the 14-month fallow, as part of the 2-year crop rotation winter wheat–fallow. Mean values for three locations in Colorado, central Great Plains, from 1988 to 1995 with zero-tilled soils (Farahani *et al.*, 1998).

Early period (96 days)				Overwinter period (203 days)				Late period (122 days)				Total period (421 days)			
15 July–19 October				19 October–11 May				12 May–12 September				15 July–12 September			
P (mm)	ΔS (mm)	PSE[a] (%)	E (mm day^{-1})	P (mm)	ΔS (mm)	PSE[a] (%)	E (mm day^{-1})	P (mm)	ΔS (mm)	PSE[a] (%)	E (mm day^{-1})	P (mm)	ΔS (mm)	PSE[b] (%)	E (mm day^{-1})
204	22	11	1.9	188	111	59	0.6	261	–8	–3	2.2	653	125	19	1.3

[a]Percentage of soil water storage of precipitation received during individual periods within the 14-month fallow.
[b]The PSE of the total period corresponds to the fallow efficiency.

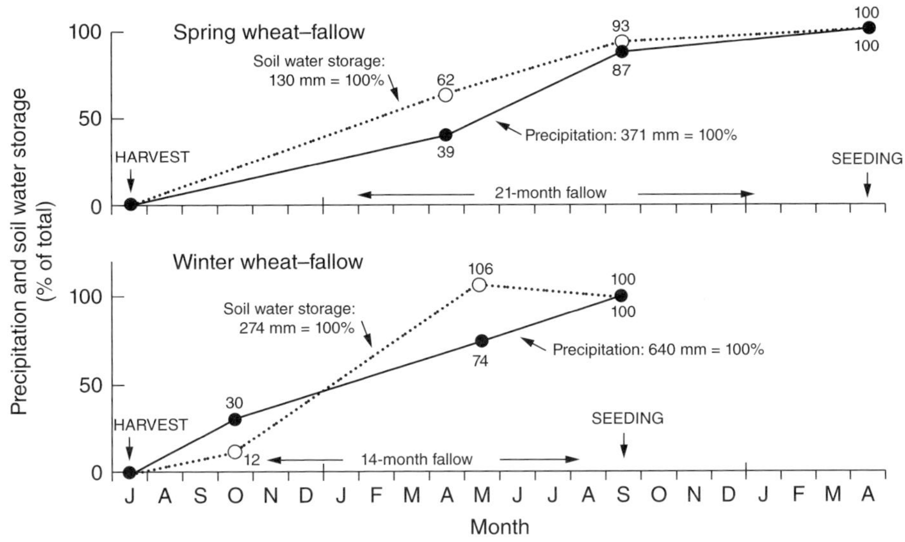

Fig. 18.1. Precipitation and soil water storage in the Great Plains, USA, with spring wheat–fallow rotation (Sidney, Montana, 1981–1985) in the northern plains and winter wheat–fallow rotation (North Platte, Nebraska, 1963–1966) in the central plains. The soils are managed by zero-tillage (data from Farahani *et al.*, 1998).

Table 18.2. The effect of conventional tillage (CT) and zero-tillage (ZT) systems on soil water storage under wheat–fallow cropping at Akron, Colorado in the central Great Plains. Precipitation (P), change in soil water storage (ΔS), precipitation storage efficiency (PSE), for three specific periods of the 14-month fallow. All values are means over 10 years (1996–2006) based on three replicates per year. Residues were incorporated in CT tillage but remained on the surface in ZT. (Based on Nielsen and Vigil, 2010.)

Period	Duration	Precipitation (P) (mm)	Change in storage (ΔS) (mm)		Precipitation storage efficiency (PSE) (%)	
			CT	ZT	CT	ZT
First summer	July–September (~82 days)	158	41	60	23	35[a]
Autumn to spring	October–April (~212 days)	118	38	94	32	81[a]
Second summer	May–September (~143 days)	262	33	34	11	12
Total period[b]	July–September (~435 days)	539	111	188	20	35[a]

[a]Difference from CT significant at $P = 0.01$.
[b]Based on annual records published by the authors.

summer crop that uses the summer rainfall together with the soil water stored during the winter fallow after wheat. According to Table 18.1 storage can amount to 133 mm (22 + 111 mm), and summer rain may add another 261 mm, giving a total of 394 mm for use by the summer crop. The choice of crop to grow, or even whether to sow a crop could be determined by how much water is extractable at the time of seeding in early summer (late May–June) (Nielsen *et al.*, 2011). Such a system is referred to as *opportunity or flexible cropping*. A possible crop for an extended rotation is maize, which follows the fallow–wheat sequence resulting in a 3-year rotation appropriate to the central and northern Great Plains with its warm summers. Another rotation is wheat–sorghum–fallow, which is more appropriate to the southern Great Plains with its hot summer months. By introducing transpiration-efficient C_4 summer crops, rotations are not only intensified but also diversified. Including additional crops results in more extended rotations having a 4-year cycle, which have been tested in field experiments. Long- or short-season crops were selected to replace the fallow. In addition to maize and sorghum, longer-season crops included sunflower (*Helianthus annuus*) and safflower (*Carthamus tinctorius*) whereas examples of short-season crops were proso millet (*Panicum miliaceum*), dry pea and pinto bean (*Phaseolus vulgaris*) together with forage crops spring triticale (X *Triticosecale*), pea, foxtail millet (*Setaria italica*) and spring oilseed rape. These additional crops in the rotation use part of the water and therefore can limit wheat yield according to seasonal weather conditions (Nielsen *et al.*, 2011; Aiken *et al.*, 2013). The differences in crop water use, especially those associated with long- or short-season summer crops, has the consequence of increasing variability in wheat yields. To some extent the impact of variability on producers is offset by taking account of the net return from all the crops included in the rotation.

The possibilities and challenges inherent in adopting more continuous cropping in a climate with erratic summer rainfall have been investigated in the central Great Plains. One study was at Colby, Kansas (Aiken *et al.*, 2013) and the other at Akron, Colorado (Nielsen *et al.*, 2011). Using eight 3-year fixed rotations, Aiken *et al.* (2013) compared continuous cropping with a system having two crops and an 11-month fallow between 2002 and 2008. After winter wheat, which was the principal crop, all rotations contained a feed grain crop, maize or sorghum, followed by fallow or an oilseed crop; spring oilseed rape, soybean or sunflower. Precipitation varied greatly between years and seasons over the course of the experiment, resulting in three 'winters' (September–February) and five 'springs' (March–June) being significantly drier and two 'winters' and 'springs' being wetter than the 30 year average. However, even in the wettest 'springs' the rainfall was substantially less than half the 677 mm long-term average open pan evaporation from April to June. Compared with the fallow system, whatever the feed grain or the oilseed crop, continuous cropping resulted in less water being extractable from the soil by winter wheat, which adversely affected its above-ground dry matter, grain production and water use efficiency (WUE) – based either on grain or above-ground biomass (Aiken *et al.*, 2013). On average, more water was extractable following the short-season oilseed rape than the long-season sunflower, with water storage after soybean being intermediate. Net returns calculated across the entire crop rotation were positive, when wheat followed a fallow, and were not influenced by the feed grain crop. With oilseed rape replacing the fallow the net returns were less but still positive. With soybean and sunflower the net returns were negative (Aiken *et al.*, 2013). From this study we can conclude that fixed rotations with fallow excluded will not contribute to the fundamental solution of better precipitation use.

A more complex experiment was designed by Nielsen *et al.* (2011), in which the use of opportunity-cropping systems (OCs) was tested against fixed rotations. The OCs all included wheat in the choice of crops available for sowing in the cropping sequence (Table 18.3), and they included forage crops, which can be grown as cash crops. Because most of the above-ground forage biomass is harvested, these crops have the potential of an economic WUE larger than in grain crops. For comparison traditional 2-year wheat–fallow systems were grown. They were either conventionally tilled with residues incorporated (conventional tillage, CT in Table 18.3), or zero-tilled (ZT in Table 18.3). Additional treatments included a 3-year rotation of wheat–maize–fallow and a double-cropping wheat–millet rotation; both of these used zero-tillage. Also all OCs were established using zero-tillage (Table 18.3). For OC-1 there were only four crops available for selection. All the other three OCs had the same ten crops available for selection (Table 18.3). At each sowing time, crops were chosen on the basis of potential water availability, which took account of the extractable water content in the soil profile to a depth of 1.65 m plus an anticipated precipitation of 100% (OC-2), 70% (OC-1, OC-3) or 130% (OC-4) of long-term average seasonal values. Anticipation of more than average rainfall would encourage the growing of longer-season summer crops, whereas less than average rainfall would tend to result in the adoption of short-season summer cropping or the establishment of a fallow, if the previous crop had been grown under drought conditions. The actual amount of water measured in the profile was reduced according to literature values indicating the ability of individual crops to explore the full depth of the profile over their expected lifetime (Nielsen *et al.*, 2011). OC-4, which had the largest expected water availability (extractable water in the profile plus 130% of average rainfall) yielded poorly in all 5 years, and in these years there was insufficient water to grow wheat. This was because rainfall was consistently less than the long-term average, so the crops selected turned out to be inappropriate under the prevailing weather pattern. OC-3, which also had a large choice of crops available but could only count on 70% of average rainfall plus water extractable from the profile after the preceding crop, included a long fallow in a drought year. Therefore total production for the rotation was somewhat small, particularly because water was not sufficient in any year to grow wheat. However, WUE based on the economic yield of all crops in this rotation was 1.04 kg m^{-3} (cf. Table 13.1) because of the inclusion of two forage crops (foxtail millet and forage pea). This WUE was not significantly different from that of OC-2, which had crops selected on the expectation of 100% of average rainfall. In OC-2 severe drought prevented formation of a harvestable yield of the long-season summer crop, sunflower. WUE for OC-1, with the limited choice of crops, was restricted because of the decision to include a long fallow under intense drought. Under double cropping with wheat and proso millet WUE was similar to that for fallow systems (Table 18.3). The precipitation use efficiency, a measure of the effectiveness of a cropping system to use the rainfall in each season, was greatest for OCs 1 and 2 but there was little difference between any of the other systems. Average wheat production was 20% greater in zero-tillage than in conventional tillage and OC-2 achieved a similar wheat production to zero-tillage. Only OC-2, zero-tillage (ZT) and conventional tillage (CT) gave positive net returns and there were no statistically significant differences between these cropping systems (Table 18.3). With the technical

Table 18.3. Water use efficiency, precipitation use efficiency and net return over 5 years for four opportunity-cropping systems, three fallow systems and a double-cropping system for the central Great Plains at Akron, Colorado. Crops available for selection and those chosen are identified (Nielsen *et al.*, 2011).

Cropping system	Crop choice in addition to wheat	Actual cropping sequence	Water use efficiency – economic yield (kg m^{-3})	Precipitation use efficiency[a] (kg m^{-3})	Net return over 5 years (US\$ ha^{-1})
Opportunity cropping					
OC-1: ZT	Maize, proso and foxtail millet	Foxtail millet, wheat, maize, fallow, wheat	0.96[B]	0.72[A]	–73[C]
OC-2: ZT	Maize, proso and foxtail millet, sunflower, soybean, oilseed rape, dry and forage pea, summer triticale	Wheat, sunflower, maize, forage pea, foxtail millet	1.15[A]	0.74[A]	146[A]
OC-3: ZT	As for OC-2	Dry pea, foxtail millet, fallow, forage pea, foxtail millet	1.04[A]	0.39[CD]	–491[D]
OC-4: ZT	As for OC-2	Oilseed rape, proso millet, sunflower, proso millet, dry pea	0.17[E]	0.01[E]	–1100[E]
Fixed fallow cropping					
Wheat: CT	–	Wheat, fallow (×3)	0.27[D]	0.34[D]	81[AB]
Wheat: ZT	–	Wheat, fallow (×3)	0.35[D]	0.43[BC]	139[A]
Wheat–maize: ZT	Maize	Wheat, maize, fallow (×2)	0.46[C]	0.46[B]	–18[C]
Double cropping					
Wheat–millet: ZT	Proso millet	Wheat, proso millet (×5)	0.27[D]	0.43[BC]	–66[C]

CT, Conventional tillage; OC, opportunity cropping; ZT, zero-tillage. Within a column, values followed by the same upper case letter are not significantly different (P = 0.05).
[a]Economic yield produced (kg) per m^3 rainwater used in evapotranspiration is a measure of the effectiveness of the cropping system in using precipitation. More than half of the annual precipitation falls in the spring and summer.

support available to the researchers, it appears that opportunistic cropping, based on the selection of crops to make best use of extractable water stored in the soil at seeding time, can provide comparable returns to the traditional wheat–fallow rotation. However, this idea of more effective precipitation use by 'cropping intensification' is as yet not generally accepted and for individual producers is questionable.

To sum it up: 'The primary challenge with reducing fallow frequency is the increase in yield variability and risk of crop failure' (Hansen *et al.*, 2012). It became obvious that for the successful establishment of winter wheat within rotations that have been extended by summer crops, a lengthy fallow before seeding wheat seems to be key. The importance of the fallow increases in proportion to how much the water supply falls short of the requirement. In any case, the benefits from replacing fallow depend greatly on the application of zero-tillage, which favours the capture of precipitation.

Cropping during the rainy season is a good strategy for directing precipitation water into plant production. But the strategy presents problems with deeply developed *Vertisols*, as the management of these soils is extremely difficult. The black-coloured soils are rich in clay (30–70%), sticky when wet but hard when dry. The clay minerals (smectites) swell and shrink with increases and decreases of water content, and give rise to deep cracking and to the 'gilgai' topography with its undulating soil surface. In India these soils are present in areas with a monsoon climate. With the onset of the monsoon rains in summer (Fig. 15.2) or just before, tillage and seedbed preparation present a difficult task for the small-scale farmer who lacks heavy, powerful machinery. Moreover, the deep cracks are considered to be essential for coping with heavy rain showers and for deep infiltration of water when the rainy period starts. However, when the cracks have closed due to swelling, the infiltrability is low. The soils easily become waterlogged and aeration becomes marginal. These are the main reasons why the soils are traditionally cropped during the dry season and fallowed during the rainy season. This traditional cropping system results in soil erosion, associated with large water loss in runoff but there is also an ineffective use of precipitation because of evaporation and deep percolation (Table 18.4). In the example given, runoff and subsoil drainage (as well as soil loss) varied dramatically with the amount of annual rainfall, whereas the post-rainy season evapotranspiration (ET) of sorghum was largely unaffected.

To make better use of precipitation and stored water the International Crops Research Institute for the Semi-Arid Tropics (ICRISAT) in Hyderabad has developed,

Table 18.4. Soil water balance components (in mm) and soil loss (t ha^{-1}) for the traditional rainy-season fallow system on Vertisols at the International Crops Research Institute for the Semi-Arid Tropics (ICRISAT), Hyderabad, India. Sorghum was grown as the post-rainy-season crop. Records are for 5 years. P, Precipitation; R, runoff; E, evaporation during the fallow rainy season; D, subsoil drainage; ET, evapotranspiration during the cropped post-rainy season (El-Swaify *et al.*, 1985).

Year	P	R	E	D	ET	Soil loss
1976	710	238	169	31	272	9.2
1977	586	53	201	15	317	1.7
1978	1117	410	185	221	301	9.7
1979	682	202	166	42	272	9.5
1980	688	166	175	47	300	4.6
1976–80	757	214	179	71	292	6.9
1976–80[a]	100	28	24	9	39	–

[a]In percentage of precipitation.

adapted and improved *crop rotations*. These rotations allow crop cultivation during the rainy season as well as during the post-rainy season. They are intended to replace the single crop, traditionally grown in the dry season. One rotation is maize, immediately followed by chickpea (*Cicer arietinum*). Both are short-season crops, grown in pure stands. Such a system with two crops per year is called '*double cropping*'. The preferred rotation, however, consists of maize or sorghum, 'intercropped' with pigeon pea or red gram (*Cajanus cajan*) (Stewart *et al.*, 1994). *Intercropping* is a form of mixed cropping in which two (or more) crops build the crop stand. Intercropping combines a long-season crop, here pigeon pea, with short-season crops, here maize and sorghum (Arnon, 1992). The advantage of intercropping is that no land has to be prepared anew for the dry-season crop. Techniques have been developed for tilling the soil ahead of the rainy season in February–March after harvesting the previous crop. A primary tillage operation produces a cloddy surface. The clods disintegrate with time and the seedbed is then prepared just ahead of the rainy season. Across the natural slope 100-cm-wide beds alternate with 50-cm-wide furrows. The furrows are good for drainage and improve soil aeration within the beds. Here in the planting zone, soil structure improves and a natural tilth develops during the course of some years. The need for intensive tillage decreases with time that the system is in operation. In the case of high intensity rains, excess water is channelled into grassed waterways (El-Swaify *et al.*, 1985). The new cropping system is quite adaptable to the particular rainfall pattern, conserves soil and moisture, increases WUE, water use and yield and reduces the year-to-year variability in yield.

Double cropping is a very effective rotation system to utilize the annual rainfall at a large scale for crop production and to avoid unproductive losses (Eqn 17.5). A prerequisite of the system is that soil and climate allow two crops to mature. Therefore climates in which cold winter temperatures restrict the growing season have a disadvantage in principle. In parts of semi-arid China, where precipitation is concentrated in the monsoon months from July to September, double cropping can be practised successfully. The success depends on the accurate timing of crop management operations, as annual rainfall may not be greater than 600 mm. Maize is seeded into dry stubble left after the wheat harvest in June. This is the end of the dry season with little precipitation, amounting roughly to only 30% of annual rainfall. With the onset of the rainy season the maize starts to germinate and emerge. At the same time the soil profile gets replenished with rainwater. By the time the maize is harvested in the middle of September soil water has been more or less fully restored (Table 18.5). Wheat is seeded in the middle of October after tillage and the application of fertilizer. The wheat crop depends more on the stored soil water than does maize as the contribution of precipitation to water use is much less (Table 18.5).

Yield variation in maize was much greater than for wheat (Table 18.5). Maize usually starts to grow when soil water is exhausted. Maize growth is therefore directly dependent on the precipitation pattern during the rainy season. In 1986, rainfall was only 169 mm during the maize season. Therefore there was little moisture recharge, actually it was negative: –6 mm (80 mm stored at harvest minus 86 mm stored at sowing). Maize yielded only 0.6 t ha^{-1}. Unlike the normal situation, the wheat that followed started on dry soil, containing only 86 mm of

Table 18.5. Water use and yield of maize and wheat grown in Gongxiang, the People's Republic of China, in a double cropping rotation. Records from October 1983 to September 1988. The average annual rainfall is 597 mm. Data are given as $\bar{x}$ (range), where $\bar{x}$ symbolizes the average for the years (Stewart *et al.*, 1994).

		Available soil water (mm (130 cm)$^{-1}$)				
Crop	Season	At sowing	At harvest	Water use (mm)	Rainfall (mm)	Grain yield (t ha^{-1})
Maize	Early June to mid-September	61 (18–95)	182 (80–244)	263 (175–305)	384 (169–482)	5.1 (0.6–7.1)
Wheat	Mid-October to end of May	217 (86–262)	39 (11–86)	353 (285–399)	175 (120–210)	5.2 (4.3–6.3)

available water. Rainfall during the 'dry' wheat season (210 mm) was greater than during the previous maize season. Moreover, wheat effectively extracted soil moisture from the 1.3-m profile, leaving just 11 mm of available water. However, the wheat crop must have been stressed, because the yield of 4.3 t ha^{-1} was the smallest one during the 5 years on record. None the less, the wheat yield was much better than the maize yield (0.6 t ha^{-1}) attained during the rainy season with exceptionally little rainfall. From this it seems that the performance of a crop, which relies on a restricted supply of soil water together with the constant but insufficient addition of precipitation, is more at risk than that for crops with access to water from what starts as a fully recharged soil profile. This is especially true when the crop (here wheat) has a long season (7 months) that covers a winter period with a smaller evaporative demand (Stewart *et al.*, 1994) than is experienced by maize during the short summer season.

Generally the practicality of crop rotations, particularly those involving double cropping, depends on how much precipitation is received, characterized by amount and temporal distribution. This is especially true for dry regions. Certainly it is not sustainable or forward-looking that irrigation is used as a prerequisite for double cropping, as in the dry North China Plain with a 'massive extraction of groundwater and a rapid decline of the groundwater table' (Zhang *et al.*, 2005, see also Fig. 17.7). Apart from irrigation, double cropping opens up the possibility of adaptation to actual conditions by *opportunity cropping*. This type of cropping requires the skill of farmers to make ad hoc decisions. They have to react flexibly to rainfall pattern and variation in soil water storage. In the Great Plains double cropping may be possible when summer rainfall is exceptionally large and allows the cultivation of proso millet (Table 18.3) or sorghum immediately following the wheat harvest. When it is too dry for double cropping but late summer rain replenishes the soil, a new wheat season can be started. When that is not possible, sorghum or maize may be planted next summer. If it is still too dry, as is common, a fallow has to be adopted. Soil water monitoring can be helpful in making that decision, as stated before.

Another idea is to organize a whole package of farm practices according to the onset of the rainy season in regions with variable Mediterranean and monsoon climates (Stewart and Steiner, 1990). The date that the rainy season starts and the total precipitation during the following season are interrelated. The earlier the start, the greater is the amount of rainfall that can be expected (Stewart, 1985). That provides the farmer with the opportunity to evaluate quite early in the season the probable conditions governing the water supply to the crop in question. The farmer can decide with somewhat more certainty on the crop, the crop cultivar (long or short season), seeding date, seed density, fertilization and other crop management factors. This concept has been named '*response farming*' (Stewart, 1985).

Catch crops and main crops can contribute positively to the water budget of subsequent crops of the rotation; for example plants that have a long *taproot* that opens inaccessible layers in the subsoil. In north-east Germany, in the region known as 'Altmark', there are coarse-textured, sandy soils that have developed a hardened B_s horizon. This is an illuvial horizon, enriched with sesquioxides (Fig. 2.1), forming an iron pan called ortstein. These podzols are not suited for potato production, as the subsoil is not penetrable to the roots of this crop. The soils, however, can be ameliorated for successful potato cultivation by adding *lupins into the rotation*. More than 100 years ago Schultz-Lupitz (1895) pioneered potato production on the light sandy soils by first cropping with blue lupin (*Lupinus angustifolius*). The lupin taproot penetrated the hardened B_s horizon and opened channels through to the subsoil for the fibrous root system of the potato to exploit. Once into the non-hardened subsoil the potato roots were able to proliferate (Fig. 18.2). Part of the ameliorative effect was due to '*biologic subsoiling*'. Another part was due to the nitrogen fixation by the legume (Schultz-Lupitz, 1895). None the less, deep

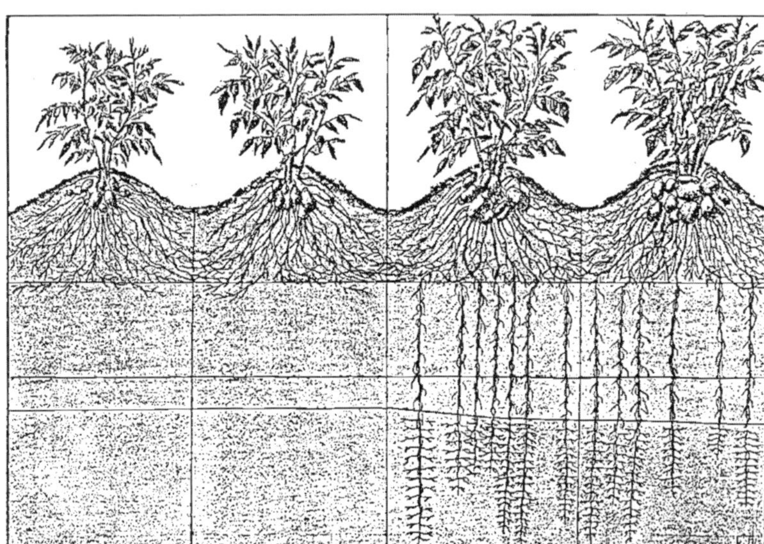

Fig. 18.2. The amelioration of sandy podzols in north-east Germany for potato production. The potato root cannot penetrate the hardened subsoil (left-hand side). After cultivation of blue lupins, however, the potato plants were able to grow their roots through the pathways created by the taproots of the lupin (right-hand side). Below the hardened horizon the potato roots spread into the subsoil (from Schultz-Lupitz, 1895).

rooting favoured water uptake, transpiration and potato production. Seepage losses were reduced and WUE increased (Eqn 17.5). A similar beneficial effect can be expected on more productive loamy-clayey soils, when the deep-rooting lucerne (Section 6.3) is included into the rotation (Gaiser *et al.*, 2012). Lucerne is able to open the subsoil by creating root channels, which increase infiltrability and the rooting of subsequent crops (Jensen and Sletten, 1965).

Greater diversity in crop rotations including '*break crops*' can be another measure to secure and increase the yield level, even under semi-arid conditions. Break crops like oilseed rape and Indian mustard (*Brassica juncea*) can break the life cycle of cereal root *diseases* in wheat. The pathogens concerned are 'take-all' (*Gaeumannomyces graminis* var. *tritici*) and 'crown rot' (*Fusarium graminearum*). In south-eastern Australia it was found that the incorporation of *Brassica* species into the rotation caused a healthier root system in wheat, as a result of which water extraction and consequently yield were increased (Angus and van Herwaarden, 2001).

The appropriate sequence of crops within the rotation can provide opportunities for the elimination of *weeds* as they provide unwelcome *competition for* the stored soil *water* (Eqn 17.5). The change between autumn- and spring-sown crops provides an effective window for weed control by cultivation. Some of the autumn-germinating weeds are killed mechanically during seedbed preparation for winter crops. The weeds that germinate later after crop establishment may be suppressed, but still survive. Spring-crop cultivation may kill more or less the entire population of such weeds. Accordingly, spring-germinating weeds can be controlled effectively by cultivation of late-seeded summer crops. Mechanical weeding, however, can be overdone with respect to water losses (Table 17.1). Reducing the number of passes with a cultivator can save soil moisture. Possibly the savings are greatest when weeds are completely controlled by herbicide application. Very effective herbicides are a prerequisite for growing crops with both reduced and zero-tillage. An economic application of a herbicide will take into account the species, the plant density and the competitive pressure of the weed population. As long as the level of weed competition has not attained limiting proportions, the costs of the herbicide application will be greater than the value of the forecasted yield reduction caused by the weeds. The critical level of competition varies considerably between the weed species. A strategy that relies upon counting the weed flora is known as 'curative' weed control and is an important contribution to sustainable agriculture. It contrasts with 'prophylactic' weed control, which is often taken as the regular and unquestioned practice in crop husbandry. But it is also true that the curative approach is much more difficult to adopt when ploughing is abandoned and tillage is reduced or the soils are zero-tilled. Under such conditions farmers may prefer the prophylactic approach to keep the weed population in check, as mechanical control as an emergency measure is more or less excluded.

18.2 Choice of Species and Cultivars

The right choice of crop species is a prerequisite for a diversified crop rotation. When putting preferred crops into sequence for a rotation, the *need for a sufficient water supply* during the season has to be considered for each one. The site conditions can be such that right from the outset a number of crop species have to be excluded from consideration, as their individual needs cannot be met by the actual water supply. In temperate climates, for instance, sugarbeet is not grown on sandy soils under rainfed conditions, as the water supply during the long summer season is insufficient. On these coarser-textured soils winter rye has a relative advantage over winter wheat, as the former grows and *develops earlier* and thus economizes on water use. For the same reason, winter barley may be superior to winter wheat on more productive soils, as the duration of the barley season is usually 2 or 3 weeks shorter. The longer wheat season can make a difference to the additional water requirement during the warm summer period. Let us suppose a daily evapotranspiration rate of roughly 2–3 mm day^{-1} (cf. Fig. 9.4) for a maturing wheat crop over the 2 weeks after barley has been harvested. Then the extra water use amounts to between 28 mm and 42 mm. That quantity may correspond in some areas to the average monthly precipitation.

In more continental climates with little summer precipitation and warmer temperatures, the field pea (*Pisum sativum*) may have a superior performance to faba bean because the *length of the season* is usually shorter and hence the required water supply is less for pea. Faba bean is a pulse with a longer season, adapted to temperate season climates that provide a well-balanced water supply without severe stress. In contrast, sunflower is a plant that is much more suited to a continental climate with warm to hot summers. Unlike pea, sunflower is not a drought escaper but much more of a water spender (Fig. 1.1), rather like lucerne. But unlike the taprooted lucerne, sunflower is a crop with a deeply penetrating fibrous root system (Fig. 6.3B). In Kansas sunflower depleted water from a loess soil down to 3.3 m depth at the end of the season with a maximum rooting depth of 3.03 m (Stone *et al.*, 2002). Sunflower as an oil crop and lucerne as a fodder plant maintain their importance in warm and dry climates because of their ability to *acquire water* from *deep soil* layers. Oilseed rape is a cool-season crop and best adapted to temperate climates, especially in the case of the winter form. Although a taprooted plant (Fig. 6.3A), it needs a continuous water supply, either provided by sufficient soil water storage or by regular rainfall. Winter-grown oilseed rape has a very *prolonged season*. In the UK and Germany it may be in the ground for up to 12 months. Growing over a long period under cool temperatures with a small vapour pressure deficit qualifies the crop is an efficient water user, having a relatively large transpiration efficiency (Eqn 17.5).

In principle, *winter-grown cereals* use water more efficiently than *spring-sown* crops (Fig. 9.5). They have already completed part of their growth and development when summer forms are sown. And that early part of dry matter production is accomplished under quite favourable gas exchange conditions at the leaf surface, where CO_2 is taken in and H_2O vapour is released. Winter cereals cover the soil early with their leaves (barley > rye > wheat). This way they reduce soil evaporation. They use the radiation for assimilation during the winter months, even when the temperatures are limiting. Winter-grown cereals are superior not only in shoot development but also in root formation. The root system develops earlier and extends generally to deeper soil layers than that of spring-sown crops. Therefore they can minimize evaporation and percolation losses much better than can the spring-sown forms. All these properties help to explain why winter cereals have the ability to produce comparatively heavier and more reliable yields (Table 15.1).

Winter barley is superior to winter wheat in the rate that it shades the soil surface due to its faster development. Another reason is the greater *leaf area ratio* and *specific leaf area* (see Section 16.1).

Cultivars with a predominantly horizontal *leaf position* can intercept a greater proportion of radiation and can suppress soil evaporation more effectively than cultivars with leaves that are more erect. These differences can be important at early stages of development and may decide the outcome of interspecific competition. The planophile leaf inclination is especially advantageous when the radiation load is small (Section 10.2). After the winter solstice the radiation increases during crop development at locations of higher latitude (Section 15.1). With the greater radiation intensity, energy can be used for biomass production more effectively when the radiation is intercepted by leaves that are erect, since by that time the leaf area index has increased considerably (Fig. 9.1). Shading from upright leaves can still be high, thus keeping evaporation to a low level. Cultivars with a leaf inclination that changes during the season may be significant in some crops, such as small-grain cereals, maize and sugarbeet. These cultivars may be significant because they use water efficiently (Eqn 17.5). They can reduce unproductive water losses and transform radiation effectively into dry matter production.

In summary, there seem to be some *essential traits* that have been accentuated during the process of crop selection. These traits make up those cultivars that are better adapted for cultivation in areas where water is a limiting factor in crop production. These traits are: (i) rapid development; (ii) early root and leaf growth (but later in the season a sparse rather than too luxuriant a canopy); (iii) short stature; (iv) deep root system; (v) relatively long post-anthesis period for maturation; (vi) high translocation rate of storage material; (vii) large harvest index (HI); and (viii) a short season. A more retarded development of individual plants within the crop stand rather than early luxurious growth seems to be important. A restricted capacity to tiller or branch is also beneficial. In Australia, for instance, strains of wheat have been bred that cannot tiller at all. However, these lines had no chance of being accepted for commercial crop production because of the higher risk in the crop establishment phase. In the case of adverse conditions, for instance if field emergence is impaired because of dryness in the seedbed, they may fail completely. Normal cultivars will compensate by enhanced tillering when the soil eventually gets wet, and thus are able to fill the spaces within the stand.

18.3 Seeding and Stand Density

Early sowing seems to be an essential prerequisite for the maximum use of environmental factors that determine growth and yield, like radiation and water. In *warm climates* seeding date will be dictated by the onset of the *rainy season* (Fig. 15.2). Some areas, such as West Africa with a tropical savannah climate, experience a bimodal rainfall distribution, allowing for two distinct cropping seasons (Cotonou, Fig. 15.2). In climates with very *cold winter* temperatures, successful planting of winter crops may be impossible because of winter-kill. Under such harsh conditions efforts are made to sow spring crops very early to get crop growth started as soon as possible. To that end spring crops can be sown in autumn. Germination is delayed because the seeds are dormant. The *dormancy* is broken during winter under the influence of chilling temperatures, and seeds will germinate in spring. In areas with a more *temperate climate*, cultivation of winter crops is possible and often preferred, as the yield is generally greater than the yield of spring-sown crops (Table 15.1). In this climate early sowing of both winter and spring crops allows more favourable utilization of transpirational water in addition to the greater use of the radiation. The more effective use of the water with respect to biomass production occurs during the cooler part of the year, when the vapour pressure deficit of the air is small. Moreover, early canopy and root development, and smaller unproductive water losses by evaporation and subsoil drainage (Section 9.3) leave more water for transpiration in absolute terms, and thereby increase the biomass yield attainable (Eqn 17.5). Early spring seeding of *delicate crops*, however, may result in damage by late frost. For instance, sugarbeet growers have to weigh the risk of frost damage against the advantage of increased sugar yield due to a longer season. That contrasts with the cultivation of warm season crops like sorghum in the dry steppe climate. Early seeding may induce slow plant canopy development and slow coverage of the soil surface, giving rise to increased water losses by evaporation compared with later seeding (Stewart and Steiner, 1990).

In the temperate climate of Germany, early seeding may improve WUE and increase grain yield of *winter-sown cereals*. This is especially true when water is yield limiting, as it is in some eastern areas where the annual precipitation is no more than 450–500 mm. For example, late seeding reduced all yield components in winter wheat, whereas in winter barley a decreasing number of grains per ear was offset somewhat by a small increase in grain weight (Table 18.6). This evidence suggests that late seeding caused some water stress early during stem elongation, and probably more in wheat than in barley, but there was also some stress later in the season during grain filling (Section 13.3).

In regions where water supply is insufficient for lush vegetative growth, the *stand density* has to be adjusted for this shortcoming. Reducing the stand density (the number of plants per unit area) may increase the *water supply per plant*. But at the same time there is a risk associated with sparse plant densities for *row crops* like sugarbeet, maize, sorghum or cotton, as the soil is left uncovered by the *foliage* for a longer period than with greater plant densities. Also *root exploration* is retarded. Consequently, in stands of low density a considerable percentage of stored soil water may be lost by unproductive *evaporation*. When the soil surface dries between plant rows after first stage evaporation (Section 5.2), radiation creates sensible heat and the raised temperature increases the vapour pressure deficit of the air near the plant canopy. The clothesline effect is responsible for excessive transpiration of individual plants (Section 12.2). The 'productive' part of water use by transpiration is added to by an 'unproductive' part of leaf water loss. That causes transpiration efficiency to decrease (Ehlers, 1991). These relationships may be the basic idea for defining an optimum plant density in *sugarbeet* production for the dry areas in eastern Germany with its annual precipitation of only about 450 mm. The recommendation is that crop density should not be less than eight plants m^{-2}. It is even better to adjust the density to nine or ten plants m^{-2} and to pay particular

Table 18.6. Grain yield and yield components in winter barley and winter wheat as influenced by seeding date. Bernburg an der Saale, eastern Germany. The soil is a Phaeozem derived from loess (Kratzsch, 1993a).

Crop	Seeding date	Grain yield (t ha^{-1})	Number of ears m^{-2}	Grains per ear	Grain weight (mg)	Yield per ear (g)
Winter barley	15 September	8.1	526	34.6	44.3	1.53
	25 September	7.5	512	33.2	44.3	1.47
	5 October	7.2	508	31.3	45.5	1.42
Winter wheat	5 October	8.6	570	32.8	46.2	1.51
	25 October	8.2	572	30.1	45.7	1.38
	15 November	7.6	548	28.8	44.5	1.28

attention to achieving uniform spacing (Heßland, 1993). Basically it turns out that this is the same as the recommendation for beet cultivation provided in the more humid areas of Germany. One reason for recommending a relatively high plant density is that some water shortage and *stress* in the individual plants appears to be *desirable* during the later season (August–September) and that is promoted by the intraspecific competition among sugarbeet plants. Water shortage and reduced nitrate uptake hinders the regrowth of leaves, which could act as competing secondary sinks for assimilates. In the absence of regrowth, the sucrose synthesized within the existing leaves (the source) is translocated into the beet (the primary sink).

Optimizing planting density for row crops is a question of the compromise between the water supply per plant over the entire season and water loss by soil evaporation. The question can only be answered for a specific crop in a given climate. In the West African Sahel, field experiments with *pearl millet* showed that within a particular year under the given rainfall the biomass yield and the grain yield were dependent on planting density. At a planting density of 0.5 'hills' m^{-2} (one hill contains three plants), biomass yield was smallest in 2 out of 3 years, and grain yield was always the smallest. At the greatest density of 2 hills m^{-2}, the largest biomass yield was produced only once in a relatively wet year with 545 mm annual precipitation, and it also produced the largest grain yield once in a dry year with only 260 mm precipitation (albeit that this yield was very small in absolute terms). In 2 out of 3 years with adequate rainfall the greatest grain yield was attained with a *medium* planting density of 1 hill m^{-2} (Payne, 1997). Within a year and for a given annual rainfall, crop management (planting density, phosphorus (P) and nitrogen (N) fertilizer addition, choice of cultivar) dictated the yield level. In contrast, however, the seasonal evapotranspiration for the year proved to be relatively constant (Fig. 18.3). These results resemble very much those shown in Fig. 12.2. This is consistent with our conclusion that crop management can result in large variations in WUE because of effects on dry matter production rather than on evapotranspiration.

For *row crops* we can draw the conclusion that in addition to crop species, climate and the actual water supply determine the appropriate value of plant density. The rule that generally applies is: *the drier the situation the smaller the*

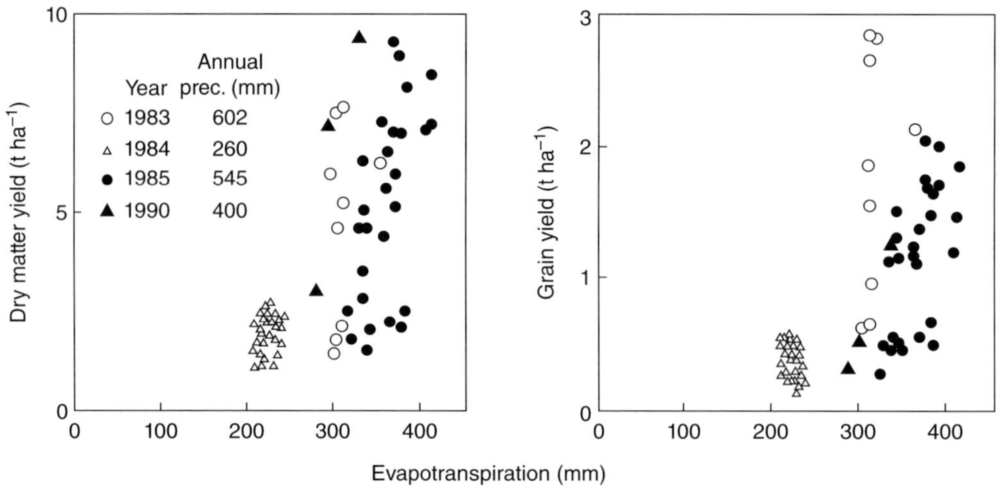

Fig. 18.3. Relationship between seasonal evapotranspiration and above-ground dry matter yield or grain yield, respectively, for pearl millet grown in Niger, Sahel, West Africa. The soil was a sandy loam to loamy sand. Yield depended on rainfall of the year, but varied greatly depending on inputs for crop management. The evapotranspiration also varied with annual rainfall. But in contrast with yield, water use was quite conservative irrespective of cultural practices (based on Payne, 1997).

optimum stand density. The problem is that in dry areas particularly the amount of rainfall and therefore the wetness of the season can only be predicted with a low level of probability. In this respect the concept of response farming (Section 18.1) may possibly offer an advantage. As more optimum conditions of water supply prevail, stand density can be increased to maximize radiation interception. Within this framework it seems that *low planting density encourages soil evaporation whereas high planting density endangers the water supply per plant*, especially at later stages of development.

When water is limiting, the recommendation for reducing the plant density is correct not only for row crops, but also for more densely seeded *small-grain cereals* and some other cool-season crops like oilseed rape. The early growth of these crops takes place during the cooler part of the season when water loss by evaporation will be relatively small. As temperature rises in spring, the soil is shaded by the growing canopy. Sowing date and *seeding density* have to be adjusted in such a way that, on the one hand, enough biomass will be produced in the cooler and more humid season with a relatively low saturation deficit. On the other hand, enough soil moisture has to be preserved for root water uptake during the period of grain filling (Gregory, 1989b). As a rule it turns out that seeding density has to be less in drier regions compared with more humid ones.

When there is little rainfall during the latter part of the growing season and evaporative demand remains high, as in some semi-arid to arid regions, crops have to rely much more on the amount that has been stored in the soil in extractable form to meet their need for water. In such a situation where soil water is not replenished during the season, plants have to be quite economical with the water right from the onset of growth. For that reason, in eastern Australia wheat genotypes have been selected with *xylem* vessels that are particularly narrow in diameter. The constricted vessels have an increased *axial resistance* to water flow, greater than in normal genotypes. In fact, in normal wheat the axial resistance is relatively small within the conducting network of the plant (Section 8.1). In conjunction with a more sparsely developed root system the new plant types have restricted water uptake relative to traditional strains that behaved more in the manner of water spenders (Passioura, 1972). At the end the water savers produced the greater yield.

Water saving during the *preflowering period* is very important in regions where water is scarce. But, on the other hand, biomass production must not be constrained too severely by limiting the water use. A limited period of grazing by sheep, goats or cattle is one approach to reducing transpiration during the vegetative stage. Although extending the duration of the crop cycle, increased availability of water during the reproductive phase and grain fill can ensure equal or even improved yields (Virgona *et al.*, 2006). Stems and leaves as well as spike and spikelets form the basis for later yield formation during the grain-filling period. Being faced with the need for yield optimization, the obvious solution is to: (i) reduce seeding rate; (ii) avoid overstressing tillering; (iii) promote strong individual culms with a long period of spike formation; and (iv) allow for large-sized ears with a large number of florets. Compared with row crops, soil evaporation and unproductive transpiration will contribute much less to unproductive water losses as the plant density is still high and row closure is completed when the season is still cool. In winter crops, rows close earlier than in spring crops. In the case of a reduced number of shoots per plant, water and nutrients can be provided to the individual shoot comparatively easily. Such a strategy of reduced stand density with an emphasis on strengthening the single culm provides for the possibility that sufficient residual soil moisture will remain for the grain-filling period. Part of the assimilates to be included in the seeds has to be translocated from other parts of the plant, where the components had been stored earlier (Section 13.3). The greater part, however, is synthesized during the *postflowering period*. The green flag leaf contributes significantly to assimilate production, but production also takes place in all the other organs that are still photosynthetically active. These consist of the residual areas of older leaves that have not yet turned yellow, green parts of the culm, the glumes and awns. Awns can be found in barley, rye, oat and durum wheat (*Triticum durum*), and in a few cultivars of common wheat (*Triticum aestivum*). The importance of *green leaf area* during the postflowering period for grain formation is shown in Fig. 18.4. A greater leaf area duration (integral of green leaf area with time, Eqn 9.5) causes a greater cumulative transpiration, a greater CO_2 assimilation (not measured) and a larger grain yield. Such a response may also be expected from plants with the 'stay green' trait in maize and sorghum (Section 16.1), possibly with stomates restricting leaf transpiration.

Reduced seeding rates had a favourable effect on grain yield formation in *winter wheat* grown in the dry areas of eastern Germany when precipitation was particularly small (about 60 mm) in the months from May to July (Fig. 18.5, curve 1). In years with little rainfall, cloud cover tends to be small, but radiation tends to be large (Fig. 15.3) and there is a high potential evapotranspiration (Fig. 9.4A).

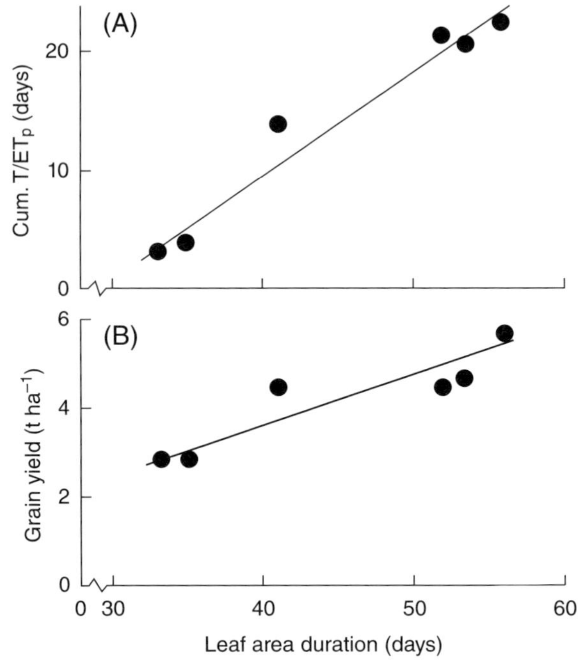

Fig. 18.4. Influence of the green leaf area after flowering on transpirational water use and grain yield in oat. (A) The cumulative transpiration (T) over the average daily potential evapotranspiration rate (ET_p) within the period from anthesis to maturity as a function of leaf area duration after anthesis. (B) The grain yield as related to leaf area duration after anthesis. Oat was grown on loess-derived Luvisols near Göttingen, Germany over several years (after Ehlers, 1991).

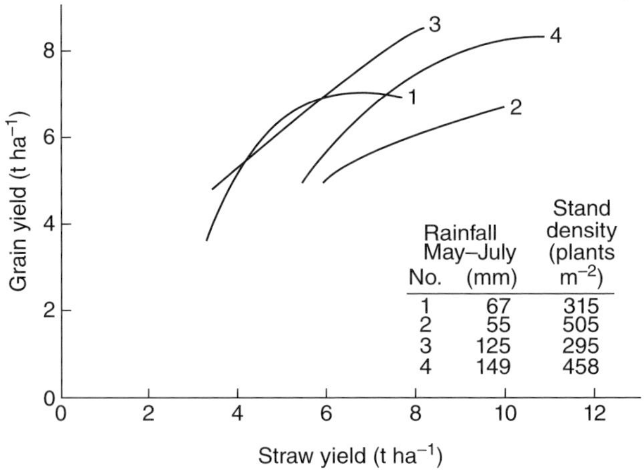

No.	Rainfall May–July (mm)	Stand density (plants m^{-2})
1	67	315
2	55	505
3	125	295
4	149	458

Fig. 18.5. Relationship between harvested vegetative biomass (straw) and grain yield in winter wheat, depending on rainfall and stand density during individual years. The yield level was changed by nitrogen fertilization. The soil is a Phaeozem derived from loess in the dry region of Bernburg an der Saale, Germany (after Waloszczyk, 1991; Kratzsch, 1993b).

Under these conditions the cumulative transpiration determines the grain yield (cf. Table 15.1). An increased *plant density* of wheat due to a higher seeding rate under conditions of inadequate rainfall produced more straw, but barely any of the grain yield (Fig. 18.5, curve 2) achieved by a sparse stand (curve 1). In years with average rainfall (about 130–140 mm from May to July), again the higher stand density produced more straw (curve 4), but the maximum grain yield only approached that of the smaller stand density (curve 3). In any case, a large production of straw (stimulated by N fertilizer) formed the basis of increased grain yield when plant density was not reduced (curves 2 and 4). The HI was unfavourably small. With a reduced plant density on the other hand, heavy grain yields were achieved with less straw (curves 1 and 3) and a better HI. Therefore, in this dry temperate region with reduced or average early summer rainfall, a smaller seeding rate proved to be a reliable crop management strategy for a heavy grain yield. Note that in the low rainfall situation the grain yield hardly showed any response to high N applications, whereas straw was very responsive (curve 1). It was concluded (Waloszczyk, 1991) that a lower plant density was the basis for *yield stability*. The reason for the better yield stability is probably the larger percentage of extractable water, remaining for use over the period of grain fill.

When water supply is limited, one option is to reduce the number of plants per unit area. An additional possibility that can be considered is to modify the *plant distribution* within the field with a sparse stand (Arnon, 1992). At a given plant density (related to the number of plants per field) the local density within the plant row can still be greater if the row spacing is widened. A similar effect may be achieved by double-row planting, leaving a wider space between the twin rows. Increasing the plant density within the rows supports the intraspecific competition. Soil moisture within the wide row spacings is saved for a time. Development of an individual plant is retarded, as the competition for nutrients, radiation and water increases. However, in the long run deeper rooting may be stimulated, thereby making the water that is stored in deeper horizons accessible to the plants during later stages of the season. Also the remaining soil water from the inter-rows may support the plants late in the season, when explored by lateral root growth. This favourable effect of row spacing on *postponing root water uptake* until the period when the economic yield components are formed has to be balanced against the more unfavourable level of losses through soil evaporation that could be taking place during the early part of the season (Eberbach and Pala, 2005). In any particular situation the question has to be answered specifically, whether the manipulation of the crop stand by row spacing will adversely affect the amount of water that is accessible to the plant roots, especially late in the season, and how any modification will affect the yield level.

18.4 Fertilizer Application

Fertilizer becomes necessary when the supply of macro- and micronutrients in the soil is insufficient or is depleted to the extent that it cannot meet the plant's need. Fertilizer application is an effective tool for increasing dry matter production and yield (Chapter 12). Generally, better nutrient supply will increase *water use efficiency* (WUE) by reducing unproductive water losses (Eqn 17.5) because well-nourished crop stands develop an extensive *canopy*, thereby controlling soil evaporation. Fertilizer also encourages an intense and deep *root system* for greater water extraction from the soil (Fig. 18.6). This includes taking water from deeper layers when these can be explored by roots (Gliemeroth, 1951; Angus and van Herwaarden, 2001). Moreover, intense rooting in shallow soil layers (Fig. 6.4) dries the soil (Figs 9.7 and 9.8) between individual plants and contributes separately from the canopy cover effect in further reducing (Fig. 5.10) soil evaporation (Section 9.7).

Crop stands that suffer from nutrient deficiency are restricted in growth. Yields are normally less than in crop stands well supplied with nutrients. Severe deficiency can cause not only WUE to decrease, but even transpiration efficiency (TE) may decline (Fig. 12.5). In countries with a highly developed agriculture, however, such severe nutrient deficiencies seem to be eliminated as long as extreme impacts of soil erosion are not taken into account. Even so, when TE is of no concern in these countries, poor fertilizer practice may nevertheless affect WUE.

Application of fertilizer stimulates shoot and root growth (Fig. 18.6). In the case where the soil water supply is limited and not adequately augmented by regular rainfall during the season, stimulated growth and greater plant water use are of no benefit if the water resources become depleted before the crop matures. It is therefore the rule that the *rate of fertilizer application has to be adjusted to both the quantity and the timeliness of the water supply* (Section 12.1).

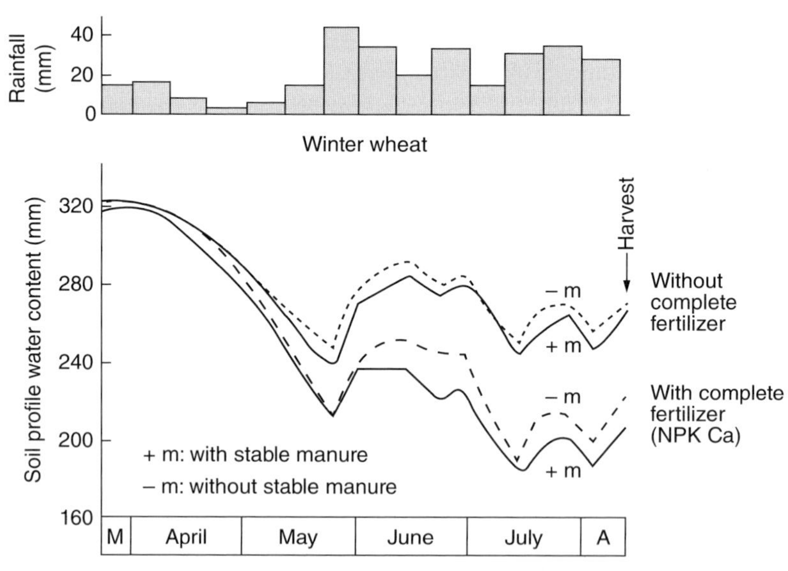

Fig. 18.6. Change in soil water stored in a silty Luvisol near Bonn, Germany, during the main part of the season with winter wheat. Mineral fertilization (NPKCa) and solid manure application (m) supported root water extraction from the soil (cf. Table 12.3). The depth of the soil profile was 1 m, the original loess being deposited on the gravel terrace of the river Rhine. The rainfall pattern is indicated at the top of the figure (after Schulze and Schulze-Gemen, 1957).

The basic *requirements* of crops for *major nutrients* like potassium (K), P, calcium (Ca), magnesium (Mg) and sulfur (S) have to be provided as fertilizer, perhaps even at seeding if soil analysis indicates a deficiency. N, on the other hand, is frequently applied in split doses during vegetative growth. The idea is to guarantee an unrestricted development of the single plant under given circumstances. In northern Syria with a Mediterranean climate, the biomass and grain yield of barley can be greatly improved by *early fertilizer application* of N and P (Brown *et al.*, 1987). Rapid crop development greatly reduced evaporation losses. Transpiration was favoured by greater shoot and root growth. Under the conditions in Syria, fertilizer improved WUE (Eqn 17.5) and TE, as more biomass developed during the winter months when the vapour pressure deficit was small (Cooper *et al.*, 1987).

When N limits yield, deficiency may be met by application of mineral fertilizer or by inclusion of a *legume* into a crop–fallow rotation. In some semi-arid regions of Australia the traditional wheat–fallow system has been replaced by a wheat–*Medicago* species rotation in combination with reduced tillage. In addition to beneficial effects on soil tilth, soil organic matter, general fertility and erosion control, the level of *mineralizable nitrogen* in the soil is enhanced. These improvements cause wheat yields to increase significantly. This system, which is intimately linked to sheep farming, has now been introduced in Algeria and Tunisia and also appears to be promising for improving soil productivity there (Smith and Elliott, 1990).

N supply has to be adjusted to the plant's requirements. Fertilizer application may increase root length density, rooting depth and water extraction (Fig. 18.6), but *excessive nitrogen supply* can result in a negative feedback. With spring wheat, grown on loamy soils under conditions of 'limited rainfall' in western Minnesota, USA, the relatively high N rate of 134 kg ha^{-1} decreased rather than increased total root length, rooting depth and water depletion in deep layers compared with the control with 67 kg N ha^{-1} (Comfort *et al.*, 1988). High N supply may shift the *allocation of assimilates* within the plant, favouring shoot mass, but to the detriment of root development. This adverse effect of high N applications may not only have consequences for water uptake during the later stages of growth but may also increase the potential for the *leaching of nitrate*, particularly any still remaining in deeper soil layers from fertilizer applied to previous crops.

In contrast to the positive effect that appropriate applications of fertilizer can have on water dynamics in temperate climates (Fig. 18.6), adverse relationships may develop in dry climates. When water supply is limited due to restricted precipitation, high evaporative demand and low available field capacity (FC_{av}) (Table 9.2), *disproportionate* applications of *N fertilizer* may lead to *yield reduction*. This is because N availability strongly favours biomass production, thereby encouraging water stress during the later part of the season (Frederick and Camberato, 1995a,b). Large soil N and post-anthesis drought may cause *premature ripening* in dryland wheat, an effect called '*haying-off*' by farmers in south-eastern Australia. Hayed-off wheat plants grow relatively tall, but are dark coloured. Grains are pinched or shrivelled and contain a high protein concentration. Symptoms for haying-off have been found in Mediterranean countries affected by the sirocco wind as well as in New Zealand, North America and Russia (van Herwaarden *et al.*, 1998). In south-eastern Australia a steppe climate prevails that is characterized by relatively mild winter temperatures, high summer temperatures and rainfall throughout the year. Here haying-off does occur, but it has never been observed within the wheat belt of Western Australia with a Mediterranean climate.

In the southern part of New South Wales, Australia, the N effect on haying-off has been tested at two contrasting locations (Fig. 18.7). Ginnindera is located near Canberra, Australian Capital Territory, 600 m above sea level (asl), with 706 mm mean annual rainfall and about 1400 mm Class A pan (Fig. 5.6) evaporation year^{-1}. The other site, Wagga Wagga, some 250 km west of Canberra and 210 m asl, is drier with 579 mm rain and about 1650 mm pan evaporation annually. N application stimulated *tillering* and *biomass at anthesis* at both sites (van Herwaarden *et al.*, 1998), but increases in grain yield were only obtained at Ginnindera, whereas in Wagga Wagga *grain yield decreased* with N fertilizer application (Fig. 18.7), a characteristic feature of haying-off. Yield components like ear density and number of grains per ear were increased at both locations, but the *grain weight* was greatly reduced at the drier site in Wagga Wagga and resulted in a decrease in HI. N application (240 kg N ha^{-1}) increased *evapotranspiration* (ET) from 456 mm up to 476 mm in Ginnindera, whereas in Wagga Wagga (200 kg N ha^{-1}) ET *decreased* from 384 mm to 374 mm at the expense of post-anthesis ET. Causes of hayed-off wheat are *excess water use before anthesis* and a *reduced water extraction* from the subsoil *thereafter* (cf. Section 16.1). The latter effect may possibly be related to a strong *competition* for the allocation of only a small quantity of *assimilates*, produced after anthesis under conditions of water shortage encouraged by N fertilizer application. This competition could be taking place between ears (a sink with an increasing number of grains per ear (Fig. 18.7)) and roots (a sink spreading in deep soil). Another cause is an inadequate *retranslocation* of pre-anthesis assimilates from stem and leaf into the grains to compensate for the small amount of post-anthesis assimilation (van Herwaarden *et al.*, 1998). Application and subsequent uptake of N favours the allocation of assimilates to structural tissue, and therefore induces a potential shortage of soluble carbohydrates for retranslocation (Angus and van Herwaarden, 2001).

Large N applications did not increase early water stress in wheat, even at Wagga Wagga, the drier site. The decline in HI and in grain weight (Fig. 18.7) suggests that at this location – in addition to water shortage associated with the general climatic conditions – the N effect caused strong additional *water stress* only at or after anthesis during grain filling (cf. Fig. 13.7). Between anthesis and maturity net biomass production ranged from 3.8 t ha^{-1} to 4.9 t ha^{-1} at Ginnindera and from 1.0 t ha^{-1} to 0.1 t ha^{-1} at Wagga Wagga for control crops and those with highest N application rate, respectively. Whereas at Ginnindera 24% (99 mm) of rain in the growing season fell after anthesis, this part was only 7% (23 mm) at Wagga Wagga (van Herwaarden *et al.*, 1998). One approach to preventing haying-off has been to delay the decision to apply N fertilizer until wheat crops begin the phase of stem elongation, when the likely availability of water for reproductive growth and grain fill, based on local weather patterns, is more certain (Angus, 2001). Passioura and Angus (2010) argue that this tactic contributed to significant increases in the production of wheat in southern Australia.

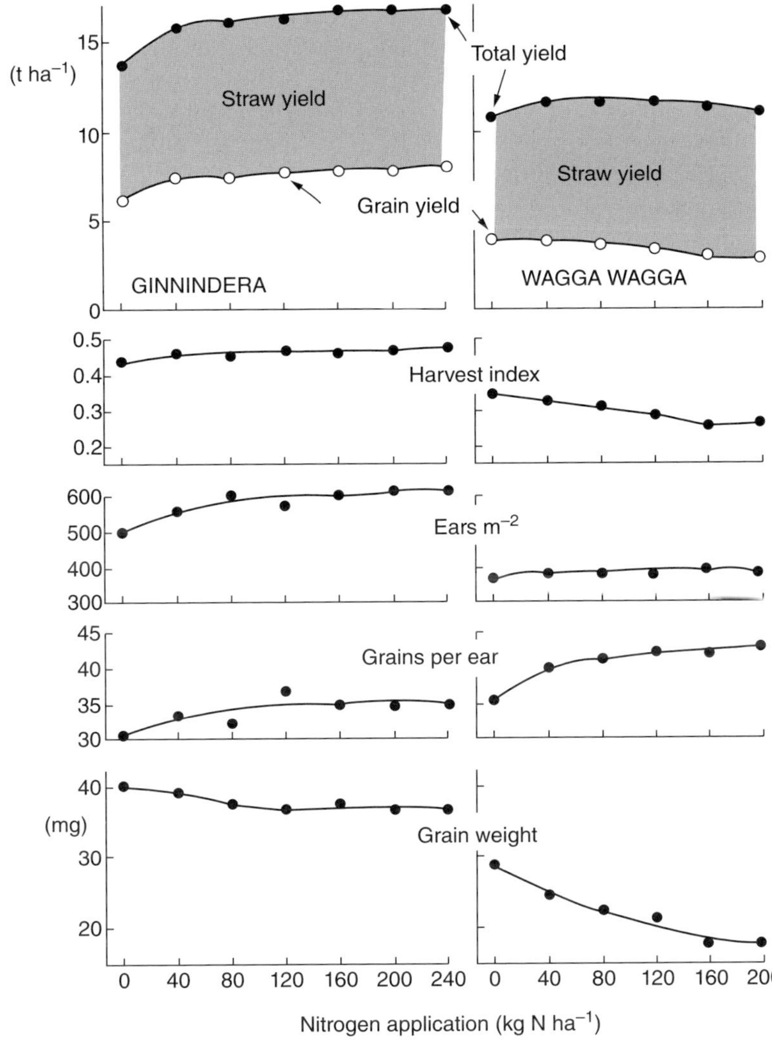

Fig. 18.7. Effect of nitrogen (N) application on total and grain yield of wheat, straw yield by difference and yield components at two locations in south-eastern Australia. Wheat grown at Wagga Wagga shows typical signs of 'haying-off'. See text for details (based on van Herwaarden *et al.*, 1998).

The N effect in Wagga Wagga clearly differs from that observed in barley from northern Syria, which was mentioned earlier (Brown *et al.*, 1987). The different outcome may be due to the *contribution of stored soil water to seasonal ET*, being only about 10–20% in Syria, but 20–40% in the drier parts of eastern Australia (Gregory, 1989b). In Syria with its Mediterranean climate, precipitation is concentrated in the winter months (Fig. 15.2) and largely coincides with the main cropping season, when the CO_2–H_2O gas exchange conditions for biomass production are favourable (Box 11.1). In the semi-arid eastern Australian steppe climate, precipitation is distributed throughout the year. Apparently, here vegetative growth is dependent on current rainfall but plants also rely very much on stored soil water for grain fill. When there is little water stored, and if precipitation fails to materialize during the grain-filling period and evaporative demand is high, then root growth restrictions by poorly adjusted N application may cause dramatic yield reductions through 'haying-off'.

19 Irrigation

19.1 Need, Concerns, Problems

World population is still increasing rapidly, although the rate has declined somewhat from the peak in 1988 of 88 million additions year^{-1} (Fig. 19.1). At the end of 1999 the population reached the 6 billion mark (UN Population Division, 1999) and on 31 October 2011 it exceeded 7 billion (US Census Bureau, 2013). Over the next few years the increase is projected to be about 76 million people annually (US Census Bureau, 2013). This growth in population is only slightly less than adding the current population of Germany every year but considerably more than the current number of people in the UK. One projection by the United Nations is that the population of the world will reach 9.6 billion people by 2050 (UN, 2013). The majority (95%) of this growth will be concentrated in *developing countries*. All the people will need food, fuel and fibre together with water for drinking, hygiene and, in many dry areas, for irrigation too.

The earth's land area covers roughly 13 billion ha (excluding Antarctica, major water bodies and rivers). That is around 29% of the total earth's surface. Approximately one-tenth of the land area, namely 1.4 billion ha, is arable land (FAO, 2010). The rest is permanent grassland (3.3 billion ha), forest and savannah (4 billion ha) or is infertile because of salinization or aridity. On average the arable land area per person now amounts to just 0.2 ha.

Just as world population has increased, especially since the middle of the 20th century, so has the land *area* under *irrigation* (Fig. 19.1). In the year 2000 the irrigated area covered about 270 million ha, or about 18% of the arable land area, and by 2011 it reached about 310 million ha – nearly 21% of arable land area (FAOSTAT, 2014). From 1960 onwards, the irrigated area per person has stayed nearly constant at around 0.045 ha (Fig. 19.1). Although the irrigated area is only 20–21% of the arable area, it produces 33% of global crop production, 44% of cereals (Portmann *et al*, 2010) and more than 40% of human food (Fereres

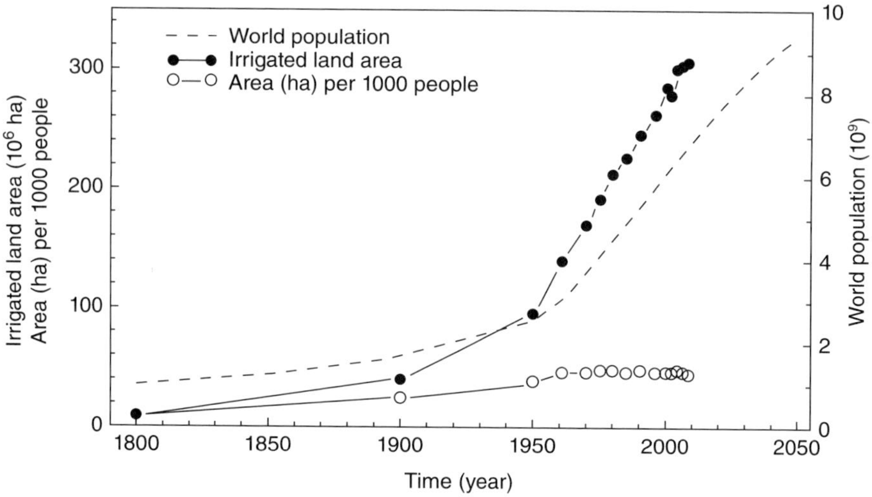

Fig. 19.1. Growth of world population, irrigated area and irrigated land per person during the last two centuries (after Howell, 2001). In 2011 world population reached 7 billion people (US Census Bureau, 2013). The irrigated area increased to ~310 × 10^6 ha. From both numbers an irrigated area is derived of 44 ha per 1000 people or 440 m^2 per person.

and Soriano, 2007). More than 70% of the irrigated area is concentrated in Asia (Table 19.1). In the densely populated South-east Asia with its monsoon climate (Fig. 15.2), flooded or paddy rice (*Oryza sativa*) serves as the prevailing crop and staple food. In contrast, the combined irrigated areas of Africa, South America and Australia amount to less than 10% of the total irrigated area (Table 19.1).

Given the fact of a steadily increasing world population, expansion of the irrigated area appears to be essential. On the other hand, the present-day irrigated land area is at risk in many parts of the world because of *environmental problems*. These problems are linked to the inadequate quality of the irrigation water, caused by high *sodium and salt* contents. Soils may also be unsuitable for irrigation because of texture, slope and inherent salt content. Those with extremely high salt content are called Solonchak (from Russian *sol* = salt), whereas a Solonetz soil is characterized by the predominance of alkaline sodium. Some other problems are related to the technology of water conduction in canals or to the application and uniform distribution of irrigation water in the field. In addition, the ability to remove excess water and salts can be limited if soil drainage and solute leaching are impaired. Deep aquifers serving as a water source for irrigation may become depleted and dry up. Overuse of groundwater and lowering of water levels have been observed worldwide in major aquifers, such as in the High Plains in Central USA, in Arabia, in the Upper Ganges and the North China Plain (cf. Section 17.3). The short-sighted excessive use of ancient water resources – sometimes called 'groundwater mining' – inevitably leads to the final and sometimes disastrous end of an irrigation-based agriculture. Where groundwater gets depleted under land located close to shorelines, saline sea water can penetrate into the coastal strip, infiltrating the former freshwater resources. When this happens, the freshwater supplies needed for humans are likely to be impaired, not only those used

Table 19.1. Irrigated area by continent as a percentage of total irrigated area (ICID-CIID, 2012).

Continent	Irrigated area (%)
Africa	5
North and Central America	12
South America	3
Asia	72
Australia	1
Europe	7
World	100 (~ 310 × 10^6 ha)

for irrigation but also the resources used for domestic and industrial purposes. It is estimated that currently from *40 to 50 million ha of irrigated land* may already *have been degraded* by *waterlogging, salinization and sodication* (Howell, 2001). Salinization of irrigated soils mainly occurs in dry regions. It is associated with inadequate application of water so that unused fertilizer and salts present in the source water or salts from subsoil horizons accumulate within the rooting zone. Sodication is the build up of exchangeable sodium in the soil, and most frequently results from saline groundwater entering the rooting zone, because irrigation or other change in land use causes a rise in the level of the water table.

Let us imagine conditions where there is the technology and social and natural resources favourable for irrigation management and for allowing sustainable crop production without adverse effects to the environment. Let us further suppose that the yield level is higher than the level attained under rainfed agriculture. Even so the water use efficiency (WUE) may be less with irrigation than it is without. Reasons for this unsatisfactory result may be an abundant but uncontrolled water supply that is associated with high unproductive water losses (Eqn 17.5). Another reason may be the failure to make full use of measures like the application of fertilizer and pesticide that support and maintain crop yields. Having adopted irrigation to improve the yield-limiting conditions associated with water deficiency, additional crop management practices are essential to improve yield and WUE (Tables 12.1 and 12.2).

Hence for the development of a successful, economically and environmentally sound irrigation practice, many aspects have to be considered. These include technological issues such as supply, delivery and distribution of irrigation water, but also those that deal with collection, conduction, re-use and final storage of surplus water which drains through the soil and is often enriched by salts. These engineering aspects are beyond the scope of this book, but some of the pertinent literature is identified in the list of references. However, we will briefly review early irrigation-based civilizations and then concentrate on the principles of irrigation that are linked to crop management and have the aim of increasing yields and WUE.

19.2 Tapping Water – the Basis of Early Civilizations

Irrigated agriculture presents a real contrast to *rainfed agriculture*. Irrigation is the human-based, artificial application of water to the soil for the purpose of facilitating plant growth and the development of yield. Irrigation is both a science and an art, and has a long tradition in many areas of the world. A principal prerequisite of irrigation is of course the availability of water, which can be drawn from *groundwater* or *flowing surface water*. The groundwater must be replenished by precipitation or it will be gradually exhausted, as has happened already to many aquifers in dry regions that were simply a store of ancient water (Section 19.1).

A famous, well-understood technique for using groundwater is the *qanat* system, which was developed by *Persians* in prehistoric times. Even today the system is employed in Iran and in many countries of the Near East and North Africa. It is also still in operation in the Chinese region of Siang-Kiang (Arnon, 1992). The qanat is an artificial underground gallery, which collects the groundwater in alluvial fans at the elevated base of mountains. From the fans the water flows through the aquifer by gravity and emerges at the soil surface some distance, up to several kilometres, from the foot of the mountain. This apparent 'upwelling' is possible because the underground aqueduct is less sloping than the land surface. The horizontal construction of the tunnel was accomplished by excavating from a series of vertical access shafts, sunk in a line but at some distance from each other.

If groundwater or flowing water sources are not available and precipitation is limited and erratic, even this meagre resource can still be used for field crop production. Rainwater from larger areas can be concentrated in smaller areas by collecting runoff water behind dams and storing it in soils, terraces or aquifers. Using this technique it is advantageous to deal with soils that exhibit a small rather than a large infiltrability so as to induce runoff. In arid regions a small infiltrability may be caused by low levels of organic matter, by surface slaking, by surface-exposed stones or by non-wettability of the soil (cf. Section 2.2, banded vegetation and Box 17.1). The technique of '*water harvesting*' has been developed by the Nabatean people in the Negev desert of ancient *Palestine*. Rainwater is diverted from hills and mountains to the valley floors, sometimes in the form of heavy torrents. Here the water is effectively conserved by storage in the loess deposits that fill the valleys. This ancient system enabled a flourishing agriculture in an area with less than 100 mm of annual rainfall as early as a few hundred years BC (Evenari *et al.*, 1961; Evenari, 1987). With increasing costs and even the failure of some irrigation systems, interest in water harvesting has been re-awakened in many semi-arid and arid regions over recent years. In one form or another it is currently practised in Australia, the USA, India and in some countries bordering the Mediterranean Sea (Arnon, 1992).

The presence of *flowing surface water* was of course the most obvious basis for the early development of some famous *advanced civilizations* in various parts of the world. Even earlier than 2000 BC, irrigated rice was already being grown in *China*, in the valleys of the Yangtse Kiang and the Huang-He. Dams and canals provided for the supply and removal of water. The Mohenjo-Daro culture flourished even earlier on the banks of the river Indus in the northern part of the *Indian subcontinent*, and was based on irrigated agriculture. Irrigation also supported progressive civilizations in *Mesopotamia*, in the basin formed between the Euphrates and Tigris rivers, as well as in *Egypt* along the River Nile. Even Egyptian agriculture of the present day is totally dependent on water from the Nile for irrigation. In

Mesopotamia, however, irrigated agriculture went into decline following an enduring golden age because of increasing problems with silting-up and salinization. In the *New World*, too, irrigation systems were developed by indigenous peoples based on river water, for instance along the rivers Gila, Colorado and Rio Grande. In Chihuahua, Mexico, a water control system was developed on the Rio Casas Grande in the middle of the 11th century, enabling irrigation and drainage (Achtnich, 1980; Arnon, 1992). Similar achievements have been associated with the ancient Inca people of Peru.

The presence of water is a natural precondition for irrigated agriculture. Either the system is totally based on water derived from rivers and underground aquifers for irrigation or it relies to a greater or lesser extent on soil water, which will just be topped up by irrigation water in periods when soil water becomes short. Another precondition, the wise and economic application of irrigation water, has to be met when the aim is large yields and high WUE. This precondition also aims at keeping the irrigation system sound for future generations and avoiding environmental hazards.

19.3 Water Requirement of Crops

'By a cruel stroke of nature, the water requirements of crops are greatest in arid and semiarid regions, while the supply by rainfall is least' (Hillel, 1990). It is the large vapour pressure deficit of the air that forces the enormous water consumption by crop stands and the considerable water use per unit of crop yield. The scarcity of water combined with the dryness of the air is the reason why, in these regions, irrigation has to act as a fundamental source of water supply 'to let the desert bear fruit' (Evenari, 1987). In contrast, irrigation takes on the role of merely supplementing the soil water supply in more temperate regions with larger and more reliable precipitation.

The *water requirement* is the unrestricted water use of a crop stand at a location without any shortages or limitations, and includes transpiration and soil evaporation. In other words it is the amount needed for a crop to grow without water stress. Of course the daily water requirement varies during the season. The total water requirement of the crop is determined at the end of the season.

Based on what has been extensively covered in previous chapters, we can say that the water requirement of crops depends on both the *internal and the external conditions of the crop*. The *potential evapotranspiration* (ET_p) is to a large extent an external term. It not only depends on solar radiation but also on the drying effect of the wind and of the vapour pressure deficit of the air. On the other hand, ET_p can be modified by some crop characteristics like stand density, leaf area index (LAI) and plant height, which all contribute to the roughness of the crop (Section 8.2). To overcome all these specific crop peculiarities affecting ET_p, experts of the Food and Agriculture Organization of the United Nations (FAO) proposed to calculate the water requirement of a hypothetical crop as a standard for irrigation purposes. It is the *reference crop evapotranspiration rate* ET_o (Allen *et al.*, 1998), explained in Section 8.2.

The larger the ET_p or ET_o, the greater is the water requirement of the plant stand or the hypothetical crop, respectively. To meet the water requirement, the daily transpirational water use by the crop under consideration must never be limited by water uptake, which could result in severe water stress and stomatal closure. This objective can be attained using irrigation. Meeting water requirements for a large evapotranspiration (ET) through irrigation implies that the water supplied has to be monitored precisely to keep the soil matric potential and soil moisture content constantly at a high level. This level of soil moisture will be greater than one that can be tolerated when ET is small (Figs 8.6 and 8.7).

In addition to external conditions that influence ET_p or ET_o, the *internal characteristics of a particular crop species* also play an important role in governing the water requirement of actual stands in the field. During germination and seedling growth, LAI and transpiration (T) are small. Under drying conditions in these early stages, T (Fig. 9.4) and ET (Figs 9.5 and 9.6) will not meet ET_p or ET_o. Frequent rainfall at this time can, however, stimulate evaporation (E), causing ET to attain the level of ET_p or ET_o (Allen *et al.*, 1998). At later stages with complete *canopy development* (Fig. 9.1), T or ET, respectively, and hence the actual water requirement, will approach ET_o (ET_p) or may even exceed it, as shown for ET_p (Figs 9.4, 9.5 and 9.6) if water extraction is not restricted by inadequate root growth or soil moisture. Finally, towards the end of the season, when the crop matures, ages and dries, the water requirement again falls below ET_o (ET_p) (Section 9.2). Since the water requirement varies during the season, the *cumulative seasonal water requirement* will not match the cumulative ET_o (ET_p), even when the soil water supply to the root system is never restricted during the growing period. In many environments the total water requirement amounts to 60–90% of seasonal ET_p (Hillel, 1990) or ET_o but in humid climates with regular rainfall it can exceed cumulative ET_p (Goss *et al.*, 1984b; Box 8.1). As a value between 60% and 90% is consistent with figures in Section 9.2, we are led to the conclusion that the crops grown in the humid-temperate climates of the UK and Germany did not experience severe, long-lasting periods of water stress.

Broadly speaking, the aim of irrigation is to allow an unrestricted water supply to crops, at least when the primary task is to achieve heavy yields. Therefore the determination of the cumulative water requirement is a first but important step in planning the total amount of irrigation water necessary for a given area. The quantity will depend on the specific field crop, soil and climate, and has to be determined individually for typical sites in the region. Daily water requirement changes during the season with the stage of crop development. More recently the term 'crop water requirement' has been replaced by the simple term *crop evapotranspiration* ET_c (Allen *et al.*, 1998). Numerically both expressions are the same, but the requirement stresses the need for water supply, whereas the evapotranspiration points to soil water loss to the atmosphere, always under the condition that the water supply to roots is not limited. In the following the term crop evapotranspiration ET_c will be used in connection with the reference crop evapotranspiration ET_o. Doing so we have to accept that we have clearly to distinguish (and not to confuse) between three special forms of evapotranspiration, between ET_p, ET_o and ET_c. For clear-cut examples we will still make use of the expression 'water requirement' (Bos *et al.*, 2010) when necessary.

A first step to quantify the *daily crop evapotranspiration* ET_c (or the *daily crop water requirement*) is to determine ET_o. The ET_o can either be measured in the field (in the form of ET_p), taking the actual water use of a well-watered reference crop like grass (Penman, 1948; Section 8.2) or lucerne, or it can be calculated by applying the FAO form of the Penman–Monteith equation. In addition to potential evapotranspiration (ET_p) or preferably to the reference crop evapotranspiration (ET_o) an empirical *crop-specific coefficient* (K_c) is also needed that reflects the changing transpirational water use of the cultivated field crop during the season. This coefficient includes possible soil evaporation losses that will occur mainly during the early period of crop growth. Data on K_c have been evaluated for many important crops (Doorenbos and Pruitt, 1977; Doorenbos and Kassam, 1979) and nowadays are available from extensive tables (Hargreaves and Merkley, 1998; Allen *et al.*, 1998). The idea of tabulating K_c values is that because the coefficients are more or less universal, they are applicable under various site conditions.

The *crop evapotranspiration* can be calculated on the basis of daily values (mm day^{-1}) for individual periods of the season (mm per season) by use of the equation developed by Doorenbos and Pruitt (1977) for crop water requirement (W_{rc}).

$$W_{rc} = K_c \cdot ET_p \tag{19.1}$$

But now we write, according to Allen *et al.* (1998):

$$ET_c = K_c \cdot ET_o \tag{19.2}$$

For example, during initial growth (from seeding to 10% ground cover) K_c varies between 0.20 and 0.40 for wheat and oat and from 0.20 to 0.50 for maize. In this early phase, intermittent rainfall can increase the variation due to Stage 1 evaporation (Fig. 5.7). During rapid growth (the stage of stem elongation and the associated period of peak water use) K_c increases to between 1.05 and 1.20 for the three crops, and late in the season K_c drops to 0.20–0.30 for wheat (0.20–0.25 for oat) and is between 0.35 and 0.60 for maize (Hargreaves and Merkley, 1998).

ET_c cumulated over the season provides a good estimate of the total quantity of water actually transpired and evaporated (and therefore needed) in growing a crop without water shortage. Whenever soil water supply restricts plant growth during the season, either right from the beginning or during the later stages of development, dry matter production and yield formation are likely to be impaired. Under these circumstances seasonal crop ET will not reach seasonal ET_c. This is when irrigation can assist the farmer in his goal of attaining the potential yield per area. That yield is achievable under the given climatic conditions of the area and with best crop management, when water does not limit the production.

19.4 Timing and Adjusting the Application of Water

The seasonal crop evapotranspiration (ET_c) or water requirement fixes the total amount of water necessary for unrestricted growth and yield formation. ET_c will be dependent largely on weather conditions and plant characteristics, and will usually increase with the length of the growing season. The water requirement can be satisfied by the supply of soil water, which depends on rainfall occurring prior to and during the growing season, either alone or in combination with irrigation. Hence the question that needs answering is: 'How much water and when should it be supplied by irrigation to avoid growth restrictions and yield loss?'

Based on the soil–plant–water relations discussed in this book, the following general principles are listed that apply to irrigation management.

Plant considerations

1. The actual need for water is crop specific and depends on the stage of plant development, characterized for instance by LAI.
2. Plants may suffer considerably from water stress causing yield loss when shortage is experienced during stress-sensitive stages of development. Stress reduces the source capacity of stems and leaves for assimilate production. It also reduces the sink capacity of the plant organs important for storing the reserves that may possibly contribute to harvestable yield.
3. Water shortage may be a desirable condition during later stages of development as it may sustain maturation, increase the quality of the harvestable product and restrict the sink capacity of leaves.
4. Slight water shortage early in the season can encourage root growth to depth, thus enhancing the ability of the plant to explore deeper layers for soil water and nutrients.

Climate considerations

5. The larger the ET_o or ET_p, the greater is the need for water. The amount and frequency of irrigation increase with the aridity of the site.
6. The smaller the amount of water stored in the soil profile at the beginning of the season, the greater is the need for irrigation water.

Soil considerations

7. Extraction of soil water by plant roots depends on the 'available field capacity' and the 'effective rooting depth'.
8. The available field capacity is determined mainly by soil characteristics. Nevertheless, at many locations the drained upper limit may be achieved only after the application of irrigation water.
9. The effective rooting depth, too, is primarily dependent on soil properties. But the depth also varies with the crop species and the actual degree of soil water storage.

When should *irrigation start*? The beginning is fixed by the need to avoid severe yield depressions. At which soil water content will crops start to reduce dry matter accumulation and yield? Generally it may be stated that roughly half or even two-thirds (Turner, 1990, 2001) of the extractable water in the rooting zone must be depleted before crops start to react negatively by stomatal closure. This rough limit can change when ET_p varies. With increasing ET_p the limit is shifted to larger water contents and vice versa. For the temperate climate of northern Germany it was found that approximately *50% of the extractable water* in the rooted profile had to be maintained for an unrestricted water supply to roots and for unrestricted evapotranspiration (Fig. 19.2). When this limit is approached, irrigation should start. On sandy soils, however, a slightly higher percentage should be maintained (Fig. 19.2). This is also generally true when growing tender crops such as fruits, strawberries (*Fragaria* × *ananassa*) for example, or vegetables such as lettuce (*Lactuca sativa*) (Roth and Werner, 2000).

The control of irrigation by invoking the idea of extractable water (W_{extr}) is straightforward but nevertheless can give rise to some difficulties. W_{extr} relies on the effective rooting depth ($z_{r\,eff}$). This $z_{r\,eff}$ is the maximum rooting depth of a crop that is effective in the removal of water from the soil profile (Fig. 9.12). The quantity of water that can actually be extracted may be smaller at earlier stages of growth, when the rooting front is not at maximum depth. These reservations become more significant, the greater the values of ET_p and K_c for these early stages and the smaller the available field capacity (FC_{av}) of the soil.

After the rainy or winter season (or irrigation) crops usually start to deplete the filled water storage of the soil when W_{extr} is at maximum. The transpiring crop depletes W_{extr} as a function of time. The remaining quantity of extractable water in the soil profile can be calculated from the daily water balance. The balance

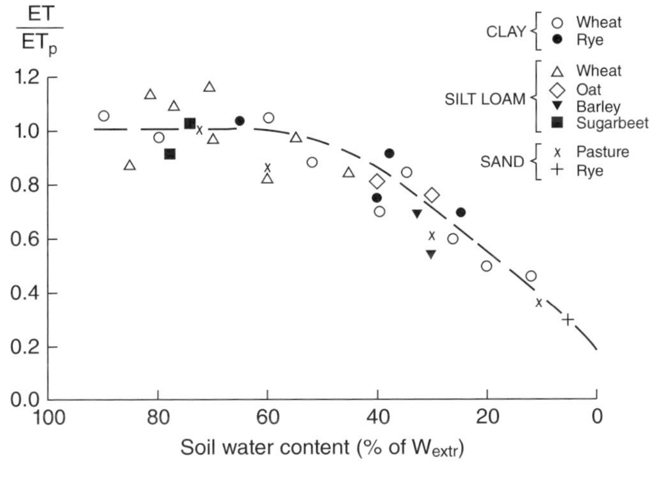

Fig. 19.2. The ratio of actual evapotranspiration rate (ET) to potential evapotranspiration rate (ET_p), known as the relative evapotranspiration, as a function of soil water content. The soil water content, defined as the extractable soil water (W_{extr}), is fixed by the available field capacity and the effective rooting depth. The relationship was obtained with soils of different textural class and with a variety of crops in Lower Saxony, Germany (after Renger *et al.*, 1974).

is based on the record of climatic variables (Section 5.1) and later modified by taking the crop coefficient K_c (Eqn 19.1) into consideration. This *climatic water balance* (CWB) records the daily difference between precipitation (P) and potential evapotranspiration (ET_p) and accumulates the differences over the season:

$$CWB = P - ET_p \tag{19.3}$$

For irrigation control a much better approximation of the quantity of residual soil moisture is obtained by taking the water requirement of a specific crop (W_{rc}) into consideration. Replacing ET_p by W_{rc} results in an equation for determining the *climatic water deficit* (CWD) during growth stages of irrigated crops (Roth and Werner, 2000):

$$\begin{aligned} CWD &= P + IR - W_{rc} \\ &= P + IR - (K_c \cdot ET_p) \end{aligned} \tag{19.4}$$

According to Allen *et al.* (1998) we replace W_{rc} and ET_p to arrive at Eqn 19.5:

$$\begin{aligned} CWD &= P + IR - ET_c \\ &= P + IR - (K_c \cdot ET_o) \end{aligned} \tag{19.5}$$

Before irrigation (IR) starts, P and ET_c are recorded day by day. As long as ET_c is greater than P within a given period, the CWD is negative. A highly negative deficit can be reduced by the addition of rainwater (P) or by application of irrigation (IR). Nowadays, computer programs are available to calculate ET_o or ET_p, select K_c for the growing crop and to balance the amount of extractable water in the rooted profile. In many countries professional agencies have been founded to advise on irrigation management.

In regions where rainfall may top up the soil water content unexpectedly, the advice is not to fill up the soil to 100% of W_{extr} with irrigation water, but to leave it at 80% or so. The goal is always to keep part of the soil reservoir free for the *storage of rainwater*. This is essential for effective transpirational water use and avoiding water loss by deep drainage or surface runoff.

The moment to start irrigation can be ascertained directly by monitoring the soil water content or the soil water potential (Box 5.1). Both *time-domain reflectometry (TDR)* and *tensiometers* can be used to control the beginning and end of water application. Tensiometer systems require at least one tensiometer, or a set of instruments, to be installed with the cup in a soil layer with high root length density and with a large potential water uptake rate (Eqn 9.2). For example the base of the tilled topsoil layer is a suitable location. The cups of a second group of tensiometers are inserted deeper in the soil profile, such as at the lower depth of the rooted zone near the rooting front, where water uptake is smaller (Fig. 19.3). The upper set of tensiometers controls the start of irrigation. That happens when the soil water tension and the tensiometric reading fall below a preset tension value due to root water extraction. When irrigation starts, a wetting zone moves down the soil (Fig. 5.13). The zone passes the installation depth of the upper cups and arrives sometime later at the depth of the lower cups. When at this lower depth

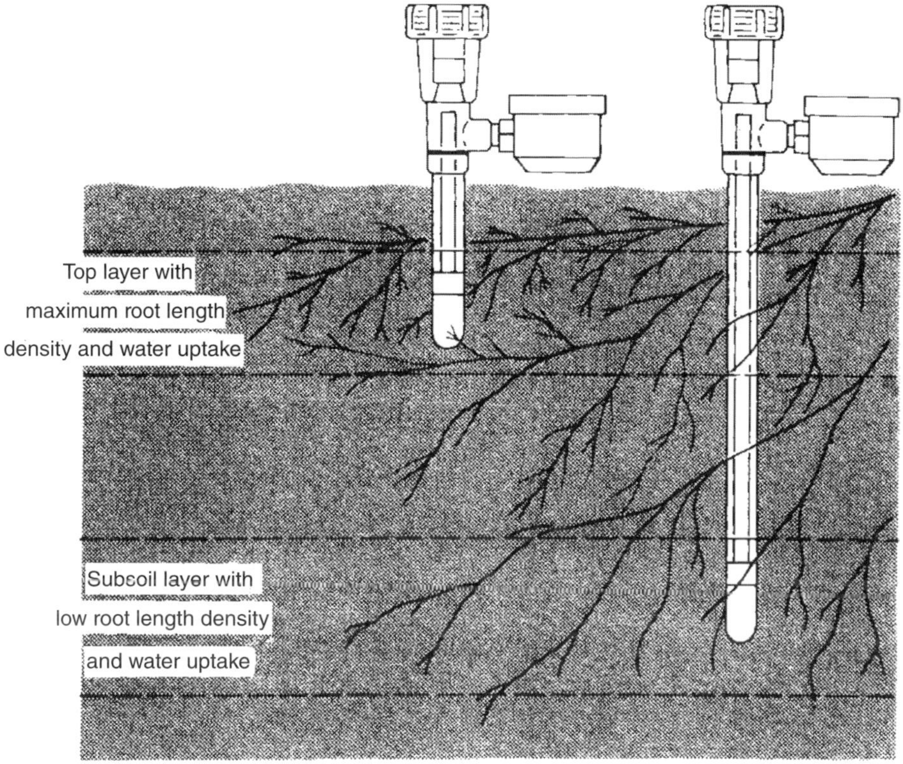

Fig. 19.3. Controlling start and termination of irrigation by use of tensiometers installed at different soil depths (based on Taylor and Ashcroft, 1972).

the tension has dropped to field capacity (FC) (0.1 bar), or better to a slightly higher value (0.2 bar), the water supply is terminated. Using this kind of tensiometric control, large drainage losses and possible losses by leaching of nitrate or of soluble agrochemicals, such as pesticides, can be avoided. Terminating irrigation before attaining FC leaves some extra storage capacity for rainwater.

The time to start irrigation can also be determined by checking the crop plant itself. The *leaf water potential* can be monitored preferably during the early morning hours by use of the pressure chamber apparatus (Section 14.1). Also the *leaf temperature* at noon as compared to air temperature is a good and easy indicator of plant water stress.

The benefit of irrigation depends on climatic, crop and soil variables. A main soil characteristic for determining the additional yield attainable by irrigation is the amount of *extractable water* in the rooted profile (W_{extr}). With winter wheat and sugarbeet, one unit of irrigation water resulted in a steeper yield increase, the less water was stored in the soil in form of W_{extr} (Fig. 19.4). In the Chernozem the attainable yield increase was much less than in the sandy Cambisol. Such a result may be typical where soil water storage contributes to a greater extent in yield development, such as in climates with a rainy period or a fallow season. In dry

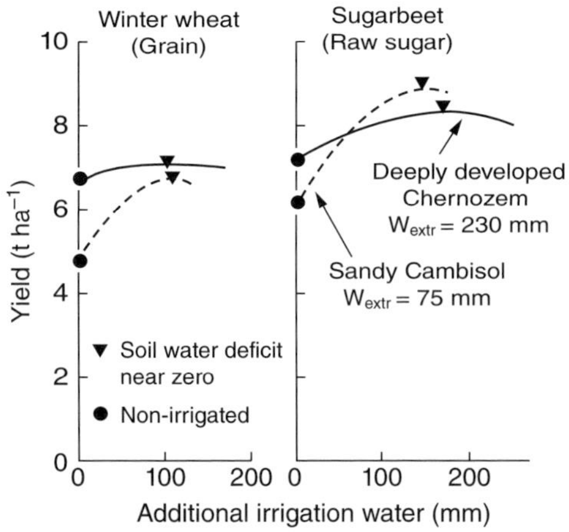

Fig. 19.4. Yield of winter wheat and sugarbeet as a function of the quantity of added irrigation water. The two soils, a Chernozem in the Thuringian Basin and a Cambisol in Mark Brandenburg, Germany, differed in the amount of extractable water (W_{extr}). In one case, irrigation was controlled with the goal of maintaining the climatic water deficit (CWD) of the soil near zero (see triangle symbol). In the other cases, irrigation was settled at 50%, 75%, 125% and 150% of irrigation water given for keeping CWD near zero, thus including dry and very wet conditions. One of the treatments was left non-irrigated at natural rainfall conditions (closed circles). The results are averages obtained over several years (after Roth and Kachel, 1989).

climates the response curves will be more similar in shape as soil water storage contributes less and less to crop growth and yield. But it is also true that irrigation control on more coarse-textured soils relies much more upon fine-tuning than is required on silty soils, rich in W_{extr} like Chernozems.

19.5 Efficient Water Use

The principal aim of irrigation in crop production is to supplement the soil water storage that relies on rainfall. *Supplemental irrigation* will assist farmers in avoiding adverse effects of plant water stress effects on crop yield. These negative effects may be especially harsh if plants suffer from water shortage during periods when particular yield components are being formed (Section 13.3). From this general statement we may conclude that irrigation is most important in areas: (i) where precipitation is limited and unpredictable; (ii) where soils have only stored a small amount of extractable water (W_{extr}); and (iii) where tender crops with high market value are grown. In dry regions, irrigation will be the *main and most important source* of water supply to crops for yield formation.

Supplemental irrigation minimizes growth restrictions due to water shortage. Optimizing one production factor (here water) calls for larger expenditures on some other factors, which become yield limiting. These inputs include fertilizer (Fig. 12.1), herbicide, fungicide and insecticide. Also a change to a high-yielding cultivar with greater leaf area, a longer period of growth after anthesis and increased harvest index might be a necessary consequence, if improvement in all the growth factors is to be exploited to the full. In combination with an *intensified crop management*, irrigation can assist in increasing yield and *water use efficiency* (WUE) in the same way (Fig. 19.5).

After a high level of grain yield has been attained, WUE does not increase any further but tends to level off (Fig. 19.5). The Kastanozem, a Pullman clay loam, retains a large quantity of extractable water (Musick *et al.*, 1994). W_{extr} is about 250 mm of water (cf. Table 9.2). Therefore one might speculate that beyond a yield level of 8 t ha^{-1} the grain-related WUE should decrease again because of increased unproductive water losses (Eqn 17.5). Also one could speculate that with a soil that stores less W_{extr} than the Kastanozem in Texas, the decrease in WUE would occur at a comparatively smaller yield, especially when applying flood irrigation with an irrigation depth of about 100 mm as was the case in Texas. With flooding the losses due to drainage might have been much greater in soils of small W_{extr}.

In dry areas water is scarce and therefore the irrigation water is valuable and costly. When the price for water increases as the availability of water decreases, a farmer may find it more profitable to strive for a high WUE than for large yield per unit area of land (Fig. 19.5). In any case he will avoid a very large yield associated with a decreased WUE resulting from marked water losses (Eqn 17.5) to the

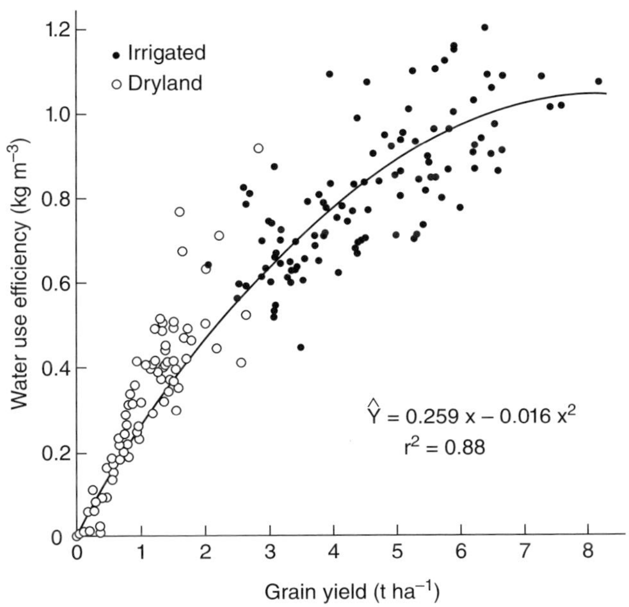

Fig. 19.5. Grain yield (12.5% moisture) and grain-related water use efficiency of irrigated and non-irrigated (dryland) winter wheat, grown on a Kastanozem in the semi-arid southern High Plains, Texas, USA. The irrigated wheat was 'adequately' fertilized with nitrogen and phosphorus. The dryland wheat received no fertilizer (after Musick *et al.*, 1994).

environment. A large WUE consistent with environmental objectives might become the most important goal in irrigated agriculture. To make this assessment, the yield (either of biomass or harvestable product) attainable by irrigation (Y_i) is compared with the dryland yield (Y_d). The difference in yield, $Y_i - Y_d$, is related to the amount of irrigation water (IR) expended in addition to soil water use (Eqn 19.6):

$$IR_{WUE} = \frac{Y_i - Y_d}{IR} \qquad (19.6)$$

Water use efficiency of irrigation water (IR_{WUE}) will be greater than WUE of dryland crops (Table 19.2) when irrigation water application avoids crop water stress, increases the yield and reduces the contribution of unproductive water loss from evaporation, runoff and drainage in total water use. As a result, a greater proportion of the soil water supply will be available for transpiration. In other words, irrigation must not contribute greatly to water loss but has to be applied according to crop water requirements. Nevertheless, at large irrigation rates, IR_{WUE} will be reduced (Table 19.2). Then the question arises whether the rare and expensive irrigation water should be applied in full to a single field (318 mm in Table 19.2) or should be distributed between two fields, each receiving about half of the water for the full irrigation rate (152 mm in Table 19.2). In the first situation the farmer can harvest 10.48 t maize grain from a fully irrigated field of, say, 1 ha size, plus 4.20 t grain from 1 ha of non-irrigated land. That makes a total harvest of 14.68 t grain from 2 ha. In the second situation the 2-ha area receiving limited irrigation will produce 15.44 t maize grain (2 × 7.72 t). From this example we can appreciate that *deficit irrigation* may make better use of a unit of irrigation water than full irrigation.

Of course, the outcome of experiments like this one from North Carolina will change with soil and climate. In the case where rainfall has contributed to a comparatively large value of W_{extr}, the total amount of irrigation water that is at the farmer's disposal must be reduced or can be distributed over a larger area. Otherwise a yield reduction can be induced (Fig. 19.4), possibly due to insufficient aeration. Moreover, unproductive water losses will increase. When, however, soil water supply is limited because of a 'dry year', the concentration of the irrigation

Table 19.2. Effect of irrigation regime on maize grain yield, water use efficiency (WUE) and water use efficiency of irrigation water (IR_{WUE}) in North Carolina, USA. The soil is a clay loam, an Acrisol. Sprinkler irrigation was scheduled using tensiometers, installed at 30 cm depth. The irrigation rate was 8 mm h^{-1}. The data present averages over 4 years for the ploughed treatment (based on Wagger and Cassel, 1993).

Irrigation treatment	Irrigation water (mm)	Total water use (mm)	Grain yield (t ha^{-1})	WUE (kg m^{-3})	IR_{WUE} (kg m^{-3})
Non-irrigated	–	412	4.20	1.02	–
Limited[a]	152	563	7.72	1.37	2.32
Full[b]	318	735	10.48	1.43	1.97

[a]Irrigation from 2 weeks prior to tasselling to 2 weeks after silking.
[b]Irrigation from 5 weeks after planting to plant maturity.

water to a smaller area can be advantageous to avoid severe stress and crop failure (cf. Section 18.1, response farming). Weather statistics on the probability of rainfall affecting the likelihood of a deficit in the climatic water balance record may be helpful in making a decision on how much irrigation to apply per unit area (Renger *et al.*, 1974). Usually the likelihood that within a season a soil water deficit is reached and exceeded will decrease with the severity of the deficit. Smaller deficits occur more often than severe deficits. The goal to be pursued is how to distribute a given amount of irrigation water and receive the greatest benefit in yield production over the years with seasons that vary in precipitation and potential evapotranspiration.

We mentioned the term 'deficit irrigation' and will explain the term more fully, as it has become an important aspect in irrigation strategy (Fereres and Soriano, 2007; Geerts and Raes, 2009). In dry areas, where crop husbandry depends primarily on irrigation, all endeavours are aimed at transforming each drop of water into economic yield. When irrigation water supply per area land is insufficient to satisfy ET_c (Eqn 19.2), neither the potential yield nor the maximum WUE will likely be attained. The farmer has to cultivate crops under water-limited conditions and this calls for compromise in the application of water. In supplemental irrigation, practised in more humid regions, the main goal is to sustain yield at a level attainable without water shortage. In deficit irrigation, applied in more arid regions, the goal is to apply a limited quantity of water to the crop, accepting a yield depression, but as small as possible. All measures of crop husbandry have to be adjusted to save water (Chapters 17 and 18) and to apply it efficiently. In particular, attention has to be directed to the variable sensitivity of crops during different growth stages (Section 13.3). Water stress is therefore allowed in periods when the expected yield depression is tolerable.

19.6 Irrigation Methods

Whatever irrigation method is employed, water should always be applied according to actual *plant water needs*. Water should never be applied to a prescribed schedule and volume, which may end in water delivery rotations fixed by sequence and time from field to field and farmer to farmer, often settled by archaic laws. Water distribution and application on *command* does not meet the crop water requirement, wastes water by unproductive losses like evaporation, runoff and deep drainage and may threaten the environment and crop production. For example, it can lead to salt accumulation. Restricting irrigation to periods of soil water deficits and plant water needs is an important step into the direction of economizing on water use in crop production.

Old methods of irrigation apply the water on the soil surface by flooding. These methods are simple and do not require sophisticated techniques. But this carries an inherent danger of large, uncontrolled water loss. Nowadays, modern methods are available that minimize unproductive water loss. They were developed with the goal of feeding the irrigation water directly to the plant root system. They require a special technique and an acquired skill of handling the system. But they make full use of irrigation water and at the same time reduce environmental problems. These modern techniques can therefore be applied beneficially in regions where: (i) water is rare and expensive; (ii) saline soils have to be cropped; (iii) irrigation water is salty; or (iv) the crops produced have a fairly good market price.

Surface irrigation 'feeds' the soil in total, but modern methods 'nourish' individual crop plants.

The most important methods are summarized in Table 19.3 and will be briefly described below.

Surface methods are still widespread and common in many countries, not just in developing countries. Surface methods account for more than 95% of the irrigated land worldwide (Arnon, 1992) and still for 51% of irrigation in the USA (Howell, 2001). In Europe, this approach makes almost no significant contribution (Roth and Werner, 2000). Hillel (cited by Arnon, 1992) defined surface irrigation as 'the process of introducing a stream of water at the head of a field and allowing gravity and hydrostatic pressure to spread the flow over the surface throughout the field'.

For a uniform application of water over the soil surface, a distribution system consisting of ditches and pipelines is required. The discharge from the supply system is controlled by devices that often include outflow meters and pressure regulators. The soil surface must be levelled (today by laser technique) but with a slight uniform slope to direct the spread of water over the field. As it spreads, the water infiltrates into the soil. Water ponds when the infiltrability is low. The challenge for all surface methods is to distribute the water evenly over the field and thereby achieve a more-or-less spatially uniform wetting of the soil profile.

The simplest method is *flooding*. The water is conveyed to the field in ditches or channels, where it is released to flood the soil downslope 'in a somewhat haphazard manner' (Kruse *et al.*, 1990). In some cases large fields get flooded from several ditches established along contour lines.

In *border strip irrigation* the area to be flooded is divided into smaller strips, bordered by dykes parallel to the slope. Often close-growing crops like lucerne, pasture or cereal are seeded in close rows aligned across the slope. The row direction facilitates the spread of water over the width of the bordered field. The width of the strips is adapted to the working width of the machinery. Application rates may be up to 100 mm (1000 m^3 ha^{-1}).

Basin irrigation means that water is applied to level basins enclosed by dykes. It is an old irrigation system used in ancient Egypt and Mesopotamia to flood the plain adjacent to the big rivers. The basin is filled quickly with water that will pond. Ponding assists in a spatially even infiltration into the soil. There is no runoff because of the dyke enclosure. Therefore soils of low infiltrability get uniformly

Table 19.3. Overview of the most important irrigation methods.

Surface methods	Sub-irrigation	Sprinkler irrigation	Micro-irrigation
Controlled flooding	Tile drains	Static sprinkler	Drip or trickle
Border strip irrigation		Linear-move	Sub-surface drip
Basin irrigation		Centre pivot	
Furrow irrigation		Rain gun	

wetted. For row crops, level furrows are constructed. These are supplied by surface water from channels bordering each side of the furrows. This system is called *basin-furrow irrigation*.

Furrow irrigation is used for annual row crops, for trees and some vine crops (Kruse *et al.*, 1990), grown on slightly sloping land. When the water supply rate cannot meet the requirements for an even distribution of water in bordered strip fields, the water flow can still be sufficient to irrigate furrows uniformly. The furrows are filled at the upper end of the field, for instance by siphon tubes. The slope of the furrows has to be adjusted to the steepness of the terrain. The steeper the slope of the terrain, the more adjustment is needed to let the furrows follow the contour with a gradient of 0.2–0.6% (Arnon, 1992). Nevertheless, as the furrows are kept free of vegetation, some erosion may take place. At the end of some furrows containing soil of low infiltrability, water may accumulate which then gets lost partly by evaporation.

It is inherent to surface irrigation systems that, in addition to drainage and runoff losses, water is lost directly by evaporation. In *sub-irrigation*, however, evaporative losses are small. Plants are supplied by water from a water table, the level of which is controlled by regulating the water level in ditches or tile drain lines. The plants extract the water they need from the capillary fringe of the water table. Sub-irrigation is not very widespread. It is a mode used to irrigate organic soils in areas of high precipitation with intermittent dry periods. The ditches then often serve the double purpose of water supply and water drainage. Sub-irrigation has also been investigated as a method of reducing the loss of fertilizer nitrogen in areas with hot summers but having a high risk of leaching in autumn and spring (Tan *et al.*, 1999).

Sprinkler or overhead irrigation mimics water application by natural rainfall in the form of drops or droplets spraying through the air. For sprinkler use, water has to be pressurized. The application rate is normally greater than the water uptake rate by crops. Therefore most of the irrigated water is stored in the soil. Another consequence is that irrigation water is not applied continuously but at discrete time intervals. From this it follows that the sprinklers, either with or without the conveyance pipes and the couplings, have to be transported from field to field unless they are permanently installed, but only used as needed. Fully portable or movable systems need less investment costs related to the total irrigated area, but operating costs are higher because of the demand for human labour.

Static sprinkler systems are used mainly in gardens or orchards, for instance in the form of swing-arm sprinklers. *Portable systems* are much more common for field crop production. Simple systems are lines of sprinklers that are moved by hand. *Mobile systems* are self-propelled and move continuously over the field. They include lateral or *linear-move* sprinklers that roll slowly over the entire rectangular field, propelled by water pressure. In North America the *centre pivot* system is in widespread use. The sprinkler line is up to 500 m long and rotates in a circle continuously around the centre pivot point, where the line gets filled with water. One rotation takes from 1 to 4 days. The discharge rate from the sprinkler line is proportional to the rate of travel: less in the innermost part of the irrigated field than at the outer edge (Arnon, 1992). This way the application rate (1 l m^{-2} soil surface ≡ 1 mm) is maintained throughout the circular field. The system is more or less self-contained and automatically driven. The system has been modified for water application in form of drops at low pressure – the *low-energy precision-application* system (LEPA).

In some European countries like Germany, *mobile rain guns* have been introduced to irrigate various crops including cereals. The guns require water at high pressure for emitting the jet that is distributed to both sides by a slewing pipe. At the same time as water is applied, the flexible hose is rolled up on to a drum, which pulls the gun from one side of the field to the other. Self-propulsion is achieved through the water pressure. The application rate of the gun is greater and the application precision within the irrigated area is less than with other sprinkler systems. After one passage of the field the gun has to be moved into a new position and adjusted according to the width of the trajectory. The width can be up to 50–60 m. Taking account of the overlapping of the jet spray from one passage to the next, the irrigated width extends out to 100 m for a field that can be several hundred metres long.

Water application depth is usually less with overhead than with surface irrigation methods. Basically, surface application of water requires some levelling of the land, the accuracy depending on the method. Sprinkler irrigation methods, however, allow some undulation of the land surface. These features apply even more to *micro-irrigation* systems. With these systems water is supplied at low pressure and more or less continuously to the plant and its root system. The rate is very much directed towards the water requirement of crops. Micro-irrigation systems are built from lateral plastic lines extending over the field. These are equipped with emitters that discharge the water at low rates. In trickle or *drip irrigation*, water is emitted at the soil surface, in principle to serve individual plants. Water release from tubes on top of the soil facilitates maintenance of the equipment. Regular inspection seems to be needed to prevent clogging of hoses and emitters. Notwithstanding these requirements, in modern *sub-surface drip irrigation systems*, hoses are buried below the soil surface and plants are supplied with water from sub-surface emitters, greatly reducing evaporative loss. The discharge rates are similar to those from drip irrigation. Both systems of micro-irrigation are used in vineyards, orchards and row crops. These systems are extremely effective for water saving. They were developed from the necessity to deal with the extreme shortage of water in the Arava desert of southern Israel (Arnon, 1992).

As water is supplied directly to the plant according to its daily requirement, and as the soil between plants can be left more or less unirrigated, losses by evaporation or runoff are small and losses by deep drainage can be nil. At the same time, leaching beyond the root zone of chemicals like nitrate is effectively controlled (Cassel Sharmasarkar *et al.*, 2001). Soil properties like texture, available and extractable water (Fig. 19.4) are of minor importance with this method of immediate water supply (Bresler, 1977). As the soil gets wetted in spots near the plants and not over the whole area, few weed plants germinate and weed competition is small. Whenever necessary, fertilizers and herbicides can be mixed into the irrigation water. The combination of micro-irrigation and fertilizer has been referred to as 'fertigation' (Bar-Yosef, 1999). As the soil in the immediate vicinity of the plant is kept at a high matric potential, some salt content of irrigation water can be tolerated – the allowable solute concentration depending on the crop. Nevertheless, the occasional leaching out of salts from the soil might be necessary. Generally, salinization by deep percolation of irrigation water and by capillary rise afterwards is not a problem with micro-irrigation.

20 Epilogue

For the vast majority of human beings life depends on crop production and livestock farming. Both these rely on the availability of productive soils and suitable sources of water. The provision of fertile soil and supplies of good quality water to meet the needs of an increasing world population is directly endangered by human activity. They are further threatened by climate change, as is food production. We have seen that even relatively small differences in the distribution of rain events can result in feast on the one hand or famine on the other. Both the inability to predict the weather over a season and the consequences of changes in climate make it difficult to identify the best course of action to ensure stable yields. The complexity of the global environment has made people blind to all the consequences of their daily activities. None the less, the more science enhances our knowledge and understanding the more certain we become that interfering with the natural balance adversely affects the terrestrial ecosystem. And the earth is certainly the basis for human existence and its prospects for the future. In this book we have shown how a sound understanding of the interactions between plants, soils and climate can lead to practices that enhance agricultural productivity and optimize the effective use of resources.

We conclude with a poem by an unknown author. The poem symbolizes humility, love and commitment to the natural environment and all its creatures, without any sense of personal advantage but for the welfare of mankind.

Earth Pledge

I pledge allegiance to the earth,
and all its sacred parts;
its water, land, and living things,
and all its human hearts.
I pledge allegiance to all life.
and promise I shall care.
to love and cherish all its gifts.
with people everywhere.

References

Abdul-Ghani, M. (1979) Estimation of daily soil heat flux during the spring season in a perennial forage crop from climatological data. MSc dissertation, University of Guelph, Ontario, Canada.

Abdul-Jabbar, A.S., Lugg, D.G., Sammis, T.W. and Gay, L.W. (1985) Relationships between crop water stress index and alfalfa yield and evapotranspiration. *Transactions of the American Society of Agricultural Engineers* 28, 454–461.

Acevedo Hinojosa, E. (1975) The growth of maize (*Zea mays* L.) under field conditions as affected by its water relations. Dissertation, University of California, Davis, California. University Microfilms International, Ann Arbor, Michigan.

Achtnich, W. (1980) *Bewässerungslandbau.* Ulmer, Stuttgart, Germany.

Ahuja, L.R., Barnes, B.B. and Rojas, K.W. (1993) Characterization of macropore transport studied with the ARS root zone water quality model. *Transactions of the American Society of Agricultural Engineers* 36, 369–380.

Ahuja, L.R., Reddy, V.R., Saseendran, S.R. and Yu, Q. (eds) (2008) *Response of Crops to Limited Water: Understanding and Modeling Water Stress Effects on Plant Growth Processes.* American Society of Agronomy, Madison, Wisconsin.

Aiken, R.M., O'Brien, D.M., Olson, B.L. and Murray, L. (2013) Replacing fallow with continuous cropping reduces crop water productivity of semiarid wheat. *Agronomy Journal* 105, 199–207.

Ainsworth, E.A. and Long, S.P. (2005) What have we learned from 15 years of free-air CO_2 enrichment (FACE)? A meta-analytic review of the responses of photosynthesis, canopy properties and plant production to rising CO_2. *New Phytologist* 165, 351–372.

Al-Khafaf, S., Wierenga, P.J. and Williams, B.C. (1978) Evaporative flux from irrigated cotton as related to leaf area index, soil water, and evaporative demand. *Agronomy Journal* 70, 912–917.

Allen, R.G., Pereira, L.S., Raes, D. and Smith, M. (1998) Crop Evapotranspiration: Guidelines for Computing Water Requirements. Food and Agriculture Organization of the United Nations (FAO) Irrigation and Drainage Paper 56. FAO, Rome.

Andreae, B. (1977) *Agrargeographie. Strukturzonen und Betriebsformen in der Weltlandwirtschaft.* Walter de Gruyter, Berlin.

Andreini, M.S., Parlange, J.-Y. and Steenhuis, T.S. (1990) A numerical model for preferential solute movement in structured soils. *Geoderma* 46, 193–208.

Angers, D.A. and Eriksen-Hamel, N.S. (2008) Full-inversion tillage and organic carbon distribution in soil profiles: a meta-analysis. *Soil Science Society of America Journal* 72, 1370–1374.

Angus, J.F. (2001) Nitrogen supply and demand in Australian agriculture. *Australian Journal of Experimental Agriculture* 41, 277–288.

Angus, J.F. and van Herwaarden, A.F. (2001) Increasing water use and water use efficiency in dryland wheat. *Agronomy Journal* 93, 290–298.

Arbeitsgruppe Bodenkunde (1982) *Bodenkundliche Kartieranleitung*, 3rd edn. Schweizerbart'sche Verlagsbuchhandung, Stuttgart, Germany.

Arnon, I. (1992) *Agriculture in Dry Lands. Principles and Practice*. Elsevier, Amsterdam.

Askenasy, E. (1895) Über das Saftsteigen. *Botanisches Centralblatt* 62, 237–238.

Asrar, G., Hipps, L.E. and Kanemasu, E.T. (1984) Assessing solar energy and water use efficiencies in winter wheat: a case study. *Agricultural and Forest Meteorology* 31, 47–58.

Asseng, S., Ritchie, J.T., Smucker, A.M.J. and Robertson, M.J. (1998) Root growth and water uptake during water deficit and recovering in wheat. *Plant and Soil* 201, 265–273.

Austin, R.B. (1999) Yield of wheat in the United Kingdom: recent advances and prospects. *Crop Science* 39, 1604–1610.

Austin, R.B., Ford, M.A. and Morgan, C.L. (1989) Genetic improvement in the yield of winter wheat: a further evaluation. *Journal of Agricultural Science, Cambridge* 112, 295–301.

Babé, A., Lavigne, T., Séverin, J.-P., Nagel, K.A., Walter, A., Chaumont, F., Batoko, H., Beeckman, T. and Draye, X. (2012) Repression of early lateral root initiation events by transient water deficit in barley and maize. *Philosophical Transactions of the Royal Society B* 367, 1534–1541.

Baeumer, K. (1992) *Allgemeiner Pflanzenbau*, 3rd edn. Ulmer, Stuttgart, Germany.

Bakker, B., van der Ploeg, M.J., de Rooij, G.H., Hoogendam, C.W., Gooren, H.P.A., Huiskes, C., Koopal, L.K. and Kruidhof, H. (2007) New polymer tensiometers: measuring matric pressures down to the wilting point. *Vadose Zone Journal* 6, 196–202.

Bange, G.G.J. (1953) On the quantitative explanation of stomatal transpiration. *Acta Botanica Neerlandia* 2, 255–296.

Barber, S.A. (1995) *Soil Nutrient Bioavailability: a Mechanistic Approach*, 2nd edn. Wiley, New York.

Barley, K.P. (1954) Effects of root growth and decay on the permeability of a synthetic sandy loam. *Soil Science* 78, 205–210.

Bar-Yosef, B. (1999) Advances in fertigation. *Advances in Agronomy* 65, 1–77.

Basch, G., Kassam, A., Friedrich, T., Santos, F.L., Gubiani, P.I., Calegari, A., Reichert, J.M. and dos Santos, D.R. (2012) Sustainable soil water management systems. In: Lal, R. and Stewart, B.A. (eds) *Soil Water and Agronomic Productivity*, Advances in Soil Science. CRC Press, Boca Raton, Florida, pp. 229–287.

van Bavel, C.H.M. (1966) Potential evaporation: the combination concept and its experimental verification. *Water Resources Research* 2, 455–467.

Baver, L.D., Gardner, W.H. and Gardner, W.R. (1972) *Soil Physics*, 4th edn. Wiley, New York.

Beale, C.V., Morison, J.I.L. and Long, S.P. (1999) Water use efficiency of C_4 perennial grasses in a temperate climate. *Agricultural and Forest Meteorology* 96, 103–115.

Beese, F., Horton, R. and Wierenga, P.J. (1982) Growth and yield response of chile pepper to trickle irrigation. *Agronomy Journal* 74, 556–561.

Begon, M., Harper, J.L. and Townsend, C.R. (1991) *Ökologie – Individuen*, Populationen und Lebensgemeinschaften. Birkhäuser, Basel, Switzerland.

Benecke, P. and van der Ploeg, R.R. (1976) Quantifizierung des zeitlichen Verhaltens der Wasserhaushaltskomponenten eines Buchen-und eines Fichtenaltholzbestandes im Solling mit Hilfe bodenhydrologischer Methoden. *Verhandlungen der Gesellschaft für Ökologie* 5, 3–16.

Benjamin, J.G., Mikha, M., Nielsen, D.C., Vigil, M.F., Calderón, F. and Henry, W.B. (2007) Cropping intensity effects on physical properties of a no-till silt loam. *Soil Science Society of America Journal* 71, 1160–1165.

Bennie, A.T.P. and Botha, F.J.P. (1986) Effect of deep tillage and controlled traffic on root growth, water-use efficiency and yield of irrigated maize and wheat. *Soil and Tillage Research* 7, 85–95.

Berisso, F.E., Schjønning, P., Keller, T., Lamandé, M., Etana, A., de Jonge, L.W., Iversen, B.V., Arvidsson, J. and Forkman, J. (2012) Persistent effects of subsoil compaction on pore size distribution and gas transport in a loamy soil. *Soil and Tillage Research* 122, 42–51.

Bettex, A. (1960) *Welten der Entdecker*. Droemer-Knaur, München.

Beven, K. and Germann, P. (1982) Macropores and water flow in soils. *Water Resources Research* 18, 1311–1325.

Beven, K. and Germann, P. (2013) Macropores and water flow in soils revisited. *Water Resources Research* 49, 3071–3092.

Bhalerao, R.P., Eklöf, J., Ljung, K., Marchant, A., Bennett, M.J. and Sandberg, G. (2002) Shoot-derived auxin is essential for early lateral root emergence in *Arabidopsis* seedlings. *The Plant Journal* 29, 325–332.

Bierhuizen, J.F. and Slatyer, R.O. (1965) Effect of atmospheric concentration of water vapour and CO_2 in determining transpiration–photosynthesis relationships of cotton leaves. *Agricultural Meteorology* 2, 259–270.

Blanke, M. (1998) CHIPs und TIPs – ein Übersichtsreferat über Water Channel. *Gartenbauwissenschaft* 63, 133–137.

Blanke, M.M. and Bower, J.P. (1991) Small fruit problem in *Citrus* trees. *Trees* 5, 239–243.

Blum, A. (2011) *Plant Breeding for Water-Limited Environments*. Springer, New York.

Blüthgen, J. (1966) *Allgemeine Klimageographie*, 2nd edn. Walter de Gruyter, Berlin.

Böhm, J. (1889) Ursache des Saftsteigens. *Berichte der Deutschen Botanischen Gesellschaft* 7, 46–56.

Böhm, J. (1893) Capillarität and Saftsteigen. *Berichte der Deutschen Botanischen Gesellschaft* 11, 203–212.

Böhm, W. (1974) Mini-rhizotrons for root observations under field conditions. *Zeitschrift für Acker- und Pflanzenbau* 140, 282–287.

Böhm, W. (1978) Untersuchungen zur Wurzelentwicklung bei Winterweizen. *Zeitschrift für Acker- und Pflanzenbau* 147, 264–269.

Böhm, W. (1979) *Methods of Studying Root Systems*. Ecological Studies, 33. Springer, Berlin.

Böhm, W. (1997) *Biographisches Handbuch zur Geschichte des Pflanzenbaus*. K.G. Saur, München.

Böhm, W. and Köpke, U. (1977) Comparative root investigations with two profile wall methods. *Zeitschrift für Acker- und Pflanzenbau* 144, 297–303.

Bond, J.J. and Willis, W.O. (1969) Soil water evaporation: surface residue rate and placement effects. *Soil Science Society of America Proceedings* 33, 445–448.

Boone, F.R. (1988) Weather and other environmental factors influencing crop responses to tillage and traffic. *Soil and Tillage Research* 11, 283–324.

Boote, K.J. and Ketring, D.L. (1990) Peanut. In: Stewart, B.A. and Nielsen, D.R. (eds) *Irrigation of Agricultural Crops*. Agronomy Series No. 30. American Society of Agronomy, Madison, Wisconsin, pp. 675–717.

Borel, C., Frey, A., Marion-Poll, A., Tardieu, F. and Simonneau, T. (2001) Does engineering abscisic acid biosynthesis in *Nicotiana plumbaginifolia* modify stomatal response to drought? *Plant, Cell and Environment* 24, 477–489.

Bos, M.G., Kselik, R.A.L., Allen, R.G. and Molden, D.J. (2010) *Water Requirements for Irrigation and the Environment*. Springer, Dordrecht, The Netherlands.

Bouma, J. (1981) Soil morphology and preferential flow along macropores. *Agricultural Water Management* 3, 235–250.

Bouyoucos, G.J. and Mick, A.H. (1940) An electrical resistance method for the continuous measurement of soil moisture under field conditions. Technical Bulletin No. 172. Michigan Agricultural Experimental Station, East Lansing, Michigan.

Brady, N.C. (1990) *The Nature and Properties of Soils*, 10th edn. Macmillan, New York.

Bresler, E. (1977) Trickle-drip irrigation: principles and application to soil-water management. *Advances in Agronomy* 29, 343–393.

Briggs, L.J. and Shantz, H.L. (1913a) The water requirements of plants. I. Investigations in the Great Plains in 1910 and 1911. United States Department of Agriculture, Bureau of Plant Industry, Bulletin No. 284, Washington, DC.

Briggs, L.J. and Shantz, H.L. (1913b) The water requirements of plants. II. A review of literature. United States Department of Agriculture, Bureau of Plant Industry, Bulletin No. 285, Washington, DC.

Briggs, L.J. and Shantz, H.L. (1914) Relative water requirements of plants. *Journal of Agricultural Research* 3, 1–65.

Briggs, L.J. and Shantz, H.L. (1917) A comparison of the hourly evaporation rate of atmometers and free water surfaces with the transpiration rate of *Medicago sativa*. *Journal of Agricultural Research* 9, 277–292.

Brito, I., Carvalho, M., Alho, L. and Goss, M.J. (2014) Managing arbuscular mycorrhizal fungi for bioprotection: Mn toxicity. *Soil Biology and Biochemistry* 68, 78–84.

Brown, R.H. and Byrd, G.T. (1996) Transpiration efficiency, specific leaf weight, and mineral concentration in peanut and pearl millet. *Crop Science* 36, 475–480.

Brown, S.C., Keatinge, J.D.H., Gregory, P.J. and Cooper, P.J.M. (1987) Effects of fertilizer, variety and location on barley production under rainfed conditions in Northern Syria. 1. Root and shoot growth. *Field Crops Research* 16, 53–66.

Buckland, S.T., Campbell, C.D., Mackie-Dawson, L.A., Horgan, G.W. and Duff, E.I. (1993) A method for counting roots observed in minirhizotrons and their theoretical conversion to root length density. *Plant and Soil* 153, 1–9.

Bürcky, K. (1991) Einfluß des Stickstoffangebotes auf Substanzbildung, Erntegewicht und Qualität der Zuckerrübe unter besonderer Berücksichtigung von Temperatur und Globalstrahlung in Gefäßversuchen. *Journal of Agronomy and Crop Science* 167, 341–349.

Bürcky, K. (1993) Einfluß des Stickstoffangebotes auf Wasserverbrauch und Stoffproduktion der Zuckerrübe in Gefäßversuchen. *Journal of Agronomy and Crop Science* 171, 153–160.

Byrd, G.T. and May, P.A. II (2000) Physiological comparisons of switchgrass cultivars differing in transpiration efficiency. *Crop Science* 40, 1271–1277.

Caldwell, M.M. and Richards, J.H. (1989) Hydraulic lift: water efflux from upper roots improves effectiveness of water uptake by deep roots. *Oecologia* 79, 1–5.

Caldwell, M.M., Dawson, T.E. and Richards, J.H. (1998) Hydraulic lift: consequences of water efflux from roots of plants. *Oecologia* 113, 151–161.

Campbell, G.S. (1985) *Soil Physics with BASIC.* Elsevier, Amsterdam.

Carminati, A., Moradi, A., Vetterlein, D., Vontobel, P., Lehmann, E., Weller, U., Vogel, H.-J. and Oswald, S.E. (2010) Dynamics of soil water content in the rhizosphere. *Plant and Soil* 332, 163–176.

Carminati, A., Schneider, C.L., Moradi, A.B., Zarebanadkouki, M., Vetterlein, D., Vogel, H.-J., Hildebrandt, A., Weller, U., Schüler, L. and Oswald, S.E. (2011) How the rhizosphere may favor water availability to roots. *Vadose Zone Journal* 10, 988–998.

Cassel Sharmasarkar, F., Sharmasarkar, S., Miller, S.D., Vance, G.F. and Zhang, R. (2001) Assessment of drip and flood irrigation on water and fertilizer efficiencies for sugarbeets. *Agricultural Water Management* 46, 241–251.

Castiglioni, P., Warner, D., Bensen, R.J., Anstrom, D.C., Harrison, J., Stoecker, M., Abad, M., Kumar, G., Salvador, S., D'Ordine, R., Navarro, S., Back, S., Fernandes, M., Jayaprakash Targolli, J., Dasgupta, S., Bonin, C., Luethy, M.H. and Heard, J.E. (2008) Bacterial RNA chaperones confer abiotic stress tolerance in plants and improved grain yield in maize under water-limited conditions. *Plant Physiology* 147, 446–455.

Chandler, P.M., Munns, R. and Robertson, M. (1993) Regulation of dehydrin expression. In: Close, T.J. and Bray, E.A. (eds) *Plant Responses to Cellular Dehydration during Environmental Stress.* American Society of Plant Physiologists, Rockville, Maryland, pp. 159–166.

Chaudhuri, U.N., Kirkham, M.B. and Kanemasu, E.T. (1990) Carbon dioxide and water level effects on yield and water use of winter wheat. *Agronomy Journal* 82, 637–641.

Claassen, M.M. and Shaw, R.H. (1970a) Water deficit effects on corn. I. Vegetative components. *Agronomy Journal* 62, 649–652.

Claassen, M.M. and Shaw, R.H. (1970b) Water deficit effects on corn. II. Grain components. *Agronomy Journal* 62, 652–655.

Claupein, W. (1993) Stickstoffdüngung und chemischer Pflanzenschutz in einem Dauerfeldversuch und die Ertragsgesetze von Liebig, Liebscher, Wollny and Mitscherlich. *Journal of Agronomy and Crop Science* 171, 102–113.

Cleland, E.E., Chuine, I., Menzel, A., Mooney, H.A. and Schwartz, M.D. (2007) Shifting plant phenology in response to global change. *Trends in Ecology and Evolution* 22, 357–365.

Close, T.J. (1996) Dehidrins: emergence of a biochemical role of a family of plant dehydration proteins. *Physiologia Plantarum* 97, 795–803.

Cohen, Y., Fuchs, M. and Green, G.C. (1981) Improvement of the heat pulse method for determining sap flow in trees. *Plant, Cell and Environment* 4, 391–397.

Cole, J.S. and Mathews, O.R. (1923) Use of water by spring wheat on the Great Plains. United States Department of Agriculture, Bureau of Plant Industry, Bulletin No. 1004, Washington, DC.

Comfort, S.D., Malzner, G.L. and Busch, R.H. (1988) Nitrogen fertilization of spring wheat genotypes: influence on root growth and soil water depletion. *Agronomy Journal* 80, 114–120.

Condon, A.G., Richards, R.A. and Farquhar, G.D. (1987) Carbon isotope discrimination is positively correlated with grain yield and dry matter production in field-grown wheat. *Crop Science* 27, 996–1001.

Condon, A.G., Richards, R.A., Rebetzke, G.J. and Farquhar, G.D. (2002) Improving intrinsic water use efficiency and crop yield. *Crop Science* 42, 122–131.

Condon, A.G., Richards, R.A., Rebetzke, G.J. and Farquhar, G.D. (2004) Breeding for high water-use efficiency. *Journal of Experimental Botany* 55, 2447–2460.

Cooper, P.J.M. (1983) Crop management in rainfed agriculture with special reference to water use efficiency. In: International Potash Institute (ed.) *Nutrient Balances and the Need for Fertilizers in Semi-arid and Arid Regions.* Proceedings of the 17th Colloquium of the International Potassium Institute in Rabat and Marrakech, Morocco, pp. 63–79.

Cooper, P.J.M., Gregory, P.J., Keatinge, J.D.H. and Brown, S.C. (1987) Effects of fertilizer, variety and location on barley production under rainfed conditions in northern Syria. 2. Soil water dynamics and crop water use. *Field Crops Research* 16, 67–84.

Cowan, I.R. (1982) Regulation of water use in relation to carbon gain in higher plants. In: Lange, O.L., Nobel, P.S., Osmond, C.B. and Ziegler, H. (eds) *Physiological Plant Ecology II.* Springer, Berlin, pp. 589–614.

Cramer, M.D., Hawkins, H.-J. and Verboom, G.A. (2009) The importance of nutritional regulation of plant water flux. *Oecologia* 161, 15–24.

Dalton, J. (1802a) Experiments and observations to determine whether the quantity of rain and dew is equal to the quantity of water carried off by the rivers and raised by evaporation; with an enquiry into the origin of springs. *Memoirs of the Proceedings of the Literary and Philosophical Society of Manchester* 5, *Part* 2, 346–372.

Dalton, J. (1802b) Experimental essays on the constitution of mixed gases; on the force of steam or vapor from water and other liquids in different temperatures, both in a Torricellian vacuum and in air; on evaporation and on the expansion of gases by heat. *Memoirs of the Proceedings of the Literary and Philosophical Society of Manchester* 5, *Part* 2, 535–602.

Danneberg, G., Latus, C., Zimmer, W., Hundeshagen, B., Schneider-Poetsch, H.J. and Bothe, H. (1992) Influence of vesicular-arbuscular mycorrhiza on phytohormone balances in maize (*Zea mays* L.). *Journal of Plant Physiology* 141, 33–39.

Davies, W.J. and Zhang, J. (1991) Root signals and the regulation of growth and development of plants in drying soil. *Annual Review of Plant Physiology and Plant Molecular Biology* 42, 55–76.

Denmead, O.T. and Shaw, R.H. (1962) Availability of soil water to plants as affected by soil moisture content and meteorological conditions. *Agronomy Journal* 54, 385–390.

Dixon, H.H. and Joly, J. (1895) The path of the transpiration current. *Annals of Botany (London)* 9, 416–419.
Dixon, R.M. and Peterson, A.E. (1971) Water infiltration control: a channel system concept. *Soil Science Society of America Proceedings* 35, 968–973.
Donald, C.M. and Hamblin, J. (1976) The biological yield and harvest index of cereals as agronomic and plant breeding criteria. *Advances in Agronomy* 28, 361–405.
Doorenbos, J. and Kassam, A.H. (1979) Yield Response to Water. Food and Agriculture Organization of the United Nations (FAO) Irrigation and Drainage Paper 33. FAO, Rome.
Doorenbos, J. and Pruitt, W.O. (1977) Crop Water Requirements. Food and Agriculture Organization of the United Nations (FAO) Irrigation and Drainage Paper 24. FAO, Rome.
Drüge, U. and Schönbeck, F. (1992) Effect of vesicular-arbuscular mycorrhizal infection on transpiration, photosynthesis and growth of flax (*Linum usitatissimum* L.) in relation to cytokinin levels. *Journal of Plant Physiology* 141, 40–48.
van Duin, R.H.A. (1955) Tillage in relation to rainfall intensity and infiltration capacity of soils. *Netherlands Journal of Agricultural Science* 3, 182–191.
Duynisveld, W.H.M. and Strebel, O. (1983) *Entwicklung von Simulationsmodellen für den Transport von gelösten Stoffen in wasserungesättigten Böden und Lockersedimenten.* Texte 17. Umweltbundesamt, Berlin.
Eastin, J.D., Dickinson, T.E., Krieg, D.R. and Maunder, A.B. (1983) Crop physiology in dryland agriculture. In: Dregne, H.E. and Willis, W.O. (eds) *Dryland Agriculture.* Agronomy Series No. 23. American Society of Agronomy, Madison, Wisconsin, pp. 333–364.
Eberbach, P. and Pala, M. (2005) Crop row spacing and its influence on the partitioning of evapotranspiration by winter-grown wheat in Northern Syria. *Plant and Soil* 268, 195–208.
Eck, H.V. (1988) Winter wheat response to nitrogen and irrigation. *Agronomy Journal* 80, 902–908.
Edmeades, G.O. (2012) Progress in achieving and delivering drought tolerance in maize – an update. In: James, C. (ed.) *Global Status of Commercialized Biotech/GM Crops: 2012. ISAAA Brief No.* 44. International Service for the Acquisition of Agri-Biotech Applications, Ithaca, New York, pp. 239–272.
Ehlers, W. (1975a) Einfluß von Wassergehalt, Struktur und Wurzeldichte auf die Wasseraufnahme von Weizen auf Löß-Parabraunerde. *Mitteilungen der Deutschen Bodenkundlichen Gesellschaft* 22, 141–156.
Ehlers, W. (1975b) Observations on earthworm channels and infiltration on tilled and untilled loess soil. *Soil Science* 119, 242–249.
Ehlers, W. (1976) Evapotranspiration and drainage in tilled and untilled loess soil with winter wheat and sugarbeet. *Zeitschrift für Acker- und Pflanzenbau* 142, 285–303.
Ehlers, W. (1977) Measurement and calculation of hydraulic conductivity in horizons of tilled and untilled loess-derived soil, Germany. *Geoderma* 19, 293–306.
Ehlers, W. (1978) Wassergehalts- und Wasserspannungsmessungen im Felde zur Bilanzierung des Bodenwasserhaushalts. *Mitteilungen der Deutschen Bodenkundlichen Gesellschaft* 26, 115–132.
Ehlers, W. (1986) Warum sind Ackerbohnen so dürreempfindlich? *Mitteilungen der Deutschen Landwirtschaftsgesellschaft, DLG* 101, 558–567.
Ehlers, W. (1989) Transpiration efficiency of oat. *Agronomy Journal* 81, 810–817.
Ehlers, W. (1991) Leaf area and transpiration efficiency during different growth stages in oats. *Journal of Agricultural Science, Cambridge* 116, 183–190.
Ehlers, W. (1993) Kann Bodenbearbeitung zum Schutz des Bodens vor Verdichtung und Erosion beitragen? *Mitteilungen der Deutschen Bodenkundlichen Gesellschaft* 72, 1447–1450.
Ehlers, W. (1997) Zum Transpirationskoeffizienten von Kulturpflanzen unter Feldbedingungen. *Pflanzenbauwissenschaften* 1, 97–108.
Ehlers, W. and Claupein, W. (1994) Approaches towards conservation tillage in Germany. In: Carter, M.R. (ed.) *Conservation Tillage in Temperate Agroecosystems.* Lewis Publishers, Boca Raton, Florida, pp. 141–165.

Ehlers, W. and van der Ploeg, R.R. (1976) Evaporation, drainage and unsaturated hydraulic conductivity of tilled and untilled fallow soil. *Zeitschrift für Pflanzenernährung und Bodenkunde* 139, 373–386.

Ehlers, W., Khosla, B.K., Köpke, U., Stülpnagel, R., Böhm, W. and Baeumer, K. (1980a) Tillage effects on root development, water uptake and growth of oats. *Soil and Tillage Research* 1, 19–34.

Ehlers, W., Edwards, W.M. and van der Ploeg, R.R. (1980b) Runoff-controlling hydraulic properties of erosion susceptible grey-brown podzolic soils in Germany. In: de Boodt, M. and Gabriels, D. (eds) *Assessment of Erosion*. Wiley, Chichester, UK, pp. 381–391.

Ehlers, W., Köpke, U., Hesse, F. and Böhm, W. (1983) Penetration resistance and root growth of oats in tilled and untilled loess soil. *Soil and Tillage Research* 3, 261–275.

Ehlers, W., Goss, M.J. and Boone, F.R. (1986a) Einfluß der Bodenbearbeitung auf Bodenwasserhaushalt, Durchwurzelung und Wasserentzug. *Kali-Briefe (Büntehof)* 18, 107–125.

Ehlers, W., Müller, U. and Grimme, K. (1986b) Untersuchungen zum Wasserhaushalt der Ackerbohne II. Wasserpotential, stomatäre Leitfähigkeit und Ertragsbildung. *Kali-Briefe (Büntehof)* 18, 189–210.

Ehlers, W., Goss, M.J. and Boone, F.R. (1987) Tillage effects on soil moisture, root development and crop water extraction. *Plant Research and Development* 25, 92–110.

Ehlers, W., Hamblin, A.P., Tennant, D. and van der Ploeg, R.R. (1991) Root system parameters determining water uptake of field crops. *Irrigation Science* 12, 115–124.

van Eimern, J. and Häckel, H. (1979) *Wetter- und Klimakunde*, 3rd edn. Ulmer, Stuttgart, Germany.

Ellis, F.B. and Barnes, B.T. (1980) Growth and development of root systems of winter cereals grown after different tillage methods including direct drilling. *Plant and Soil* 55, 283–295.

El-Swaify, S.A., Pathak, P., Rego, T.J. and Singh, S. (1985) Soil management for optimized productivity under rainfed conditions in the semi-arid tropics. *Advances in Soil Science* 1, 1–64.

Enquete-Kommission 'Schutz der Erdatmosphäre' of the German Bundestag (1994) *Schutz der Grünen Erde. Klimaschutz durch umweltgerechte Landwirtschaft und Erhalt der Wälder*. Economica Verlag, Bonn, Germany.

Enquete-Kommission 'Schutz der Erdatmosphäre' of the German Bundestag (1995) *Mehr Zukunft für die Erde – Nachhaltige Energiepolitik für dauerhaften Klimaschutz*. Economia Verlag, Bonn, Germany.

Evenari, M. (1987) *Und die Wüste trage Frucht*. Ein Lebensbericht. Bleicher, Gerlingen, Germany.

Evenari, M., Shanan, L., Tadmor, M. and Aharoni, Y. (1961) Ancient agriculture in the Negev. *Science* 133, 976–996.

FAO (Food and Agriculture Organization of the United Nations) (2010) Statistical Yearbook 2010. Available at: http://www.fao.org/docrep/015/am081m/am081m00.htm (accessed 26 June 2014).

FAOSTAT (2014) FAOSTAT Database. Available at: http://faostat3.fao.org/faostat-gateway/go/to/download/G1/GT/E (accessed 26 June 2014).

Farahani, H.J., Peterson, G.A. and Westfall, D.G. (1998) Dryland cropping intensification: a fundamental solution to efficient use of precipitation. *Advances in Agronomy* 64, 197–223.

Farquhar, G.D. and Richards, R.A. (1984) Isotopic composition of plant carbon correlates with water-use efficiency of wheat genotypes. *Australian Journal of Plant Physiology* 11, 539–552.

Feddes, R.A., Kowalik, P.J. and Zaradny, H. (1978) *Simulation of Field Water Use and Crop Yield*. Wageningen Centre for Agricultural Publishing and Documentation, Wageningen, The Netherlands.

Federal Office of Statistics (Statistisches Bundesamt) (1983–1992) *Land- und Forstwirtschaft, Fischerei*. Fachserie 3, Reihe 3. *Landwirtschaftliche Bodennutzung und pflanzliche Erzeugung*. W. Kohlhammer, Stuttgart; ab 1988 Metzler-Poeschel, Stuttgart, Germany.

Fenner, S. (1995) Wirkung und Nachhaltigkeit mechanischer Lockerung von Krumenbasisverdichtungen unter Wendepflug- und Mulchwirtschaft. PhD dissertation, University of Göttingen, Cuvillier, Göttingen, Germany.

Fereres, E. and Soriano, M.A. (2007) Deficit irrigation for reducing agricultural water use. *Journal of Experimental Botany* 58, 147–159.

Fischer, R.A. (2007) Understanding the physiological basis of yield potential in wheat. *Journal of Agricultural Science* 145, 99–113.

Fitter, A.H. (1985) Functioning of vesicular-arbuscular mycorrhizas under field conditions. *New Phytologist* 99, 257–265.

Fitter, A.H. and Hay, R.K.M. (1981) *Environmental Physiology of Plants*. Academic Press, London.

Flindt, R. (2000) *Biologie in Zahlen*, 5th edn. Spektrum, Heidelberg.

Föhse, D., Claassen, N. and Jungk, A. (1991) Phosphorus efficiency of plants. II. Significance of root radius, root hairs and cation–anion balance for phosphorus influx in seven plant species. *Plant and Soil* 132, 261–272.

Frederick, J.R. and Camberato, J.J. (1995a) Water and nitrogen effects on winter wheat in the southeastern coastal plain. I. Grain yield and kernel traits. *Agronomy Journal* 87, 521–526.

Frederick, J.R. and Camberato, J.J. (1995b) Water and nitrogen effects on winter wheat in the southeastern coastal plain. II. Physiological responses. *Agronomy Journal* 87, 527–533.

French, R.J. and Schultz, J.E. (1984a) Water use efficiency in wheat in a Mediterranean-type environment. I. The relation between yield, water use and climate. *Australian Journal of Agricultural Research* 35, 743–764.

French, R.J. and Schultz, J.E. (1984b) Water use efficiency in wheat in a Mediterranean-type environment. II. Some limitations to efficiency. *Australian Journal of Agricultural Research* 35, 765–775.

French, B.K., Long, I.F. and Penman, H.L. (1973a) Water use by farm crops I. Test of the neutron meter on barley, beans and sugar beet, 1970. *Rothamsted Experimental Station, Report for 1972, Part 2*. Lawes Agricultural Trust, Harpenden, UK, pp. 5–42.

French, B.K., Long, I.F. and Penman, H.L. (1973b) Water use by farm crops II. Spring wheat, barley, potatoes (1969); potatoes, beans, kale (1968). *Rothamsted Experimental Station, Report for 1972, Part 2*. Lawes Agricultural Trust, Harpenden, UK, pp. 43–61.

French, B.K., Long, I.F. and Penman, H.L. (1973c) Water use by farm crops III. Bare soil, short turf and crops in rotation, 1962 to 1967, 1971. *Rothamsted Experimental Station, Report for 1972, Part 2*. Lawes Agricultural Trust, Harpenden, UK, pp. 62–85.

Gaiser, T., Perkons, U., Küpper, P.M., Puschmann, D.U., Peth, S., Kautz, T., Pfeifer, J., Ewert, F., Horn, R. and Köpke, U. (2012) Evidence of improved water uptake from subsoil by spring wheat following lucerne in a temperate humid climate. *Field Crops Research* 126, 56–62.

Gall, H., Burck, M., Klitsch, S. and Zachow, B. (1994) Ergebnisse der Lysimetermessungen in Groß-Lüsewitz aus pflanzenbaulicher Sicht. *Wasser und Boden* 46, 64–68.

Gallagher, J.N. and Biscoe, P.V. (1978) Radiation absorption, growth and yield of cereals. *Journal of Agricultural Science, Cambridge* 91, 47–60.

Gardner, F.P., Pearce, R.B. and Mitchell, R.L. (1985) *Physiology of Crop Plants*. The Iowa State University Press, Ames, Iowa.

Gardner, W.R. (1960) Dynamic aspects of water availability to plants. *Soil Science* 89, 63–73.

Gäth, S., Meuser, H., Abitz, C.-A., Wessolek, G. and Renger, M. (1989) Determination of potassium delivery to the roots of cereals plants. *Zeitschrift für Pflanzenernährung und Bodenkunde* 152, 143–149.

Geerts, S. and Raes, D. (2009) Deficit irrigation as an on-farm strategy to maximize crop water productivity in dry areas. *Agricultural Water Management* 96, 1275–1284.

German Weather Service (Deutscher Wetterdienst) (1982–1998) *Monatlicher Witterungsbericht 30–46*. Zentralamt, Offenbach, Germany.

German Weather Service (Deutscher Wetterdienst) (1999–2000) *Witterungsreport 1–2*. Zentralamt, Offenbach, Germany.

Gerwitz, A. and Page, E.R. (1974) An empirical mathematical model to describe plant root systems. *Journal of Applied Ecology* 11, 773–782.

Gholipoor, M., Prasad, P.V.V., Mutava, R.N. and Sinclair, T.R. (2010) Genetic variability of transpiration response to vapor pressure deficit among sorghum genotypes. *Field Crops Research* 119, 85–90.

Gill, K.S. (1994) Efficient plant types. In: Virmani, S.M., Katyal, J.C., Esweran, H. and Abrol, I.P. (eds) *Stressed Ecosystems and Sustainable Agriculture*. Oxford and IBH Publishing Co., New Delhi, pp. 307–332.

Gimeno, V., Fernandez-Martinez, J.M. and Fereres, E. (1989) Winter planting as a means of drought escape in sunflower. *Field Crops Research* 22, 307–316.

Gliemeroth, G. (1951) Der Einfluß von Düngung auf den Wasserentzug der Pflanzen aus den Unterbodentiefen. *Zeitschrift für Pflanzenernährung, Düngung und Bodenkunde* 52, 21–41.

Goss, M.J. (1991) Consequences of the effects of roots on soil. In: Atkinson, D. (ed.) *Plant Root Growth – an Ecological Perspective*. Blackwell Scientific Publications, Oxford, pp. 171–186.

Goss, M.J. and Ehlers, W. (2009) The role of lysimeters in the development of our understanding of water dynamics in ecosystems. *Soil Use and Management* 25, 213–223.

Goss, M.J. and Kay, B.D. (2005) Soil aggregation. In: Wright, S.F. and Zobel, R.W. (eds) *Roots and Soil Management: Interactions between Root and the Soil*. Agronomy Monograph No. 48. American Society of Agronomy, Crop Science Society of America and Soil Science Society of America, Madison, Wisconsin, pp. 163–180.

Goss, M.J. and Russell, R.S. (1980) Effects of mechanical impedance on root growth in barley (*Hordeum vulgare* L.) III. Mechanism of the response of roots to mechanical stress. *Journal of Experimental Botany* 31, 577–588.

Goss, M.J., Harris, G.L. and Howse, K.R. (1983) Functioning of mole drains in a clay soil. *Agricultural Water Management* 6, 27–30.

Goss, M.J., Boone, F.R., Ehlers, W., White, I. and Howse, K.R. (1984a) Effects of soil management practice on soil physical conditions affecting root growth. *Journal of Agricultural Engineering Research* 30, 131–140.

Goss, M.J., Howse, K.R., Vaughan-Williams, J.M., Ward, M.A. and Jenkins, W. (1984b) Water use by winter wheat as affected by soil management. *Journal of Agricultural Science, Cambridge* 103, 189–199.

Goss, M.J., Ehlers, W. and Unc, A. (2010) The role of lysimeters in the development of our understanding of processes in the vadose zone relevant to contamination of groundwater aquifers. *Physics and Chemistry of the Earth* 35, 913–926.

Gowing, D.J., Davies, W.J. and Jones, H.G. (1990) A positive root-sourced signal as an indicator of soil drying in apple, *Malus × domestica* Borkh. *Journal of Experimental Botany* 41, 1535–1540.

Grable, A.T. (1966) Soil aeration and plant growth. *Advances in Agronomy* 18, 58–106.

Grashoff, C. (1990a) Effect of pattern of water supply on *Vicia faba* L. 1. Dry matter partitioning and yield variability. *Netherlands Journal of Agricultural Science* 38, 21–44.

Grashoff, C. (1990b) Effect of pattern of water supply on *Vicia faba* L. 2. Pod retention and filling, and dry matter partitioning, production and water use. *Netherlands Journal of Agricultural Science* 38, 131–143.

Grashoff, C. and Verkerke, D.R. (1991) Effect of pattern of water supply on *Vicia faba* L. 3. Plant water relations, expansive growth and stomatal reactions. *Netherlands Journal of Agricultural Science* 39, 247–262.

Greacen, E.L. and Oh, J.S. (1972) Physics of root growth. *Nature (London) New Biology* 235, 24–25.

Greb, B.W. (1970) Deep plowing a shallow clay layer to increase soil water storage and crop yields. Colorado State University Experiment Station, Publication No. PR-70-23, pp. 1–2.

Greb, B.W., Smika, D.E. and Welsh, J.R. (1979) Technology and wheat yields in the Central Great Plains: experiment station advances. *Journal of Soil and Water Conservation* 34, 264–268.

Greef, J.M., Hansen, F., Pasda, G. and Diepenbrock, W. (1993) Die Strahlungs-, Energie- und Kohlendioxidbindung landwirtschaftlicher Kulturpflanzen. *Berichte über Landwirtschaft* 71, 554–566.

Green, S.R. and Clothier, B.E. (1988) Water use of kiwifruit vines and apple trees by the heat-pulse technique. *Journal of Experimental Botany* 39, 115–123.

Green, S., Clothier, B. and Jardine, B. (2003) Theory and practical application of heat pulse to measure sap flow. *Agronomy Journal* 95, 1371–1379.

Gregory, P.J. (1989a) Water-use efficiency of crops in the semi-arid tropics. In: International Crops Research Institute for the Semi-Arid-Tropics (ICRISAT) (ed.) *Soil, Crop and Water Management*

Systems for Rainfed Agriculture in the Sudano-Sahelian Zone. Proceedings of an International Workshop, 7–11 January 1987. ICRISAT Sahelian Center, Niamey, Niger, pp. 85–98.

Gregory, P.J. (1989b) The role of root characteristics in moderating the effects of drought. In: Baker, F.W.G. (ed.) *Drought Resistance in Cereals.* CAB International, Wallingford, UK, pp. 141–150.

Gregory, P.J., Simmonds, L.P. and Pilbeam, C.J. (2000) Soil type, climatic regime, and the response of water use efficiency to crop management. *Agronomy Journal* 92, 814–820.

Guan, Y. and Fredlund, D.G. (1997) Use of the tensile strength of water for the direct measurement of high soil suction. *Canadian Geotechnical Journal* 34, 604–614.

Guerlac, H. (1972) Hales, Stephen. In: Gillispie, C.C. (ed.) *Dictionary of Scientific Biography*, vol. VI. Charles Scribner's Sons, New York.

Haefele, S.M., Siopongco, J.D.L.C., Boling, A.A., Bouman, B.A.M. and Tuong, T.P. (2009) Transpiration efficiency of rice (*Oriza sativa* L.). *Field Crops Research* 111, 1–10.

Hainsworth, J.M. and Aylmore, L.A.G. (1986) Water extraction by single plant roots. *Soil Science Society of America Journal* 50, 841–848.

Håkansson, I., Voorhees, W.B., Elonen, P., Raghavan, G.S.V., Lowery, B., van Wijk, A.L.M., Rasmussen, K. and Riley, H. (1987) Effect of high axle-load traffic on subsoil compaction and crop yield in humid regions with annual freezing. *Soil and Tillage Research* 10, 259–268.

Hales, S. (1727) *Vegetable Staticks.* W. and J. Innis and T. Woodward, London. Reprint in History of Science Library (1961), Macdonald, London.

Hamblin, A.P. (1981) Filter-paper method for routine measurement of field water potential. *Journal of Hydrology* 53, 355–360.

Hamblin, A.P., Tennant, D. and Perry, M.W. (1990) The cost of stress: dry matter partitioning changes with seasonal supply of water and nitrogen to dryland wheat. *Plant and Soil* 122, 47–58.

Hamza, M.A. and Anderson, W.K. (2005) Soil compaction in cropping systems. A review of the nature, causes and possible solutions. *Soil and Tillage Research* 82, 121–145.

Hanks, R.J. (1974) Model for predicting plant yield as influenced by water use. *Agronomy Journal* 66, 660–665.

Hanks, R.J. and Ashcroft, G.L. (1980) *Applied Soil Physics*. Springer, Berlin.

Hanks, R.J. and Rasmussen, V.P. (1982) Predicting crop production as related to plant water stress. *Advances in Agronomy* 35,193–215.

Hanks, R.J. and Ritchie, J.T. (eds) (1991) *Modeling Soil and Plant Systems*. American Society of Agronomy, Madison, Wisconsin.

Hanks, R.J., Gardner, H.R. and Florian, R.L. (1968) Evapotranspiration–climate relations for several crops in the Central Great Plains. *Agronomy Journal* 60, 538–542.

Hansen, N.C., Allen, B.L., Baumhardt, R.L. and Lyon, D.J. (2012) Research achievements and adoption of no-till, dryland cropping in the semi-arid US Great Plains. *Field Crops Research* 132, 196–203.

Hargreaves, G.H. and Merkley, G.P. (1998) *Irrigation Fundamentals*. Water Resources Publications, LLC, Highlands Ranch, Colorado.

Harris, P.J. (1988) Microbial transformation of nitrogen. In: Wild, A. (ed.) *Russell's Soil Conditions and Plant Growth*, 11th edn. Wiley, New York, pp. 608–651.

Hartge, K.H. and Horn, R. (1991) *Einführung in die Bodenphysik*, 2nd edn. Ferdinand Enke, Stuttgart, Germany.

Hatfield, J.L. (1990) Methods of estimating evapotranspiration. In: Stewart, B.A. and Nielsen, D.R. (eds) *Irrigation of Agricultural Crops*. American Society of Agronomy, Madison, Wisconsin, pp. 435–474.

Hatfield, J.L., Boote, K.J., Kimball, B.A., Ziska, L.H., Izaurralde, R.C., Ort, D., Thomson, A.M. and Wolfe, D. (2011) Climate impacts on agriculture: implications for crop production. *Agronomy Journal* 103, 351–370.

Hatfield, W.A. (1988) Water and anion movement in a Typic Hapludult. PhD dissertation, Clemson University, Clemson, South Carolina.

Hattendorf, M.J., Carlson, R.E., Halim, R.A. and Buxton, D.R. (1988) Crop water stress index and yield of water-deficit-stressed alfalfa. *Agronomy Journal* 80, 871–875.

Helling, C.S. and Gish, T.J. (1991) Physical and chemical processes affecting preferential flow. In: Gish, T.J. and Shrimohammadi, A. (eds) *Preferential Flow*. American Society of Agricultural Engineers, St Joseph, Michigan, pp. 77–86.

Herkelrath, W.N., Miller, E.E. and Gardner, W.R. (1977) Water uptake by plants: I. Divided root experiments. *Soil Science Society of America Journal* 41, 1033–1038.

van Herwaarden, A.F., Farquhar, G.D., Angus, J.F., Richards, R.A. and Howe, G.N. (1998) 'Haying-off', the negative grain yield response of dryland wheat to nitrogen fertilizer. I. Biomass, grain yield, and water use. *Australian Journal of Agricultural Research* 49, 1067–1081.

Heßland, F. (1993) Pflanzenbau im Trockengebiet. Anpassungsstrategien bei Sortenwahl, Saatdichte, Düngung und Pflanzenschutz im Zuckerrübenanbau. *Zuckerrübe* 42, 269–272.

Hillel, D. (1971) *Soil and Water. Physical Principles and Processes*. Academic Press, New York.

Hillel, D. (1980) *Applications of Soil Physics*. Academic Press, New York.

Hillel, D. (1990) Role of irrigation in agricultural systems. In: Stewart, B.A. and Nielsen, D.R. (eds) *Irrigation of Agricultural Crops*. Agronomy Series No. 30. American Society of Agronomy, Madison, Wisconsin, pp. 5–30.

Hillel, D. (1998) *Environmental Soil Physics*. Academic Press, San Diego, California.

Hillel, D. (2011) An overview of soil and water management: the challenge of enhancing productivity and sustainability. In: Hatfield, J.L. and Sauer, T.J. (eds) *Soil Management*. Building a Stable Base for Agriculture. American Society of Agronomy, Madison, Wisconsin, pp. 3–11.

Hillel, D. and van Bavel, C.H.M. (1976) Simulation of profile water storage as related to soil hydraulic properties. *Soil Science Society of America Journal* 40, 807–815.

Hillel, D., Krentos, V.D. and Styllianou, Y. (1972) Procedure and test of an internal drainage method for measuring soil hydraulic characteristics *in situ*. *Soil Science* 114, 395–400.

Hillel, D., van Beek, C. and Talpaz, H. (1975) A microscopic-scale model of soil water uptake and salt movement to plant roots. *Soil Science* 120, 385–399.

Hillel, D., Talpaz, H. and van Keulen, H. (1976) A macroscopic-scale model of water uptake by a non-uniform root system and of water and salt movement in the soil profile. *Soil Science* 121, 242–255.

Hlavinka, P., Trnka, M., Semerádová, D., Dubrovský, M., Žalud, Z. and Možný, M. (2009) Effect of drought on yield variability of key crops in Czech Republic. *Agricultural and Forest Meteorology* 149, 431–442.

Horgan, G.W., Buckland, S.T. and Mackie-Dawson, L.A. (1993) Estimating three-dimensional line process densities from tube counts. *Biometrics* 49, 899–906.

Horn, R., van den Akker, J.J.H. and Arvidsson, J. (eds) (2000) *Subsoil Compaction*. Advances in Geoecology 32. Catena, Reiskirchen, Germany.

Horton, R., Beese, F. and Wierenga, P.J. (1982) Physiological response of chile pepper to trickle irrigation. *Agronomy Journal* 74, 551–555.

Howell, T.A. (1990) Relationships between crop production and transpiration, evapotranspiration, and irrigation. In: Stewart, B.A. and Nielsen, D.R. (eds) *Irrigation of Agricultural Crops*. Agronomy Series No. 30. American Society of Agronomy, Madison, Wisconsin, pp. 391–434.

Howell, T.A. (2001) Enhancing water use efficiency in irrigated agriculture. *Agronomy Journal* 93, 281–289.

Howell, T.A., Schneider, A.D., Dusek, D.A., Marek, T.H. and Steiner, J.L. (1995) Calibration and scale performance of Bushland weighing lysimeters. *Transactions of the American Society of Agricultural Engineers* 38, 1019–1024.

Hsiao, T.C. (1990) Measurement of plant water status. In: Stewart, B.A. and Nielsen, D.R. (eds) *Irrigation of Agricultural Crops*. Agronomy Series No. 30. American Society of Agronomy, Madison, Wisconsin, pp. 243–279.

Huber, B. (1932) Beobachtung und Messung pflanzlicher Saftströme. *Berichte der Deutschen Botanischen Gesellschaft* 50, 89–109.

Huber, B. (1956) *Die Saftströme der Pflanzen*. Springer, Berlin.

Hubick, K.T. and Farquhar, G.D. (1989) Carbon isotope discrimination and the ratio of carbon gained to water lost in barley cultivars. *Plant Cell and Environment* 12, 795–804.

Hunsaker, D.J., Kimball, B.A., Pinter, P.J., Wall, G.W., LaMorte, R.L., Adamsen, F.J., Leavitt, S.W., Thompson, T.L., Matthias, A.D. and Brooks, T.J. (2000) CO_2 enrichment and soil nitrogen effects on wheat evapotranspiration and water use efficiency. *Agricultural and Forest Meteorology* 104, 85–105.

Hüsken, D., Steudle, E. and Zimmermann, U. (1978) Pressure probe technique for measuring water relations of cells in higher plants. *Plant Physiology* 61, 158–163.

ICID-CIID (International Commission on Irrigation and Drainage) (2012) Annual Report 2011–12. Available at: icid.org/ar_e_2011.pdf (accessed 27 June 2014).

Idso, S.B. (1982) Non-water-stressed baselines: a key to measuring and interpreting plant water stress. *Agricultural Meteorology* 27, 59–70.

Idso, S.B., Reginato, R.J., Jackson, R.D., Kimball, B.A. and Nakayama, F.S. (1974) The three stages of drying of a field soil. *Soil Science Society of America Proceedings* 38, 831–837.

Idso, S.B., Jackson, R.D. and Reginato, R.J. (1977) Remote-sensing of crop yields. *Science* 196, 19–25.

Idso, S.B., Jackson, R.D., Pinter, P.J., Reginato, R.J. and Hatfield, J.L. (1981) Normalizing the stress-degree-day parameter for environmental variability. *Agricultural Meteorology* 24, 45–55.

IPCC (Intergovernmental Panel on Climate Change) (2000) *Emissions Scenarios*. Nakicenovic, N. and Swart, R. (eds) Cambridge University Press, Cambridge.

IPCC (Intergovernmental Panel on Climate Change) (2007) Climate Change 2007: Synthesis Report. Contribution of Working Groups I, II and III to the Fourth Assessment Report of the Intergovernmental Panel on Climate Change [Core Writing Team, Pachauri, R.K. and Reisinger, A. (eds)]. IPCC, Geneva, Switzerland, 104 pp. Available at: http://www.ipcc.ch/publications_and_data/publications_ipcc_fourth_assessment_report_synthesis_report.htm (accessed 27 May 2014).

IPCC (Intergovernmental Panel on Climate Change) (2013a) Carbon Dioxide: Projected Emissions and Concentrations. Available at: http://www.ipcc-data.org/observ/ddc_co2.html (accessed 24 November 2013).

IPCC (Intergovernmental Panel on Climate Change) (2013b) Climate Change 2013: the Physical Science Basis. Contribution of Working Group I to the Fifth Assessment Report of the Intergovernmental Panel on Climate Change [Stocker, T.F., Qin, D., Plattner, G.-K., Tignor, M., Allen, S.K., Boschung, J., Nauels, A., Xia, Y., Bex, V. and Midgley, P.M. (eds)]. Cambridge University Press, Cambridge, 1535 pp. Available at: https://www.ipcc.ch/report/ar5/wg1/#.Uxdwhtyc-ao (accessed 5 March 2014).

Irmak, S., Haman, D.Z. and Bastug, R. (2000) Determination of crop water stress index for irrigation timing and yield estimation of corn. *Agronomy Journal* 92, 1221–1227.

Ishida, T., Campbell, G.S. and Calissendorff, C. (1991) Improved heat balance method for determining sap flow rate. *Agricultural and Forest Meteorology* 56, 35–48.

Ismail, A.M. and Hall, A.E. (1993) Inheritance of carbon isotope discrimination and water-use efficiency in cowpea. *Crop Science* 33, 498–503.

Janzen, H.H., Campbell, C.A., Izaurralde, R.C., Ellert, B.H., Juma, N., McGill, W.B. and Zentner, R.P. (1998) Management effects on soil C storage on the Canadian prairies. *Soil and Tillage Research* 47, 181–195.

Jardine, P.M., Wilson, G.V. and Luxmoore, R.J. (1990) Unsaturated solute transport through a forest soil during rain storm events. *Geoderma* 46, 103–118.

Jenny, H. (1933) Soil fertility losses under Missouri conditions. Missouri Agricultural Experiment Station, Bulletin No. 324. Missouri Agricultural Experiment Station, Columbia, Missouri.

Jensen, M.E. and Sletten, W.H. (1965) Effects of alfalfa, crop sequence, and tillage practice on intake rates of Pullman silty clay loam and grain yields. Conservation Research Report 1. United States Department of Agriculture, Washington, DC.

Jones, H.G. (1980) Interaction and integration of adaptive responses to water stress: the implications of an unpredictable environment. In: Turner, N.C. and Kramer, P.J. (eds) *Adaptation of Plants to Water and High Temperature Stress*. Wiley, New York, pp. 353–365.

Jones, S.B., Wraith, J.H. and Or, D. (2002) Time domain reflectometry measurement principles and applications. *Hydrological Processes* 16, 141–153.

Jungk, A.O. (2002) Dynamics of nutrient movement at the soil–root interface. In: Waisel, Y., Eshel, A. and Kafkafi, U. (eds) *Plant Roots. The Hidden Half*, 3rd edn. Marcel Dekker, New York, pp. 587–616.

Jungk, A. and Claassen, N. (1989) Availability in soil and acquisition by plants as the basis for phosphorus and potassium supply to plants. *Zeitschrift für Pflanzenernährung und Bodenkunde* 152, 151–157.

Jury, W.A., Gardner, W.R. and Gardner, W.H. (1991) *Soil Physics*. Wiley, New York.

Kage, H. and Ehlers, W. (1996) Does transport of water to roots limit water uptake of field crops? *Zeitschrift für Pflanzenernährung und Bodenkunde* 159, 583–590.

Kanemasu, E.T., Thurtell, G.W. and Tanner, C.B. (1969) Design, calibration, and field use of a stomatal diffusion porometer. *Plant Physiology* 44, 881–885.

Keatinge, J.D.H. and Cooper, P.J.M. (1983) Kabuli chickpea as a winter-sown crop in northern Syria: moisture relations and crop productivity. *Journal of Agricultural Science, Cambridge* 100, 667–680.

Keeling, R.F., Piper, S.C., Bollenbacher, A.F. and Walker, S.J. (2009) Atmospheric CO_2 values (PPMV) derived from *in situ* air samples collected at Mauna Loa, Hawaii. Available at: http://cdiac.ornl.gov/ftp/trends/co2/maunaloa.co2 (accessed 5 March 2014).

Keen, B.A. (1936) The circulation of water in the soil between the surface and the level of underground water. *Bulletin of the International Association of Scientific Hydrology* 22, 328–331.

Keener, M.E. and Kirchner, P.L. (1983) The use of canopy temperature as an indicator of drought stress in humid regions. *Agricultural Meteorology* 28, 339–349.

van Keulen, H. (1986a) Crop yield and nutrient requirements. In: van Keulen, H. and Wolf, J. (eds) *Modelling of Agricultural Production: Weather, Soils and Crops*. Centre for Agricultural Publishing and Documentation, Pudoc, Wageningen, The Netherlands, pp. 155–181.

van Keulen, H. (1986b) Plant data. In: van Keulen, H. and Wolf, J. (eds) *Modelling of Agricultural Production: Weather, Soils and Crops*. Centre for Agricultural Publishing and Documentation, Pudoc, Wageningen, The Netherlands, pp. 235–245.

Kimball, B.A. (1983) Carbon dioxide and agricultural yield: an assemblage and analysis of 430 prior observations. *Agronomy Journal* 75, 779–788.

Kimball, B.A., Kobayashi, K. and Bindi, M. (2002) Response of agricultural crops to free-air CO_2 enrichment. *Advances in Agronomy* 77, 293–368.

Kirkegaard, J.A., Lilley, J.M., Howe, G.N. and Graham, J.M. (2007) Impact of subsoil water use on wheat yield. *Australian Journal of Agricultural Research* 58, 303–315.

Klapp, E. (1962) Ertragssteigerung and Wasserverbrauch landwirtschaftlicher Kulturen. *Zeitschrift für Kulturtechnik* 3, 1–5.

Klute, A. and Dirksen, C. (1986) Hydraulic conductivity and diffusivity: laboratory methods. In: Klute, A. (ed.) *Methods of Soil Analysis. Part I. Physical and Mineralogical Methods*. Agronomy Monograph 9, 2nd edn. American Society of Agronomy, Madison, Wisconsin, pp. 687–734.

Kohnke, H. (1968) *Soil Physics*. McGraw-Hill, New York.

Könings, I. (1987) Durchwurzelung des Bodens von Hafer, Erbsen und Ackerbohnen und im Boden verbleibende Nitratmengen. Diploma thesis, Institute of Agronomy and Plant Breeding, University of Göttingen, Cuvillier, Göttingen, Germany.

Kramer, P.J. (1983) *Water Relations of Plants*. Academic Press, New York.

Kramer, P.J. (1988) Changing concepts regarding plant water relations. *Plant, Cell and Environment* 11, 565–568.

Kratzsch, G. (1993a) Anpassungsstrategien bei Sortenwahl, Saatdichte, Düngung und Pflanzenschutz im Getreidebau. In: Deutsche Landwirtschaftsgesellschaft, DLG (ed.) *Pflanzenbau im Trockengebiet*. DLG-Arbeitsunterlagen No. R/93, Frankfurt, Germany.

Kratzsch, G. (1993b) Weizen: Was tun, wenn's Wasser fehlt? *Mitteilungen der Deutschen Landwirtschaftsgesellschaft, DLG* 108(issue 8), 30–33.

Kremer, C., Stöckle, C.O., Kemanian, A.R. and Howell, T. (2008) A canopy transpiration and photosynthesis model for evaluating simple crop productivity models. In: Ahuja, L.R., Reddy, V.R.,

Saseendran, S.A. and Yu, Q. (eds) *Response of Crops to Limited Water*. Advances in Agricultural Systems Modeling Series 1. American Society of Agronomy, Madison, Wisconsin, pp.165–189.

Kruse, E.G., Bucks, D.A. and von Bernuth, R.D. (1990) Comparison of irrigation systems. In: Stewart, B.A. and Nielsen, D.R. (eds) *Irrigation of Agricultural Crops*. Agronomy Series No. 30. American Society of Agronomy, Madison, Wisconsin, pp. 475–508.

Kutilek, M. and Nielsen, D.R. (1994) *Soil Hydrology*. Catena Verlag, Cremlingen, Germany.

Kutschera, L. (1960) *Wurzelatlas mitteleuropäischer Ackerunkräuter und Kulturpflanzen*. Deutsche Landwirtschaftsgesellschaft, DLG-Verlags-GmbH, Frankfurt, Germany.

Larcher, W. (1973) *Ökologie der Pflanzen*. Ulmer, Stuttgart, Germany.

Larcher, W. (1994) *Ökophysiologie der Pflanzen*, 5th edn. Ulmer, Stuttgart, Germany.

Lascano, R.J. (2000) A general system to measure and calculate daily crop water use. *Agronomy Journal* 92, 821–832.

Laskowski, M. (2013) Lateral root initiation is a probabilistic event whose frequency is set by fluctuating levels of auxin response. *Journal of Experimental Botany* 64, 2609–2617.

Lawes, D.A., Bond, D.A. and Poulsen, M.H. (1983) Classification, origin, breeding methods and objectives. In: Hebblethwaite, P.D. (ed.) *The Faba Bean (*Vicia faba *L.)*. Butterworths, London, pp. 23–76.

Lawes, J.B. (1850) Experimental investigation into the amount of water given off by plants during their growth; especially in relation to fixation and source of their various constituents. *Journal of the Horticultural Society of London* 5, 38–63.

Lawes, J.B., Gilbert, J.H. and Warington, R. (1881) On the amount and composition of the rain and drainage-waters collected at Rothamsted. Parts I and II. The amount and composition of rainfall. *Journal of the Royal Agricultural Society of England* 17, 241–279.

Lawlor, D.W. (2013) Genetic engineering to improve plant performance under drought: physiological evaluation of achievements, limitations, and possibilities. *Journal of Experimental Botany* 64, 83–108.

Leenhardt, D., Voltz, M. and Rambal, S. (1995) A survey of several agroclimatic soil water balance models with reference to their spatial application. *European Journal of Agronomy* 4, 1–14.

Lehne, I. (1968) Zur Quantifizierung der Gesamtwirkung von Düngungsmaßnahmen (Pflanze und Boden) auf einem Sandstandort mit Hilfe der Differenzmethode. *Albrecht-Thaer-Archiv* 12, 717–731.

Levitt, J. (1980) *Responses of Plants to Environmental Stresses, Volume II. Water, Radiation, Salt, and Other Stresses*, 2nd edn. Academic Press, New York.

Lie, T.A. (1981) Environmental physiology of the legume-rhizobium symbiosis. In: Broughton, W.J. (ed.) *Nitrogen Fixation, Volume 1: Ecology*. Oxford Scientific Publications, Clarendon Press, Oxford, pp. 104–134.

Lockhart, J.A. (1965) An analysis of irreversible plant cell elongation. *Journal of Theoretical Biology* 8, 264–275.

Long, I.F. and French, B.K. (1967) Measurement of soil moisture in the field by neutron moderation. *Journal of Soil Science* 18, 149–166.

Lotze-Campen, H. and Schellnhuber, H.-J. (2009) Climate impacts and adaptation options in agriculture: what we know and what we don't know. *Journal für Verbraucherschutz und Lebensmittelsicherheit* 4, 145–150.

Lu, J., Sun, G., McNulty, S.G. and Amatya, D.M. (2005) A comparison of six potential evapotranspiration methods for regional use in the southeastern United States. *Journal of the American Water Resources Association* 41, 621–633.

Ludlow, M.M. and Muchow, R.C. (1990) A critical evaluation of traits for improving crop yields in water-limited environments. *Advances in Agronomy* 43, 107–153.

Luo, Y., Ou Yang, Z., Yuan, G., Tang, D. and Xie, X. (2003) Evaluation of macroscopic root water uptake models using lysimeter data. *American Society of Agricultural Engineers* 46, 625–634.

Luxmoore, R.J. (1981) Micro-, meso-, and macroporosity of soil. *Soil Science Society of America Journal* 45, 671–672.

MacFarling Meure, C., Etheridge, D., Trudinger, C., Steele, P., Langenfelds, R., van Ommen, T., Smith, A. and Elkins, J. (2006) Law Dome CO_2, CH_4 and N_2O ice core records extended to 2000 years BP. *Geophysical Research Letters* 33(14). Available at: http://onlinelibrary.wiley.com/doi/10.1029/2006GL026152/full (accessed 15 April 2015).

Madhu, M. and Hatfield, J.L. (2013) Dynamics of plant root growth under increased atmospheric carbon dioxide. *Agronomy Journal* 105, 657–669.

Mahdieh, M. and Mostajeran, A. (2009) Abscisic acid regulates root hydraulic conductance via aquaporin expression modulation in *Nicotiana tabacum*. *Journal of Plant Physiology* 166, 1993–2003.

Mahouachi, J., Arbona, V. and Gómez-Cadenas, A. (2007) Hormonal changes in papaya seedlings subjected to progressive water stress and re-watering. *Plant Growth Regulation* 53, 43–51.

Maraux, F. and Lafolie, F. (1998) Modeling soil water balance of a maize-sorghum sequence. *Soil Science Society of America Journal* 62, 75–82.

Marek, T., Piccinni, G., Schneider, A., Howell, T., Jett, M. and Dusek, D. (2006) Weighing lysimeters for the determination of crop water requirements and crop coefficients. *Applied Engineering in Agriculture* 22, 851–856.

Marschner, H. (1993) *Mineral Nutrition of Higher Plants*, 2nd edn. Academic Press, London.

Marshall, D.C. (1958) Measurement of sap flow in conifers by heat transport. *Plant Physiology* 33, 385–396.

Marshall, T.J., Holmes, J.W. and Rose, C.W. (1996) *Soil Physics*. Cambridge University Press, Cambridge.

Mårtensson, A.M., Rydberg, I. and Vestberg, M. (1998) Potential to improve transfer of N in intercropped systems by optimising host–endophyte combinations. *Plant and Soil* 205, 57–66.

Martin, J.H., Leonard, W.H. and Stamp, D.L. (1976) *Principles of Field Crop Production*. Macmillan, New York.

Masle, J. and Passioura, J.B. (1987) The effect of soil strength on the growth of young wheat plants. *Australian Journal of Plant Physiology* 4, 643–656.

McGowan, M. and Williams, J.B. (1980) The water balance of an agricultural catchment. I. Estimation of evaporation from soil water records. *Journal of Soil Science* 31, 217–230.

McKeague, J.A., Eilers, R.G., Thomasson, A.J., Reeve, M.J., Bouma, J., Grossman, R.B., Favrot, J.C., Renger, M. and Strebel, O. (1984) Tentative assessment of soil survey approaches to the characterization and interpretation of air–water properties of soils. *Geoderma* 34, 69–100.

Meinshausen, M., Meinshausen, N., Hare, W., Raper, S.C.B., Frieler, K., Knutti, R., Frame, D.J. and Allen, M.R. (2009) Greenhouse-gas emission targets for limiting global warming to 2°C. *Nature* 258, 1158–1162.

Mengel, K. and Kirkby, E.A. (1982) *Principles of Plant Nutrition*, 3rd edn. International Potash Institute, Bern.

Merrill, S.D., Tanaka, D.L. and Hanson, J.D. (2005) Comparison of fixed-wall and pressurized-wall minirhizotrons for fine root growth measurements in eight crop species. *Agronomy Journal* 97, 1373–1376.

Meyer, W.S. and Green, G.C. (1980) Water use by wheat and plant indicators of available soil water. *Agronomy Journal* 72, 253–257.

Millar, A.A., Gardner, W.R. and Goltz, S.M. (1971) Internal water status and water transport in seed onion plants. *Agronomy Journal* 63, 779–784.

Mogensen, V.O. (1980) Drought sensitivity at various growth stages of barley in relation to relative evapotranspiration and water stress. *Agronomy Journal* 72, 1033–1038.

Mogensen, V.O. and Jensen, H.E. (1989) The concept of stress days in modelling crop yield response to water stress. In: Plancquaert, P. (ed.) *Management of Water Resources in Cash Crops and in Alternative Production Systems*. Commission of the European Communities Report EUR 11935 EN. Commission of the European Communities, Luxembourg, pp. 13–22.

Moinuddin, Fischer, R.A., Sayre, K.D. and Reynolds, M.P. (2005) Osmotic adjustment in wheat in relation to grain yield under water deficit environments. *Agronomy Journal* 97, 1062–1071.

Monnier, G. (1965) Action des materies organiques sur la stabilite structurale des sols. *Annales Agronomiques* 16, 327–400, 471–534.

Monteith, J.L. (1957) Dew. *Quarterly Journal of the Royal Meteorological Society* 83, 322–341.

Monteith, J.L. (1965) Evaporation and environment. *Symposia of the Society for Experimental Biology* X1X, 205–234.

Monteith, J.L. (1977) Climate and the efficiency of crop production in Britain. *Philosophical Transactions of the Royal Society of London B* 281, 277–294.

Monteith, J.L. (1978) Reassessment of maximum growth rates for C_3 and C_4 crops. *Experimental Agriculture* 14, 1–5.

Monteith, J.L. (1981) Evaporation and surface temperature. *Quarterly Journal of the Royal Meteorological Society* 107, 1–27.

Monteith, J.L. (1986a) Howard Latimer Penman. 10 April 1909–13 October 1984. *Biographical Memoirs of Fellows of the Royal Society* 32, 378–404.

Monteith, J.L. (1986b) Significance of the coupling between saturation vapour pressure deficit and rainfall in monsoon climates. *Experimental Agriculture* 22, 328–338.

Monteith, J.L. (1990) Steps in crop climatology. In: Unger, P.W., Sneed, T.V., Jordan, W.J. and Jensen, R. (eds) *Challenges in Dryland Agriculture – A Global Perspective.* Texas Agricultural Experiment Station, Amarillo/Bushland, Texas, pp. 273–282.

Montgomery, E.G. and Kiesselbach, T.A. (1912) Studies in water requirement of corn. Nebraska Agricultural Experiment Station, Bulletin No. 128, Lincoln, Nebraska.

Morrison, M.J. and Stewart, D.W. (1995) Radiation-use efficiency in summer rape. *Agronomy Journal* 87, 1139–1142.

Müller, M.J. (1982) *Selected Climatic Data for a Global Set of Standard Stations for Vegetation Science.* Dr W. Junk Publishers, The Hague, The Netherlands.

Müller, R. (1988) Bedeutung des Wurzelwachstums und der Phosphatmobilität im Boden für die Phosphaternährung von Winterweizen, Wintergerste und Zuckerrüben. PhD dissertation, University of Göttingen, Göttingen, Germany.

Müller, U. (1984) Wasserhaushalt von Ackerbohne und Hafer auf Löß-Parabraunerde. PhD dissertation, University of Göttingen, Göttingen, Germany.

Müller, U. and Ehlers, W. (1986) Untersuchungen zum Wasserhaushalt der Ackerbohne I. Wurzelwachstum, Wasseraufnahme und Wasserverbrauch. *Kali-Briefe (Büntehof)* 18, 167–187.

Müller, U. and Ehlers, W. (1987) Verfahren zur Berechnung der Bodenevaporation in Pflanzenbeständen. *Zeitschrift für Kulturtechnik und Flurbereinigung* 28, 394–399.

Müller, U., Meyer, C., Ehlers, W. and Böhm, W. (1985) Wasseraufnahme und Wasserverbrauch von Ackerbohne und Hafer auf einer Löß-Parabraunerde. *Zeitschrift für Pflanzenernährung und Bodenkunde* 148, 389–404.

Müller, U., Grimme, K., Meyer, C. and Ehlers, W. (1986) Leaf water potential and stomatal conductance of field-grown faba bean (*Vicia faba* L.) and oats (*Avena sativa* L.). *Plant and Soil* 93, 17–33.

Mundel, G. (1992) Untersuchungen zur Evapotranspiration von Silomaisbeständen in Lysimetern. *Archiv für Acker- und Pflanzenbau und Bodenkunde* 36, 35–44.

Munns, R. and King, R.W. (1988) Abscisic acid is not the only stomatal inhibitor in the transpiration stream. *Plant Physiology* 88, 703–708.

Munyankusi, E., Gupta, S.C., Moncrief, J.F. and Berry, E.C. (1994) Earthworm macropores and preferential transport in a long-term manure applied Typic Hapludalf. *Journal of Environmental Quality* 23, 733–784.

Musick, J.T. and Porter, K.B. (1990) Wheat. In: Stewart, B.A. and Nielsen, D.R. (eds) *Irrigation of Agricultural Crops.* Agronomy Series 30. American Society of Agronomy, Madison, Wisconsin, pp. 597–638.

Musick, J.T., Jones, O.R., Stewart, B.A. and Dusek, D.A. (1994) Water–yield relationships for irrigated and dryland wheat in the US Southern Plains. *Agronomy Journal* 86, 980–986.

Narayanan, S., Aiken, R.M., Prasad, P.V.V., Xin, Z. and Yu, J. (2013) Water and radiation use efficiencies in Sorghum. *Agronomy Journal* 105, 649–656.

NASA Goddard Institute for Space Studies (2014) Surface Temperatures. Available at: data.giss.nasa.gov/gistemp/ (accessed 20 March 2014).

Newman, E.I. (1966) A method of estimating the total length of root in a sample. *Journal of Applied Ecology* 3, 139–145.

Newman, E.I. (1969) Resistance to water flow in soil and plant. 1. Soil resistance in relation to amount of root: theoretical estimates. *Journal of Applied Ecology* 6, 1–12.

Nickel, S.E., Crookston, R.K. and Ruselle, M.P. (1995) Root growth and distribution are affected by corn–soybean cropping sequence. *Agronomy Journal* 87, 895–902.

Nielsen, D.C. and Vigil, M.F. (2010) Precipitation storage efficiency during fallow in wheat–fallow systems. *Agronomy Journal* 102, 537–543.

Nielsen, D.C., Vigil, M.F. and Benjamin, J.G. (2011) Evaluating decision rules for dryland rotation crop selection. *Field Crops Research* 120, 254–261.

Nobel, P.S. (1974) *Introduction to Biophysical Plant Physiology*. Freeman, San Francisco, California.

Nobel, P.S. (1983) *Biophysical Plant Physiology and Ecology*. Freeman, New York.

Nultsch, W. (1991) *Allgemeine Botanik*, 9th edn. Georg Thieme, Stuttgart, Germany.

Or, D. (2001) Who invented the tensiometer? *Soil Science Society of America Journal* 65, 1–3.

van Ouwerkerk, C. (1974) Rational tillage. Semaine d'étude agriculture et environnement. *Bulletin Recherche Agronomique Gembloux*, hors séries, 695–709.

Palmer, R.E., Meisinger, J.J. and Magette, W.L. (2011) Undisturbed soil columns for lysimetry: 1. Collection, field testing, and construction. *Applied Engineering in Agriculture* 27, 379–389.

Park, R. and Epstein, S. (1960) Carbon isotope fractionation during photosynthesis. *Geochimica et Cosmochimica Acta* 21, 110–126.

Passioura, J.B. (1972) The effect of root geometry on the yield of wheat growing on stored water. *Australian Journal of Agricultural Research* 23, 745–752.

Passioura, J.B. (1977) Grain yield, harvest index and water use of wheat. *Journal of the Australian Institute of Agricultural Science* 43, 117–120.

Passioura, J.B. (1988) Root signals control leaf expansion in wheat seedlings growing in drying soil. *Australian Journal of Plant Physiology* 15, 687–693.

Passioura, J.B. (2007) The drought environment: physical, biological and agricultural perspectives. *Journal of Experimental Botany* 58, 113–117.

Passioura, J.B. and Angus, J.F. (2010) Improving productivity of crops in water-limited environments. *Advances in Agronomy* 106, 37–75.

Payne, W.A. (1997) Managing yield and water use of pearl millet in the Sahel. *Agronomy Journal* 89, 481–490.

Payne, W.A., Lascano, R.J., Hossner, L.R., Wendt, C.W. and Onken, A.B. (1991) Pearl millet growth as affected by phosphorus and water. *Agronomy Journal* 83, 942–948.

Payne, W.A., Drew, M.C., Hossner, L.R., Lascano, R.J., Onken, A.B. and Wendt, C.W. (1992) Soil phosphorus availability and pearl millet water-use efficiency. *Crop Science* 32, 1010–1015.

Penman, H.L. (1940a) Gas and vapour movements in the soil. I. The diffusion of vapours through porous solids. *Journal of Agricultural Science, Cambridge* 30, 437–462.

Penman, H.L. (1940b) Gas and vapour movements in the soil. II. The diffusion of carbon dioxide through porous solids. *Journal of Agricultural Science, Cambridge* 30, 570–581.

Penman, H.L. (1940c) Meteorological and soil factors affecting evaporation from fallow soil. *Quarterly Journal of the Royal Meteorological Society* 66, 401–410.

Penman, H.L. (1948) Natural evaporation from open water, bare soil and grass. *Proceedings of the Royal Society of London Series A* 193, 120–146.

Penman, H.L. (1950) The dependence of transpiration on weather and soil conditions. *Journal of Soil Science* 1, 74–89.

Penman, H.L. and Long, I.F. (1960) Weather in wheat: an essay in micro-meteorology. *Quarterly Journal of the Royal Meteorological Society* 86, 16–50.

Peressotti, A. and Ham, J.M. (1996) A dual-heater gauge for measuring sap flow with an improved heat-balance method. *Agronomy Journal* 88, 149–155.

Philip, J.R. (1957) The theory of infiltration: 4. Sorptivity and algebraic infiltration equations. *Soil Science* 84, 257–264.

Philip, J.R. (1966) Plant water relations: some physical aspects. *Annual Review of Plant Physiology* 17, 245–268.

Philip, J.R. (1995) Desperately seeking Darcy in Dijon. *Soil Science Society of America Journal* 59, 319–324.

Pisek, A., Larcher, W., Vegis, A. and Napp-Zinn, K. (1973) The normal temperature range. In: Precht, H., Christophersen, J., Hensel, H. and Larcher, W. (eds) *Temperature and Life*. Springer, Berlin, pp. 102–144.

van der Ploeg, R.R., Beese, F., Strebel, O. and Renger, M. (1978) The water balance of a sugar beet crop: a model and some experimental evidence. *Zeitschrift für Pflanzenernährung und Bodenkunde* 141, 313–328.

van der Ploeg, R.R., Böhm, W. and Kirkham, M.B. (1999) On the origin of the theory of mineral nutrition of plants and the law of the minimum. *Soil Science Society of America Journal* 63, 1055–1062.

van der Ploeg, M.J., Gooren, H.P.A., Bakker, G. and de Rooij, G.H. (2008) Matric potential measurements by polymer tensiometers in cropped lysimeters under water-stressed conditions. *Vadose Zone Journal* 7, 1048–1054.

Portmann, F.T., Siebert, S. and Döll, P. (2010) MIRCA2000 – global monthly irrigated and rainfed crop areas around the year 2000: a new high-resolution data set for agricultural and hydrological modeling. *Global Biogeochemistry Cycles* 24, 1–24.

Postma, J.A. and Lynch, J.P. (2011) Root cortical aerenchyma enhances the growth of maize on soils with suboptimal availability of nitrogen, phosphorus, and potassium. *Plant Physiology* 156, 1190–1201.

Quisenberry, V.L. and Phillips, R.E. (1976) Percolation of surface-applied water in the field. *Soil Science Society of America Journal* 40, 484–489.

Quisenberry, V.L., Smith, B.R., Phillips, R.E., Scott, H.D. and Nortcliff, S. (1993) A soil classification system for describing water and chemical transport. *Soil Science* 156, 306–315.

Ragab, R., Beese, F. and Ehlers, W. (1990) A soil water balance and dry matter production model: I. Soil water balance of oat. *Agronomy Journal* 82, 152–156.

Raschke, K. (1975) Stomatal action. *Annual Review of Plant Physiology* 26, 309–340.

Rauh, W. (1941) *Morphologie der Nutzpflanzen*. Verlagsbuchhandlung Quelle and Meyer, Leipzig.

Rauhe, K., Dittrich, D. and Kunze, A. (1968) Die Wirkung verschiedener langjähriger Düngungsmaßnahmen auf Pflanzenertrag und Humusversorgung des Bodens in einem Dauerversuch auf Lößlehmboden. *Albrecht-Thaer-Archiv* 12, 733–745.

Raven, J.A., Handley, L.L. and Wollenweber, B. (2004) Plant nutrition and water use efficiency. In: Bacon, M.S. (ed.) *Water Use Efficiency in Plant Biology*. Blackwell, Oxford, pp. 162–188.

Rehm, S. and Espig, G. (1976) *Die Kulturpflanzen der Tropen und Subtropen*. Ulmer, Stuttgart, Germany.

Rehm, S. and Espig, G. (1991) *The Cultivated Plants of the Tropics and Subtropics*. Verlag Josef Margraf, Weikersheim.

Reicosky, D.C. and Ritchie, J.T. (1976) Relative importance of soil resistance and plant resistance in root water absorption. *Soil Science Society of America Journal* 40, 293–297.

Renger, M. and Strebel, O. (1980) Beregnungsbedarf landwirtschaftlicher Kulturen in Abhängigkeit vom Boden. *Wasser und Boden* 32, 572–575.

Renger, M., Strebel, O. and Giesel, W. (1974) Beurteilung bodenkundlicher, kulturtechnischer und hydrologischer Fragen mit Hilfe von klimatischer Wasserbilanz und bodenphysikalischen Kennwerten. 1. Bericht: Beregnungsbedüftigkeit. *Zeitschrift für Kulturtechnik und Flurbereinigung* 15, 148–160.

Richards, L.A. (1928) The usefulness of capillary potential to soil moisture and plant investigators. *Journal of Agricultural Research* 37, 719–742.

Richards, L.A. (1941) A pressure-membrane extraction apparatus for soil solution. *Soil Science* 51, 377–386.

Richards, L.A. (1942) Soil moisture tensiometer materials and construction. *Soil Science* 53, 241–248.

Richards, R.A. (2006) Physiological traits used in the breeding of new cultivars for water-scarce environments. *Agricultural Water Management* 80, 197–211.

Richards, R.A., Lopez-Castaneda, C., Gomez-Macpherson, H. and Condon, A.G. (1993) Improving the efficiency of water use by plant breeding and molecular biology. *Irrigation Science* 14, 93–104.

Richards, S.J. (1965) Soil suction measurements with tensiometers. In: Black, C.A. (ed.) *Methods of Soil Analysis. Part 1*. Agronomy Series No. 9. American Society of Agronomy, Madison, Wisconsin, pp. 153–163.

Richter, G.L., Monshausen, G.B., Krol, A. and Gilroy, S. (2009) Mechanical stimuli modulate lateral root organogenesis. *Plant Physiology* 151, 1855–1866.

Richter, J. (1987) *The Soil as a Reactor – Modelling Processes in the Soil.* Catena Publisher, Cremlingen and Reiskirchen, Germany.

Ridley, A.M. and Burland, J.B. (1993) A new instrument for the measurement of soil moisture suction. *Géotechnique* 43, 321–324.

Ridley, A.M. and Burland, J.B. (1999) Use of tensile strength of water for direct measurement of high soil suction: discussion. *Canadian Geotechnical Journal* 36, 178–180.

Riggs, T.J., Hanson, P.R., Start, N.D., Miles, D.M., Morgan, C.L. and Ford, M.A. (1981) Comparison of spring barley varieties grown in England and Wales between 1880 and 1980. *Journal of Agricultural Science, Cambridge* 97, 599–610.

Rijtema, P.E. and Aboukhaled, A. (1975) Crop water use. In: *Research on Crop Water Use, Salt Affected Soils and Drainage in the Arab Republic of Egypt.* Food and Agriculture Organization of the United Nations (FAO), Rome, pp. 5–61.

Ritchie, J.T. (1972) Model for predicting evaporation from a row crop with incomplete cover. *Water Resources Research* 8, 1204–1213.

Ritchie, J.T. (1981) Soil water availability. *Plant and Soil* 58, 327–338.

Ritchie, J.T. (1983) Efficient water use in crop production: discussion on the generality of relations between biomass production and evapotranspiration. In: Taylor, H.M., Jordan, W.R. and Sinclair, T.R. (eds) *Limitations to Efficient Water Use in Crop Production.* American Society of Agronomy, Madison, Wisconsin, pp. 29–44.

Ritchie, J.T. and Burnett, E. (1971) Dryland evaporative flux in a subhumid climate: II. Plant influences. *Agronomy Journal* 63, 56–62.

Ritchie, J.T. and Johnson, B.S. (1990) Soil and plant factors affecting evaporation. In: Stewart, B.A. and Nielsen, D.R. (eds) *Irrigation of Agricultural Crops.* American Society of Agronomy, Madison, Wisconsin, pp. 363–390.

Rogers, H.H., Runion, G.B. and Krupa, S.V. (1994) Plant responses to atmospheric CO_2 enrichment with emphasis on roots and the rhizosphere. *Environmental Pollution* 83, 155–189.

Roth, C.H. (1992) *Die Bedeutung der Oberflächenverschlämmung für die Auslösung von Abfluß and Abtrag.* Habilitationsschrift, Reihe Bodenökologie und Bodengenese, Heft 6, Technical University of Berlin, Berlin.

Roth, D. and Kachel, K. (1989) Zusatzwasser-Ertrags-Beziehungen von Winterweizen, Sommergerste, Zuckerrüben, Kartoffeln und Welschem Weidelgras auf drei Standorten mit unterschiedlichem Bodenwasserbereitstellungsvermögen. *Archiv für Acker- und Pflanzenbau und Bodenkunde* 33, 393–403.

Roth, D. and Werner, D. (2000) Bewässerung. In: Blume, H.P. (ed.) *Handbuch der Bodenkunde.* Ecomed, Landsberg, Section 5.3.4.2.

Roth, D., Günther, R. and Roth, R. (1988) Transpirationskoeffizienten und Wasserausnutzungsraten landwirtschaftlicher Fruchtarten. 1. Mitteilung: Transpirationskoeffizienten und

Wasserausnutzungsraten von Getreide, Hackfrüchten und Silomais unter Feldbedingungen und in Gefäßversuchen. *Archiv für Acker- und Pflanzenbau und Bodenkunde* 32, 397–403.

Roth, D., Günther, R. and Knoblauch, S. (1994) Technische Anforderungen an Lysimeteranlagen als Voraussetzung für die Übertragbarkeit von Lysimeterergebnissen auf landwirtschaftliche Nutzflächen. *4. Gumpensteiner Lysimetertagung 'Übertragung von Lysimeterergebnissen auf landwirtschaftlich genutzte Flächen und Regionen'.* Bundesanstalt für alpenländische Landwirtschaft Gumpenstein, Irdning, Austria, pp. 9–21.

Rötter, R. and van de Geijn, S.C. (1999) Climate change effects on plant growth, crop yield and livestock. *Climatic Change* 43, 651–681.

Russell, E.J. (1907) Note on an apparent secular change in the Rothamsted Drain Gauges. *Journal of Agricultural Science, Cambridge* 2, 29–34.

Sadras, V.O., Grassini, P. and Steduto, P. (2011) Status of Water Use Efficiency of Main Crops. The State of Land and Water Resources (SOLAW), Thematic Report No. 7, Food and Agriculture Organization of the United Nations (FAO), Rome.

Safir, G.R., Boyer, J.S. and Gerdemann, J.W. (1972) Nutrient status and mycorrhizal enhancement of water transport in soybean. *Journal of Plant Physiology* 49, 700–703.

Savage, M.J., Ritchie, J.T., Bland, W.L. and Dugas, W.A. (1996) Lower limit of soil water availability. *Agronomy Journal* 88, 644–651.

Sawadogo, H. (2011) Using soil and water conservation techniques to rehabilitate degraded lands in northwestern Burkina Faso. *International Journal of Agricultural Sustainability* 9, 120–128.

Scheffer, F. and Schachtschabel, P. (1989) *Lehrbuch der Bodenkunde*, 12th edn. Enke, Stuttgart, Germany.

Schmid, H. (1991) Wurzelentwicklung von Zuckerrüben in verschieden texturierten Böden und die Methodik ihrer Erfassung. Diploma thesis, Institute of Agronomy and Plant Breeding, University of Göttingen, Göttingen, Germany.

Schmidt, H. (1979) Ertragsbildung von Zuckerrüben in Abhängigkeit von der Bodenbearbeitung und der N-Düngung. Diploma thesis, Institute of Agronomy and Plant Breeding, University of Göttingen, Göttingen, Germany.

Schofield, R.K. (1935) The pF of the water in soil. *Transactions of the 3rd International Congress of Soil Science* 2, 37–48.

Scholander, P.F., Hammel, H.T., Bradstreet, E.D. and Hemmingsen, E.A. (1965) Sap pressure in vascular plants. *Science* 148, 339–346.

Schönwiese, C.-D. (2013) *Klimatologie*, 4th edn. Ulmer, Stuttgart, Germany.

Schönwiese, C.-D., Rapp, J., Fuchs, T. and Denhard, M. (1993) *Klimatrend-Atlas Europa 1891–1990.* Berichte des Zentrums für Umweltforschung No. 20. ZUF-Verlag, Frankfurt/Main.

Schönwiese, C.-D., Staeger, T. and Trömel, S. (2005) Klimawandel und Extremereignisse in Deutschland. In: *Klimastatusbericht*. Deutscher Wetterdienst, Offenbach, Germany, pp. 7–17.

Schröder, J.I., Raschke, K. and Neher, E. (1987) Voltage dependence of K^+ channels in guard-cell protoplasts. *Proceedings of the National Academy of Science USA* 84, 4108–4112.

Schultz-Lupitz, A. (1895) *Zwischenfruchtbau auf leichten Böden*. Arbeiten der Deutschen Landwirtschaftsgesellschaft, Heft 7, Berlin.

Schulze, E. and Schulze-Gemen, P. (1957) Der Wasserhaushalt des Bodens im Dauerdüngungsversuch Dikopshof (1953/54). *Zeitschrift für Acker- und Pflanzenbau* 103, 22–58.

Schulze, E.-D., Hall, A.E., Lange, O.L. and Walz, H. (1982) A portable steady-state porometer for measuring the carbon dioxide and water vapour exchanges of leaves under natural conditions. *Oecologia* 53, 141–145.

von Seelhorst, C. (1902) Vegetationskästen zum Studium des Wasserhaushaltes im Boden. *Journal für Landwirtschaft* 50, 277–280.

von Seelhorst, C. and Fresenius, L. (1904) Beiträge zur Lösung der Frage nach dem Wasserhaushalt im Boden und dem Wasserverbrauch der Pflanzen. *Journal für Landwirtschaft* 52, 355–393.

von Seelhorst, C. and Müther, A. (1905) Beiträge zur Lösung der Frage nach dem Wasserhaushalt im Boden und dem Wasserverbrauch der Pflanzen. *Journal für Landwirtschaft* 53, 239–259.

Segal, E., Kushnir, T., Mualem, Y. and Shani, U. (2008) Microsensing of water dynamics and root distribution in sandy soils. *Vadose Zone Journal* 7, 1018–1026.

Senock, R.S., Ham, J.M., Loughin, T.M., Kimball, B.A., Hunsaker, D.J., Pinter, P.J., Wall, G.W., Garcia, R.L. and LaMorte, R.L. (1996) Sap flow under free-air CO_2 enrichment. *Plant, Cell and Environment* 19, 147–158.

Serraj, R., Vadez, V., Purcell, C.L. and Sinclair, T.R. (1999) Recent advances in the physiology of drought stress effects on symbiotic N_2 fixation in soybean. In: Martinez, E. and Hernández, G. (eds) *Highlights of Nitrogen Fixation Research.* Plenum Publishers, Kluwer Academic Press, New York, pp. 49–55.

Sharratt, B.S., Hanks, R.J. and Aase, J.K. (1980) Environmental factors associated with yield differences between seeding dates of spring wheat. Utah Agricultural Experiment Station, Research Report No. 92. Utah Agricultural Experiment Station, Logan, Utah.

Shaw, R.H. and Laing, D.R. (1966) Moisture stress and plant response. In: Pierre, W.H., Kirkham, D., Pisek, J. and Shaw, R. (eds) *Plant Environment and Efficient Water Use.* American Society of Agronomy, Madison, Wisconsin, pp. 73–94.

Shibles, R.M. and Weber, C.R. (1965) Leaf area, solar radiation interception and dry matter production by soybeans. *Crop Science* 5, 575–577.

Siddique, K.H.M., Belford, R.K., Perry, M.W. and Tennant, D. (1989) Growth, development and light interception of old and modern wheat cultivars in a Mediterranean-type environment. *Australian Journal of Agricultural Research* 40, 473–487.

Siddique, K.H.M., Belford, R.K. and Tennant, D. (1990a) Root:shoot ratios of old and modern, tall and semi-dwarf wheats in a Mediterranean environment. *Plant and Soil* 121, 89–98.

Siddique, K.H.M., Tennant, D., Perry, M.W. and Belford, R.K. (1990b) Water use and water use efficiency of old and modern wheat cultivars in a Mediterranean-type environment. *Australian Journal of Agricultural Research* 41, 431–447.

da Silva, A.P., Kay, B.D. and Perfect, E. (1994) Characterization of the least limiting water range of soils. *Soil Science Society of America Journal* 58, 1775–1781.

Simmons, S.R. and Jones, R.J. (1985) Contributions of pre-silking assimilate to grain yield in maize. *Crop Science* 25, 1004–1006.

Sinclair, T.R. (2011) The thousand pound gorilla in crop response to climate change. In: Hillel, D. and Rosenzweig, C. (eds) *Handbook of Climate Change and Agroecosystems: Impacts, Adaptations and Mitigation.* Columbia University and Goddard Institute for Space Studies. Imperial College Press, London, pp. 179–190.

Sinclair, T.R. and Muchow, R.C. (1999) Radiation use efficiency. *Advances in Agronomy* 65, 215–265.

Skopp, J. (1981) Comment on 'Micro-, meso-, and macroporosity of soil'. *Soil Science Society of America Journal* 45, 1244–1246.

Slatyer, R.O. (1968) *Plant–Water Relationships*, 2nd edn. Academic Press, London.

Smit, A.L., Bengough, A.G., Engels, C., van Noordwijk, M., Pellerin, S. and van de Geijn, S.C. (eds) (2000) *Root Methods. A Handbook.* Springer, Berlin.

Smith, D.M. and Allen, S.J. (1996) Measurement of sap flow in plant stems. *Journal of Experimental Botany* 47, 1833–1844.

Smith, J.L. and Elliott, L.F. (1990) Tillage and residue management effects on soil organic matter dynamics in semiarid regions. *Advances in Soil Science* 13, 69–88.

Smucker, A.J.M., McBurney, S.L. and Srivastava, A.K. (1982) Quantitative separation of roots from compacted soil profiles by the hydropneumatic elutriation system. *Agronomy Journal* 74, 500–503.

Söhne, W. (1958) Fundamentals of pressure distribution and soil compaction under tractor tires. *Agricultural Engineering* 39, 276–281, 290.

Sperry, J.S. (2000) Hydraulic constraints on plant gas exchange. *Agricultural and Forest Meteorology* 104, 13–23.

Sperry, J.S., Stiller, V. and Hacke, U.G. (2003) Xylem hydraulics and the soil–plant–atmosphere continuum: opportunities and unresolved issues. *Agronomy Journal* 95, 1362–1370.

Steiner, J.L. and Hatfield, J.L. (2008) Winds of change: a century of agroclimate research. *Agronomy Journal* 100, S132–S152.

Stewart, J.I. (1985) Response farming: a scientific approach to ending starvation and alleviating poverty in drought zones of Africa. In: Moses, Y.T. (ed.) *Proceedings of the International Conference on African Agricultural Development: Technology, Ecology and Society*, May 28-June 2, 1985, California State Polytechnic University, Pomona, California, pp. 50–52.

Stewart, B.A. and Steiner, J.L. (1990) Water-use efficiency. *Advances in Soil Science* 13, 151–173.

Stewart, B.A., Zixi, Z. and Jones, O.R. (1994) Optimizing rain water use. In: Virmani, S.M., Katyal, J.C., Eswaran, H. and Abrol, I.P. (eds) *Stressed Ecosystems and Sustainable Agriculture*. Oxford and IBH Publishing Co., New Delhi, pp. 253–265.

Stockle, C.O. and Kiniry, J.R. (1990) Variability in crop radiation-use efficiency associated with vapor-pressure deficit. *Field Crops Research* 25, 171–181.

Stone, L.R., Goodrum, D.E., Schlegel, A.J., Jaafar, M.N. and Khan, A.H. (2002) Water depletion depth of grain sorghum and sunflower in the central High Plains. *Agronomy Journal* 94, 936–943.

Strasburger, E., Noll, F., Schenck, H. and Schimper, A.F.W. (1991) *Lehrbuch der Botanik für Hochschulen*, 33rd edn. Gustav Fischer, Stuttgart, Germany.

Strebel, O. and Duynisveld, W.H.M. (1989) Nitrogen supply to cereals and sugar beet by mass flow and diffusion on a silty loam soil. *Zeitschrift für Pflanzenernährung und Bodenkunde* 152, 135–141.

Strebel, O., Böttcher, J. and Duynisveld, W.H.M. (1984) *Einfluß von Standortbedingungen und Bodennutzung auf Nitratauswaschung und Nitratkonzentration des Grundwassers*. Landwirtschaftliche Forschung, Kongreßband, Karlsruhe, pp. 33–44.

Strebel, O., Duynisveld, W.H.M. and Böttcher, J. (1989) Nitrate pollution of groundwater in Western Europe. *Agriculture, Ecosystems and Environment* 26, 189–214.

Supit, I., van Diepen, C.A., de Wit, A.J.W., Wolf, J., Kabat, P., Baruth, B. and Ludvig, F. (2012) Assessing climate change effects on European crop yields using the Crop Growth Monitoring System and a weather generator. *Agricultural and Forest Meteorology* 164, 96–111.

Swanson, R.H. (1994) Significant historical developments in thermal methods for measuring sap flow in trees. *Agricultural and Forest Meteorology* 72, 113–132.

Swanson, R.H. and Whitfield, D.W.A. (1981) A numerical-analysis of heat pulse velocity theory and practice. *Journal of Experimental Botany* 32, 221–239.

Swaraj, K., Topunov, F., Golubeva, L.I. and Kretovich, V.L. (1986) Effect of water stress on enzymatic reduction of leghemoglobin in soybean nodules. *Translated from Fiziologia Rastenii* 33, 87–92.

Taiz, L. and Zeiger, E. (1991) *Plant Physiology*. The Benjamin/Cummings Publishing Company, Redwood City, California.

Take, W.A. and Bolton, M.D. (2003) Tensiometer saturation and the reliable measurement of soil suction. *Géotechnique* 53, 159–172.

Tan, C.S. and Buttery, B.R. (1995) Determination of the water use of two pairs of soybean isolines differing in stomatal frequency using a heat balance stem flow gauge. *Canadian Journal of Plant Science* 75, 99–103.

Tan, C.S., Drury, C.F., Ng, H.Y.F. and Gaynor, J.D. (1999) Effect of controlled drainage and subirrigation on subsurface tile drainage and nitrate loss and crop yield at the farm scale. *Canadian Water Resources Journal* 24, 177–186.

Tanner, C.B. (1957) Factors affecting evaporation from plants and soils. *Journal of Soil and Water Conservation* 12, 221–227.

Tanner, C.B. (1963) Plant temperatures. *Agronomy Journal* 55, 210–211.

Tanner, C.B. (1981) Transpiration efficiency of potato. *Agronomy Journal* 73, 59–64.

Tanner, C.B. and Sinclair, T.R. (1983) Efficient water use in crop production: research or re-search? In: Taylor, H.M., Jordan, W.R. and Sinclair, T.R. (eds) *Limitations to Efficient Water Use in Crop Production*. American Society of Agronomy, Madison, Wisconsin, pp. 1–27.

Tardieu, F. (1994) Growth and functioning of roots and of root systems subjected to soil compaction. Towards a system with multiple signalling? *Soil and Tillage Research* 30, 217–243.

Tardieu, F. and Davies, W.J. (1993) Integration of hydraulic and chemical signalling in the control of stomatal conductance and water status of droughted plants. *Plant Cell and Environment* 16, 341–349.

Tardieu, F., Zhang, J., Katerji, N., Bethenod, O., Palmer, S. and Davies, W.J. (1992) Xylem ABA controls the stomatal conductance of field-grown maize subjected to soil compaction or soil drying. *Plant Cell and Environment* 15, 193–197.

Tardieu, F., Zhang, J. and Gowing, D.J.G. (1993) Stomatal control by both [ABA] in the xylem sap and leaf water status: test of a model and of alternative hypotheses for droughted or ABA-fed field-grown maize. *Plant Cell and Environment* 16, 413–420.

Taylor, H.M. and Klepper, B. (1975) Water uptake by cotton root systems: an examination of assumptions in the single root model. *Soil Science* 120, 57–67.

Taylor, S.A. and Ashcroft, G.L. (1972) *Physical Edaphology*. W.H. Freeman, San Francisco, California.

Tennant, D. (1975) A test of a modified line intersect method of estimating root length. *Journal of Ecology* 63, 995–1001.

Thom, A.S. and Oliver, H.R. (1977) On Penman's equation for estimating regional evaporation. *Quarterly Journal of the Royal Meteorological Society* 103, 345–357.

Thornthwaite, C.W. (1948) An approach toward a rational classification of climate. *The Geographical Review* 38, 55–94.

Thumma, B.R., Naidu, B.P., Cameron, D.F. and Bahnisch, L.M. (1998) Transpiration efficiency and its relationship with carbon isotope discrimination under well-watered and water-stressed conditions in *Stylosanthes scraba*. *Australian Journal of Agricultural Research* 49, 1039–1045.

Tollenaar, M. and Aguilera, A. (1992) Radiation use efficiency of an old and a new maize hybrid. *Agronomy Journal* 84, 536–541.

Topp, G.C., Davis, J.L. and Annan, A.P. (1980) Electromagnetic determination of soil water content: measurements in coaxial transmission lines. *Water Resources Research* 16, 574–582.

Topp, G.C., Davis, J.L. and Annan, A.P. (2003) The early development of TDR for soil measurements. *Vadose Zone Journal* 2, 492–499.

Triplett, G.B. Jr, van Doren, D.M. and Schmidt, B.L. (1968) Effect of corn (*Zea mays* L.) stover mulch on no-tillage corn yield and water infiltration. *Agronomy Journal* 60, 236–239.

Turner, N.C. (1981) Techniques and experimental approaches for the measurement of plant water status. *Plant and Soil* 58, 339–366.

Turner, N.C. (1986) Adaptation to water deficits: a changing perspective. *Australian Journal of Plant Physiology* 13, 175–190.

Turner, N.C. (1990) Plant water relations and irrigation management. *Agricultural Water Management* 17, 59–73.

Turner, N.C. (1993) Water use efficiency of crop plants: potential for improvement. In: Buxton, D.R., Shibles, R., Forsberg, R.A., Blad, B.L., Asay, K.H., Paulsen, G.M. and Wilson, R.F. (eds) *International Crop Science I*. Crop Science Society of America, Madison, Wisconsin, pp. 75–82.

Turner, N.C. (2001) Optimizing water use. In: Nösberger, J., Geiger, H.H. and Struik, P.C. (eds) *Crop Science: Progress and Prospects*. CAB International, Wallingford, UK, pp 119–135.

Turner, N.C. (2004) Agronomic options for improving rainfall-use efficiency of crops in dryland farming systems. *Journal of Experimental Botany* 55, 2413–2425.

Tyree, M.T. and Dixon, M.A. (1983) Cavitation events in *Thuja occidentalis* L.? *Plant Physiology* 72, 1094–1099.

Unger, P.W. (1984) Tillage and residue effects on wheat, sorghum, and sunflower grown in rotation. *Soil Science Society of America Journal* 48, 885–891.

UN (United Nations) (1992) United Nations Framework Convention on Climate Change (Article 2). Available at: http://unfccc.int/files/essential_background/background_publications_htmlpdf/application/pdf/conveng.pdf (accessed 26 June 2014).

UN (United Nations) (2013) *World Population Prospects. The 2012 Revision. Volume I: Comprehensive Tables*. United Nations, New York.

UNEP (United Nations Environmental Programme) Ozone Secretariat (2010–2011) Scientific Assessment of Ozone Depletion: 2010. Report of the 2010 Assessment of the Scientific Assessment Panel. Available at: http://ozone.unep.org/new_site/en/scientific_assessment_2010.php (accessed 6 November 2014).

UNFCCC (United Nations Framework Convention on Climate Change) (2013) National Greenhouse Gas Inventory Data for the Period 1990–2011. Available at: http://unfccc.int/resource/docs/2013/sbi/eng/19.pdf (accessed 5 March 2014).

UN (United Nations) Population Division (1999) The World at Six Billion. Available at: http://www.un.org/esa/population/publications/sixbillion/sixbilpart1.pdf (accessed 25 June 2014).

Uri, N.D. (1999) Factors affecting the use of conservation tillage in the United States. *Water, Air, and Soil Pollution* 116, 621–638.

US Census Bureau (2013) World Population. Available at: http://www.census.gov/population/international/data/idb/informationGateway.php (accessed 26 June 2014).

Vaadia, Y. and Itai, C. (1969) Interrelationships of growth with reference to the distribution of growth substances. In: Whittington, W.J. (ed.) *Root Growth*. Proceedings of the Fifteenth Easter School in Agricultural Science, University of Nottingham. Butterworths, London, pp. 65–79.

Valentin, C., d'Herbès, J.M. and Poesen, J. (1999) Soil and water components of banded vegetation patterns. *Catena* 37, 1–24.

Vandeleur, R.K., Sullivan, W., Athman, A., Jordans, C., Gilliham, M., Kaiser, B.N. and Tyerman, S.D. (2014) Rapid shoot-to-root signalling regulates root hydraulic conductance via aquaporins. *Plant, Cell and Environment* 37, 520–538.

Veen, B.W. (1982) The influence of mechanical impedance on the growth of maize roots. *Plant and Soil* 66, 101–109.

Ventura, F., Spano, D., Duce, P. and Snyder, R.L. (1999) An evaluation of common evapotranspiration equations. *Irrigation Science* 18, 163–170.

Vetter, H. and Scharafat, S. (1964) Die Wurzelverbreitung landwirtschaftlicher Kulturpflanzen im Unterboden. *Zeitschrift für Acker- und Pflanzenbau* 120, 275–298.

Vieweg, G.H. and Ziegler, H. (1960) Thermoelektrische Registrierung der Geschwindigkeit des Transpirationsstromes. *Berichte der Deutschen Botanischen Gesellschaft* 73, 221–226.

Virgona, J.M., Gummer, F.A.J. and Angus, J.F. (2006) Effects of grazing on wheat growth, yield, development, water use and nitrogen use. *Australian Journal of Agricultural Research* 57, 1307–1319.

Wagenet, R.J. (1990) Quantitative prediction of the leaching of organic and inorganic solutes in soil. *Philosophical Transactions Royal Society London B* 329, 321–330.

Wagger, M.G. and Cassel, D.K. (1993) Corn yield and water-use efficiency as affected by tillage and irrigation. *Soil Science Society of America Journal* 57, 229–234.

Waisel, Y., Eshel, A. and Kafkafi, U. (eds) (2002) *Plant Roots. The Hidden Half*, 3rd edn. Marcel Dekker, New York.

Walbot, V. and Bruening, G. (1988) Plant development and ribozymes for pathogens. *Nature* 334, 196–197.

Walker, G.K. (1986) Transpiration efficiency of field-grown maize. *Field Crop Research* 14, 29–38.

Walker, G.K. and Richards, J.E. (1985) Transpiration efficiency in relation to nutrient status. *Agronomy Journal* 77, 263–269.

Waloszczyk, K. (1991) Ergebnisse zur Ertragsbildung bei ökologisch orientiertem Winterweizenanbau auf Lößschwarzerdeböden in trockenen Lagen. *Journal of Agronomy and Crop Science* 166, 238–248.

Walter, H. (1970) *Vegetationszonen und Klima.* Ulmer, Stuttgart, Germany.

Wang, E. and Smith, C.J. (2004) Modelling the growth and water uptake function of plant root systems: a review. *Australian Journal of Agricultural Research* 55, 501–523.

Wang, Y., Zhang, X., Liu, X., Zhang, X., Shao, L., Sun, H. and Chen, S. (2013) The effects of nitrogen supply and water regime on instantaneous WUE, time-integrated WUE and carbon isotope discrimination in winter wheat. *Field Crops Research* 144, 236–244.

Warkentin, B.P. (1971) Effects of compaction on content and transmission of water in soils. In: American Society of Agricultural Engineers (ASAE) (ed.) *Compaction of Agricultural Soils.* ASAE Monograph, St Joseph, Michigan, pp. 126–153.

Wasson, A.P., Richards, R.A., Chatrath, R., Misra, S.C., Sai Prasad, S.V., Rebetzke, G.J., Kirkegaard, J.A., Christopher, J. and Watt, M. (2012) Traits and selection strategies to improve root systems and water uptake in water-limited wheat crops. *Journal of Experimental Botany* 63, 3485–3498.

Watson, K.W. and Luxmoore, R.J. (1986) Estimating macroporosity in a forest watershed by use of a tension infiltrometer. *Soil Science Society of America Journal* 50, 578–582.

Weatherley, A.B. and Dane, J.H. (1979) Effect of tillage on soil-water movement during corn growth. *Soil Science Society of America Journal* 43, 1222–1225.

Welles, J.M. and Norman, J.M. (1991) Instrument for indirect measurement of canopy architecture. *Agronomy Journal* 83, 818–825.

Wendroth, O., Ehlers, W., Hopmans, J.W., Kage, H., Halbertsma, J. and Wösten, J.H.M. (1993) Reevaluation of the evaporation method for determining hydraulic functions in unsaturated soils. *Soil Science Society of America Journal* 57, 1436–1443.

West, T.O. and Post, W.M. (2002) Soil organic carbon sequestration rates by tillage and crop rotation. *Soil Science Society of America Journal* 66, 1930–1946.

Whalley, W.R., Clark, L.J., Take, W.A., Bird, N.R.A., Leech, P.K., Cope, R.E. and Watts, C.W. (2007) A porous-matrix sensor to measure the matric potential of soil water in the field. *European Journal of Soil Science* 58, 18–25.

Whalley, W.R., Ober, E.S. and Jenkins, M. (2013) Measurement of the matric potential of soil water in the rhizosphere. *Journal of Experimental Botany* 64, 3951–3963.

Whittaker, R.H. (1975) *Communities and Ecosystems*, 2nd edn. Macmillan, London.

Wild, A. (ed.) (1988) *Russell's Soil Conditions and Plant Growth*, 11th edn. Longman, Burnt Mill, Harlow, UK.

Williams, W.A., Loomis, R.S. and Lepley, C.R. (1965a) Vegetative growth of corn as affected by population density. I. Productivity in relation to interception of solar radiation. *Crop Science* 5, 211–215.

Williams, W.A., Loomis, R.S. and Lepley, C.R. (1965b) Vegetative growth of corn as affected by population density. II. Components of growth, net assimilation rate and leaf area index. *Crop Science* 5, 215–219.

Willis, W.O. (1983) Water conservation: introduction. In: Dregne, H.E. and Willis, W.O. (eds) *Dryland Agriculture.* American Society of Agronomy, Madison, Wisconsin, pp. 21–24.

Wilson, D.R. (1985) The value of water for crop production. *New Zealand Agricultural Science* 19, 174–179.

Wilson, D.R. and Jamieson, P.D. (1985) Models of growth and water use of wheat in New Zealand. In: Day, W. and Atkin, R.K. (eds) *Wheat Growth and Modelling.* Plenum Press, New York, pp. 211–216.

Windt, A. and Märländer, B. (1994) Wurzelwachstum von Zuckerrüben unter besonderer Berücksichtigung des Wasserhaushaltes. *Zuckerindustrie* 119, 659–663.

de Wit, C.T. (1953) *A Physical Theory on Placement of Fertilizers.* Verslagen van landbouwkundige onderzoekingen, No. 59.4. Staatsdrukkerij, s'-Gravenhage.

de Wit, C.T. (1958) *Transpiration and Crop Yields.* Mededeling 59. Verslagen van landbouwkundige onderzoekingen, No. 64.6, Wageningen, The Netherlands.

de Wit, C.T. (1991) On the efficiency of resource use in agriculture. In: Böhm, W. (ed.) *Ziele und Wege der Forschung im Pflanzenbau. Festschrift für Kord Baeumer zum* 65. Geburtstag. Triade-Verlag, Göttingen, Germany, pp. 29–54.

Withers, B. and Vipond, S. (1974) *Irrigation. Design and Practice*. B T Batsford Ltd, London.

Wittmuss, H.D., Triplett, G.B. and Greb, D.W. (1973) Concepts of conservation tillage systems using surface mulches. In: Soil Conservation Society of America (ed.) *Conservation Tillage*. The Proceedings of a National Conference, Des Moines, Iowa. Soil Conservation Society of America, Ankeny, Iowa, pp. 5–12.

Witty, J.F., Keay, P.J., Frogatt, P.J. and Dart, P.J. (1979) Algal nitrogen fixation on temperate arable fields. *Plant and Soil* 52, 151–164.

Woli, P., Jones, J.W., Ingram, K.T. and Fraisse, C.W. (2012) Agricultural reference index for drought (ARID). *Agronomy Journal* 104, 287–300.

WRI (World Resource Institute) (2014) CAIT Climate Data Explorer. Available at: http://cait2.wri.org/wri (accessed 27 June 2014).

Wright, G.C., Nageswara Rao, R.G. and Farquhar, G.D. (1994) Water-use efficiency and carbon isotope discrimination in peanut under water deficit conditions. *Crop Science* 34, 92–97.

Xin, Z., Aiken, R. and Burke, J. (2009) Genetic diversity of transpiration efficiency in sorghum. *Field Crops Research* 111, 74–80.

Youngs, E.G. (1995) Developments in the physics of infiltration. *Soil Science Society of America Journal* 59, 307–313.

Zadoks, J.C., Chang, T.T. and Konzak, C.F. (1974) A decimal code for the growth stages of cereals. *Weed Research* 14, 415–421.

Zegelin, S.J., White, I. and Russell, G.F. (1992) A critique of the time domain reflectometry technique for determining field soil-water content. In: Topp, G.C., Reynolds, W.D. and Green, R.E. (eds) *Advances in Measurement of Soil Physical Properties: Bringing Theory into Practice*. Soil Science Society of America Special Publication No. 30, Madison, Wisconsin, pp. 187–208.

Zeit (2012) Der Mann, der die Wüste aufhielt (The Man Who Stopped the Desert). Available at: http://www.zeit.de/2012/49/Hunger-Sahelzone-Baeumepflanzer (accessed 8 July 2014).

Zhang, X., Chen, S., Liu, M., Pei, D. and Sun, H. (2005) Improved water use efficiency associated with cultivars and agronomic management in the North China Plain. *Agronomy Journal* 97, 783–790.

Index